STUDY GUIDE
FOR MILLER AND SCHROEER's
COLLEGE PHYSICS
SIXTH EDITION

ROBERT W. STANLEY
Late of Purdue University

DIETRICH SCHROEER
University of North Carolina at Chapel Hill

HARCOURT BRACE JOVANOVICH, PUBLISHERS
and its subsidiary, Academic Press
San Diego New York Chicago Austin Washington, D.C.
London Sydney Tokyo Toronto

Cover credit: © Photo Research International

ISBN: 0-15-511744-0
Library of Congress Catalog Card Number: 86-81636

Printed in the United States of America

Contents

Foreword

This Study Guide has been designed to help students master the material presented in *College Physics*, Sixth Edition, by Franklin Miller, Jr., and Dietrich Schroeer. Students who use this Study Guide in close conjunction with the text will become actively involved in the learning process. The many questions and problems with detailed solutions permit students to correct their errors promptly and to reinforce their learning as it occurs. Frequent opportunities are provided for students to assess their progress. Students in a Keller-plan course or other course format emphasizing individual study will find this material especially useful. With the aid of this Study Guide, independent learning of physics becomes a feasible alternative for any serious student.

To the Student

The purpose of this Study Guide is to help you learn the material presented in *College Physics*, Sixth Edition, by Franklin Miller, Jr., and Dietrich Schroeer. When properly used, this Study Guide will help you gain facility in answering questions and solving problems related to the text. The strong emphasis on quantitative problems corresponds to a similar emphasis in the textbook and in courses that utilize this and similar texts.

Your success in learning to solve problems in physics depends a great deal on how you use the materials available to you, especially the textbook and the Study Guide. Since no two students learn in exactly the same manner, there are as many ways to use this material as there are students in the course. The important thing is to develop an approach to learning physics that works best for you.

With this in mind, we offer the following guidelines, which may prove useful regardless of the specific way in which you study.

- *Consult the solutions often*. Get as much help as you need, especially when you are beginning a new topic. The first time you approach a specific problem you may have the feeling of being completely lost, of having no idea where to start. Turn back to the solution to find out what physical principles are involved and how they are applied to the case at hand. Then return to the problem and begin writing out its solution. If you get stuck, consult the solution again. The first time you attempt a problem you may have to consult the solution several times.
- *Repeat the solutions to problems that give difficulty*. Devise some scheme for marking those problems that give you difficulty the first time you tackle them. Return to these problems at a later time, repeatedly if necessary. For some of the more difficult problems, you may find it necessary to write out the solution three or four times before becoming 100 percent certain of your understanding. When you repeat a problem that you have worked on before, note whether you are able to proceed further than on the previous attempt. If you cannot get at least a little more written down the second time through, you know something is wrong and you should probably read the material again.
- *Settle for nothing less than complete mastery*. After writing out the solution to a problem once or twice, perhaps consulting the solution only once during the second attempt, you may be tempted to regard that problem as "solved" and go on to other things. But remember, during an examination you will not be given even one quick peek at the solution. If you cannot write out a complete solution fairly quickly, and with nearly total confidence in what you have done, then you have not really mastered that problem.

Within the Questions and Problems and the Solutions sections in this guide, there are frequent references to figures. Unless a specific reference is made to a textbook figure, all such references apply to the figures that appear in the Study Guide.

Finally, you will notice that when vectors appear in the figures they are designated by an arrow above the symbol, whereas when they appear elsewhere in the guide they are designated by **bold face** type. The reason is that the figures are hand drawn and it is conventional to use arrows in all handwritten mathematical work.

Robert W. Stanley
Dietrich Schroeer

CHAPTER 1

The Nature of Physics

GOALS To establish a view of physics as a basic natural science—the science that deals with fundamental aspects of energy and nonliving matter. To acquire some general notions concerning the properties of matter and the structure of the atoms of which it is composed.

OBJECTIVES After completing this chapter the student should be able to do the following:

1. State a reason for beginning the study of physics with the subject of mechanics rather than such other subjects as heat, sound, light, or electricity.
2. Name the three basic dimensional quantities in whose terms all mechanical quantities (force, energy, velocity, momentum, etc.) may be expressed.
3. Give the letter symbols associated with the three dimensional quantities named above.
4. Given that velocity has the dimensions of distance divided by time, write the dimensional equation for velocity using conventional notation.
5. Apply dimensional analysis to verify the correctness, from a dimensional point of view, of an equation involving mechanical quantities.
6. Write from memory the abbreviation and the meaning of the six most common metric prefixes.
7. Name the SI units of distance, mass, and time.
8. Define the meter in terms of the speed of light.
9. Name the basic attribute of matter that causes it to resist a change in its state of motion.
10. Name three more attributes of matter.
11. Write the definition of density in both words and symbols.
12. Use the definition of density to find any one of the quantities—mass, volume, density—when the two others are known.
13. Write the definition of specific gravity and calculate any one of the quantities—density of substance, density of water, specific gravity—when the two others are known.
14. Name the two particles of which the nucleus is composed.
15. Write the symbol for an element, iron or chlorine for example, when the mass number and atomic number are given.
16. Name at least two of the three factors that cause the variation of density among solids.

SUMMARY

In an age when geographical frontiers have ceased to exist, except perhaps for the ocean depths, and when the exploration of nearby space is limited only by its enormous cost and not by technological difficulties, it is reassuring to the adventurous to know that there are still challenging mental frontiers yet to be explored in all areas of human knowledge. This is especially true in the realm of the physical sciences, where we seem to be on the verge of new discoveries that will lead to a greatly improved understanding of the underlying structure of the universe.

There is no such thing as a specific and universally agreed-upon scientific method. But science has progressed because of its communal nature, as scientists challenge each other's discoveries in public and in the literature.

Much of our success in understanding and controlling both physical and biological phenomena results from the application of a systematic procedure called the *scientific method*. Key elements in this method are careful *observations*, the formation of a *hypothesis*, and the meticulous *testing* of the hypothesis. A hypothesis that has withstood critical testing over a period of time and that has led to new discoveries may achieve the status of a *theory* or perhaps even a *principle*.

In this course we will deal with that particular branch of natural science, called physics, that is concerned with very basic aspects of nonliving matter. Physics deals with the motions and interactions of physical objects from the smallest constituents of matter—the components of atoms and their nuclei—to the largest entities in our universe—galaxies and clusters of galaxies. Physics recognizes no limitations of scale nor any limitations of time. The physical principles that you will learn to work with were in operation before the earth was born and will remain valid for centuries to come.

Physics is a quantitative science. It deals with entities that can be measured and relationships that can be expressed in mathematical form. Physics is also a universal science. Thus physical quantities, such as mass, length, time, etc., must be expressed in well-defined and internationally accepted units. The *Systeme International* (abbreviated SI) serves this function admirably. In the SI, the units of length, mass, and time are the meter, the kilogram, and the second, respectively. These three units suffice for the study of mechanics. Later on in this course other SI units will be introduced as needed. These are the ampere (electric current), the kelvin (temperature), the mole (amount of substance), and the candela (luminous intensity).

Physical quantities have such a large range of values, from the microscopically small to the astronomically large, that it is necessary to express them in scientific notation using powers of ten.

A fundamental characteristic of matter is that any piece of it tends to resist any change in its state of motion. This attribute of matter is called *inertia*, for which the quantitative measure is *mass*. A less fundamental yet important characteristic of matter is *weight*, the attractive force that the earth exerts on any body in its vicinity. Matter is also characterized by other types of forces, including electric, magnetic, elastic, and nuclear forces.

The *density* of a substance is the mass per unit volume, usually expressed in g/cm^3 or kg/m^3 or $tonne/m^3$. The dimensionless ratio, (density of substance)/(density of water), is called *specific gravity*.

Matter is made up of atoms. Even a piece of matter so small that it can only be seen under a microscope contains a very large number of atoms. In the case of a solid, the atoms are often arranged in a highly regular array called a *crystal*. Individual atoms consist of an extremely small *nucleus* surrounded by a cloud of *electrons*. The nucleus contains positively charged *protons* and uncharged *neutrons*. The mass of the neutron is the same as that of the proton. The electron has a mass that is 1840 times less. Since the proton and electron have equal and opposite charges, a neutral atom must have equal numbers of protons and electrons. The number of protons is called the *atomic number*. The number of protons plus neutrons is the *mass number*. *Nucleon* is the general term for a nuclear particle, either proton or neutron.

The chemical properties of an atom are determined by the number and arrangement of its electrons. The number of neutrons in the nucleus has no effect on chemical properties. Atoms that have the same number of protons but different numbers of neutrons are called *isotopes*. The element hydrogen has three isotopes, ${}^1_1H$ with a single proton, ${}^2_1H$ with a proton plus a neutron, and ${}^3_1H$ with a proton plus two neutrons.

QUESTIONS AND PROBLEMS

Read Chapter 1 of the text looking especially for those items mentioned in the list of objectives. Answer the questions and solve the problems given in this section. It is advisable to check your answers immediately by referring to the Solutions section that follows. If your answer is correct you will gain confidence and will be even more likely to give that correct answer again. If your answer is wrong you can correct it at once and be less likely to make the same error in the future.

1. The three basic dimensional quantities with their corresponding symbols are: ____________ , ____________ , and ____________ .

2. Write the formula for the volume of a sphere.* What are the dimensions of r, π, 3, and V?

3. Given that force = mass × acceleration, what is the dimensional formula for force?

[force] = ____________ .

4. Pressure is defined as *force per unit area.* Give the dimensional formula for pressure.

[pressure] = ____________ .

5. Write the formula for the area of a triangle and the dimensional formula for the area of a triangle.

A = ____________ . [area] = ____________ .

6. Is the dimensional formula [area] = $[L^2]$ valid for a portion of the surface of a sphere?

7. Average acceleration may be defined as change in velocity divided by the time interval required for the change. Symbolically,

$$a_{av} = \frac{v_2 - v_1}{t_2 - t_1}$$

What is the dimensional equation for acceleration?

8. A gigatonne (metric) contains how many grams?

9. An hour contains how many nanoseconds?

10. Given that a galaxy contains about 10^{11} stars and that the mass of a galaxy is about 10^{41} kg, make an estimate of the average mass of a star. (Note that this is merely an "order of magnitude" calculation at best.)

11. Approximately two-thirds of the human body is composed of water (H_2O). Use the approximate masses in Table 1-2 in the text to obtain an estimate of the number of water molecules in the human body. (A molecule of H_2O has about one-third the mass of an iron atom.)

*The formulas for areas and volumes of simple geometric figures (triangle, cylinder, sphere, etc.) are taught in high school; you will be expected to know them.

12. List four attributes of matter.

13. Calculate the mass of a thin sheet of aluminum whose dimensions are 4 cm $\times$ 2 cm $\times$ 0.2 cm.

14. Calculate the approximate mass in kilograms of a hydrogen atom. The atomic mass of hydrogen is very close to 1.01. (A more exact value, not needed in this problem, is 1.0080.) Since the mass of the electron is about 1/200th that of the proton, the mass of the proton is nearly equal to the mass of the hydrogen atom.

15. The density of cork is approximately 0.25 g/cm^3. What is its specific gravity?

16. If the radius of the proton is 1.2×10^{-13} cm, calculate the density of the proton and, hence, the density of nuclear matter. (The neutron and the proton have approximately the same mass and volume.) Examples of approximate calculations are given in Appendix F of the text.

17. Write the dimensional equation for density.

$$[\text{density}] = [\qquad\qquad]$$

18. Approximately 9% of naturally occurring neon consists of the isotope $^{22}_{10}\text{Ne}$. How many neutrons are contained in the nucleus of an atom of $^{22}_{10}\text{Ne}$? How many electrons are contained in its electron cloud? What is the symbol for a neon isotope that contains two less neutrons? What is the atomic number of the isotope $^{21}_{10}\text{Ne}$?

19. A teaspoon of water has a volume of approximately 5 cm^3. Find the number of moles of water, H_2O, in a teaspoon and the approximate number of water molecules.

20. Compare the chemical properties of $^{23}_{11}Na$ and $^{22}_{11}Na$. Ordinary table salt, NaCl, is made of the sodium isotope $^{23}_{11}Na$. Would you expect salt to look different or taste different if it were made with $^{22}_{11}Na$?

21. Since the atoms in the solid are packed rather closely together, we can make an estimate of the size of an individual atom in the following way.

a. Given the density of a pure substance in the solid state, determine the mass of a cubic centimeter.

b. From this mass and the known atomic mass, determine the number of moles.

c. Using Avogadro's number, calculate the number of atoms in the volume of 1 cm^3.

d. The result of the above calculation allows us to calculate the *volume* occupied by a single atom. Call this volume ΔV.

e. Assuming that the atom "fills up" ΔV, its diameter is given by $d \cong \sqrt[3]{\Delta V}$.

Following this procedure, estimate the diameter of an atom of *magnesium*, a very light metal, and an atom of *uranium*, one of the heaviest metals. The required densities are given in Table 1-3 in the text, and the atomic masses are found on the inside back cover of the text. Retain only one significant figure in the final result.

Solutions

1. Distance (or length), L; mass, M; time, T.

2. $V = \frac{4}{3}\pi r^3$; $[r] = [L]$, $[V] = [L^3]$; π and 3 are dimensionless constants.

3. [force] = [mass] × [acceleration] = $[M] \times [LT^{-2}] = [MLT^{-2}]$

4. [pressure] = [force]/[area] = $[MLT^{-2}] \times [L^2]^{-1} = [ML^{-1}T^{-2}]$

5. $A = bh/2$; [area] = $[L^2]$

6. Yes. Any area has dimensions of length squared.
7. [acceleration] = [velocity]/[time] = $[LT^{-1}]/[T] = [LT^{-2}]$

8. $1 \text{ gigatonne} = (10^9 \text{ tonnes})\left(\frac{10^3 \text{ kg}}{1 \text{ tonne}}\right)\left(\frac{10^3 \text{ g}}{1 \text{ kg}}\right) = 10^{15} \text{ grams}$

9. $1 \text{ h} = (60 \text{ min})\left(\frac{60 \text{ s}}{1 \text{ min}}\right)\left(\frac{1 \text{ ns}}{1 \times 10^{-9} \text{ s}}\right) = 3.6 \times 10^{12} \text{ ns}$

10. $\frac{10^{41} \text{ kg}}{10^{11} \text{ stars}} = 10^{30} \text{ kg/star}$

11. $\frac{10^2 \text{ kg } (2/3)}{0.3 \times 10^{-25} \text{ kg/molecule}} \cong 2 \times 10^{27}$ molecules of H_2O

12. Matter has inertia, and matter exerts gravitational, electrical, and nuclear forces upon other matter.
13. Start with the definition of density and solve for the mass.

$$\text{density} = \frac{\text{mass}}{\text{volume}} \text{ or } d = \frac{m}{V} \qquad m = (\text{density})(\text{volume}) = d \cdot V$$

$$m = (2.70 \text{ g/cm}^3)(4 \text{ cm})(2 \text{ cm})(0.2 \text{ cm}) = 4.32 \text{ g}$$

14. The mass of 6.023×10^{23} atoms of hydrogen is about 1.01 g. The mass of one hydrogen atom and, hence, of one proton is approximately

$$\left(\frac{1.01}{6.02} \times 10^{-23} \text{ g}\right)\left(\frac{1 \text{ kg}}{10^3 \text{ g}}\right) = 1.67 \times 10^{-27} \text{ kg}$$

15. $\text{specific gravity} = \frac{\text{density of substance}}{\text{density of water}} = \frac{0.25 \text{ g/cm}^3}{1.0 \text{ g/cm}^3} = 0.25$

16. $d = \frac{m}{V}; V = (4/3)\pi r^3 = (4/3)\pi(1.2 \times 10^{-13} \text{ cm})^3 \cong 7 \times 10^{-39} \text{ cm}^3$

$$d = \frac{(1.7 \times 10^{-27} \text{ kg})(10^3 \text{ g/kg})}{7 \times 10^{-39} \text{ cm}^3} \cong 2 \times 10^{14} \text{ g/cm}^3$$

Since the radius of a nucleon is not well defined, the density is necessarily a rough estimate. It should be noted that a small change in the estimate of the radius, from 1.2 to 1.0 × 10^{-13} cm for example, causes a large change in the computed density, in this case from 2 × 10^{14} g/cm³ to 4 × 10^{14} g/cm³. The best way to express the result is to say that the density of a nucleon, or of nuclear matter, is "of the order of" 10^{14} g/cm³.

17. $[\text{density}] = [\text{mass}]/[\text{volume}] = [M]/[L^3] = [ML^{-3}]$

18. The nucleus of $^{22}_{10}\text{Ne}$ contains 12 neutrons and 10 protons. A neutral atom of $^{22}_{10}\text{Ne}$ has 10 electrons in its electron cloud.
    The isotope containing 10 neutrons is $^{20}_{10}\text{Ne}$.
    The atomic number of $^{21}_{10}\text{Ne}$ is 10.

19. First find the mass of H_2O in the teaspoon. $m = d \times V = (1\ \text{g/cm}^3)(5\ \text{cm}^3) = 5\ \text{g}$

    The number of moles is (5 g)/(18 g/mole) = 0.278 moles.

    The number of water molecules is

$$(6.02 \times 10^{23}\ \text{molecules/mole})(0.278\ \text{moles}) = 1.67 \times 10^{23}\ \text{molecules}$$

20. Different isotopes of an element have the same number and arrangement of electrons surrounding the nucleus. Since chemical properties of an element depend only on the electrons, and since taste depends on chemical reactions, there would be no difference in the taste of salt made from different sodium isotopes. The crystalline structure likewise depends only on the number and arrangement of the electrons and would be exactly the same for different isotopes.

21. We will follow the suggested steps. In the case of magnesium, $d = 1.75\ \text{g/cm}^3$

    a. $d = m/V;\ m = d \times V$
    $m = (1.75\ \text{g/cm}^3)(1\ \text{cm}^3) = 1.75\ \text{g}$

    b. $\text{number of moles} = \dfrac{1.75\ \text{g}}{24.3\ \text{g/mole}} = 0.072\ \text{moles}$

    c. $\text{number of atoms} = (0.072\ \text{moles})\left(6.02 \times 10^{23}\ \dfrac{\text{atoms}}{\text{mole}}\right) = 4.3 \times 10^{22}\ \text{atoms}$

    d. $\text{volume per atom} = \dfrac{1\ \text{cm}^3}{4.3 \times 10^{22}\ \text{atoms}} = 2.4 \times 10^{-23}\ \text{cm}^3 = 24 \times 10^{-24}\ \text{cm}^3$

    e. $\text{diameter} \cong \sqrt[3]{24} \times 10^{-8}\ \text{cm} \cong 3 \times 10^{-8}\ \text{cm}$

    We conclude that the diameter of a magnesium atom is of the order of 3 × 10^{-8} cm. When the same calculation is carried out for uranium, the result is

$$\text{diameter} \cong 3 \times 10^{-8}\ \text{cm}$$

    In the solid state these two atoms occupy essentially the same space.

CHAPTER 2 Kinematics—The Description of Motion

GOALS To recognize situations in which the acceleration vector is constant; to be able to apply the relationships between kinematic variables—displacement, velocity, acceleration, time—to solve for unknown quantities; and to distinguish between physical quantities that are scalars and those that are vectors, and to perform arithmetic operations on both types of quantities.

OBJECTIVES After completing this chapter the student should be able to do the following:

1. Describe three physical situations in which there is motion in a straight line at a constant acceleration.
2. Write from memory the definitions of average velocity, instantaneous velocity, average acceleration, and instantaneous acceleration, using the commonly accepted letter symbols, s, v, a, and t.
3. Given numerical values of v_1, v_2, and Δt, calculate a_{av}.
4. Given graphs of distance versus time, velocity versus time, and acceleration versus time, select the graphs that apply to uniformly accelerated motion.
5. Given a graph of velocity as a function of time, draw the corresponding graphs of s versus t and a versus t.
6. Given the graph of v versus t, determine the instantaneous acceleration from a measurement of the slope, and the distance from a measurement of the area.
7. Given a table of values of distance at known times, determine graphically whether the velocity was constant.
8. Given a table of values of velocity at known times, determine graphically whether the acceleration was constant.
9. Write from memory (in any order) the four equations of uniformly accelerated motion (see text Eqs. 2-9, 2-10, 2-11, and 2-12).
10. Describe two physical situations in which the motion is one-dimensional but *not* uniformly accelerated.
11. Write from memory the numerical value of the acceleration of gravity g in the SI system of units. (Only the first two significant figures are required.)
12. Calculate the final velocity of a falling object given the height from which it fell.
13. Calculate the time required for an object to fall through a given distance (1) starting from rest; (2) starting with an initial velocity in the direction of acceleration; and (3) starting with an initial velocity in a direction opposite to the acceleration.
14. Calculate the maximum height to which an object will rise when thrown upward with a given velocity.
15. Given the distance that an object has fallen and the time required to traverse that distance, calculate the acceleration that has taken place.
16. Write the appropriate SI units for s, v, and a.
17. Describe three physical situations in which motion takes place in two dimensions, with or without constant acceleration.
18. Describe two cases in which motion is confined to a plane and acceleration is constant.
19. Give a correct word definition of average acceleration, a vector, in terms of change in velocity.
20. Write from memory the definition of average acceleration using the commonly accepted symbols $\mathbf{a}_{av}$, $\Delta\mathbf{v}$, $\mathbf{v}$, and t.

21. Distinguish between vector and scalar quantities in the above definition.
22. Name the characteristics of a vector quantity that distinguish it from a scalar.
23. Calculate graphically the sum or difference of any two given vectors, stating both magnitude and direction of the resultant vector.
24. Calculate the average acceleration from any pair of initial and final velocities given with the corresponding time interval.
25. Resolve any given vector (e.g., force, displacement, velocity, acceleration) into perpendicular components.
26. Calculate the sum or difference of any two given vectors by the method of resolution into components.
27. Calculate the velocity of an object with respect to a fixed coordinate system when its velocity is given with respect to a second, moving, coordinate system.
28. Calculate the velocity of an object with respect to a moving coordinate system when the velocity of that coordinate system is known.
29. Calculate the horizontal range of a projectile from a knowledge of its initial velocity.
30. Calculate the resultant velocity of a projectile from a knowledge of its horizontal and vertical motions.

SUMMARY

Kinematics is that part of mechanics which deals with the motions of bodies without regard to the forces responsible for those motions. Kinematics may be further subdivided according to the path followed by the moving object. Among the many possible paths, the simplest and easiest to deal with is the straight line.

Rectilinear motion, unlike other types of motion, does not require the use of vector quantities. We begin with definitions of velocity and acceleration. The *average velocity*, v_{av}, is distance traveled divided by time. Symbolically, $v_{av} = (s - s_0)/(t - t_0)$, which may also be written in an abbreviated form, $v_{av} = \Delta s/\Delta t$. The *instantaneous velocity* is the limit of this ratio as the time interval is made smaller and smaller, $v = \lim_{\Delta t \to 0} \Delta s/\Delta t$.

Acceleration, a measure of the rate at which the velocity is changing, is defined by equations very similar to those just given for velocity, $a_{av} = (v - v_0)/(t - t_0)$, $a_{av} = \Delta v/\Delta t$, and $a = \lim_{\Delta t \to 0} \Delta v/\Delta t$. A special case of considerable importance arises when the acceleration is constant, i.e., when the velocity is changing at a constant rate. For this particular type of motion (uniform, rectilinear motion), the average velocity over any time interval will be exactly halfway between the instantaneous values at the beginning and the end of the time interval, i.e., $v = \frac{1}{2}(v_0 + v)$. As a consequence, the following four relationships hold for uniform, rectilinear motion:

$$s = \tfrac{1}{2}(v_0 + v)t \qquad v = v_0 + at$$

$$s = v_0 t + \tfrac{1}{2}at^2 \qquad v^2 = v_0^2 + 2as$$

When the motion of a body is not confined to a straight line, that motion must be described by means of *vector* quantities, mathematical entities having both *magnitude* and *direction*. Vectors may be added by a graphical method in which arrows representing the vectors are placed head to tail. The *sum*, or *resultant*, of two or more vectors is represented by an arrow drawn from the tail of the first vector to the head of the last vector. The *component* of a vector is the projection of the vector along a particular direction. It is often useful to replace a vector by its two components along a convenient set of mutually perpendicular x and y axes. The vector quantities that arise in the study of kinematics are *displacement, velocity,* and *acceleration*.

Average velocity and average acceleration are defined as follows:

$$\mathbf{v}_{av} = \frac{\Delta \mathbf{s}}{\Delta t} = \frac{\mathbf{s}_2 - \mathbf{s}_1}{t_2 - t_1} \qquad \mathbf{a}_{av} = \frac{\Delta \mathbf{v}}{\Delta t} = \frac{\mathbf{v}_2 - \mathbf{v}_1}{t_2 - t_1}$$

The motion of a projectile is generally not confined to a straight line; it is, however, confined to a plane. A characteristic of projectile motion is that the acceleration has a fixed direction and a constant magnitude, symbolically $\mathbf{a}$ = constant. The motion may be regarded as the sum of two independent motions, one along the direction of $\mathbf{a}$ and another along a direction at right angles to $\mathbf{a}$. The motion along the direction of the acceleration is uniformly accelerated motion for which the equations are those already found. The same equations with $a = 0$ apply to the motion in the perpendicular direction.

Other types of motion, particularly circular motion and oscillatory motion, will be treated in subsequent chapters.

QUESTIONS AND PROBLEMS

The questions and problems listed for this chapter, as for the remaining chapters, follow closely the order in which the material is presented in the textbook. The problems should reinforce what you have read and indicate what portions of the text need to be read more closely. The material in the textbook, including the illustrative examples, will help you to solve the problems. You should use text and Study Guide together, going back and forth frequently from one to the other.

The first two problems concern Secs. 2-1 and 2-2 in the text. Problems 3 through 15 deal with motion along a straight line, rectilinear motion, which is treated in Secs. 2-3 through 2-7. From a geometrical point of view, this is one-dimensional motion, and the equations that describe it are scalar (rather than vector) equations. Study the worked-out examples in the text until you are able to complete the solution on your own. Consult the problem solutions at the end of this chapter as often as you need to. Remember to mark the problems you miss so that you can return to them later.

1. A sound wave travels a distance of 68 m in 1/5 s. Calculate the velocity of sound in m/s.

2. The pilot of a jet is distracted for 1/3 of a second while she is making a landing approach at 390 km/h. How far will the plane travel during this brief period of inattention?

3. An arrow was shot straight upward, reached a maximum height of 20 m, and then fell back to the earth. The graph below (Fig. 2-1) gives the displacement as a function of time.

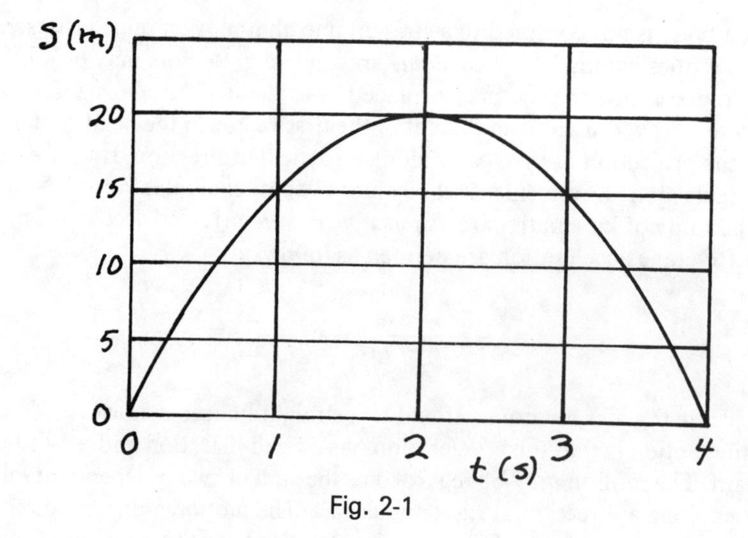

Fig. 2-1

a. Determine from this graph the average velocity of the arrow during the first two seconds of flight.

$$\left(\text{Hint: } v_{av} = \frac{s - s_0}{t - t_0} = \frac{\Delta s}{\Delta t}\right)$$

b. Determine the instantaneous velocity at the moment the arrow left the bow ($t = 0$) and at the top of its flight. (Hint: The instantaneous velocity is the slope of a line that is tangent to the curve at the point of interest.)
Record your three results below.

$v_0 =$ ______________. $v_f =$ ______________. $v_{av} =$ ______________.

c. If an object has a constant acceleration, what relation exists between v_0, v_f, and v_{av}? Does it appear from your measurements that the arrow has a constant acceleration?

d. Make a graphical determination of the velocity of the arrow at $t = 1$ s and $t = 3$ s. Then plot the points carefully on the velocity-time curve below.

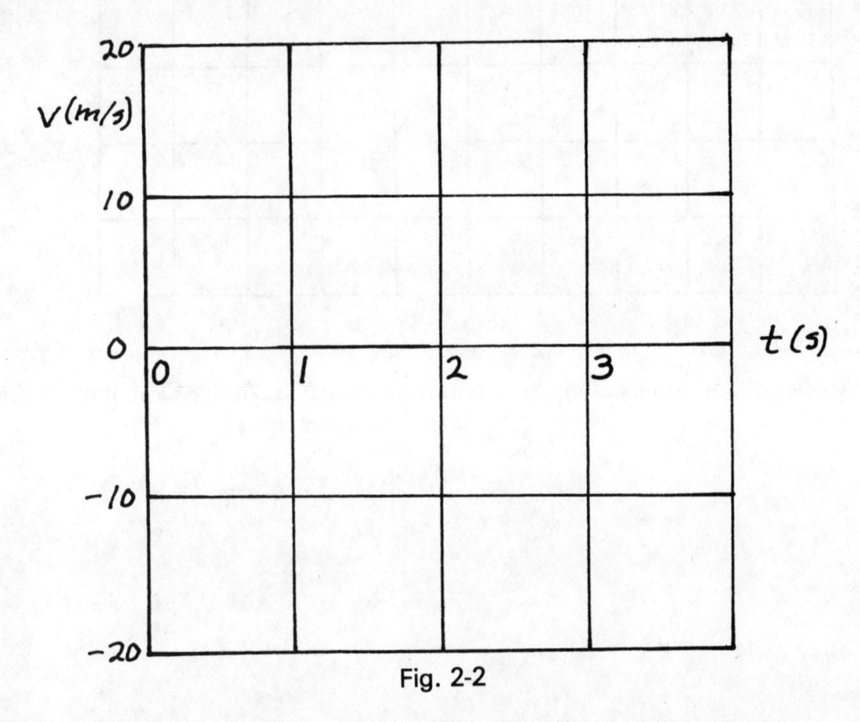

Fig. 2-2

If you have made your measurements with reasonable care, the points will fall along a straight line. Draw the best straight line through these points.

e. Write the definitions of average velocity and average acceleration.

v_{av} = ______________ . a_{av} = ______________ .

From the $v - t$ graph that you have just completed, determine the average acceleration during the time interval 1 s to 3 s.

f. From the graph determine the instantaneous acceleration of the arrow at $t = 2$ s.

g. Finally, to complete the analysis of this example of free fall, plot the acceleration as a function of time on the graph below.

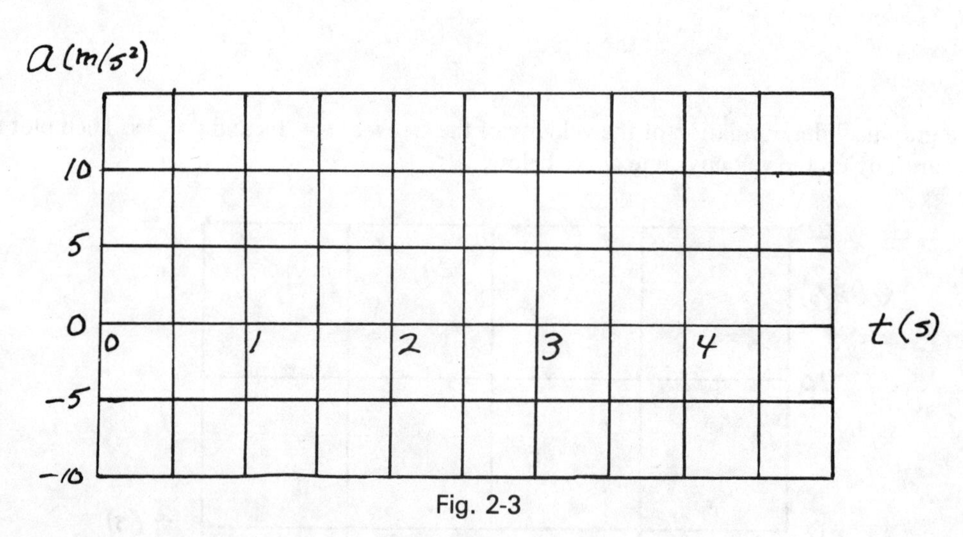

Fig. 2-3

4. An automobile accelerates uniformly along a straight road from 5 km/h to 85 km/h in 4 s. Calculate the acceleration in m/s².

5. The displacement of an automobile as a function of time is given by the graph on the left below.

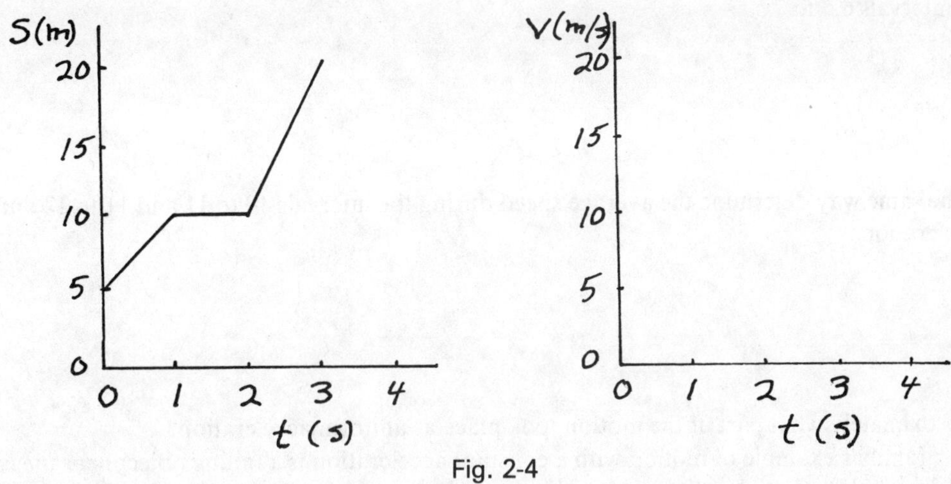

Fig. 2-4

Plot the graph of speed versus time for this automobile.

6. The displacement-time graph of an object is shown in the graph below. From this graph determine the speed of the object at 1 s and at 3 s.

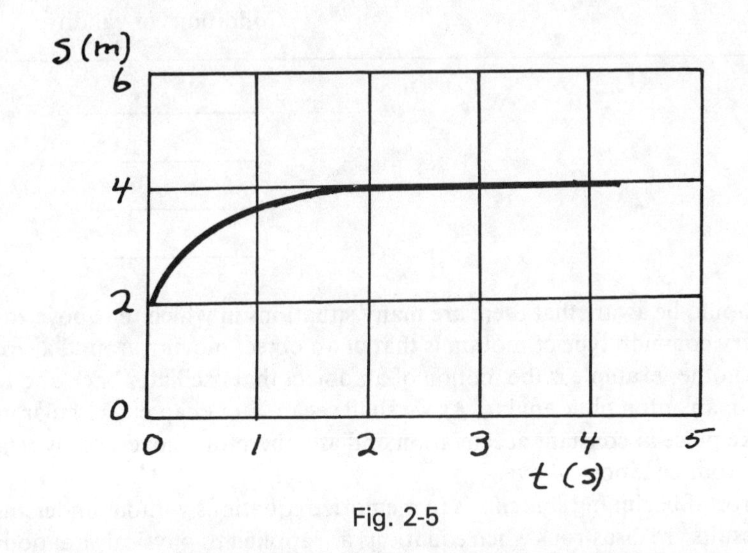

Fig. 2-5

7. A small object, moving along a straight line, was photographed at equal time intervals of 0.01 s. Its instantaneous position was recorded on film by means of a flash lamp that fired repeatedly at 0.01 s intervals. The resulting positions are shown in the drawing by means of small circles.

0 1 2 3 4 5 6 7 8 9 10 11 12

a. What was the average speed (in m/s) during the first 0.06 s? (Use a metric rule to measure the distance traveled.)

b. Did the object maintain a constant speed throughout the motion?

c. Determine the average speed during the intervals 6 to 7 and 7 to 8 and the average acceleration over the intervals 6.5 to 7.5.

d. In the same way determine the average speed during the intervals 10 to 11 and 11 to 12 and the average acceleration.

e. Approximately what part of the motion took place at uniform acceleration?

8. The most familiar example of motion with a constant acceleration is a falling object near the earth. In this case, neglecting air resistance, the acceleration has a constant magnitude of 9.8 m/s^2 (or 32 ft/s^2) and a constant direction, vertically downward.

Write, from memory if possible, the four equations that govern this type of motion *and* the conditions under which these equations are valid.

Equations	Conditions of validity
$v =$	__________
$s =$	__________
$v^2 =$	__________
$s =$	__________

The student should be aware that there are many situations in which the above equations are not valid. For example, a very common type of motion is that of an object moving around a circular path at a constant speed. Another example is the motion of an object that oscillates back and forth like a pendulum or like the piston in an automobile engine. As we shall see in later chapters, circular motion and oscillatory motion do not take place at constant acceleration and are, therefore, governed by relationships very different from the four equations above.

A common error of beginning students is to memorize equations without understanding their conditions of validity. The results are disastrous when equations are applied to physical situations in which those equations are invalid.

9. Fortunately, there are some equations that are always valid. Definitions are in this category; they require no conditions of validity. So far in this chapter, average velocity and average acceleration have been defined as well as instantaneous velocity and instantaneous acceleration. You should now be able to write these definitions from memory.

$v_{av} =$ __________. $v =$ __________.

$a_{av} =$ __________. $a =$ __________.

10. At 3:15 P.M. a motorboat was observed moving due east at 12 m/s; at 3:17 P.M. its velocity was 8 m/s eastward. Calculate the magnitude and direction of the acceleration that took place. (Note that the motorboat slowed down; Δv was negative.)

11. A small object is dropped from a high building and falls freely. Assume that air resistance is negligible.
    a. What is the instantaneous velocity at the end of 2 s?

    b. What is the velocity at the end of 3 s?

    c. What is the average velocity over the interval $t = 2$ s to $t = 3$ s?

    d. What distance did the object travel during the interval $t = 3$ s to $t = 4$ s?

    e. How long did it take the falling object to reach the speed of 48 m/s?

    f. How long did it take to fall a distance of 48 m?

12. The water level in a well is 46 m below the ground. If the speed of sound is 340 m/s, what time interval elapses between the moment of releasing a stone and hearing its splash?

13. A speedboat increases its speed at the rate of 2 m/s^2. How long does it take for the speed to increase from 7 m/s to 15 m/s? How far does it travel during this time?

14. A small rock is released at a height of 78.4 m above the ground.
   a. How many seconds will elapse before the rock reaches the ground?

   b. What will the speed of the rock be after three seconds of free fall?

   c. How long will it take for the rock to reach a speed of 19.6 m/s?

   d. Through what distance will the rock fall during the interval 1.0 s to 2.0 s?

   e. Suppose the rock is thrown *upward* from the height of 78.4 m and that its initial velocity is 20 m/s. How long will it take for the rock to reach the ground this time? (Hint: Directions become important in this problem. In particular $\mathbf{v}_0$ and $\mathbf{a}$ have opposite directions. You will probably find it easier to solve this problem in two steps. First find the velocity of the rock just before it strikes the ground. Then find the total time from $v = v_0 + at$.)

15. A ball is thrown upward with an initial speed of 12 m/s.
   a. Determine the magnitude and direction of the ball's velocity 1 s after being thrown.

   b. Determine the magnitude and direction of the velocity 2 s after being thrown.

   c. Calculate the maximum height reached by the ball.

d. Calculate the time required for the ball to reach its maximum height and the time required to return to the point from which it was thrown.

* * * * * * *

The remaining problems in this chapter deal with motion in two dimensions. It thus becomes important to take into account the vectorial nature of kinematic quantities. Displacement, velocity, and acceleration are all vectors. A brief introduction to vector addition is given in Chapter 2, Secs. 8 and 10. Vector subtraction is treated in Chapter 7, Sec. 2. Additional practice will be provided in the problems that follow. Study Secs. 2-8 through 2-11, then do the following problems.

16. The two arrows in Fig. 2-6 represent successive displacements of an object. The first displacement, $\mathbf{s}_1$, is 12 m to the north. The second displacement, $\mathbf{s}_2$, is 16 m to the east.
    a. Make a drawing to show how these vectors are to be added. Label the two displacements and their sum.

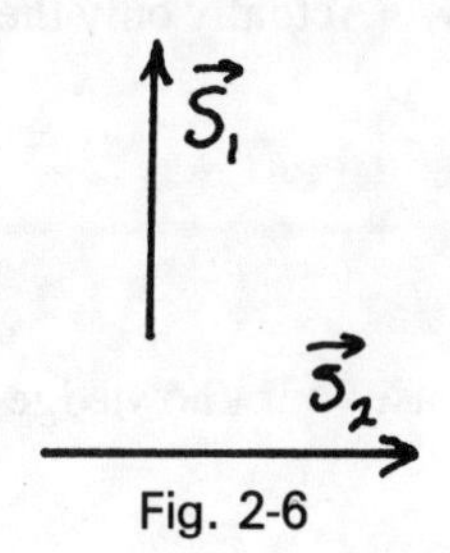

Fig. 2-6

b. Find the magnitude of the resultant displacement.

c. Find the direction of the resultant displacement. Give the angle between **s**, the resultant, and $\mathbf{s}_1$.

17. A motorboat is heading due north in a river that flows toward the west. The speed of the boat with respect to the water is 3.46 m/s. As a result of the flow of the river, the resultant velocity of the boat with respect to the shore is 4.0 m/s at an angle of 30° west of north. The velocities of the river, $\mathbf{v}_r$, and of the boat, $\mathbf{v}_b$, both with respect to the shore, are shown by the arrows below.

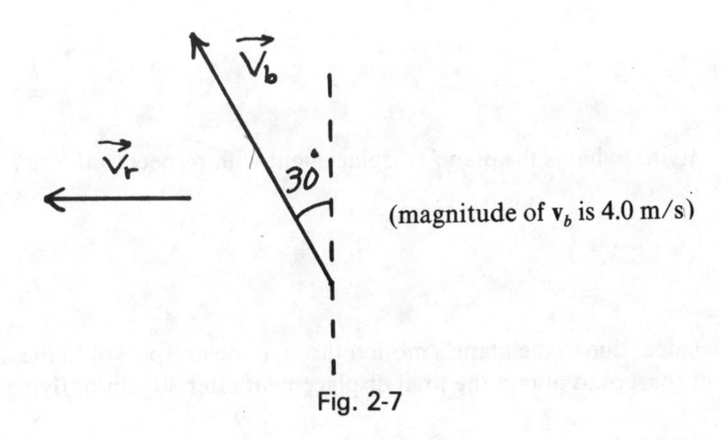

Fig. 2-7

a. Using $\mathbf{v}_{bw}$ to represent the velocity of the boat with respect to the water, write down the equation that expresses the relationship between these three vectors.

b. Make a scale drawing that gives a graphical picture of this relationship. Show clearly which vector is the sum of the other two.

c. Starting with the vectorial equation that you have found in part a, solve algebraically for the unknown vector, $\mathbf{v}_r$. (Actually only the magnitude of $\mathbf{v}_r$ is unknown, the direction being given.)

d. Finally, use your knowledge of trigonometry to determine the speed at which the river is flowing.

18. A light plane has a heading of due east and an air speed of 200 km/h. There is a steady wind from the north of 40 km/h. As a result of the wind the plane is blown off its course (which was to have been due east).

a. What is the resultant velocity with respect to the ground? Give both magnitude and direction.

b. After 30 min of flying, by how many kilometers is the pilot off course?

c. After 30 min of flying, what is the plane's displacement with respect to the air?

d. Add the displacement due to the plane's motion through the air (part c) to the displacement due to the motion of the air (part b) to obtain the total displacement after 30 min of flying.

e. In what direction should the pilot head the plane so that its resultant motion will be due east?

19. An airplane is in a steep dive at an angle of 30° from the vertical. Its speed is 200 m/s. If the sun is directly overhead, how fast is the plane's shadow moving along the ground?

20. A stone is thrown at an angle of 30° with respect to the horizontal. It leaves the origin with an initial speed of 20 m/s, follows a parabolic path and lands at $x = R$ as shown in Fig. 2-8.

a. What is the acceleration of the stone, (1) just after it is thrown and (2) when it reaches the highest point of its path (point A on the diagram)?

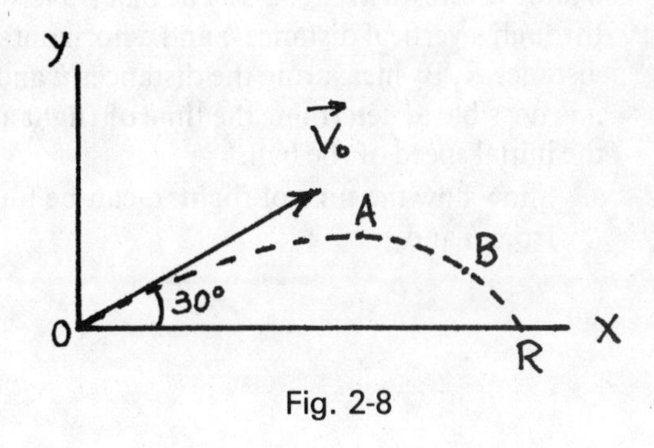

Fig. 2-8

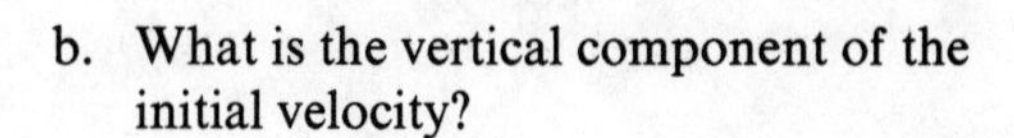

b. What is the vertical component of the initial velocity?

c. What is the vertical component of the velocity when the stone reaches the highest point of its trajectory?

d. How long does it take the stone to reach the highest point?

e. What is the maximum height reached by the stone?

f. What is the horizontal component of the velocity of the stone as it passes through point B?

g. What is the horizontal distance traveled (the distance R in the diagram)?

h. What is the magnitude of the velocity of the stone just before it strikes the ground?

21. A plane, having an air speed of 100 km/h, is headed due east. The pilot notices, however, that he is traveling exactly northeast and that he covers 70.7 km on the ground in 30 min. What is the magnitude and direction of the velocity of the wind that is blowing him off course? (Hint: It will be helpful to make a drawing showing how $\mathbf{v}_a$, the velocity of air, and $\mathbf{v}_{pa}$, the velocity of the plane with respect to the air, are related to $\mathbf{v}_p$, the resultant velocity of the plane.)

22. A steel ball is fired horizontally from the top of a table as shown in Fig. 2-9. The ball travels through a vertical distance h and a horizontal distance R. By measuring the distances h and R it is possible to determine the time of flight and the initial speed of the ball.

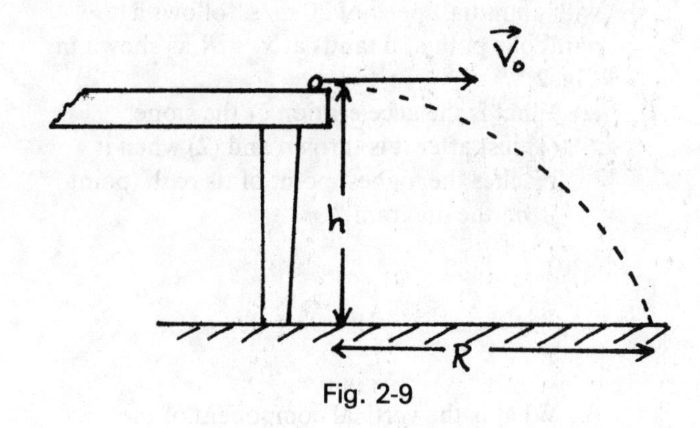

Fig. 2-9

a. Show how the time of flight, t, can be found from h and g.

b. Show how the initial speed, v_0, can be found from the range, R, and the time of flight, t.

c. Combine the two results above to get a single expression giving v_0 as a function of R, g, and h.

d. Calculate v_0 for the case $h = 1$ m, $R = 3$ m.

e. What is the magnitude of the velocity of the ball just before it strikes the floor? (Use $h = 1$ m, $R = 3$ m.)

Solutions

1. $$v_0 = \frac{\text{distance}}{\text{time}} = \frac{s}{t}$$

$$v_0 = \frac{68 \text{ m}}{0.2 \text{ s}} = 340 \text{ m/s}$$

2. $$s = v_0 t = (390 \text{ km/h})\left(\frac{1 \text{ h}}{3600 \text{ s}}\right)\left(\frac{10^3 \text{ m}}{1 \text{ km}}\right)\left(\frac{1}{3} \text{ s}\right) = 36 \text{ m}$$

3a. $v_{av} = \dfrac{s - s_0}{t - t_0} = \dfrac{20\text{ m} - 0\text{ m}}{1\text{ s} - 0\text{ s}} = 10\text{ m/s}$

b. First draw the tangent at $t = 0$.
The slope of the tangent is

$$v_0 = \text{slope} \cong \frac{21\text{ m} - 0\text{ m}}{1\text{ s} - 0\text{ s}} = 21\text{ m/s}$$

The tangent drawn to the curve at $t = 2$ s is a horizontal line whose slope is zero.

$$v_0 \cong 21\text{ m/s} \qquad v_f = 0\text{ m/s} \qquad v_{av} = 10\text{ m/s}$$

c. If $a =$ constant, then

$$v_{av} = \frac{v_f + v_0}{2}$$

The measured average velocity $1/2(0 + 21\text{ m/s}) = 10.5\text{ m/s}$ is quite close to the average value of 10 m/s, and because the $s - t$ curve is quite smooth, it appears that the acceleration was indeed constant for this flight.

d. The slopes of the tangents at $t = 1$ s and $t = 3$ s are shown in Fig. 2-10.

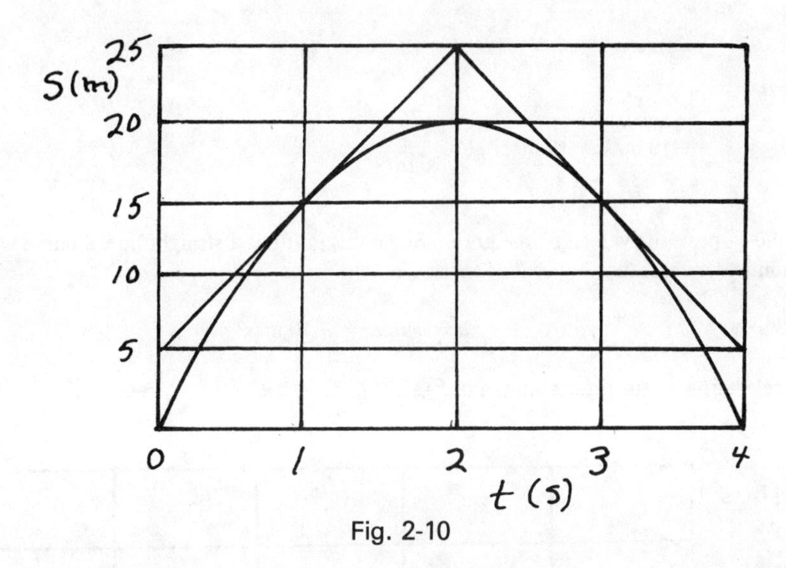

Fig. 2-10

$$v_1 = \frac{25\text{ m} - 5\text{ m}}{2\text{ s} - 0\text{ s}} = 10\text{ m/s} \quad \text{and} \quad v_2 = \frac{5\text{ m} - 25\text{ m}}{4\text{ s} - 2\text{ s}} = -10\text{ m/s}$$

(Your answers may be a little higher or a little lower, depending on how you have drawn the tangents.) The plot of the arrow's velocity as a function of time is shown in Fig. 2.11.

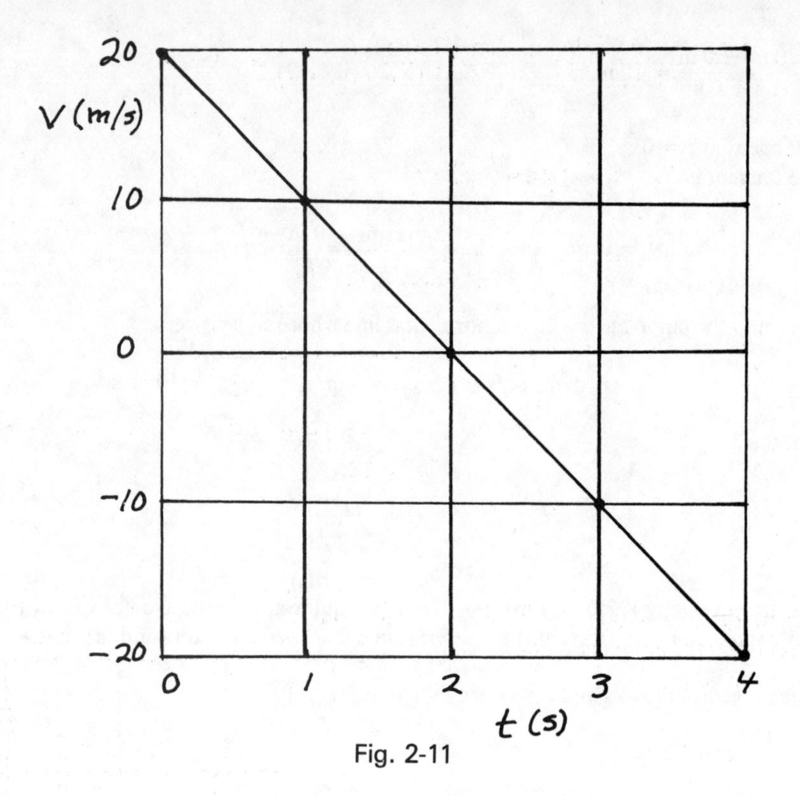

Fig. 2-11

e. $v_{av} = \dfrac{s - s_0}{t - t_0}$

$$a_{av} = \frac{v - v_0}{t - t_0}; \qquad a_{av} = \frac{-10\ \text{m/s} - 10\ \text{m/s}}{3\ \text{s} - 1\ \text{s}} = -10\ \text{m/s}^2$$

f. Acceleration is the slope of the velocity-time graph. Since this graph is a straight line, a curve whose slope is constant, the acceleration must be constant.

$$a = \text{slope} = -10\ \text{m/s}^2$$

g. The graph of acceleration versus time is shown in Fig. 2-12.

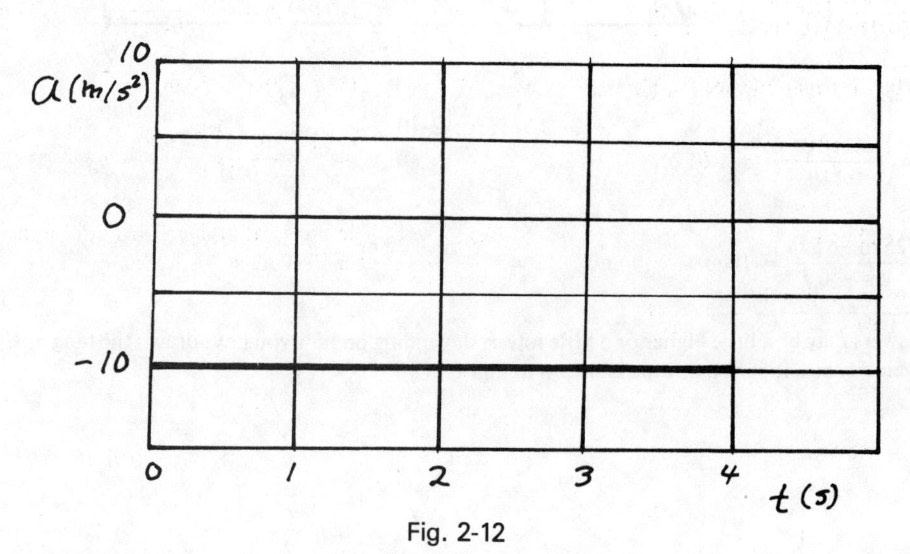

Fig. 2-12

4. $a = \frac{\Delta v}{\Delta t} = \frac{v - v_0}{t - t_0} = \frac{85 \text{ km/h} - 5 \text{ km/h}}{4 \text{ s}} = 20 \frac{\text{km/h}}{\text{s}} \left(\frac{10^3 \text{ m}}{1 \text{ km}}\right)\left(\frac{1 \text{ h}}{3600 \text{ s}}\right) = 5.6 \text{ m/s}^2$

5.

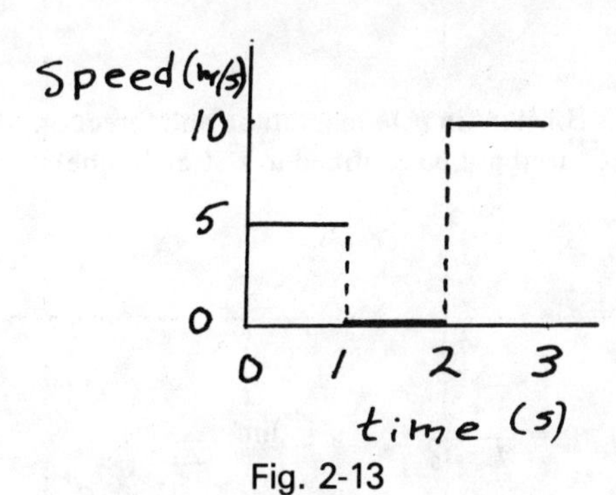

Fig. 2-13

6. The speed (or magnitude of velocity) is the slope of the tangent to the curve at the point in question (see Fig. 2-14). At $t = 1$ s the slope is

$$v_1 = \frac{5.0 \text{ m} - 2.8 \text{ m}}{3 \text{ s}} \cong 0.7 \text{ m/s}$$

At $t = 3$ s the slope is zero. Thus, $v_3 = 0$ m/s.

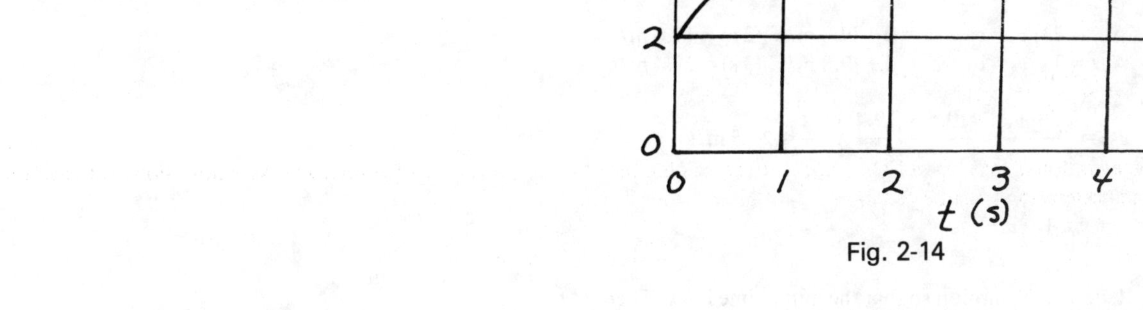

Fig. 2-14

7a. $v_{av} = \frac{s}{t} = \frac{3.6 \text{ cm}}{6(0.01 \text{ s})}\left(\frac{1 \text{ m}}{100 \text{ cm}}\right) = 0.6 \text{ m/s}$

b. No; after 0.06 s it began to speed up.

c. $v_{av}(6 \text{ to } 7) = \frac{1.05 \times 10^{-2} \text{ m}}{0.01 \text{ s}} = 1.05 \text{ m/s}$

$v_{av}(7 \text{ to } 8) = 1.25 \text{ m/s}$

$a_{av} = \frac{1.25 \text{ m/s} - 1.05 \text{ m/s}}{0.01 \text{ s}} = 20 \text{ m/s}^2$

d. $v_{av}(10 \text{ to } 11) = \frac{2.05 \times 10^{-2} \text{ m}}{0.01 \text{ s}} = 2.05 \text{ m/s}$

$v_{av}(11 \text{ to } 12) = 2.50 \text{ m/s}$

$a_{av} = \frac{2.50 \text{ m/s} - 2.05 \text{ m/s}}{0.01 \text{ s}} = 45 \text{ m/s}$

e. The acceleration increased during the interval 0.06 s to 0.12 s.

8. Equations — Conditions of validity

$$\left.\begin{aligned} v &= v_0 + at \\ s &= v_0t + \frac{1}{2}at^2 \\ v^2 &= v_0^2 + 2as \\ s &= \frac{1}{2}(v + v_0)t \end{aligned}\right\}$$

if **a** is constant in *both* magnitude and direction (the path of the object need not be confined to a straight line)

9. $v_{av} = \dfrac{s - s_0}{t - t_0}$ $\quad v = \lim\limits_{\Delta t \to 0} \dfrac{\Delta s}{\Delta t}$ $\quad a_{av} = \dfrac{v - v_0}{t - t_0}$ $\quad a = \lim\limits_{\Delta t \to 0} \dfrac{\Delta v}{\Delta t}$

10. $a = \dfrac{\Delta v}{\Delta t}$

$$a = \frac{v_2 - v_1}{\Delta t} = \frac{8\text{ m/s} - 12\text{ m/s}}{120\text{ s}} = -0.033\text{ m/s}^2$$

The minus sign indicates that the acceleration has a direction opposite to that of the velocities, i.e., **a** is westward.

11. Free fall takes place at a constant, downward acceleration of 9.8 m/s^2. Thus the four equations of uniformly accelerated motion are applicable. It will be convenient to take the downward direction as positive. The initial velocity is zero.

a. At $t = 2$ s, $v_2 = v_0 + at = 0 + (9.8\text{ m/s}^2)(2\text{ s}) = 19.6\text{ m/s}$
b. At $t = 3$ s, $v_3 = v_0 + at = 0 + (9.8\text{ m/s}^2)(3\text{ s}) = 29.4\text{ m/s}$

c. $v_{av} = \dfrac{v_2 + v_3}{2} = \dfrac{(19.6 + 29.4)\text{ m/s}}{2} = 24.5\text{ m/s}$

d. $s = v_0t + \dfrac{1}{2}at^2$

Redefine the motion so that the initial time is 3 s. Then, at $t = 3$ s,

$v_0 = 29.4\text{ m/s}$

The displacement at $t = 4$ s will be

$$s = v_0t + \frac{1}{2}at^2 = (29.4\text{ m/s})(1\text{ s}) + \frac{1}{2}(9.8\text{ m/s}^2)(1\text{ s})^2 = 29.4\text{ m} + 4.9\text{ m} = 34.3\text{ m}$$

This problem can also be solved by finding the distance traveled in 4 s and in 3 s and subtracting.

$$s_4 = 0 + \frac{1}{2}(9.8\text{ m/s}^2)(4\text{ s})^2 = 78.4\text{ m}$$

$$s_3 = 0 + \frac{1}{2}(9.8\text{ m/s}^2)(3\text{ s})^2 = 44.1\text{ m}$$

Thus the distance traveled during the fourth second is

$$s = 78.4\text{ m} - 44.1\text{ m} = 34.3\text{ m}$$

e. $v = v_0 + at;\quad v_0 = 0$

$$t = \frac{v}{a} = \frac{48 \text{ m/s}}{9.8 \text{ m/s}^2} = 4.9 \text{ s}$$

f. $s = v_0t + \frac{1}{2}at^2;\quad v_0 = 0$

$$t^2 = \frac{2s}{a} = \frac{2(48 \text{ m})}{9.8 \text{ m/s}^2} = 9.80 \text{ s}^2$$

$$t = 3.13 \text{ s}$$

12. Let t_1 be the time required for the stone to fall to the surface of the water. Let t_2 be the time required for the sound of the splash to reach the listener's ear. The total time from dropping the stone to the splash is $t_1 + t_2$.

$$s = \frac{1}{2}gt^2 \qquad t_1 = \sqrt{2s/g}$$

$$t_1 = \sqrt{2(46 \text{ m})/(9.8 \text{ m/s}^2)} = 3.06 \text{ s}$$

Since sound travels at a constant speed,

$$s = vt \qquad t_2 = s/v$$

$$t_2 = 46 \text{ m}/(340 \text{ m/s}) = 0.14 \text{ s}$$

Thus the total time is $3.1 \text{ s} + 0.14 \text{ s} \cong 3.2 \text{ s}$.

13. $a = 2 \text{ m/s}^2$

$$a = \frac{v_2 - v_1}{\Delta t};\ \Delta t = \frac{v_2 - v_1}{a} = \frac{15 \text{ m/s} - 7 \text{ m/s}}{2 \text{ m/s}^2} = 4 \text{ s}$$

$$s = v_0t + \frac{1}{2}at^2; \qquad s = (7 \text{ m/s})(4 \text{ s}) + \frac{1}{2}(2 \text{ m/s}^2)(4 \text{ s})^2 = 28 \text{ m} + 16 \text{ m} = 44 \text{ m}$$

The distance can also be calculated from the average velocity.

$$s = v_{av}t = \frac{1}{2}(7 + 15) \text{ m/s } (4 \text{ s}) = (11 \text{ m/s})(4 \text{ s}) = 44 \text{ m}$$

14. The acceleration during free fall is constant in both magnitude and direction. Thus the four equations of this chapter are applicable.

a. $s = v_0t + \frac{1}{2}at^2;\quad v_0 = 0 \qquad t = \sqrt{\frac{2s}{a}} = \sqrt{\frac{2(78.4 \text{ m})}{9.8 \text{ m/s}^2}} = 4.0 \text{ s}$

b. $v = v_0 + at;$

$v_0 = 0 \qquad v = at = (9.8 \text{ m/s}^2)(3 \text{ s}) = 29.4 \text{ m/s}$

c. Since $v_0 = 0$, $t = v/a \quad t = \frac{19.6 \text{ m/s}}{9.8 \text{ m/s}^2} = 2.0 \text{ s}$

d. Find how far the rock travels during two seconds and then subtract the distance that it covers during the first second.

From $t = 0$ to $t = 2$ s: $s_2 = 0 + \frac{1}{2}(9.8 \text{ m/s}^2)(2 \text{ s})^2 = 19.6 \text{ m}$

From $t = 0$ to $t = 1$ s: $\quad s_1 = 0 + \frac{1}{2}(9.8\text{ m/s}^2)(1\text{ s})^2 = 4.9\text{ m}$

Hence, from $t = 1$ s to $t = 2$ s: $\quad \Delta s = s_2 - s_1 = 19.6\text{ m} - 4.9\text{ m} = 14.7\text{ m}$

e. We will first follow the suggested two-step method.
Let the downward direction be positive (see Fig. 2-15).

$v^2 = v_0^2 + 2as \quad v_0$ is upward, thus negative.
$v^2 = (-20\text{ m/s})^2 + 2(9.8\text{ m/s}^2)(78.4\text{ m})$
$v^2 = 1937\text{ m}^2/\text{s}^2$
$v = 44.0\text{ m/s}$

Now the time can be found for the entire distance.

$$t = \frac{v - v_0}{a} = \frac{44.0\text{ m/s} - (-20\text{ m/s})}{9.8\text{ m/s}^2} = 6.53\text{ s}$$

This problem can also be solved in one step, but
it then requires the solution of a quadratic equation.

$$s = v_0 t + \frac{1}{2}at^2$$

$$78.4\text{ m} = (-20\text{ m/s})t + \frac{1}{2}(9.8\text{ m/s}^2)t^2$$

Fig. 2-15

Dividing by 4.9 m/s² gives

$$t^2 - (4.08\text{ s})t - 16\text{ s}^2 = 0$$

The solution is

$$t = \frac{4.08 \pm \sqrt{(4.08)^2 + 64}}{2} = 6.53\text{ s}$$

15. We take the upward direction to be positive. Then, $v_0 = +12$ m/s; $a = -9.8$ m/s².
This is one-dimensional motion with a constant (downward) acceleration. Thus, $v = v_0 + at$

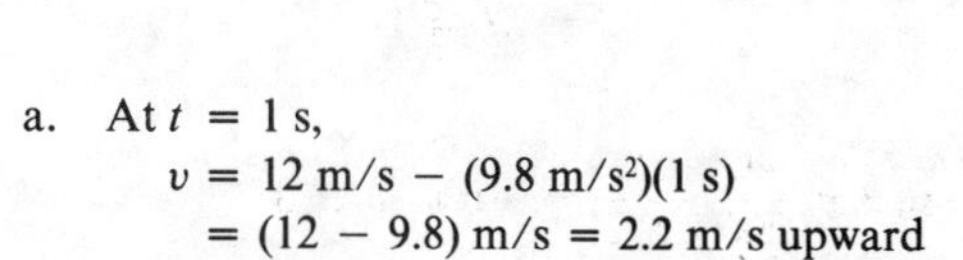

a. At $t = 1$ s,
$v = 12\text{ m/s} - (9.8\text{ m/s}^2)(1\text{ s})$
$= (12 - 9.8)\text{ m/s} = 2.2$ m/s upward

b. At $t = 2$ s,
$v = 12\text{ m/s} - (9.8\text{ m/s}^2)(2\text{ s}) = -7.6\text{ m/s}$
or 7.6 m/s downward

c. $v^2 = v_0^2 + 2as;\ s = \dfrac{v^2 - v_0^2}{2a}$

$$s = \frac{0^2 - (12\text{ m/s})^2}{2(-9.8\text{ m/s}^2)} = 7.35\text{ m}$$

d. $v = v_0 + at;\ t = \dfrac{v - v_0}{a}$

$$t = \frac{0 - 12\text{ m/s}}{9.8\text{ m/s}^2} = 1.22\text{ s}$$

time to reach top = time to return

[Note: The minus sign means "downward" since we chose the upward direction to be positive. It would be redundant (and incorrect) to write −7.6 m/s downward.]

16a.

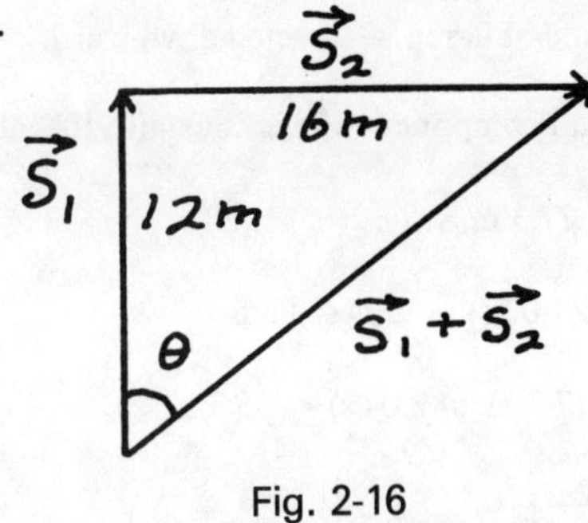

Fig. 2-16

b. $s = \sqrt{s_1^2 + s_2^2} = \sqrt{144 + 256} = \sqrt{400} = 20$ m

c. $\tan\theta = \dfrac{16\text{ m}}{12\text{ m}} = 1.33;\ \theta = 53°$

17a. $\mathbf{v}_b = \mathbf{v}_{bw} + \mathbf{v}_r$

b.

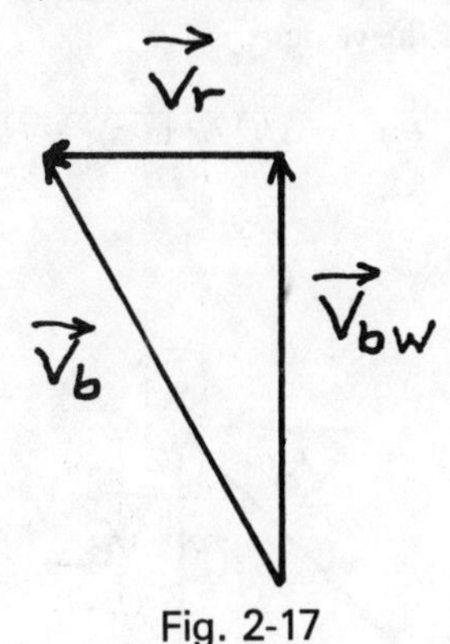

Fig. 2-17

c. $\mathbf{v}_r = \mathbf{v}_b - \mathbf{v}_{bw}$

d. $v_r^2 = v_b^2 - v_{bw}^2 = 16.0\text{ m}^2/\text{s}^2 - 120\text{ m}^2/\text{s}^2 = 4.0\text{ m}^2/\text{s}^2$

$v_r = 2.0$ m/s

18a. We begin with a vector diagram:

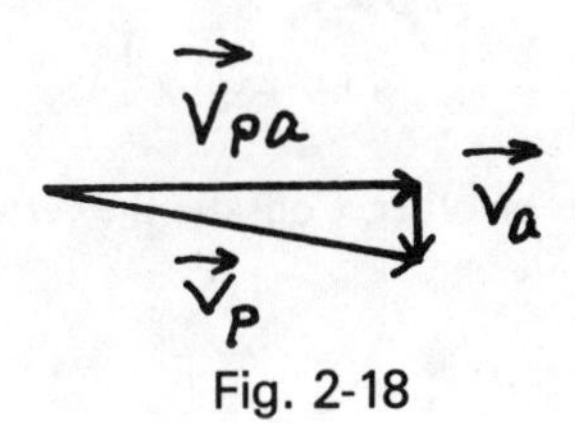

Fig. 2-18

$\mathbf{v}_p = \mathbf{v}_{pa} + \mathbf{v}_a$

$v_p = \sqrt{(200\text{ km/h})^2 + (40\text{ km/h})^2} = 204$ km/h

b. $s_a = (0.5\text{ h})(40\text{ km/h}) = 20$ km

c. $s_{pa} = (0.5\text{ h})(200\text{ km/h}) = 100$ km

d. Displacements, like velocities, are vectors

$\mathbf{s}_R = \mathbf{s}_{pa} + \mathbf{s}_a$

$s_R = \sqrt{(100\text{ km})^2 + (20\text{ km})^2} = 102$ km

e.

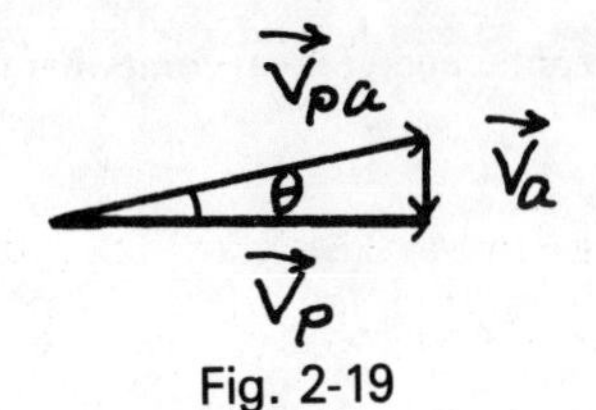

Fig. 2-19

$\tan\theta = \dfrac{40\text{ km/h}}{200\text{ km/h}} = 0.2;\ \theta = 11.3°$

19.

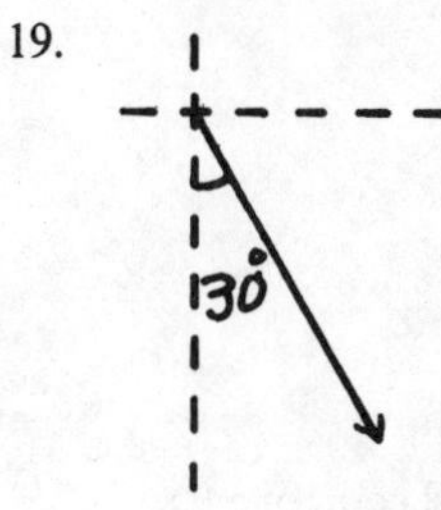

Fig. 2-20

$v_x = v\sin 30°$

$= (200\text{ m/s})(0.5) = 100$ m/s

20a. The acceleration of the stone is the same at all points of its path with a magnitude of 9.8 m/s² directed downward.

b. $v_{y0} = v_0 \sin 30° = (20\ \text{m/s})(0.5) = 10\ \text{m/s}$

c. $v_y = 0$ at the highest point of the trajectory.

d. $v_y = v_{y0} + at; \quad t = \dfrac{v_y - v_{y0}}{a}$

$$t = \frac{(0 - 10)\ \text{m/s}}{-9.8\ \text{m/s}^2} = 1.02\ \text{s}$$

e. $v_y^2 = v_{y0}^2 + 2as$

$$s = \frac{v_y^2 - v_{y0}^2}{2a} = \frac{0 - (10\ \text{m/s})^2}{2(-9.8\ \text{m/s}^2)} = 5.10\ \text{m}$$

f. The horizontal component of **v** is constant. Its value is

$$v_0 \cos 30° = 17.3\ \text{m/s}$$

g. The time is 2(1.02 s) = 2.04 s. Thus

$$R = v_x t = (17.3\ \text{m/s})(2.04\ \text{s}) = 35.3\ \text{m}$$

h. $v_y^2 = v_{y0}^2 + 2as$

Since the vertical displacement s is zero, $v_y = \sqrt{v_{y0}^2} = v_{y0}$. In this case the negative sign must be taken. Thus, $v_y = -10$ m/s. The resultant velocity is $\mathbf{v} = \mathbf{v}_x + \mathbf{v}_y$. The magnitude of the velocity is

$$v = \sqrt{v_x^2 + v_y^2} = \sqrt{(17.3\ \text{m/s})^2 + (10\ \text{m/s})^2} = 20\ \text{m/s}$$

21. The resultant velocity of the plane is the sum of the velocity of the plane with respect to the air $\mathbf{v}_{pa}$ and the velocity of the air $\mathbf{v}_a$. Symbolically, $\mathbf{v}_p = \mathbf{v}_{pa} + \mathbf{v}_a$. (Note that the angle of $\mathbf{v}_a$ in Fig. 2-21 is not given but that of $\mathbf{v}_p$ is.) The magnitude of $\mathbf{v}_p$ is (70.7 km)/(0.5 h) = 141.4 km/h. The eastward component of $\mathbf{v}_p$ is $v_p \cos 45° = (141.4\ \text{km/h})(0.707) = 100\ \text{km/h}$. This is exactly the magnitude of $\mathbf{v}_{pa}$. Thus we can see that $\mathbf{v}_a$ must have an eastward component of zero. In other words, the wind must be blowing due north. Since the triangle is now known to be a right triangle, the unknown side may be found from the Pythagorean theorem.

$$v_a = \sqrt{v_p^2 - v_{pa}^2} = \sqrt{(141.4)^2 - (100)^2} = 100\ \text{km/h}$$

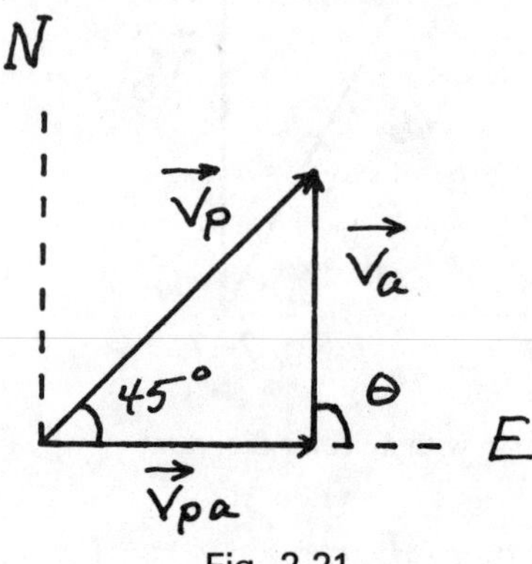

Fig. 2-21

22a. The horizontal and vertical components of the motion are independent of each other. Consider the vertical component first.

$$v_{y0} = 0$$
$$h = v_{y0}t + \tfrac{1}{2}at^2 = \tfrac{1}{2}gt^2$$

$$t = \sqrt{\frac{2h}{g}}$$

b. The horizontal motion takes place at the constant speed, v_0.

$$v_0 = v_x = \frac{R}{t}$$

c. $v_x = \dfrac{R}{\sqrt{2h/g}} = R\sqrt{g/2h}$

d. $v_x = (3\ \text{m})\sqrt{(9.8\ \text{m/s}^2)/(2\ \text{m})} = 6.6\ \text{m/s}$

e. The velocity of the ball just before striking the floor is the vector sum of its horizontal and vertical components.

$$\mathbf{v} = \mathbf{v}_x + \mathbf{v}_y$$
$$v_y = v_{y0} + gt = 0 + g\sqrt{2h/g} = \sqrt{2gh} = \sqrt{2(9.8\ \text{m/s}^2)(1\ \text{m})} = 4.43\ \text{m/s}$$
$$v = \sqrt{v_x^2 + v_y^2} = \sqrt{(6.6\ \text{m/s})^2 + (4.43\ \text{m/s})^2} = 7.95\ \text{m/s}$$

REVIEW TEST FOR CHAPTERS 1 AND 2

1. Which of the following combinations of units have the dimensions of density?

a. $kg \cdot m^3$

b. $kg \cdot m^2$

c. $\frac{kg}{m^2}$

d. $\frac{kg}{m \cdot s}$

e. $\frac{kg}{m^3}$

f. $\frac{m^3}{kg}$

2. Aluminum has a specific gravity of 2.7. Calculate the mass of a cube of aluminum 0.2 m on a side.

3. The isotope of carbon that is often used in the dating of archeological finds is carbon-14, whose symbol is $^{14}_{6}C$. How many protons are contained in the nucleus of this atom?

4. Given that 2.0×10^{23} atoms of the same kind have a mass of 60 g, calculate the mass of one atom of this substance.

5. Calculate the number of molecules of H_2O contained in a spherical droplet 1 mm in diameter. (One mole of H_2O has a mass of 18 g.)

6. Figure T-1 shows the velocity of an automobile as a function of time. What was the distance traveled during the first 3 seconds?

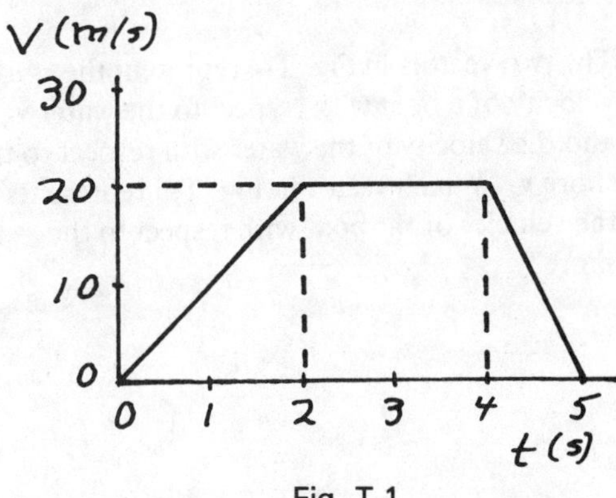

Fig. T-1

7. What was the average acceleration of the automobile in the previous problem during the interval 3 s to 5 s?

8. A small cart, illustrated in Fig. T-2, rolls without friction down an inclined plane. Starting from rest and moving with a constant acceleration, it covers the first 50 cm in 2 s. How much time is required to traverse the second 50 cm?

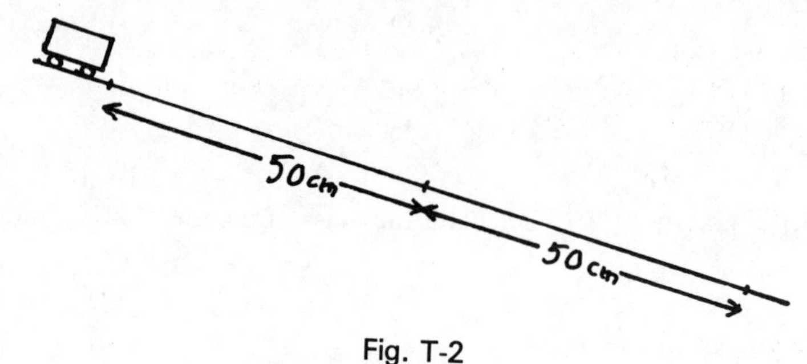

Fig. T-2

9. An object is moving along a straight line with an acceleration that is constant but not zero. Which of the velocity-time graphs below corresponds to this motion?

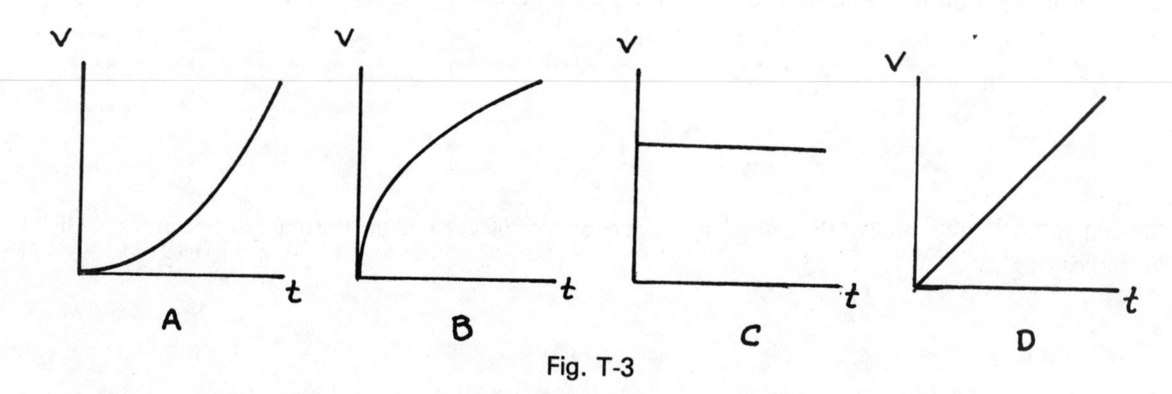

Fig. T-3

10. A small rock having a mass of 50 g is thrown upward at a speed of v_0. A second rock having twice the mass of the first is thrown upward at a speed of $3v_0$. The smaller rock reaches its maximum height after 2.0 s. How long does it take for the larger rock to reach its maximum height?
   a. 2.0 s
   b. 4.0 s
   c. 1.0 s
   d. 6.0 s
   e. 0.67 s

11. The two vectors in Fig. T-4 represent the velocity of a boat with respect to the water $\mathbf{v}_b$ and the velocity of the water with respect to the shore $\mathbf{v}_w$. Which vector in Fig. T-5 represents the velocity of the boat with respect to the shore?

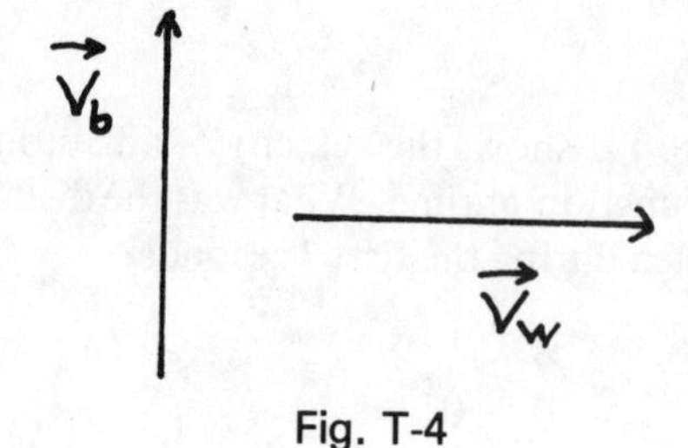

Fig. T-4

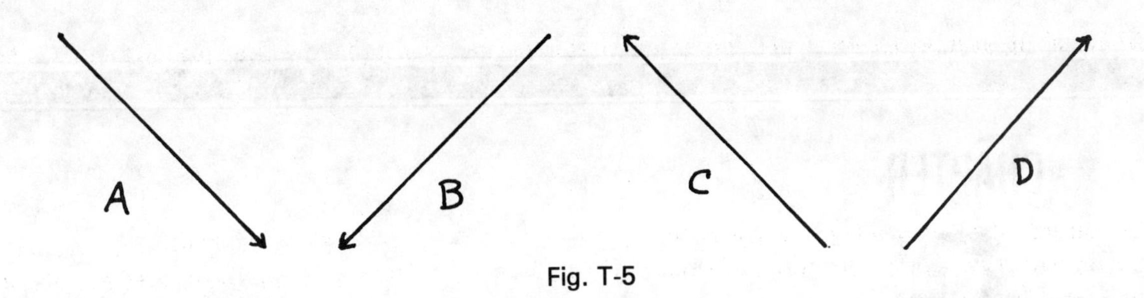

Fig. T-5

12. A jet plane is making a steep takeoff from an east-west runway on a still day. With the sun directly overhead, the plane's shadow is observed to have a speed of 433 km/h when the plane is climbing at an angle of 30°. What is the plane's air speed?

13. An arrow is shot at an angle of 37° with respect to the horizontal as shown in Fig. T-6. The initial velocity v_0 has a magnitude of 20 m/s. What is the magnitude of the velocity of the arrow when it reaches the highest point of its flight?

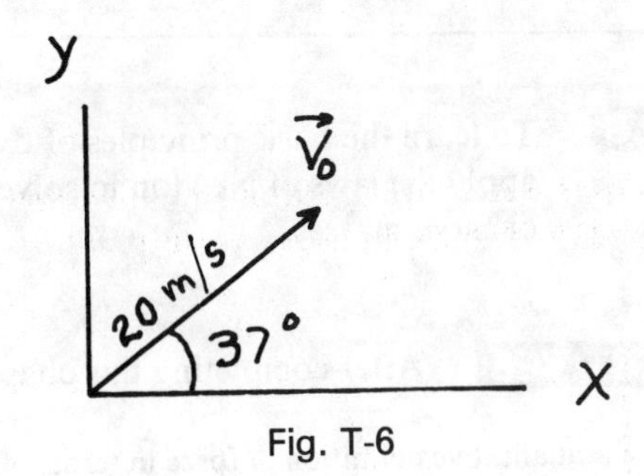

Fig. T-6

14. Two rocks are thrown upward simultaneously with different initial speeds, the faster having twice the initial speed of the slower. It takes 2 s for the slower rock to return to the ground. How long does it take the faster rock to return?

15. A marble is dropped from a third story window to the sidewalk below. At the same moment, a second marble is thrown horizontally from the same height and lands on the sidewalk across the street, a horizontal distance of 20 m. If the time of flight of the first marble was 1.5 s, what was the time of flight of the second?

CHAPTER 3

Dynamics

GOALS

To learn the basic principles of dynamics as embodied in the laws of Newton, and to learn to apply the laws of Newton to solve simple problems in mechanics involving systems containing one or more masses.

OBJECTIVES

After completing this chapter the student should be able to do the following:

1. Write a qualitative definition of force in terms of the acceleration of a material body.
2. State correctly that equilibrium is the condition of constant velocity, which includes zero velocity as a special case.
3. Write a complete statement of Newton's second law of motion such as that given in Chapter 3 of the text. The statement must relate an acceleration to an unbalanced force (or net force) and the mass of the body involved. Magnitude and direction must be explicitly mentioned.
4. Given the force that a body A exerts on a second body B by any means whatsoever, describe completely the force that the second body exerts upon the first.
5. Write the equation that represents symbolically Newton's second law of motion using the standard symbols $\Sigma \mathbf{F}$ (or net $\mathbf{F}$), m, and $\mathbf{a}$.
6. Calculate the weight of a body whose mass is given, and vice versa.
7. Calculate any of the three quantities net F, m, or a when the other two are given or when information is supplied from which the other two quantities may be calculated. (Typical of such problems are examples 3-1, 3-2, 3-4, and 3-5 in the text.)
8. Write the ratio equation using the symbols net F, W, a, and g.
9. Calculate the acceleration of an object acted upon by a given force when the weight of the object is given in SI units. (See examples 3-4 and 3-5 in the text.)
10. Calculate the tension in a cord connecting two parts of a system. (See Example 3-7 in the text.)

SUMMARY

Dynamics is that part of mechanics that deals with the *causes* of motion which are, in fact, *forces*. To put it more exactly, as Newton did so well in his *Principia* in 1687, a net force acting on a body causes that body to be accelerated, to experience a change in its velocity. The magnitude of the acceleration depends on the mass of the body. The exact relationship, called Newton's second law of motion, is written net $\mathbf{F} = m\mathbf{a}$, where net $\mathbf{F}$ is the vector sum of *all* forces acting on an object whose mass is m. This vector equation makes two statements concerning these quantities: (1) The magnitude of the acceleration produced is equal to the magnitude of the resultant force divided by the mass, and (2) the direction of the acceleration is the same as that of the resultant force.

If net $\mathbf{F} = 0$, then it follows that $\mathbf{a} = 0$ which implies in turn that $\mathbf{v}$ = a constant. Thus an object that is subject to no net force must either be at rest ($v = 0$) or moving in a straight line at a constant speed. This statement is called Newton's first law of motion. It is a special case of the more general second law.

Newton's third law of motion states that the mutual interaction of two objects has an important symmetry property, namely, if a body "*A*" exerts a force on a second body "*B*," then body "*B*" must exert an equal and opposite force on body "*A*." This law holds for all types of forces and for bodies of any mass whatsoever.

In the system of units preferred for scientific work, the SI system, forces are expressed in newtons (N) and masses in kilograms (kg). The force with which we are most familiar is the *gravitational* attraction that the earth exerts on bodies in its vicinity. If this gravitational force, called *weight*, is the only force acting on an object, the acceleration produced will be the acceleration of gravity, g. The relationship between weight $\mathbf{W}$ and mass m is $\mathbf{W} = mg$.

QUESTIONS AND PROBLEMS

Introduction and Sec. 3-1

1. A hockey puck is sliding across the ice in a straight line at a constant speed with a negligible frictional force acting on it. This motion can be described as $\mathbf{v}$ = ______________, or as $\mathbf{a}$ = ______________. Since the puck has an acceleration of zero, we know that $\Sigma\mathbf{F}$ is also zero. Thus the puck is in a state of ______________.
2. A small body is acted upon by two forces, $\mathbf{F}_1$ and $\mathbf{F}_2$. The body is moving in a straight line at a constant speed. $\mathbf{F}_1$ has a magnitude of 5 N toward the east. Describe the force $\mathbf{F}_2$.

   $|\mathbf{F}_2|$ = ______________.

   The direction of $\mathbf{F}_2$ is ______________.
3. What is a force? To say that a force is a push or a pull is not very helpful. There is much more content in the definition: A force is that which can cause the ______________ of a material body.

Secs. 3-2—3-5

4. A certain body is acted upon by only two forces, $\mathbf{F}_1$ and $\mathbf{F}_2$, which satisfy the equation $\mathbf{F}_1 = -\mathbf{F}_2$. Describe the motion of this body. Could this body be at rest?

5. A force of 10 N causes a low-friction cart to accelerate at the rate of 0.5 m/s². When the first force is replaced by a second, an acceleration of 1.5 m/s² is observed. What was the magnitude of the second force? Do the calculation without finding the mass of the cart.

6. Figure 3-1 shows a spaceship in orbit around the moon. The moon exerts a force **A** on the lunar orbiter as indicated on the diagram. Draw another arrow, **B**, to represent the force that the spaceship exerts upon the moon. What statements can you make about these two forces?

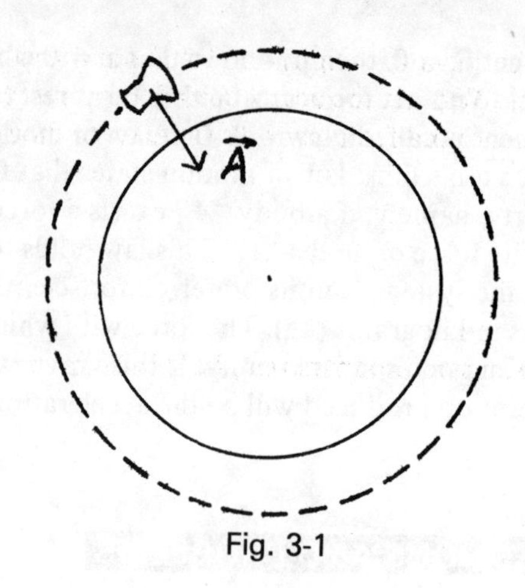

Fig. 3-1

7. A block of weight **W** is resting upon a horizontal surface as shown in Fig. 3-2. According to Newton's third law "every action has an equal and opposite reaction." What is the force that may be described as the "reaction" associated with **W**? On what body does it act?

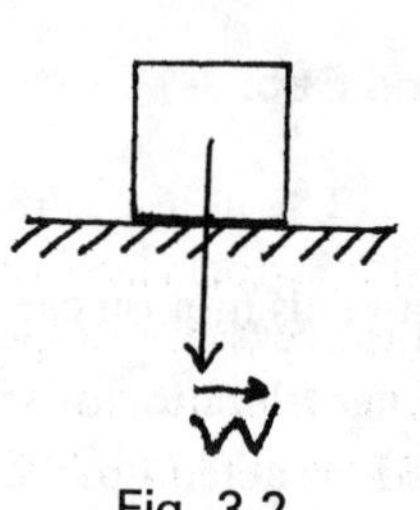

Fig. 3-2

8. Since the block in Prob. 7 is in equilibrium (i.e., **a** = 0) there must be a second force, call it **N**, such that **W** + **N** = 0. What is this force **N**? On what body does it act?

9. A single force **F**, acting on a mass of 2 kg, produces an acceleration of 3 m/s². What is the magnitude of this force?

10. The mass of a falling coconut is 2.0 kg. At a certain point during its fall the upward force of air resistance becomes 11.6 N. What is the acceleration of the coconut at that point? Make a drawing of the coconut with all of the forces that act on it. (Check the solution for this step before proceeding.)

11. An elevator weighs 4×10^5 N but the cable which supports it exerts an upward force of 5×10^5 N.
   a. What is the acceleration? Give direction as well as magnitude. (Hint: The mass of the elevator may be found from $m = W/g$. A force diagram will be very helpful.)

   b. Starting from rest, how far will this elevator rise in 2 s?

Sec. 3-6

In this section Newton's laws of motion are applied to systems consisting of two or more parts. In solving such problems it is necessary to specify clearly what the system consists of. The second law of motion, $\Sigma\mathbf{F} = m\mathbf{a}$, applies to a system of bodies, but one must remember that $\Sigma\mathbf{F}$ includes all forces acting on the system or any part of it due to the action of bodies *outside* the system. These are called *external forces*. Forces exerted by one body in a system on another body within the system are equal and opposite according to the third law of Newton. Such forces occur only in pairs and thus cancel out. Let us look at an example. Consider two pucks on a horizontal surface connected by a massless string (Fig. 3-3). Their masses are indicated by m_1 and m_2. Some external agent (not shown) is exerting a force **F** to the right. Friction with the surface produces small forces $\mathbf{f}_1$ and $\mathbf{f}_2$ which oppose the motion.

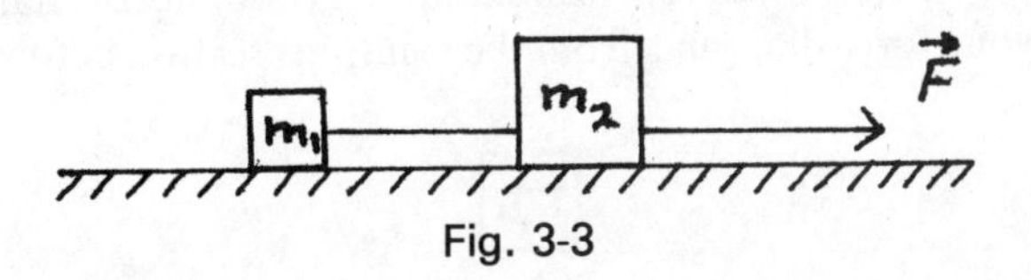

Fig. 3-3

Let us choose as our system the two masses and the string connecting them. This is illustrated in Fig. 3-4. The horizontal surface is outside the system. The two frictional forces are therefore external forces and must be included in $\Sigma\mathbf{F}$. There is also a force $\mathbf{T}_2$ pulling to the left on m_2 due to the tension in the string, and a force $\mathbf{T}_1$ pulling to the right on m_1. Since the string is massless, it serves only to transmit the forces $\mathbf{T}_1$ and $\mathbf{T}_2$. Thus $\mathbf{T}_2$ is in fact the force that m_1 exerts on m_2, and $\mathbf{T}_1$ is the force that m_2 exerts on m_1. By Newton's third law, these forces are equal and opposite; their sum is zero. Thus, taking the positive direction toward the right,

$$\Sigma F = F - f_1 - f_2$$

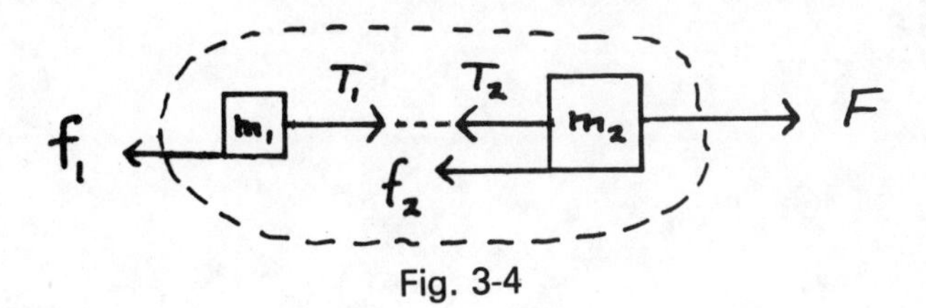

Fig. 3-4

This is the total external force acting on the system. Since the system contains two masses, its *total* mass is $m_1 + m_2$.

$$F - f_1 - f_2 = (m_1 + m_2)a$$

The three blocks pictured in Fig. 3-5 are connected by massless strings and are free to move without friction on a horizontal surface. A force **F** acts on the first block thus causing all three to be accelerated toward the left.

Write your answers to all questions in the space provided.

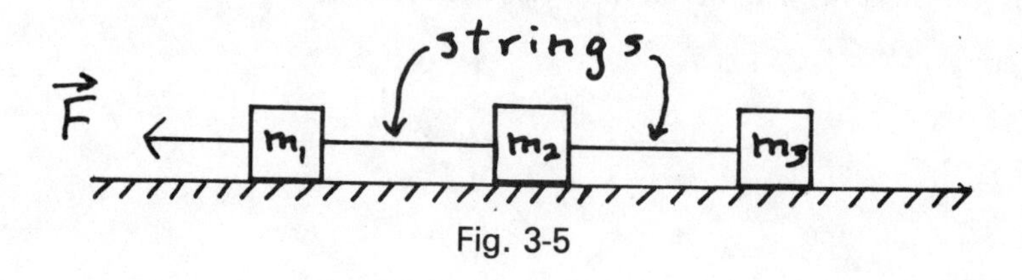

Fig. 3-5

12. Place arrows on this drawing to represent all *horizontal* forces acting on m_2. Label these forces.

Fig. 3-6

13. Place arrows on the drawing in Fig. 3-7 to represent all forces (both horizontal and vertical) acting on m_1. Verify the correctness of your force diagrams from the solutions section before proceeding.

Fig. 3-7

14. Taking the "system" under consideration to be the two masses m_2 and m_3 (Fig. 3-8), make a force diagram showing all external forces (in whatever direction) acting on this system. Do *not* include any internal forces, that is, forces that one part of the system may exert on another part of the system.

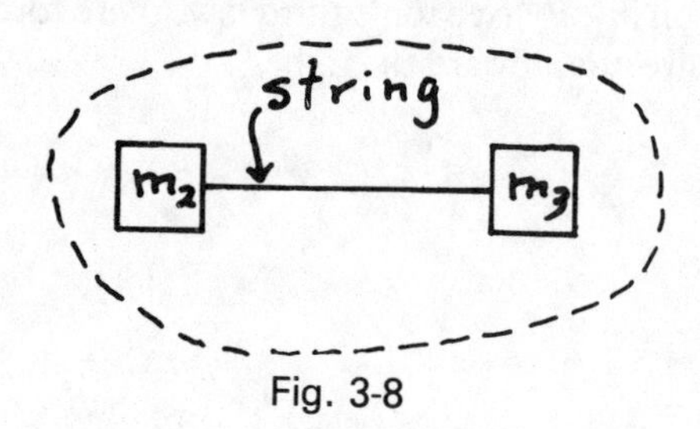

Fig. 3-8

15. Now consider a still different "system" consisting of all three blocks. Make a force diagram showing all *horizontal, external* forces acting on this system.

16. Suppose that the force **F** is known and that the three masses are given. What law of motion will allow you to calculate the acceleration of one of the blocks, say m_2? Name the law and write down a mathematical expression of it.

The law of motion involved is ______________________________.

The mathematical expression of this law is ______________________________.

Since the motion under discussion is one-dimensional, we need deal only with horizontal components of forces. The vertical forces add up to zero and may be left out of our consideration.

17. To what system should the second law of motion be applied in order to find the acceleration of m_2? (There are six possibilities to be considered: m_1, m_2, m_3, m_1 and m_2, m_2 and m_3, and all three masses together.)

The system to be chosen consists of ______________________________.

18. What is the advantage of the choice that you have just made? Why would a different choice of the "system" lead to an equation that could not be solved?

19. The magnitude of F is 12 N and each mass is 1.0 kg. Calculate the acceleration of m_2.

$a =$ ______________.

20. We proceed now to the second part of the problem, the determination of the force of tension in the connecting string. The acceleration of each block is now known. To what "system" should the law of motion now be applied so that the tension in the string between m_1 and m_2 can be determined? Describe the "system" and state the reasons for your choice.

The second law of motion should be applied to a system consisting of ______________________________.

This system is chosen because ______________________________

______________________________.

21. Make a force diagram for the system just described. Show only the horizontal forces.

22. Apply Newton's second law of motion to this system and solve for the unknown force of tension.

$T_1 =$ ______________.

23. Following the same procedure, determine the force of tension in the string connecting m_2 and m_3. Make the necessary force diagram.

$T_2 =$ ______________.

24. We will now consider a more realistic case in which friction is present. We suppose that a frictional force of 1.0 N is acting on each of the three blocks, somewhat impeding their motion. The force F is 12 N as before and the masses are unchanged.

a. Make a complete force diagram for a properly chosen system which will allow the unknown acceleration to be determined.

b. Apply Newton's second law of motion to the system and solve for the acceleration.

$a =$ ______________ .

c. Choose a different system that will allow the tension in the string between m_1 and m_2 to be determined. Make a force diagram for this system.

d. Apply Newton's second law of motion to the system and solve for the unknown tension.

$T_1 =$ ______________ .

e. Finally, carry out the complete calculation to determine the other unknown tension. Include the force diagram.

$T_2 =$ ______________ .

* * * * * * *

25. The two objects in Fig. 3-9 have the same mass and the same weight, 5 N in each case. The magnitude of the upward force **F** acting on object 1 due to the supporting cord is 20 N. Determine the acceleration of each object and the tension in the connecting cord.

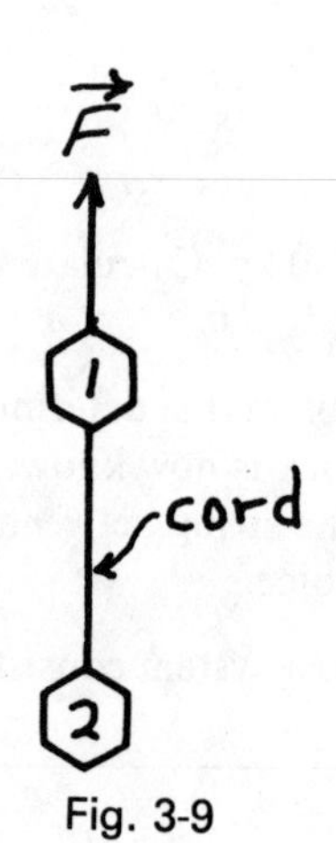

Fig. 3-9

Make a drawing of the system to which the second law of motion should be applied in order to determine the acceleration. Indicate in your drawing all external forces acting on this system. Solve for the unknown acceleration showing all necessary details.

$a =$ ______________ .

26. Make a drawing of the system shown in Fig. 3-9 to which the second law of motion should be applied in order to determine the unknown tension. Show all forces acting on this system. Calculate the unknown tension.

$T =$ ______________ .

You should now be able to solve the following problems without difficulty. If you need further help, study the worked-out examples.

27. A block whose mass is 3 kg is to be pulled along a horizontal table top by means of a string passing over a pulley (see Fig. 3-10). The string is also attached to a second block of mass 2 kg, which hangs vertically. A small peg keeps the 3-kg block from sliding prematurely. The pulley is frictionless.

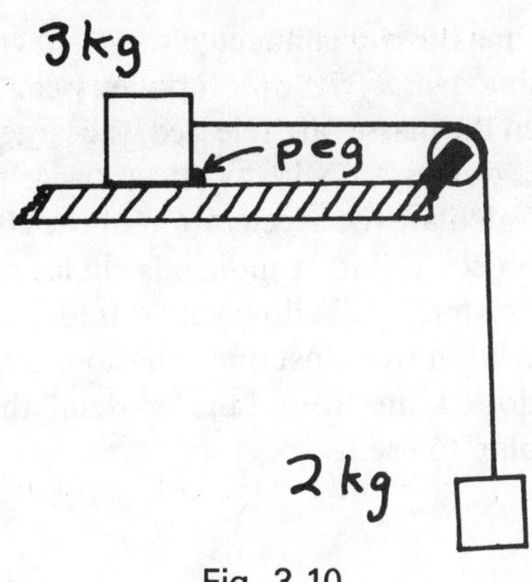

Fig. 3-10

a. What is the tension in the string when the peg is in place thus preventing any motion?

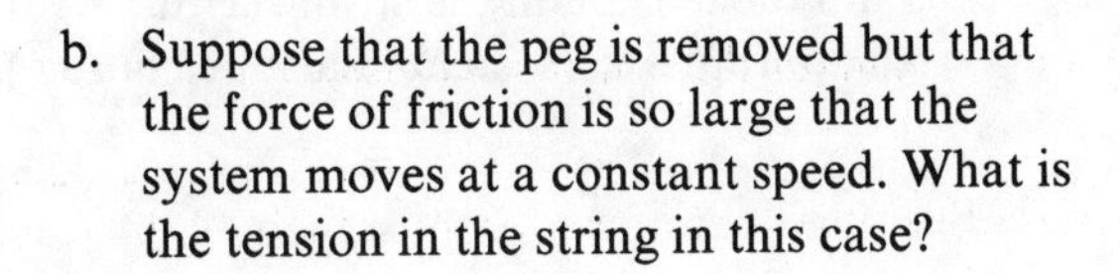

b. Suppose that the peg is removed but that the force of friction is so large that the system moves at a constant speed. What is the tension in the string in this case?

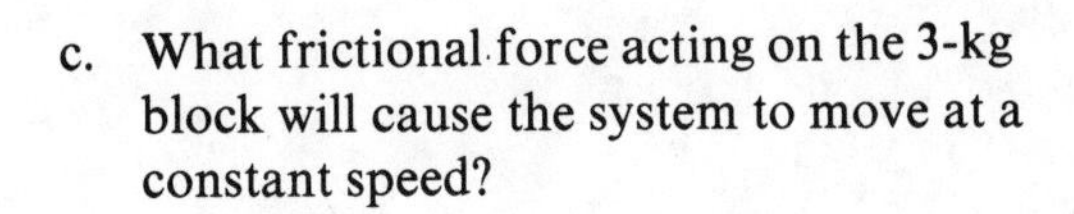

c. What frictional force acting on the 3-kg block will cause the system to move at a constant speed?

d. Assume now that there is no friction between the sliding block and the horizontal surface. What is the acceleration of the system?

e. Starting from rest, how long will it take for the hanging block to fall 2 m?

f. Under the condition of no friction, what will be the tension in the string?

g. Supposing that the force of friction between the 3-kg block and the horizontal surface is 9.8 N, what will be the acceleration and the tension in the string?

28. Two masses are connected to a light string passing over a frictionless pulley (see Fig. 3-11). When the masses are released, the larger one will move downward with a constant acceleration.

a. Calculate the acceleration of the 200-g object. (Hint: A judicious choice of the "system" will allow you to find the acceleration from just one equation. Draw a dotted line around the "system" that you plan to use.)

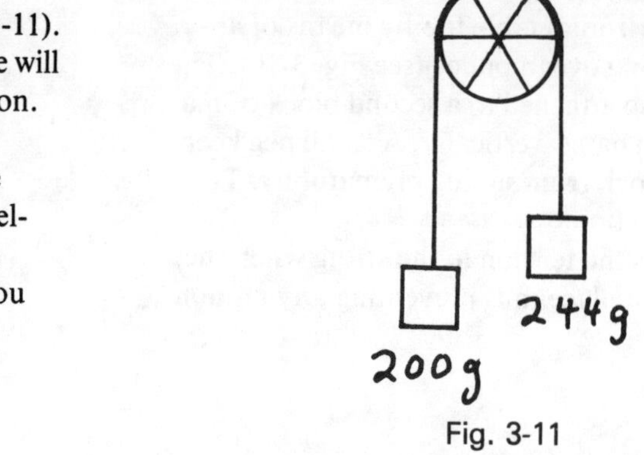

Fig. 3-11

b. Through what distance must the 244-g object fall, starting from rest, to attain a speed of 0.8 m/s?

c. Find the tension in the string. (Hint: Again an appropriate "system" must be chosen.)

Solutions

1. constant; zero; equilibrium
2. Since $\mathbf{a} = 0$, it follows that net $\mathbf{F} = 0$. Thus $\mathbf{F}_1 + \mathbf{F}_2 = 0$ or $\mathbf{F}_2 = -\mathbf{F}_1$. The unknown force has the same magnitude as $\mathbf{F}_1$ but the opposite direction. Thus $|\mathbf{F}_2| = 5$ N. The direction of $\mathbf{F}_2$ is toward the west.
3. acceleration
(If you wrote "motion" or "velocity," read Sec. 3-1 again. It is important to understand that motion does *not* imply the presence of a force; it is change of motion, i.e., acceleration, that requires a force.)
4. Since $\Sigma\mathbf{F} = \mathbf{F}_1 + \mathbf{F}_2 = 0$, we know that the body on which these forces act is in equilibrium, that is to say *it is moving in a straight line at a constant speed* or *it is at rest.*
5. For a given object net F/a is a constant. Thus the magnitudes of the vectors are related as follows:

$$\frac{F_1}{a_1} = \frac{F_2}{a_2}$$

Inserting the known quantities,

$$\frac{10\text{ N}}{0.5\text{ m/s}^2} = \frac{F_2}{1.5\text{ m/s}^2}; \; F_2 = 30\text{ N}$$

6. Since the force **B** acts on the moon as a whole, we place it at the center of the moon. These two forces are equal and opposite. They are represented by arrows of equal length, shown in Fig. 3-12.

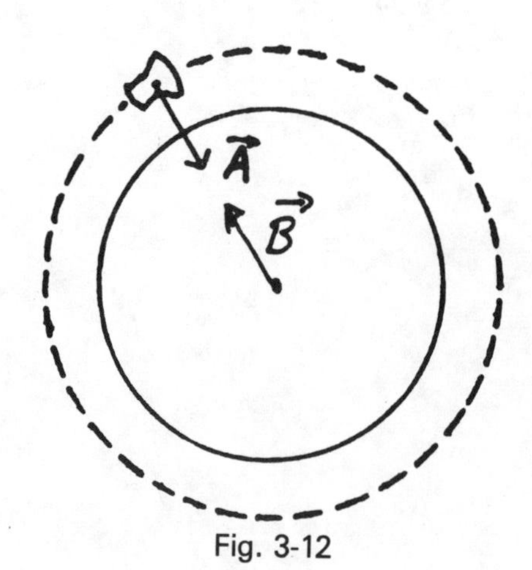

Fig. 3-12

7. The force **W** is the gravitational attraction exerted *by* the earth *on* the block. Its reaction is the equal and opposite force exerted *by* the block *on* the earth. This reaction to **W** should be drawn at the center of the earth.

8. The required force must be acting upward so as to balance **W**. This force, **N**, must be an upward force of the table top *against the block*, as illustrated in Fig. 3-13.

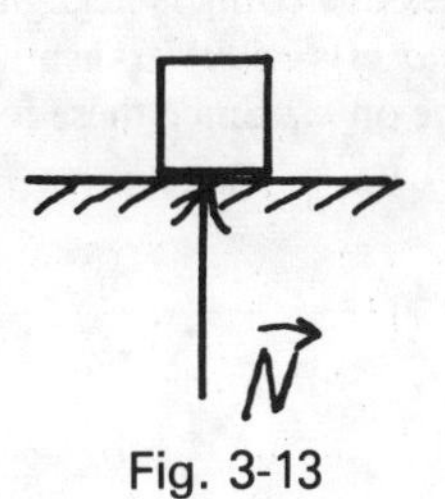

Fig. 3-13

9. $F = ma = (2\text{ kg})(3\text{ m/s}^2) = 6\text{ N}$

10. The forces are shown in Fig. 3-14. Calculate the weight, mg.

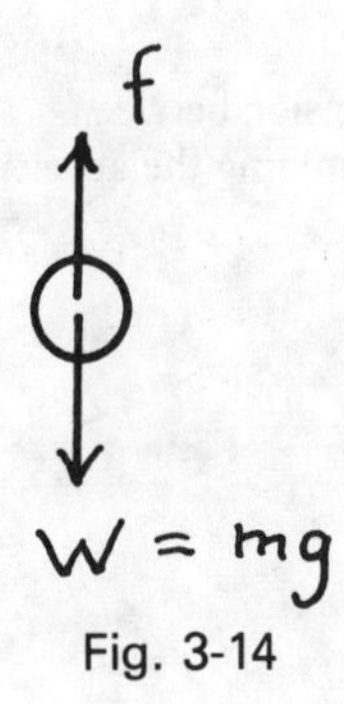

Fig. 3-14

$$mg = (2\text{ kg})(9.8\text{ m/s}^2) = 19.6\text{ N}$$

Calculate the net force. Let the positive direction be downward.

$$\text{net } F = mg - f = 19.6\text{ N} - 11.6\text{ N} = 8.0\text{ N}$$

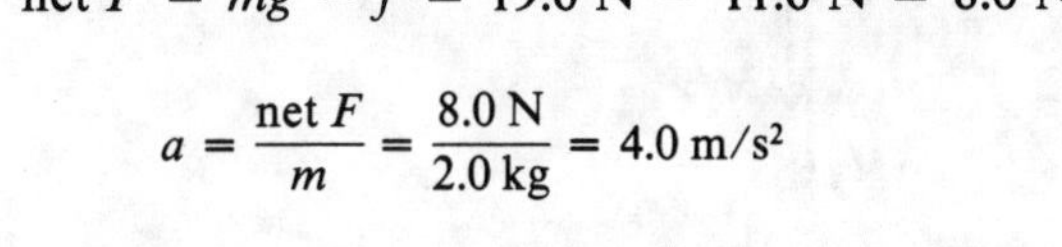

$$a = \frac{\text{net } F}{m} = \frac{8.0\text{ N}}{2.0\text{ kg}} = 4.0\text{ m/s}^2$$

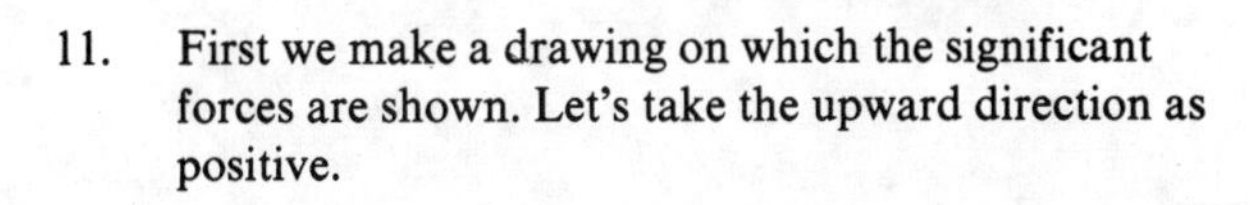

11. First we make a drawing on which the significant forces are shown. Let's take the upward direction as positive.

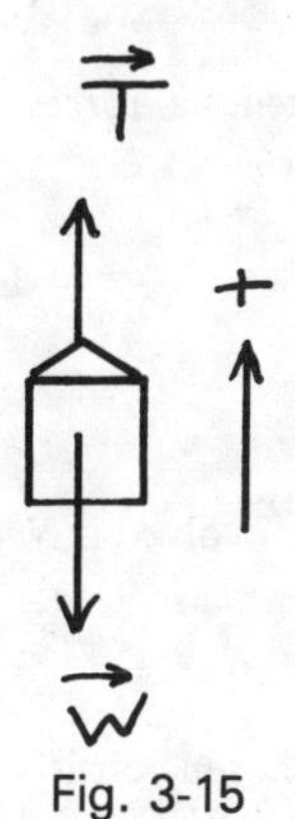

Fig. 3-15

a. net **F** = **T** + **W**

This is one-dimensional motion so we deal only with the vertical components.

$$\text{net } F = 5 \times 10^5\text{ N} - 4 \times 10^5\text{ N} = 1 \times 10^5\text{ N}$$

From Newton's second law,

$$\text{net } F = ma$$

$$a = \frac{\text{net } F}{m} = \frac{\text{net } F}{W/g} = \frac{(\text{net } F)\,g}{W}$$

$$a = \frac{1 \times 10^5\text{ N}}{4 \times 10^5\text{ N}}(9.8\text{ m/s}^2) = 2.45\text{ m/s}^2$$

b. Since the acceleration is constant, we may use the relation

$$s = v_0 t + \tfrac{1}{2}at^2$$

Putting $v_0 = 0$ and $a = 2.45\text{ m/s}^2$,

$$s = \tfrac{1}{2}(2.45\text{ m/s}^2)(2\text{ s})^2 = \mathbf{4.9\ m}$$

12. Since two strings are attached to m_2 there are two forces of tension acting on it. The vertical forces are omitted below.

Fig. 3-16

13. Since the mass of a string is negligible, the tension in a string has the same value everywhere. T_1 represents the force that m_1 exerts on m_2 (transmitted by the string). According to Newton's third law, m_2 must exert an equal and opposite force on m_1. Since these forces have the same magnitude, we represent their common magnitude by the single symbol T_1.

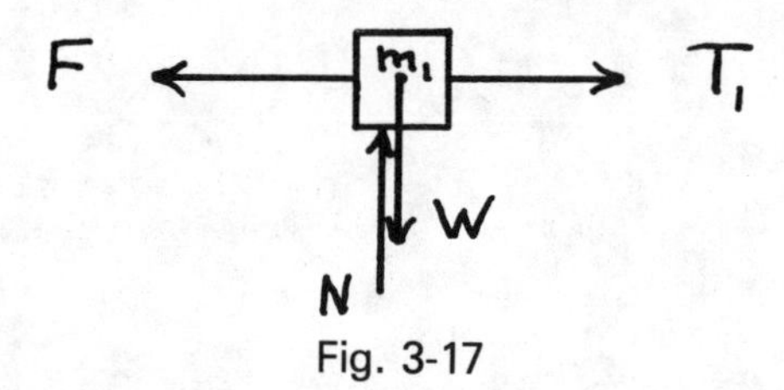

Fig. 3-17

14. The forces of tension between m_2 and m_3 are internal forces and are omitted from the diagram below since they play no role in determining the acceleration of the "system."

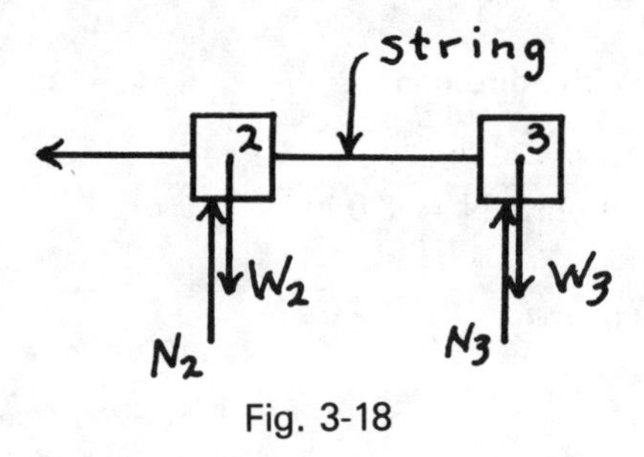

Fig. 3-18

15. Remember that frictional forces are being neglected.

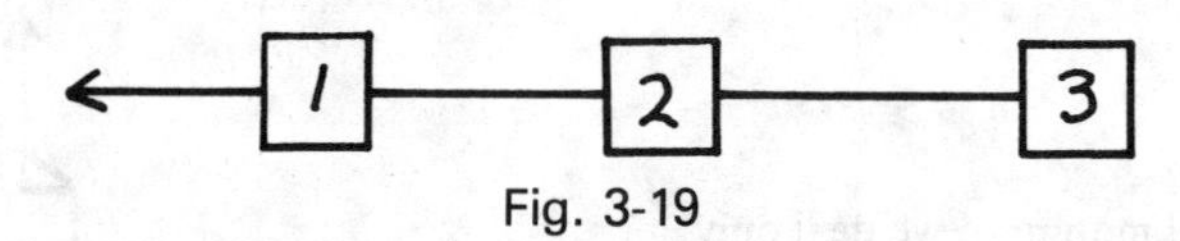

Fig. 3-19

16. The law of motion involved is *Newton's second law of motion.*

$$\text{net } \mathbf{F} = m\mathbf{a} \quad \text{or} \quad \Sigma\mathbf{F} = m\mathbf{a}$$

17. The system to be chosen consists of *all three masses together*.
18. The advantage of this choice is that no unknown forces are acting on the system. The external force **F** is known. The internal forces, $\mathbf{T}_1$ and $\mathbf{T}_2$, are unknown. If any other choice were made, the equation, net $\mathbf{F} = m\mathbf{a}$, would contain at least one unknown force plus an unknown acceleration.
19. net $F = ma$ (m is the *total* mass of the system)

$$12 \text{ N} = (3.0 \text{ kg})a; \; a = 4 \text{ m/s}^2$$

20. m_1 alone. The system is chosen because *the only unknown is the force* T_1*, which we wish to calculate.*

21.

12 N ← m_1 → T_1

Fig. 3-20

22. net $F = ma$; $12 \text{ N} - T_1 = (1.0 \text{ kg})(4 \text{ m/s}^2)$
$T_1 = 12 \text{ N} - 4 \text{ N} = 8 \text{ N}$

23.

8 N ← m_2 → T_2

Fig. 3-21

net $F = ma$; $8 \text{ N} - T_2 = (1.0 \text{ kg})(4 \text{ m/s}^2)$
$T_2 = 8 \text{ N} - 4 \text{ N} = 4 \text{ N}$

24a.

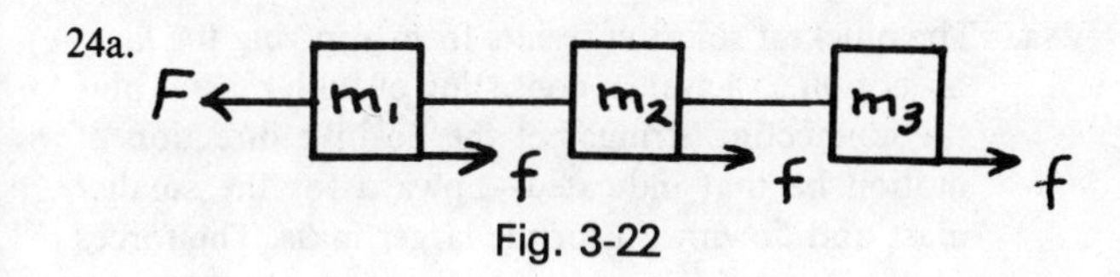

Fig. 3-22

b. net $F = F - 3f = 12\text{ N} - 3\text{ N} = 9\text{ N}$
net $F = ma$; $9\text{ N} = (3.0\text{ kg})(a)$
$a = 3\text{ m/s}^2$

c.

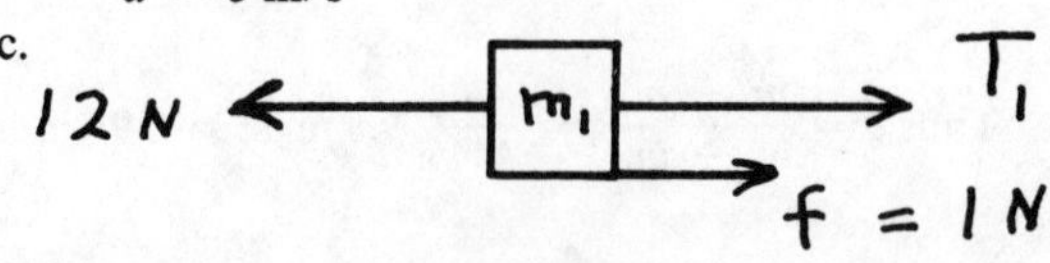

Fig. 3-23

d. net $F = ma$; $12\text{ N} - T_1 - 1\text{ N} = (1.0\text{ kg})(3\text{ m/s}^2)$
$T_1 = 11\text{ N} - 3\text{ N} = 8\text{ N}$

e.

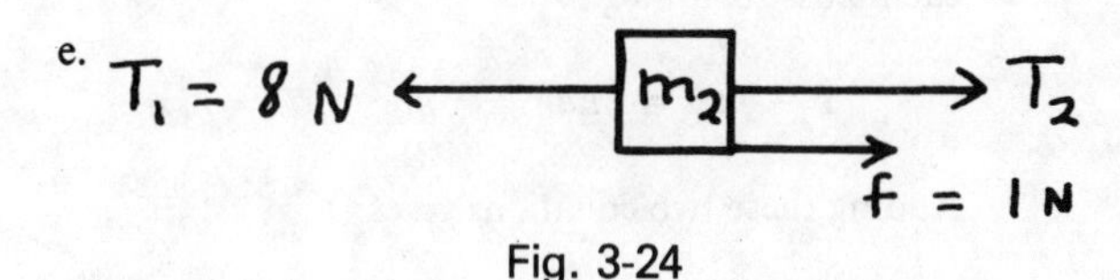

Fig. 3-24

$8\text{ N} - T_2 - 1\text{ N} = (1.0\text{ kg})(3\text{ m/s}^2)$
$T_2 = 7\text{ N} - 3\text{ N} = 4\text{ N}$

25. net $F = ma$

$F - W_1 - W_2 = (m_1 + m_2)a$

$$a = \frac{20\text{ N} - 10\text{ N}}{W_1/g + W_2/g}$$

$$a = \frac{10\text{ N}}{10\text{ N}} 9.8\text{ m/s}^2$$

$a = 9.8\text{ m/s}^2$
This acceleration is upward.

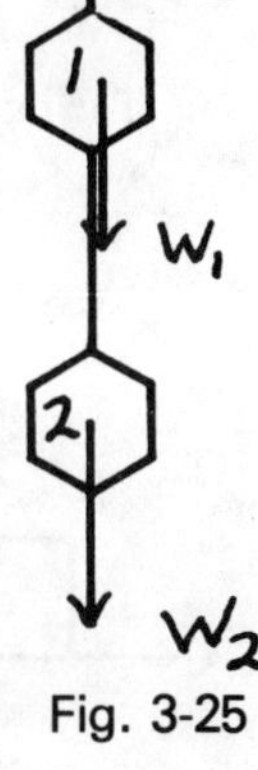

Fig. 3-25

26. net $F = ma$

$$T - W_2 = m_2 a = \frac{W_2}{g} a$$

$$T - 5\text{ N} = \frac{5\text{ N}}{g} a = 5\text{ N}$$

$T = 10\text{ N}$

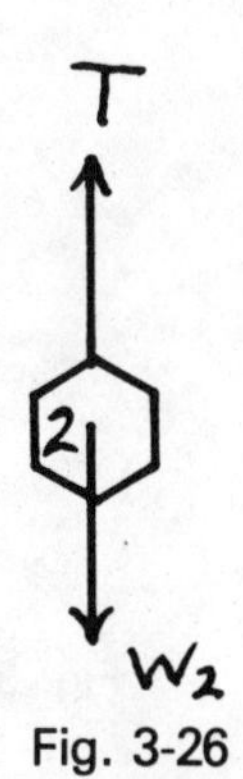

Fig. 3-26

27a. Since v has the constant value of zero, the acceleration is also zero. The upward force of the string on the hanging block will be exactly equal to the downward force of gravity. Thus,

$$T = mg = (2\text{ kg})\left(9.8\frac{\text{m}}{\text{s}^2}\right) = 19.6\text{ N}$$

b. Here again the velocity is constant. Thus the acceleration is zero and the net force is zero.

$$\text{net } F = mg - T = 0$$

Thus,

$$T = mg = 19.6\text{ N}$$

c. Since the 3 kg block is in equilibrium, the force of friction must be equal in magnitude to the force exerted by the string.

$$f = 19.6\text{ N}$$

f ← □ → T = 19.6 N

Fig. 3-27

d. The acceleration can be found by applying Newton's second law to a "system" consisting of both blocks plus the connecting string. This system has a mass of 5 kg and is subject to an external force of (2 kg) (9.8 m/s²) = 19.6 N. The two forces of tension are equal and opposite; they cancel each other.

$$\text{net } F = ma$$

$$(2\text{ kg})\left(9.8\frac{\text{m}}{\text{s}^2}\right) = (2\text{ kg} + 3\text{ kg})\,a$$

$$a = \left(\frac{2}{5}\right)\left(9.8\frac{\text{m}}{\text{s}^2}\right) = 3.92\frac{\text{m}}{\text{s}^2}$$

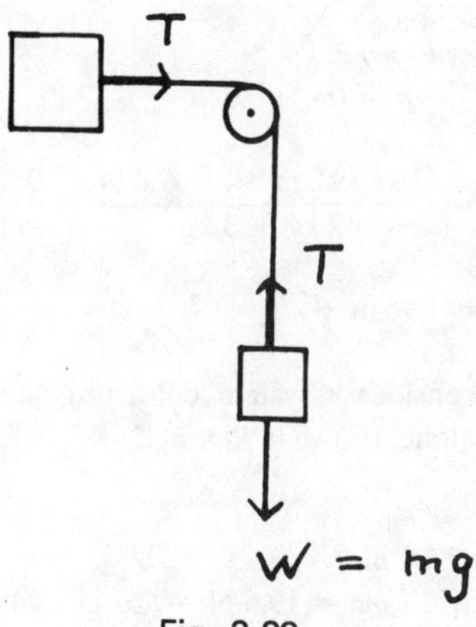

Fig. 3-28

e. This is uniformly accelerated motion.

$$s = v_0 t + \frac{1}{2}at^2$$

Since $v_0 = 0$, $t = \sqrt{\frac{2s}{a}}$

$$t = \sqrt{\frac{2(2\text{m})}{3.92 \text{ m/s}^2}} = 1.01 \text{ s}$$

f. The acceleration is known from part d. The tension can be found by applying the second law of motion to the 3-kg block alone or to the 2-kg block alone:

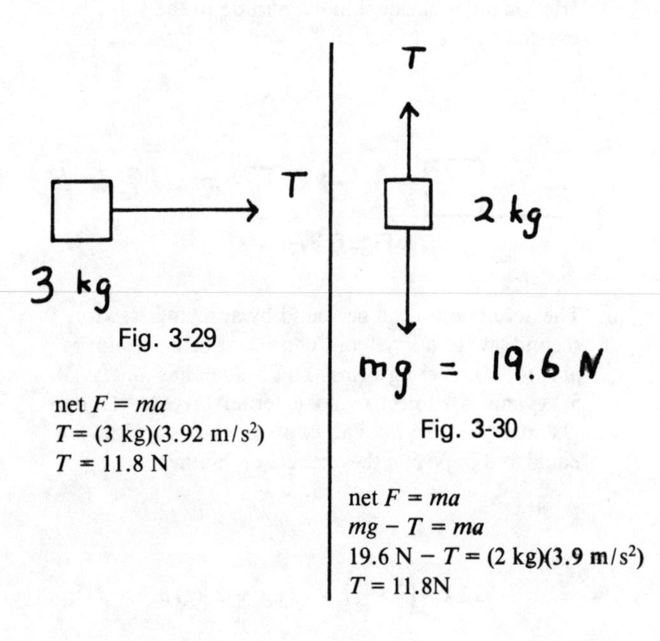

Fig. 3-29

net $F = ma$
$T = (3 \text{ kg})(3.92 \text{ m/s}^2)$
$T = 11.8$ N

Fig. 3-30

net $F = ma$
$mg - T = ma$
$19.6 \text{ N} - T = (2 \text{ kg})(3.9 \text{ m/s}^2)$
$T = 11.8\text{N}$

g. We consider again a system consisting of both blocks, shown in Fig. 3-31.

net $F = ma$
$W - f = (m_1 + m_2)a$

$$a = \frac{(2 \text{ kg})(9.8 \text{ m/s}^2) - 9.8 \text{ N}}{2 \text{ kg} + 3 \text{ kg}} = \frac{9.8 \text{ N}}{5 \text{ kg}}$$

$a = 1.96 \text{ m/s}^2$

Now consider a system consisting of the hanging block alone, shown in Fig. 3-32.

net $F = ma$
$W - T = ma$
$T = W - ma = 19.6 \text{ N} - (2 \text{ kg})(1.96 \text{ m/s}^2)$
$T = 15.7$ N

28a. The quickest solution results from applying the law of motion to a system consisting of both masses plus the connecting string. Let the positive direction of motion be that indicated—upward for the smaller mass and downward for the larger mass. The forces of tension are internal forces; they cancel out.

net $F = ma$
$m_2 g - m_1 g = (m_1 + m_2)a$

Thus $a = \frac{m_2 - m_1}{m_2 + m_1} g = \frac{44 \text{ g}}{444 \text{ g}} 9.8 \text{ m/s}^2 = 0.97 \text{ m/s}^2$

Although it takes a little longer, the same result can be obtained by applying Newton's second law to each mass separately.

$$T - m_1 g = m_1 a \qquad m_2 g - T = m_2 a$$

Adding these two equations gives

$$T - m_1 g + m_2 g - T = m_1 a + m_2 a$$

Hence

$$a = \frac{m_2 - m_1}{m_2 + m_1} g$$

b. This is uniformly accelerated motion. Thus,

$$v^2 = v_0^2 + 2as$$

The initial speed is zero. Solve for the displacement s.

$$s = \frac{v^2}{2a} = \frac{(0.8 \text{ m/s})^2}{2(0.97 \text{ m/s}^2)} = 0.33 \text{ m}$$

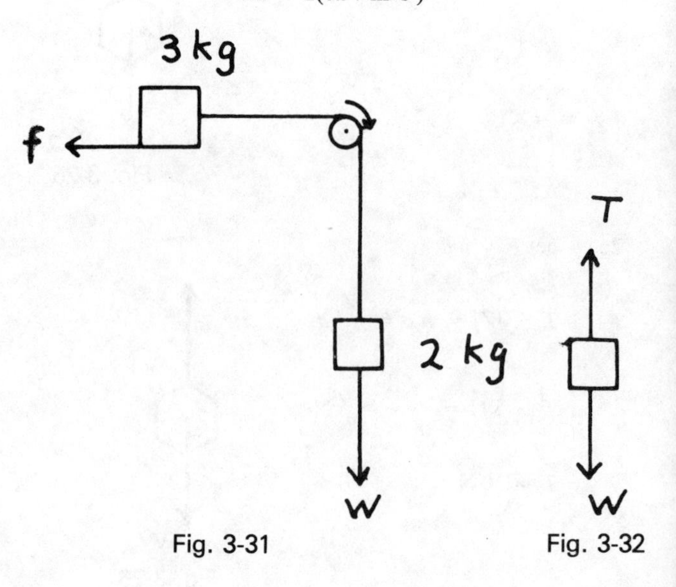

Fig. 3-31

Fig. 3-32

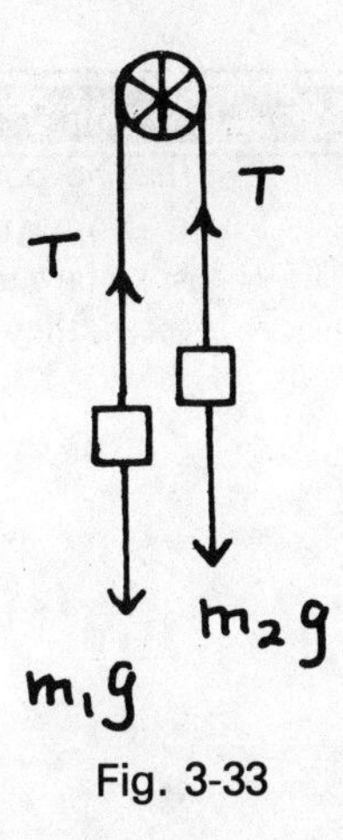

Fig. 3-33

c. The tension can be found by isolating either m_1 or m_2. The force diagrams are given in Fig. 3-34.

$$T - m_1 g = m_1 a$$
$$T = m_1 g + m_1 a = (0.2\ \text{kg})(9.8\ \text{m/s}^2 + 0.97\ \text{m/s}^2)$$
$$T = 2.15\ \text{N}$$

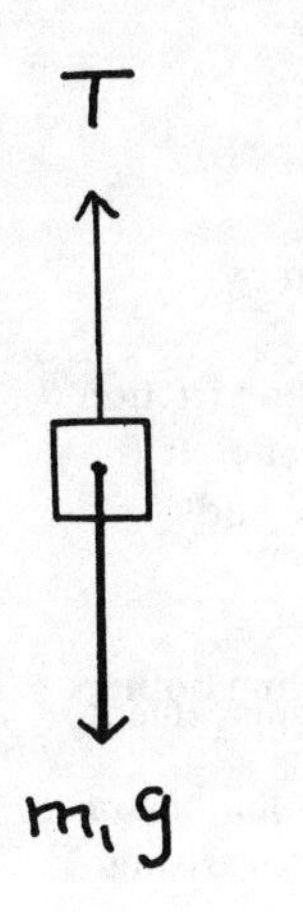

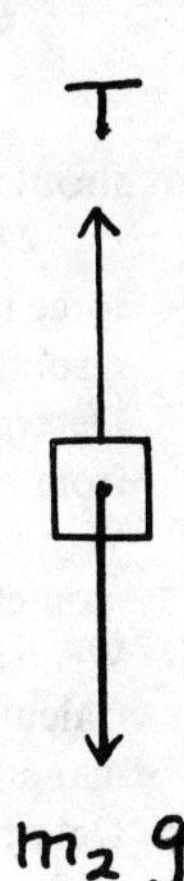

Fig. 3-34

CHAPTER 4

Statics

GOALS To learn how to apply the two conditions of equilibrium to determine unknown forces or torques, and to know the laws of friction and to be able to apply them to simple situations.

OBJECTIVES After completing this chapter the student should be able to do the following:

1. Write in symbolic form, using standard symbols, the two conditions of equilibrium.
2. Write a definition of *center of gravity* and determine the location of the center of gravity for simple bodies (uniform rod, block, separated point masses).
3. Determine whether a particle is in equilibrium under the action of known forces. (Does $\Sigma\mathbf{F} = 0$?)
4. Determine an unknown force required to maintain equilibrium of a particle. ($\mathbf{A} + \mathbf{B} + \mathbf{X} = 0$; find $\mathbf{X}$.)
5. State correctly the relationship between a frictional force and the normal force.
6. Calculate the force of static friction between two objects in contact even when F_s is less than its maximum value.
7. Calculate the force of sliding friction when the normal force is not given directly but must be found by resolution of forces into components.
8. Determine the coefficients of static and sliding friction from measurements made on an inclined plane.
9. Write a definition of *torque* in terms of force and moment arm.
10. Write a definition of the *lever arm* of a force.
11. Calculate the torque of a given force specifying both its magnitude and direction.
12. Determine whether a body is in rotational equilibrium under the action of known torques, or torques arising from known forces. (Does $\Sigma\tau = 0$?)
13. Determine simultaneously unknown forces and unknown torques (or moment arms, or angles) for a body in equilibrium. (Examples 4-11 and 4-12 in the text.)

SUMMARY

When the sum of all forces acting on a small body is zero, the acceleration of the body is also zero, a consequence of Newton's second law of motion. The condition of zero acceleration, called *equilibrium*, applies equally well to a body at rest and to a body moving in a straight line at a constant speed. For a point particle the condition of equilibrium is net $\mathbf{F} = 0$ or $\Sigma\mathbf{F} = 0$. If all but one of the forces is given, the unknown force can be found through vector arithmetic.

Two objects in physical contact necessarily exert forces on each other. This force can be resolved into two components, one parallel to the surface of contact and a second perpendicular to this surface. These two components are called the *frictional* force and the *normal* force.

If one body is sliding along the other, the frictional force F_k will have a well-defined value given by the equation $F_k = \mu_k N$ where N is the magnitude of the normal force and μ_k is a constant that depends only on the nature of the surfaces in contact. The frictional force in this case is called *kinetic* friction. If the two bodies are not sliding but are at rest with respect to each other, the frictional force F_s may have any value up to a maximum, $\mu_s N$. This relationship is written $F_s \leq \mu_s N$ where μ_s is a constant that depends only on the nature of the two surfaces.

An extended body, as opposed to a point particle, may rotate as well as translate. For each type of motion there is a corresponding condition of equilibrium. These equations of equilibrium are: $\Sigma\mathbf{F} = 0$, for translational motion, and $\Sigma\tau = 0$, for rotational motion, where the torque, τ, is the product of the force and the moment arm.

Solving of problems in equilibrium usually involves finding unknown forces. For this purpose, the forces are resolved into components along conveniently chosen mutually perpendicular x and y axes, and the conditions of equilibrium are written in scalar form: $\Sigma F_x = 0$, $\Sigma F_y = 0$, $\Sigma\tau = 0$.

The weight of a body is a force directed toward the center of the earth. Although every part of the object is attracted downward, the net effect is that of a single force **W** acting at a point in the body called its *center of gravity*.

QUESTIONS AND PROBLEMS

Secs. 4-1—4-3 It is assumed in this chapter and in those that follow that you know how to add and subtract vectors both graphically and by resolution into components. These skills are absolutely necessary. Review Secs. 2-8 and 2-10 of the text if necessary.

1. A small body whose weight is 3.0 N is subject to the two forces illustrated below plus its own weight. All three forces are in a vertical plane. Is the body in equilibrium?

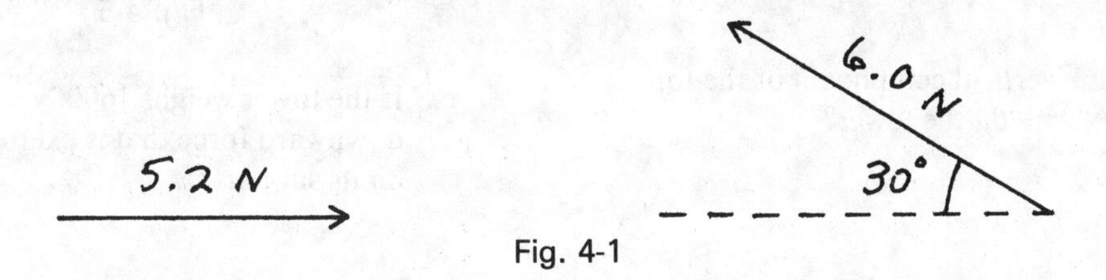

Fig. 4-1

2. A point particle is acted upon by two known forces, given below. Determine a third force, $\mathbf{F}_3$, such that the particle will be in equilibrium. Both magnitude and direction must be specified.

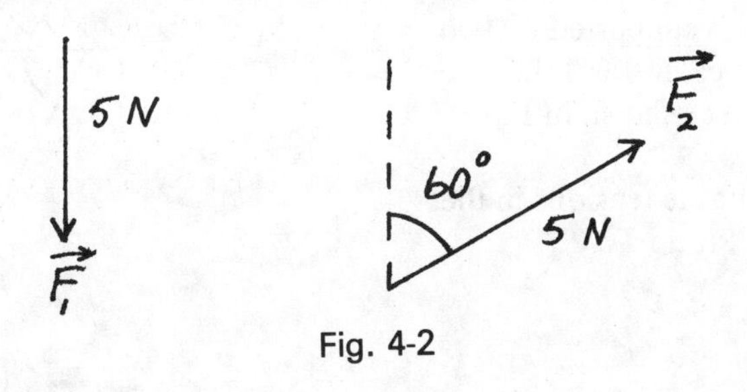

Fig. 4-2

$F_3 =$ ______________. The direction of $\mathbf{F}_3$ is ______________.

3. Find the magnitude of **F** such that the block will be in equilibrium on this inclined plane. The weight of the block is 10 N. (Assume that the friction between the block and plane is zero.)

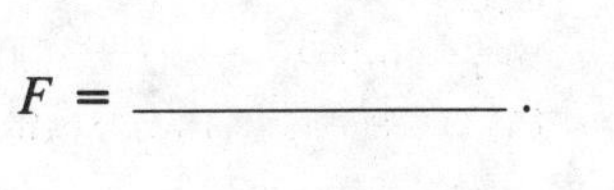

$F =$ ______________.

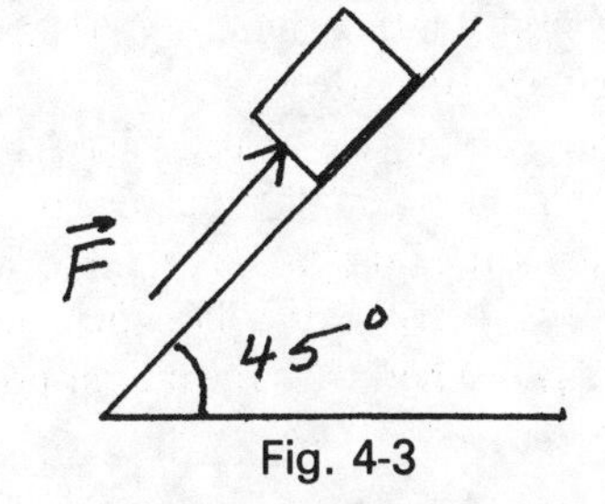

Fig. 4-3

4. The concrete ball of a wrecker shown in Fig. 4-4 weighs 20 kN. It is pushed away from its vertical position by a force F of 34.6 kN.

   a. Find the angle θ between the cable and the vertical.

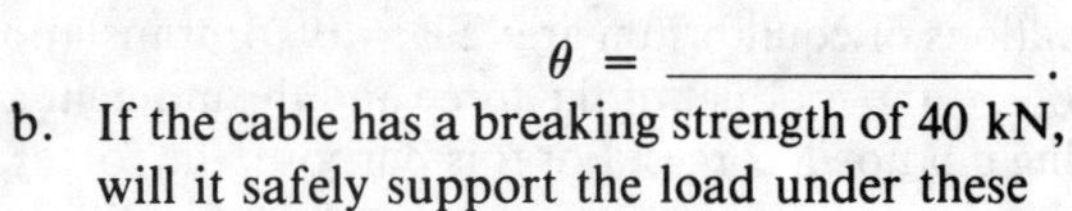

θ = ______________.

   b. If the cable has a breaking strength of 40 kN, will it safely support the load under these conditions? Explain.

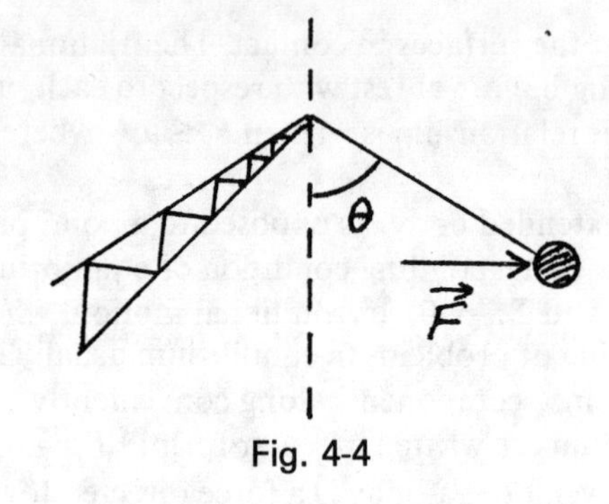

Fig. 4-4

5. The radio transmitting tower in Fig. 4-5 has a guy wire that is 6 m long. The guy wire is anchored to the ground at a point 3 m from the base of the tower. The tension in the wire is 800 N.

   a. What is the horizontal component of the force that the guy wire exerts on the tower?

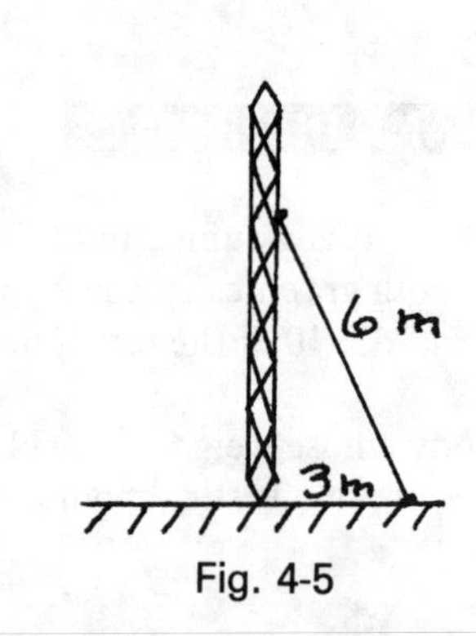

Fig. 4-5

   b. What is the vertical component of the force of the guy wire on the tower?

   c. If the tower weighs 1600 N what is the total downward force that is exerted by the tower on its support?

6. An object weighing 120 N is supported by two cables, one fastened to the ceiling and the second fastened to the wall as shown in Fig. 4-6.

   a. Let T_1 and T_2 represent the tensions in the two cables. What is the ratio T_1/T_2?

   b. Calculate the tension in each cable.

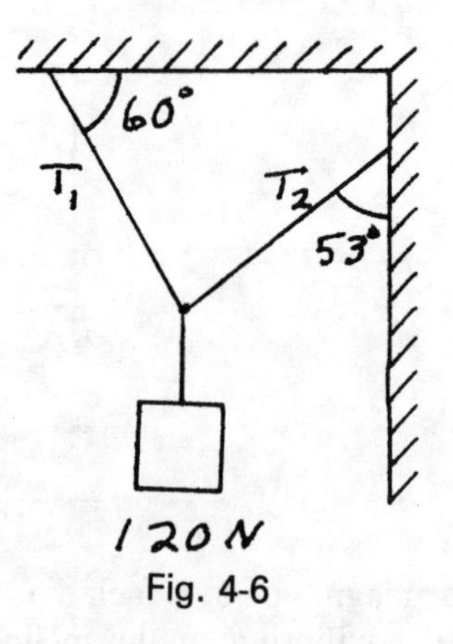

Fig. 4-6

Sec. 4-4 Friction arises when two surfaces are in contact. The frictional force, which is always parallel to the surface of contact, depends on the force with which one of the bodies pushes against the other. The latter force is *perpendicular* to the surface of contact and is therefore called the *normal* force. In certain simple cases the normal force may have the same magnitude as the weight of one of the bodies. *In general, however, the normal force,* $\mathbf{F}_n$, *is not equal to the weight,* **W**, *either in direction or in magnitude.* The examples that follow will serve to illustrate this important point.

Consider a block at rest on an inclined surface as shown in Fig. 4-7. The downward force of gravity is shown by the arrow labeled **W**. Since the block is in equilibrium there must be an upward force of exactly the right magnitude to counterbalance **W**. This upward force on the block must be supplied by the inclined surface as shown in Fig. 4-8.

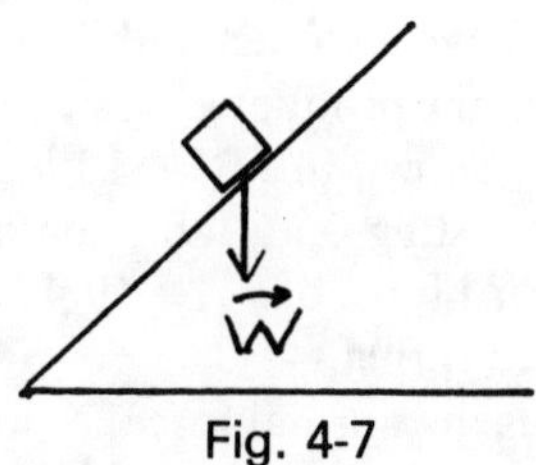

Fig. 4-7

Friction between the block and the surface on which it rests clearly plays an important role. If the surface is too slippery the block will slide downward with an ever-increasing velocity. Any detailed analysis of the situation requires that the force **F** be resolved into two components, one parallel to the plane and one perpendicular to it. The component parallel to the surface is called the frictional force $\mathbf{F}_s$ or $\mathbf{F}_k$ for static and sliding friction; the component perpendicular to the surface is called the normal force, $\mathbf{F}_n$ (or **N**). The frictional force depends on the nature of the surfaces. If there is an oil film present, for example, the frictional force may be very small. The normal force, on the other hand, does not depend on the nature of the surfaces.

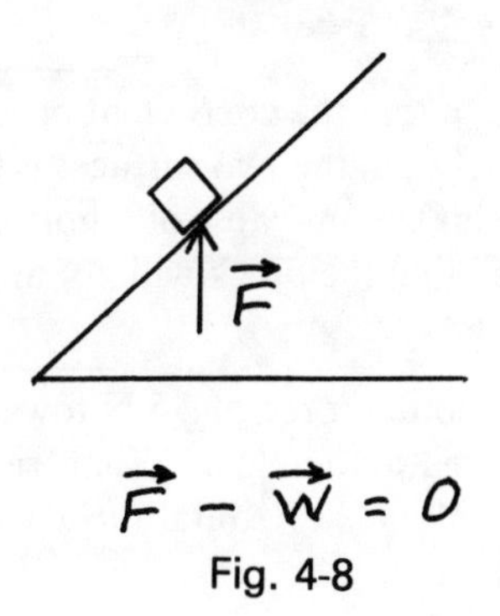

Fig. 4-8

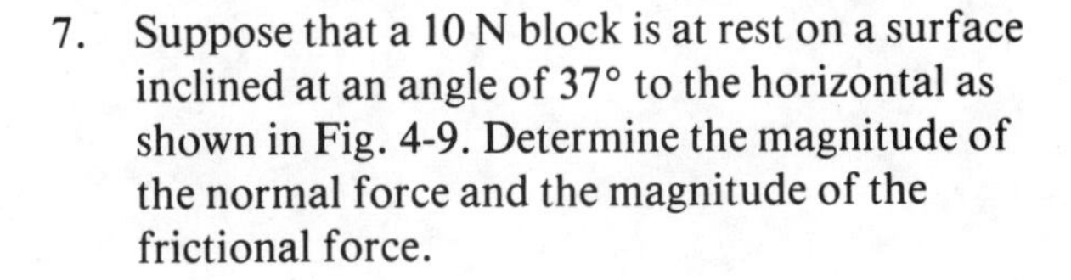

7. Suppose that a 10 N block is at rest on a surface inclined at an angle of 37° to the horizontal as shown in Fig. 4-9. Determine the magnitude of the normal force and the magnitude of the frictional force.

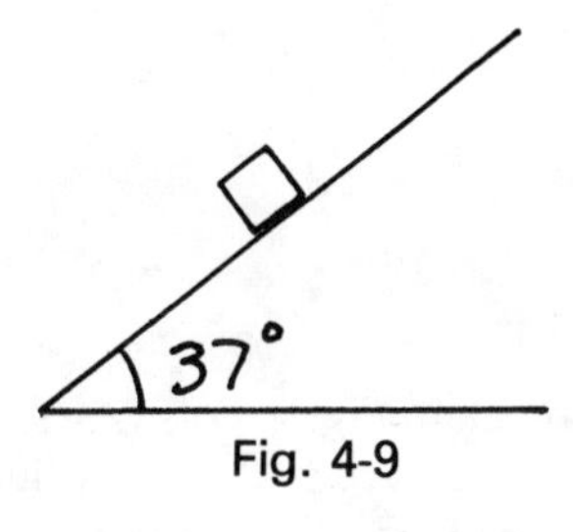

Fig. 4-9

F_n = ________________.

F_s = ________________.

The frictional force is proportional to the normal force, the constant of proportionality being dependent upon the nature of the two surfaces that are in contact. It is necessary to make a distinction between the kinetic case—the block in motion—and the static case. If the block is moving, the frictional force has a well-defined value given by the equation

$$F_k = \mu_k F_n$$

On the other hand, if the block is at rest the frictional force may have any value up to a maximum $\mu_s F_n$. This relationship is written

$$F_s \leq \mu_s F_n$$

8. Returning to the 10-N block on the 37° inclined plane, suppose that μ_s is known to be 0.80.
   a. What is the *maximum possible* force of static friction that could be exerted on the block?

   maximum value of $F_s = \mu_s F_n =$ ________.

   b. What is the *actual* force of friction exerted by the surface on the block?

   $F_s =$ ________.

As we have just seen in this example, the force of static friction present in a particular case cannot be found from the product $\mu_s F_n$ since the latter gives only the theoretical maximum value of f_s.

9. The object shown in Fig. 4-10 has a mass of 3 kg. It is resting on a table top and is subject to an upward force of $F_{sp} = 19.4$ N produced by an extended spring.

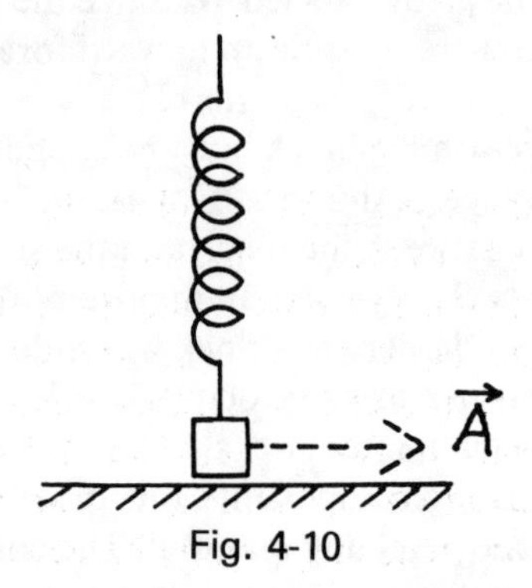

Fig. 4-10

   a. What is the normal force of the horizontal surface against the block?

   $F_n =$ ________.

   b. Suppose that the coefficient of static friction between the two surfaces is 0.5. What is the maximum value of a horizontal force **A** such that the block will not move?

   $A_{max} =$ ________.

   c. A horizontal force of 2.5 N toward the right is applied to the block. What frictional force will the horizontal surface exert on the block?

   $F_s =$ ________.

   d. Suppose that the coefficient of kinetic friction between these two surfaces is 0.3. What will be the acceleration of the block when a horizontal force of 6.0 N is applied to the block?

   $a =$ ________.

10. Describe one special physical situation in which the weight and the normal force have the same magnitude.

11. The 8-N block in Fig. 4-11 is free to move along a horizontal surface under the influence of an applied force **F** parallel to the horizontal surface. The coefficients of friction are 0.25 and 0.40. Complete all missing entries in the accompanying table.

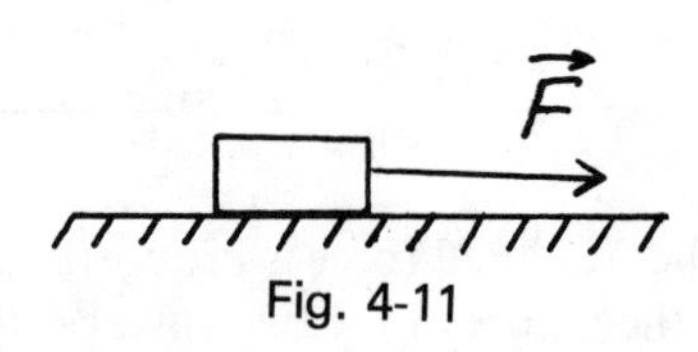

Fig. 4-11

F	State of Motion	Frictional Force *F*	Acceleration
1.0 N	at rest		
2.5 N			
	const. non-zero speed		zero
	moving		1.0 m/s^2

12. Figure 4-12 shows a 10-kg block of steel being pulled along a horizontal surface at a constant speed by a horizontal force **F**.

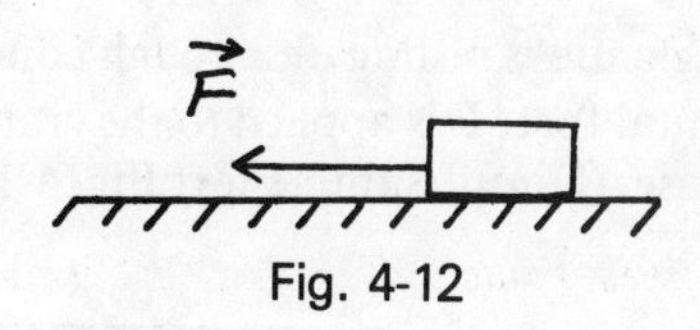

Fig. 4-12

a. Calculate the force of friction between the steel block and the wood surface on which it is sliding. Coefficients of friction are given in Table 4-1 in the text.

b. What is the magnitude of **F**?

13. The same block is being pulled along the same surface by a force **F**, which makes an angle of 30° with the horizontal. Find the magnitude of **F**. (Hint: The change from a horizontal force to an inclined force considerably complicates the problem. You can get help from Ex. 4-4 in the text.) Draw your own force diagram in the space provided and then fill in the table of horizontal and vertical components. All forces may be considered to act at a common point, the center of the block.

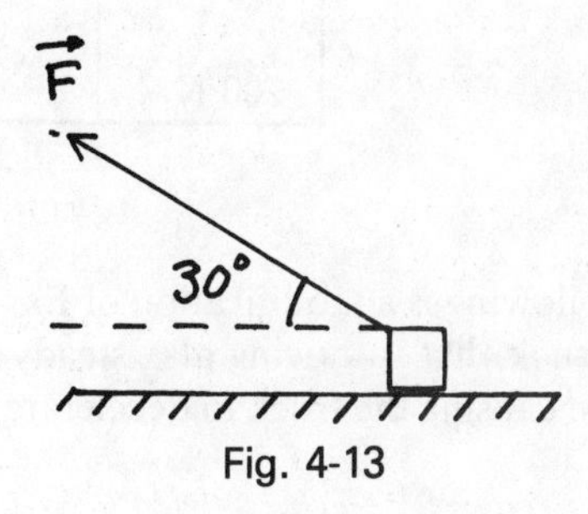

Fig. 4-13

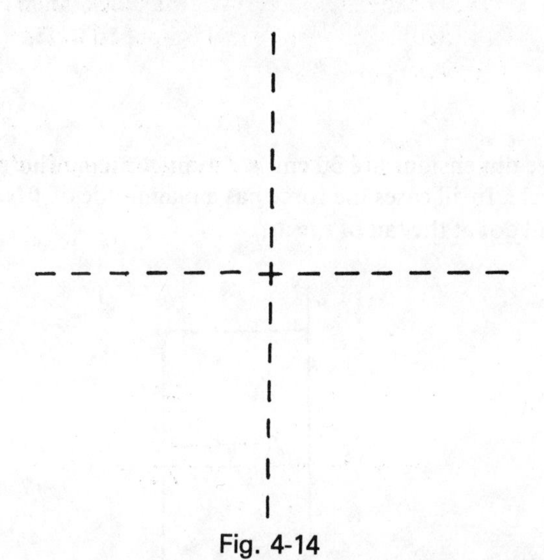
Fig. 4-14

Force	*x*-components	*y*-components
F		
W		
$\mathbf{F}_k$		
$\mathbf{F}_n$		

Check your components against the solutions section before proceeding. Next, apply the conditions of equilibrium, $\Sigma F_x = 0$ and $\Sigma F_y = 0$, to get two equations containing the unknowns. Remember that the frictional force depends on the normal force, the force that the horizontal surface exerts against the block.

14. A 25 kg crate is resting on a rough concrete floor. The coefficients of friction are $\mu_s = 0.7$ and $\mu_k = 0.4$. A horizontal force **F** is applied to the crate; several values of it are given. Determine the frictional force for each case. (Hint: Examine text Fig. 4-12.) Fill in all the missing entries in the table below.

F	Frictional Force F	Is ΣF = 0?
25 N		yes
50 N		
100 N		
200 N		

15. The following is a modification of Ex. 4-7 in the text. An iron casting of mass 2000 kg is being transported by a truck that is moving at a steady velocity of 0.3 m/s (see text Fig. 4-13). The driver applies his brakes, causing the truck to decelerate at a rate of 4 m/s². Will the casting slide forward?

Secs. 4-5—4-8

16. The drawing shows a force **F** acting on a rigid body whose dimensions are 30 cm × 40 cm × 1 cm. The axis of rotation to be used is indicated by a small open circle. In all cases the force has a magnitude of 1 N. The point of application of the force is indicated by a solid dot at the tail of the arrow.

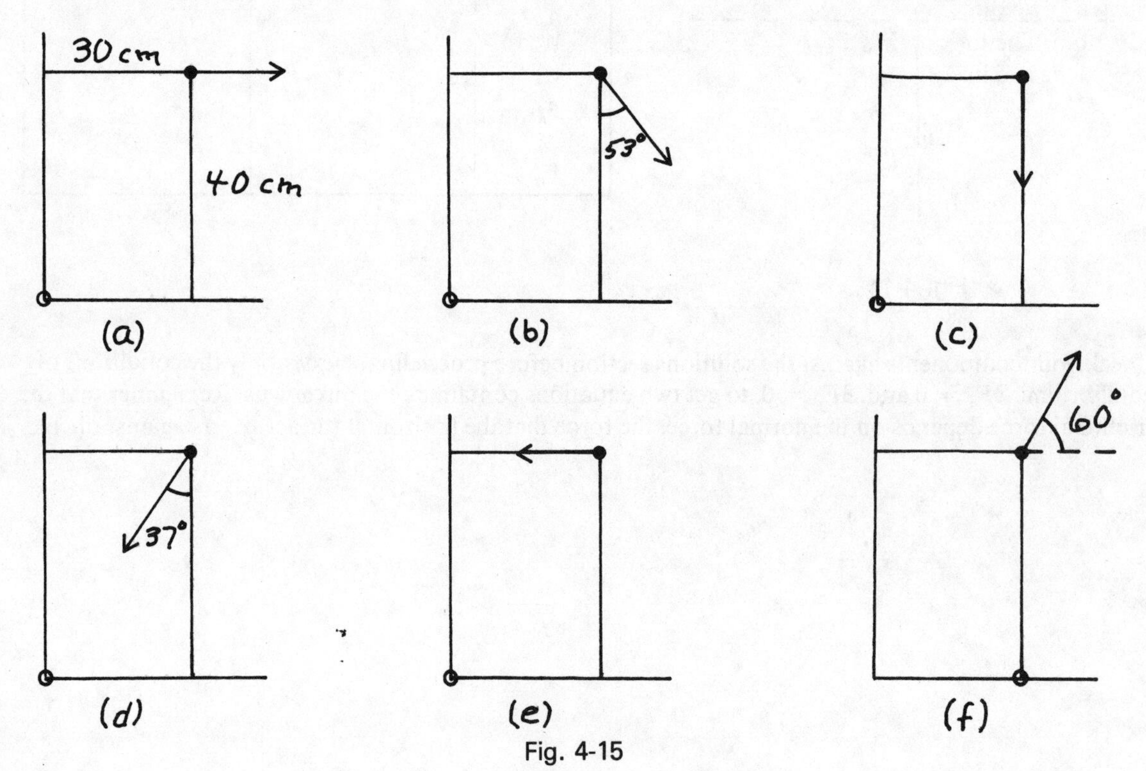

Fig. 4-15

Draw a dashed line to represent the line of action in each case. Draw a light solid line to represent the lever arm in each case. (The lever arm is the perpendicular distance from the axis of rotation to the line of action of the force.) If you have drawn the lever arm correctly (check it!) you should have no difficulty in calculating the torque for each of these six cases. Tabulate your results below. Note that counterclockwise torques are taken to be positive.

Figure	Lever arm	Torque
a		
b		
c		
d		
e		
f		

17. A rigid bar is subjected to three forces as shown in Fig. 4-16. The point of application of each force is shown by the tail of the vector. The magnitude of the forces are $A = 3$ N, $B = 4$ N, and $C = 5$ N. **A** is vertical and **B** is horizontal. The direction of **C** is shown in the diagram. The total length of the rod is 62.5 cm.
    a. Is the rod in translational equilibrium? (That is, is $\Sigma \mathbf{F} = 0$? It will be necessary to resolve the forces into components.)

    b. Is the rod in rotational equilibrium? (Hint: It will be necessary to calculate the torques around some convenient axis of rotation and determine whether $\Sigma \tau = 0$.

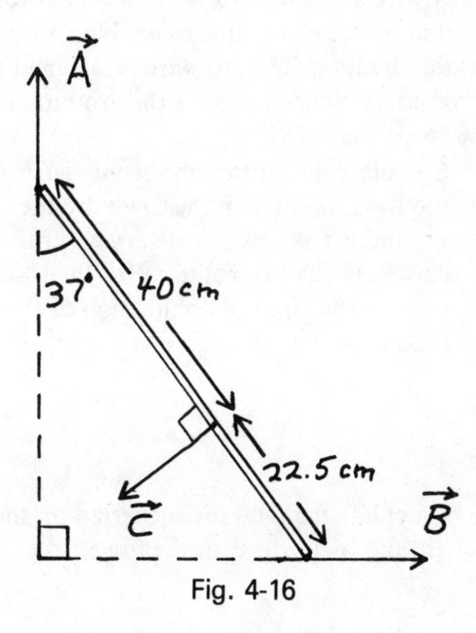

Fig. 4-16

18. A load weighing 5 N is to be supported by a force **F** applied at the end of a lever. The fulcrum is 6 cm to the right of the load and the unknown force is applied at a point 30 cm from the load as shown below.

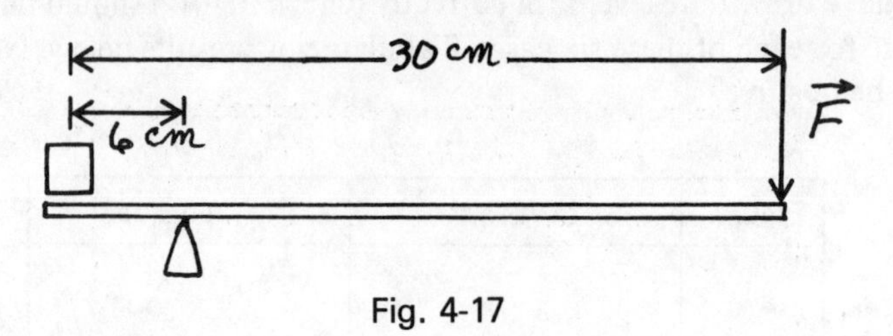

Fig. 4-17

a. What force is necessary to keep the lever in equilibrium? (Note that the fulcrum also exerts a force on the lever. A judicious choice of the axis of rotation greatly simplifies the solution.)

b. What is the force on the fulcrum when the lever is in equilibrium?

19. A high-power line is supported by a tower with a guy wire as shown in Fig. 4-18. The force exerted by the power line is 500 N in a horizontal direction. The guy wire is fastened to the tower at a point 50 m above the ground. The tower weighs 507 N.

a. Calculate the tension in the guy wire. (Since the force exerted on the tower by the ground is unknown, it is very helpful to choose the axis of rotation for the torques to be at the point of application of this unknown force.)

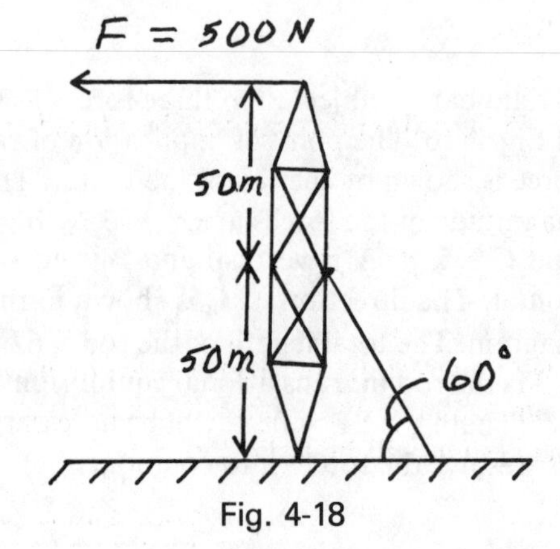

Fig. 4-18

b. Calculate the total force exerted on the ground by the base of the tower.

20. Two small weights W_1 and W_2 are connected by a light, rigid rod whose weight is very much less than W_1 and W_2. The length of the rod, illustrated below, is L.

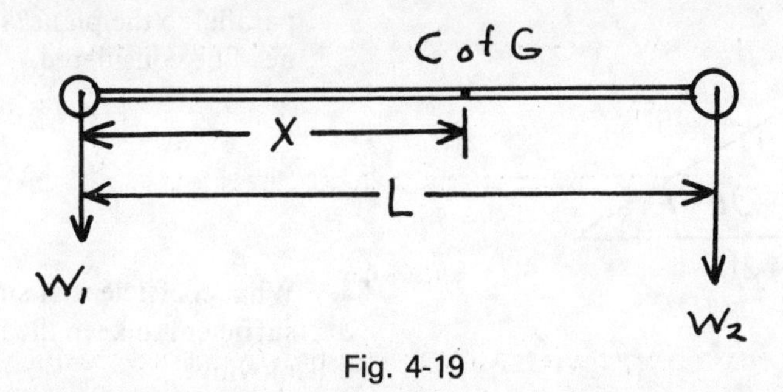

Fig. 4-19

a. Find the center of gravity. How far is it from the weight W_1? Since numerical values of W_1, W_2, and L are not given, you will have to find an algebraic expression for x in terms of W_1, W_2, and L.

b. If $W_1 = 2$ N, $W_2 = 6$ N, and $L = 40$ cm, what is the numerical value of x?

c. Where should a support be placed so that the two masses are balanced?

21. Two long rectangular pieces of wood are glued together as shown in Fig. 4-20. Their dimensions are 1 cm × 1 cm × 30 cm and 1 cm × 1 cm × 60 cm. Find the point at which a fulcrum should be placed so that the system will be balanced.

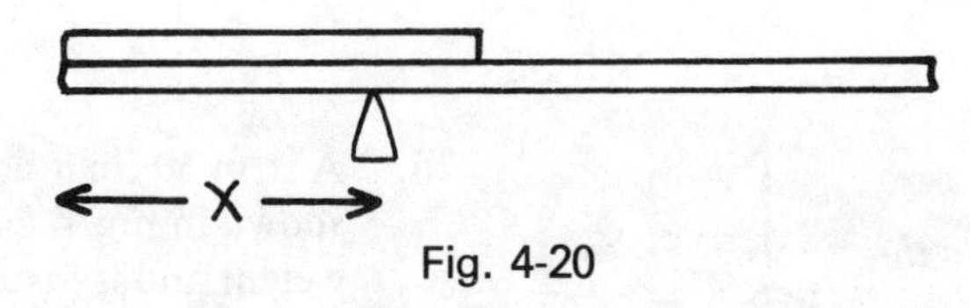

Fig. 4-20

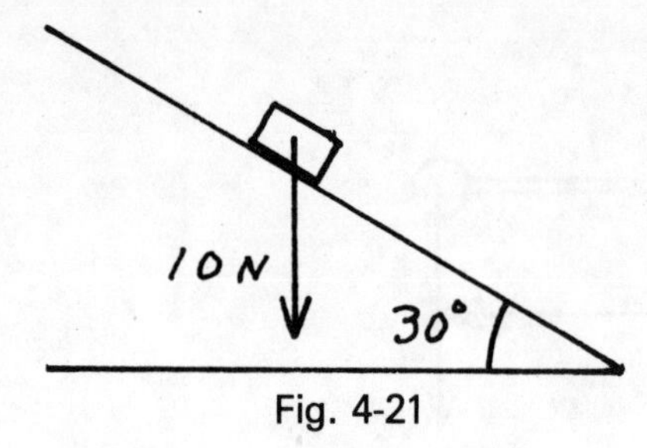

Fig. 4-21

22. What is the component of force on the mass parallel to the plane? (Only gravitational forces need be considered.)

23. What coefficient of static friction will be just sufficient to keep the mass in Fig. 4-21 from moving?

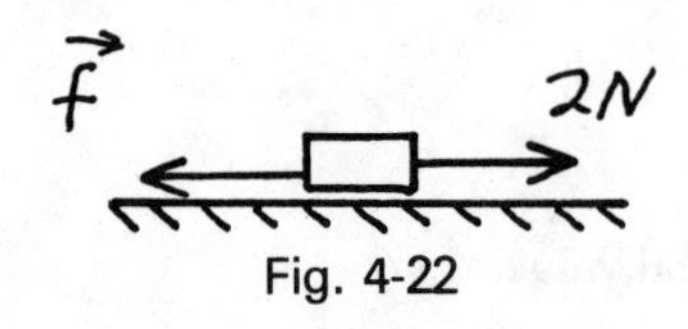

Fig. 4-22

24. The block in Fig. 4-22 weighs 10 N. It is subject to an external, horizontal force of 2 N. The coefficients of friction between the two surfaces are 0.3 and 0.4. What is the magnitude of the force of friction **F**?

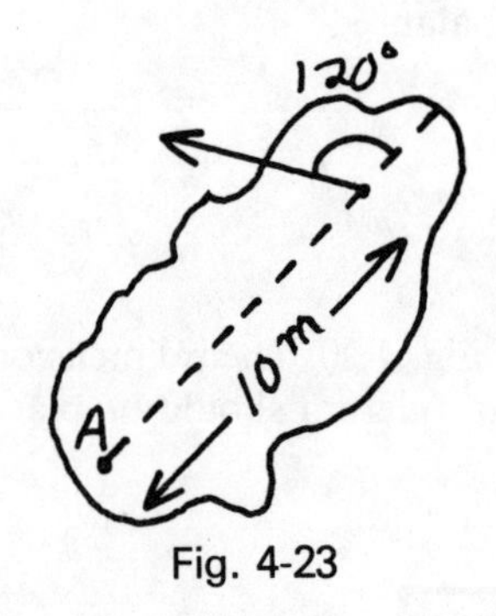

Fig. 4-23

25. An object of irregular shape is free to rotate without friction about the axis *A* shown in Fig. 4-23. A force of 12 N is applied at the point indicated. What is the magnitude of the torque?

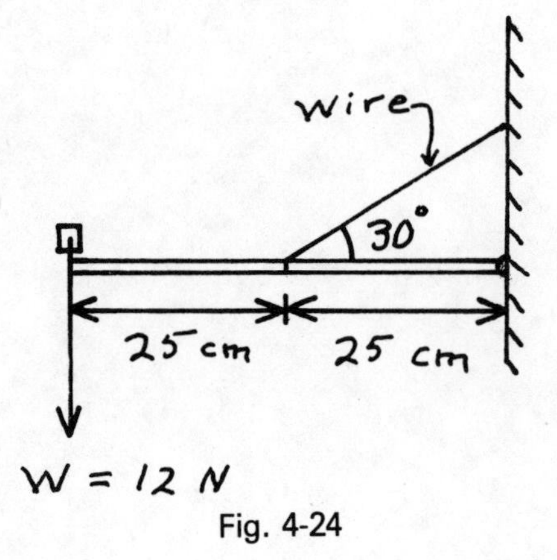

Fig. 4-24

26. A strut 50 cm long is supporting a 12-N load as shown in Fig. 4-24. The strut has negligible weight and is fastened to the wall with a frictionless hinge. What is the tension in the wire?

27. The two weights in Fig. 4-25 are connected by a rod whose weight may be neglected. The distance from the larger object to the center of gravity of the system is designated by x in the drawing. What is the value of x?

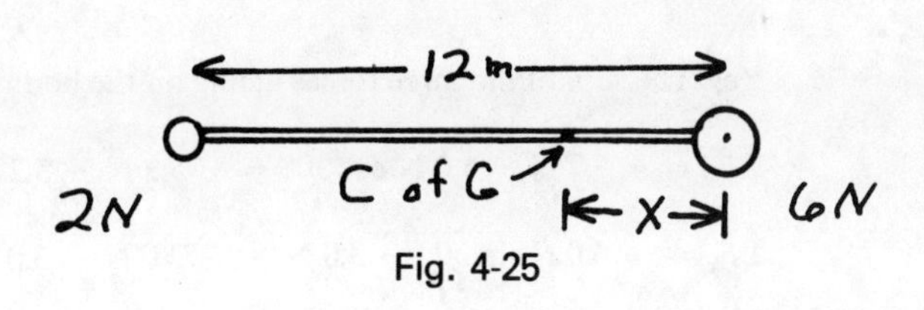

Fig. 4-25

28. The following solution to a problem in statics contains a single but fatal error. On some exams 50% of the students have made this mistake. Examine the solution carefully before answering the questions.
Problem: A weight of 50 N is hanging from the midpoint of a rod of length L. The rod is supported by a hinge and a wire as shown in Fig. 4-26. Find the tension in the wire.

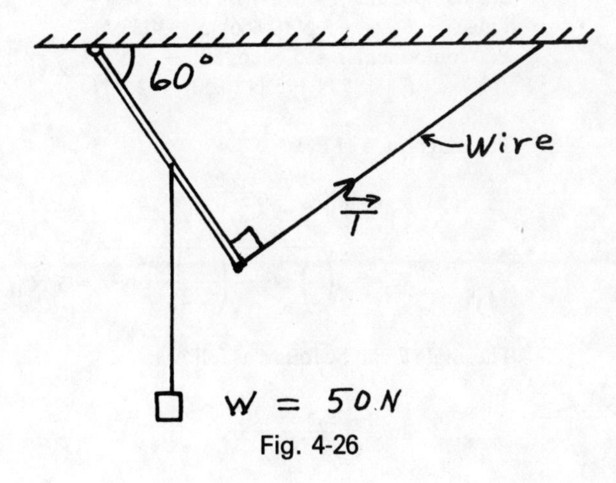

Fig. 4-26

Often-proposed solution: The system is in equilibrium, therefore

$$\Sigma\tau = 0 \quad \text{where} \quad \tau = F \times \text{lever arm}$$

Taking the axis of rotation at the hinge

$$\tau_1 = -50\,\text{N}(L/2)$$
$$T(L) - 50\,\text{N}(L/2) = 0$$
$$T = 50\,\text{N}(1/2) = 25\,\text{N}$$

a. This result is incorrect. What error has been made?

b. How can you avoid making this mistake?

c. Correct the solution and find the correct value of the tension.

$T =$ ______________

Solutions

1. Yes. The sum of the three forces acting on the body is zero. This can be shown by adding up the x and y components.

$$\Sigma F_x = +5.2\text{ N} - 6.0\text{ N}\cos 30° = \quad 5.2\text{ N} - 5.2\text{ N} = 0$$

$$\Sigma F_y = +6.0\text{ N}\sin 30° - 3.0\text{ N} = \quad 3.0\text{ N} - 3.0\text{ N} = 0$$

2. Let the third force be designated by $\mathbf{F}_3$ with components F_x and F_y.

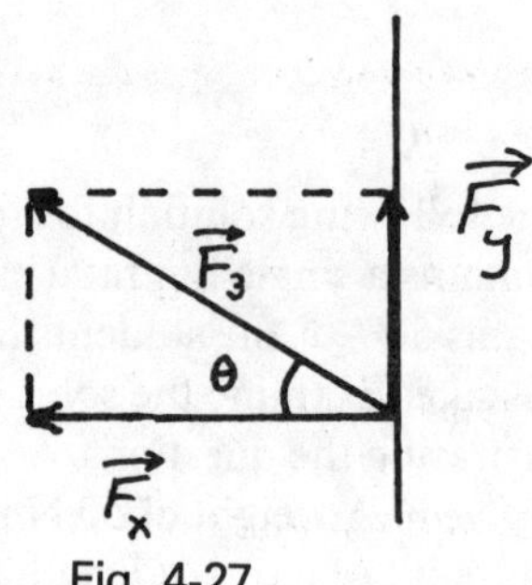

Fig. 4-27

$\Sigma(x\text{-components}) = (5\text{ N})(\sin 60°) + F_x = 0$
Thus $\quad F_x = -5\text{ N}(0.866) \cong -4.33\text{ N}$
$\Sigma(x\text{-components}) = 5\text{ N}\cos 60° - 5\text{ N} + F_y = 0$
$\quad F_y = 5\text{ N} - 5\text{ N}(0.50) = 2.5\text{ N}$

The magnitude of $\mathbf{F}_3$ is

$$F_3 = |\mathbf{F}_3| = \sqrt{F_x^2 + F_y^2}$$

$$F_3 = \sqrt{\left(\frac{5}{2}\right)^2\text{N}^2 + \left(\frac{(\sqrt{3})5}{2}\right)^2\text{N}^2} = 5\text{ N}$$

The angle θ can be found as follows:

$$\sin\theta = \frac{2.5\text{ N}}{5\text{ N}} = 0.5$$

Thus $\quad \theta = 30°$

This problem can be solved more easily by making use of symmetry. Since the two given forces are equal, their resultant must be halfway between them. Thus the angle φ must be $\frac{1}{2}(120°) = 60°$. It then is evident that the three forces make equal angles with one another. Since two of them have the same magnitude, so must the third.

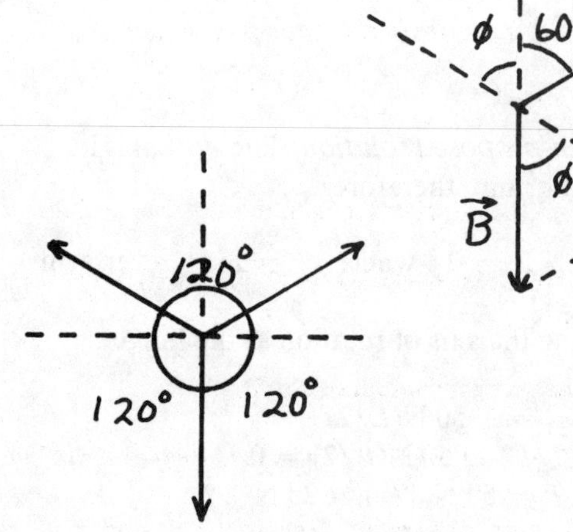

Fig. 4-28

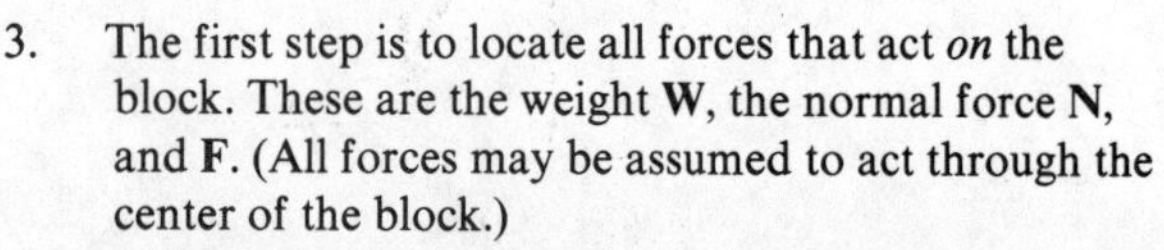

3. The first step is to locate all forces that act *on* the block. These are the weight **W**, the normal force **N**, and **F**. (All forces may be assumed to act through the center of the block.)
   The second step is to choose a convenient coordinate system. In this case we choose the x-axis to be parallel to the plane. Then the force **N** has no x-component. Using the condition of equilibrium,

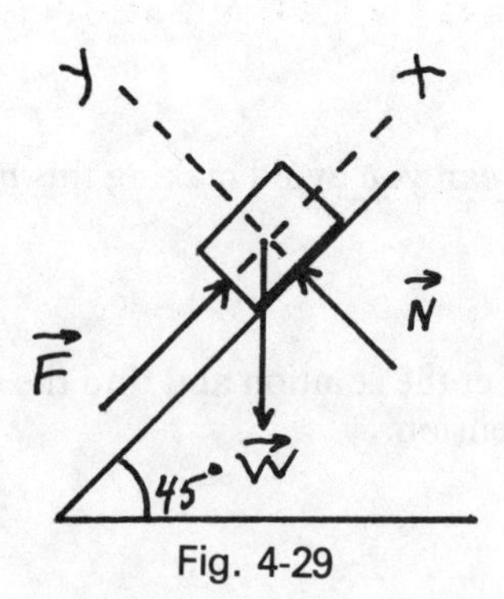

Fig. 4-29

$\Sigma(x\text{-components}) = F - W\sin 45° = 0$
$F = (10\text{ N})(0.707) \cong 7.1\text{ N}$

4. *All* forces acting *on* the object must be indicated in a force diagram. **T** is the tension in the cable, **W** is the weight. Using the conditions of equlibrium,

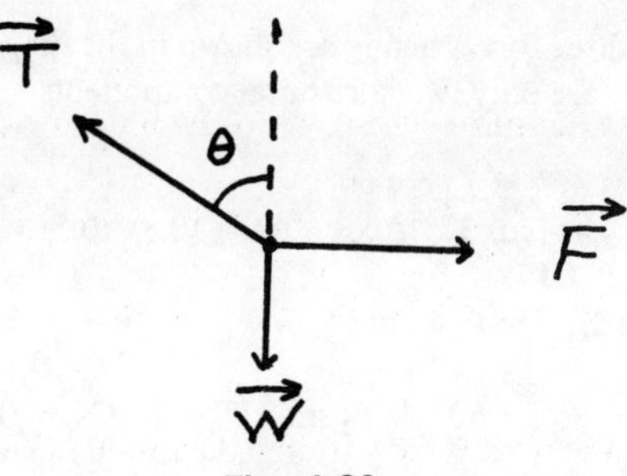

Fig. 4-30

$$\Sigma(x\text{-components}) = F - T_x = 0$$
$$\Sigma(y\text{-components}) = T_y - W = 0$$

Thus $T_x = F = 34.6 \text{ kN}$
$T_y = W = 20.0 \text{ kN}$

The unknown force **T** is the vector sum of the two components.

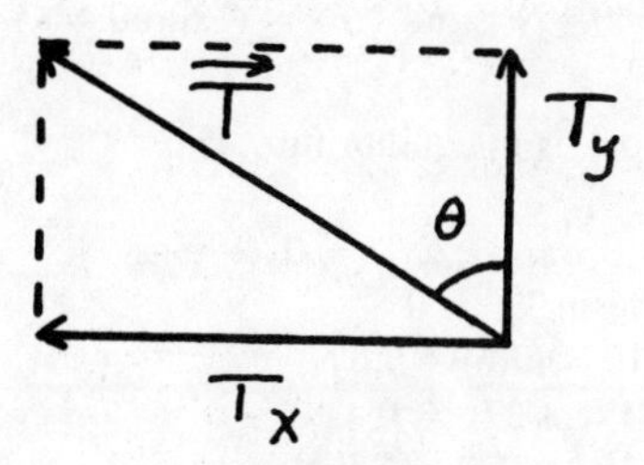

Fig. 4-31

a. $\tan \theta = \frac{T_x}{T_y} = \frac{34.6}{20.0}$

$\tan \theta = 1.73; \quad \theta = 60°$

b. The magnitude of the resultant is given by

$$T = \sqrt{T_x^2 + T_y^2} = 40 \text{ kN}$$

The cable is in danger of breaking!

5. First find the angle θ from the space diagram.

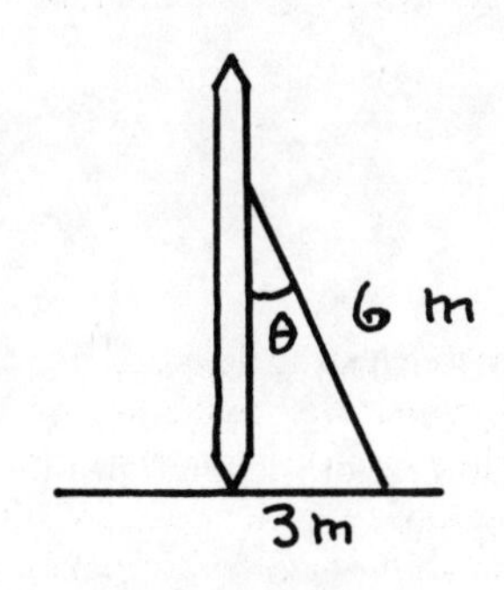

$\sin \theta = \frac{3 \text{ m}}{6 \text{ m}} = 0.5 \quad \theta = 30°$

a. Find the horizontal component of the tension.

$T_x = \text{T} \sin 30° = 800 \text{ N } (0.50) = 400 \text{ N}$

b. The vertical component of **T** is

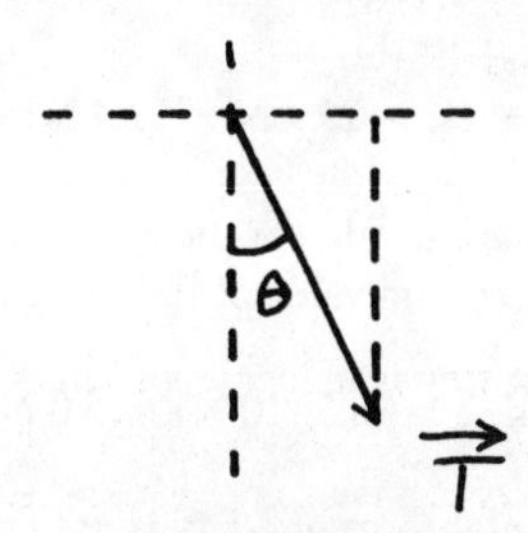

Fig. 4-32

$T_y = -T \cos \theta = -800 \text{ N } (0.866) = 693 \text{ N (downward)}$

c. The total downward force on the ground is

$F = T_y + w = 693 \text{ N} + 1600 \text{ N} = 2293 \text{ N}$

6a. The three forces acting are shown in the diagram. There are only two horizontal components.

$T_2 \cos 37° = T_1 \cos 60°$
$T_1/T_2 = (\cos 37°)/(\cos 60°) = (0.8)/(0.5) = 1.6$

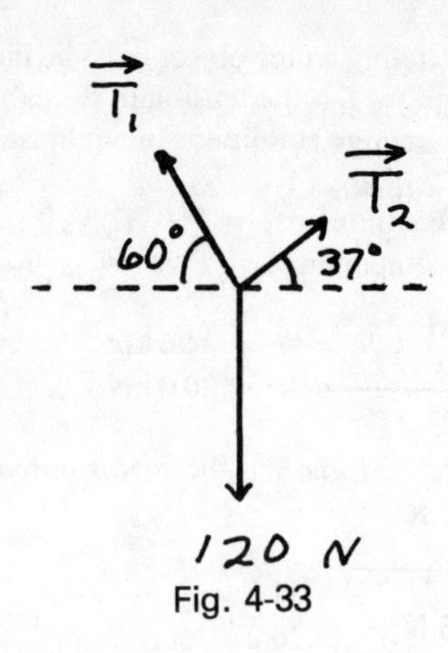

Fig. 4-33

b. Since $\Sigma F_y = 0$, we have

$$T_1 \sin 60° + T_2 \sin 37° - 120\text{ N} = 0$$

Eliminating T_1,

$$1.6\, T_2\,(0.866) + T_2\,(0.6) = 120\text{ N}$$
$$T_2 = 60.4\text{ N} \quad \text{and} \quad T_1 = (1.6)(60.4\text{ N}) = 96.6\text{ N}$$

7. The block is in equilibrium. Thus

$\Sigma F_x = 0$ and $\Sigma F_y = 0$
$F_s - W \sin 37° = 0$
$F_s = (10\text{ N})(0.6) = 6.0\text{ N}$
$F_n - W \cos 37° = 0$
$F_n = (10\text{ N})(0.8) = 8.0\text{ N}$

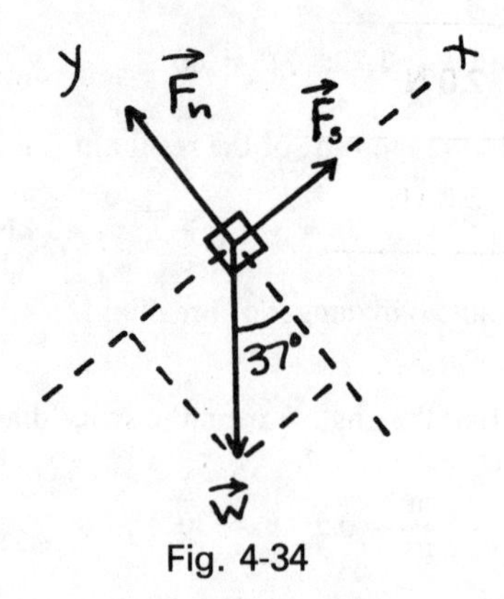

Fig. 4-34

8a. $\mu_s F_n = (0.8)(8.0\text{ N}) = 6.4\text{ N}$

b. The actual force of friction is just that force parallel to the plane that will keep the block in equilibrium.

$$F_s = 6.0\text{ N}$$

9a. $F_n + F_{sp} - W = 0$
$F_n = W - F_{sp} = 29.4\text{ N} - 19.4\text{ N} = 10.0\text{ N}$

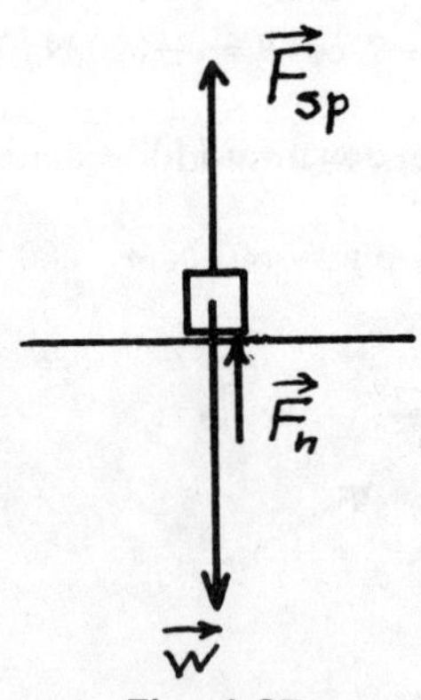

Fig. 4-35

b. The maximum value of the horizontal force is equal to the maximum value of the static friction.

$$A_{\text{max}} = \mu_s F_n = (0.5)(10\text{ N}) = 5.0\text{ N}$$

c. A horizontal force of 2.5 N is not enough to start the block moving; it will remain at rest. Thus

$$\Sigma F_x = 0;\ A - F_s = 0;\ F_s = A = 2.5\text{ N}$$

d. $$F_k = \mu_k F_n = (0.3)(10\text{ N}) = 3.0\text{ N}$$

$$\Sigma F_x = ma;\ a = \frac{\Sigma F_x}{m} = \frac{6.0\text{ N} - 3.0\text{ N}}{3\text{ kg}} = 1.0\text{ m/s}^2$$

10. If an object is on a horizontal surface and if the object is subject only to horizontal forces, then the magnitude of the normal force is equal to the magnitude of the weight.

11. $F_k = \mu_k F_n = (0.25)(8\ \text{N}) = 2\ \text{N}$
$F_s \leq \mu_s F_n = (0.40)(8\ \text{N}) = 3.2\ \text{N}$

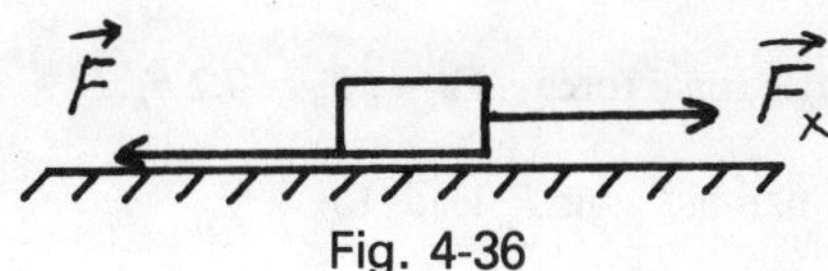

Fig. 4-36

F_x	State of Motion	Frictional Force F	Acceleration
1.0 N	at rest	1.0 N	zero
2.5 N	at rest	2.5 N	zero
2.0 N	const. non-zero speed	2.0 N	zero
2.8 N	moving	2.0 N	1.0 m/s^2

$$\text{net } F_x = ma = \left(\frac{8\ \text{N}}{9.8\ \text{m/s}^2}\right) 1.0\ \text{m/s}^2 = 0.82\ \text{N}$$

$$F_x - F_k = ma;\ F_x = ma + F_k$$
$$F_x = 0.82\ \text{N} + 2.0\ \text{N} = 2.8\ \text{N}$$

12a. The forces are in equilibrium in the vertical direction. Thus the magnitude of the normal force is $N = mg = 98\ \text{N}$. The frictional force is given by

$$F_k = \mu N = (0.2)(98\ \text{N}) = 19.6\ \text{N}$$

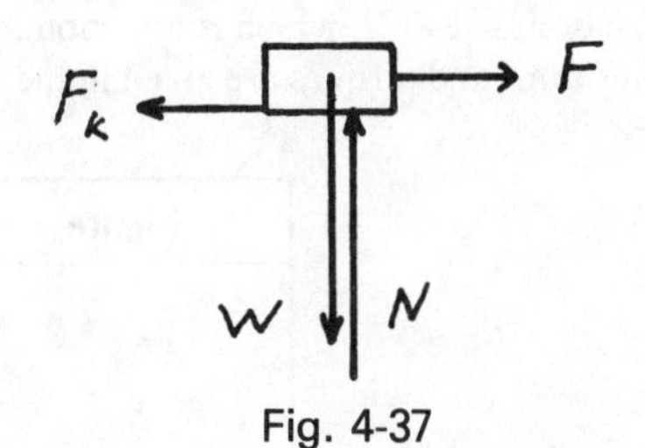

Fig. 4-37

b. Since the horizontal forces are also in equilibrium,

$$F = F_k = 19.6\ \text{N}$$

13. There are four forces acting on the block. Their components are given in the table below.

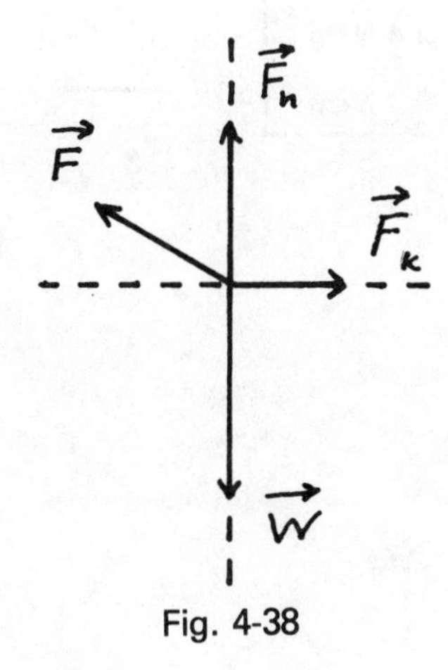

Fig. 4-38

Force	x-component	y-component
F	$-(\sqrt{3}/2)F$	$F/2$
W	0	-98 N
F_k	F_k	0
F_n	0	F_n

$\Sigma F_x = 0$ gives $F_k = (\sqrt{3}/2)F$
$\Sigma F_y = 0$ gives $F/2 + F_n = 98$ N

The frictional force is $F_k = \mu F_n = 0.2 F_n$.

Eliminating F_k and F_n leads to

$$F/2 + 5(\sqrt{3}/2)F = 98 \text{ N}$$

$$F = \frac{98 \text{ N}}{4.83} = 20.3 \text{ N}$$

14. This problem checks your understanding of the idea that $\mu_s N$ gives only the *maximum* value of F_s. Any value less than or equal to $\mu_s N$ is possible.

F	Frictional force F	Is $\Sigma F = 0$?
25 N	25 N	yes
50 N	50 N	yes
100 N	100 N	yes
200 N	98 N*	no

*Crate is no longer at rest.

15. The frictional force required to cause the casting to decelerate is −8000 N (toward the left). The maximum force of static friction is 7840 N, which is not enough. The casting will slide forward.
16. The lever arms and torques are given in the table.

Figure	Lever arm	Torque
a	40 cm	-0.4 N·m
b	50 cm	-0.5 N·m
c	30 cm	-0.3 N·m
d	0 cm	0.0 N·m
e	40 cm	+0.4 N·m
f	20 cm	-0.2 N·m

17. Let us select a horizontal x-axis and a vertical y-axis. The components of the forces appear in the table.

a. Since both x-components and y-components add up to zero the rod is in translational equilibrium.

b. Take the axis of rotation at the upper end of the rod.

Force	X-comp	Y-comp
A	0	+3 N
B	+4 N	0
C	-4 N	-3 N
ΣF	0	0

The torque due to **C** is

$$\tau_C = -C(40 \text{ cm}) = -2.0 \text{ N}\cdot\text{m}$$

The moment arm of force **B** is

$$(62.5 \text{ cm})(\cos 37°) = (5/8)\text{m} \times (4/5) = 0.5 \text{ m}$$

Thus the torque is

$$\tau_B = +B(0.5 \text{ m}) = 2.0 \text{ N}\cdot\text{m}$$

This gives $\Sigma\tau = 0$. The rod is in rotational equilibrium.

18a. Choose the axis of rotation at the fulcrum. The torque due to the load is (5 N)(6 cm) = 30 N·cm. The force **F** produces a torque of F(24 cm) in the opposite direction. Thus

$$F(24 \text{ cm}) = 30 \text{ N}\cdot\text{cm}$$
$$F = 1.25 \text{ N}$$

b. The force on the fulcrum is 6.25 N, the sum of the two downward forces.

19a. Taking torques about the foot of the tower,

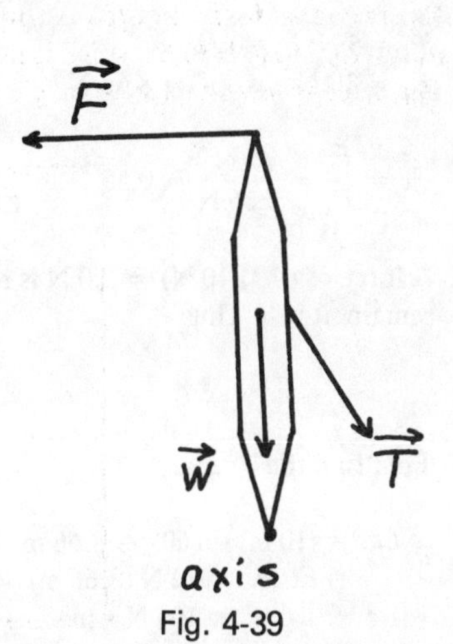

Fig. 4-39

$$\tau_1 = (500 \text{ N})(100 \text{ m}) = 5 \times 10^4 \text{ N}\cdot\text{m}$$
$$\tau_2 = -T(50 \text{ m})(\sin 30°) = -T \times 25 \text{ m}$$

Since $\Sigma\tau = 0$,

$$T \times 25 \text{ m} = 5 \times 10^4 \text{ N}\cdot\text{m}$$
$$T = 2 \times 10^3 \text{ N}$$

b. The forces must be resolved into components.

Force	X-comp	Y-comp
F	− 500 N	0
W	0	− 507 N
T	+ 1000 N	− 1732 N
Sum	+ 500 N	− 2239 N

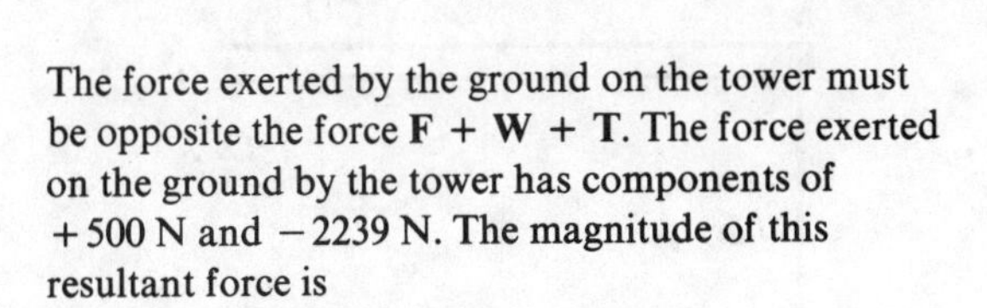

The force exerted by the ground on the tower must be opposite the force **F** + **W** + **T**. The force exerted on the ground by the tower has components of +500 N and −2239 N. The magnitude of this resultant force is

$$\sqrt{(500)^2 + (2239)^2} = 2294 \text{ N}$$

20. Place the axis of rotation at the center of gravity. Taking torques about this point leads to

a. $W_1x = W_2(L - x) = W_2L - W_2x$

$$x = \frac{W_2}{W_1 + W_2}L$$

b. $x = \frac{6\text{ N}}{8\text{ N}} 40\text{ cm} = 30\text{ cm}$

c. The support should be placed at the center of gravity.

21. Let the balance point be located a distance x from the line of action of W. Taking torques about this point,

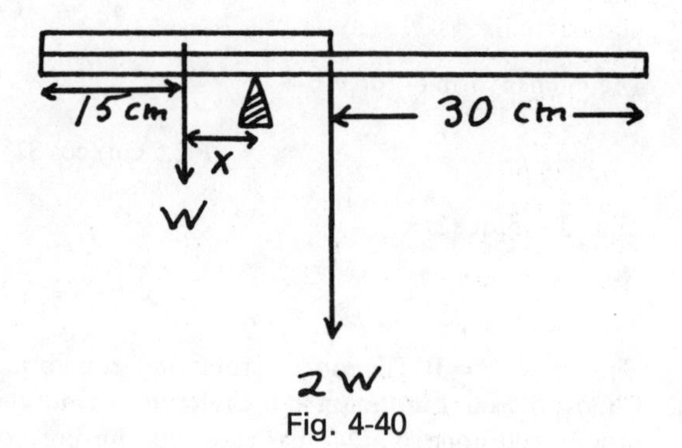

Fig. 4-40

$Wx = (2W)(15\text{ cm} - x) = 2W(15\text{ cm}) - 2Wx$
$3x = 30\text{ cm};\ x = 10\text{ cm}$

Thus the balance point is 25 cm from the left end.

22. The component parallel to the plane is

$$(10\text{ N})(\sin 30°) = 5\text{ N}$$

23. $F_s = 5\text{ N};\ F_s = \mu F_n$
$F_n = 10\text{ N}\cos 30° = 8.66\text{ N}$

$$\mu = \frac{F_s}{F_n} = \frac{5\text{ N}}{8.66\text{ N}} = 0.58$$

24. A force of (0.3)(10 N) = 3.0 N is required to keep the block moving. Since the applied force is only 2 N the block will remain at rest. Thus

$$\Sigma F_x = 2\text{ N} - F_s = 0;\quad F_s = 2\text{ N}$$

25. First find the lever arm.

$$l.a. = (10\text{ m})\sin 60° = 8.66\text{ m}$$
$$\tau = F(l.a.) = 12\text{ N}\ (8.66\text{ m})$$
$$= 104\text{ N}\cdot\text{m}$$

26. Take the axis of rotation at the pivot. The counterclockwise torque is

$$\tau_1 = 12\text{ N}\ (50\text{ cm})$$

The clockwise torque is

$$\tau_2 = -T\ (25\text{ cm}\sin 30°)$$

Since $\Sigma\tau = \tau_1 + \tau_2 = 0$, we have

$$T(25/2)\text{cm} = 12\text{ N}\ (50)\text{ cm}$$

$$T = 48\text{ N}$$

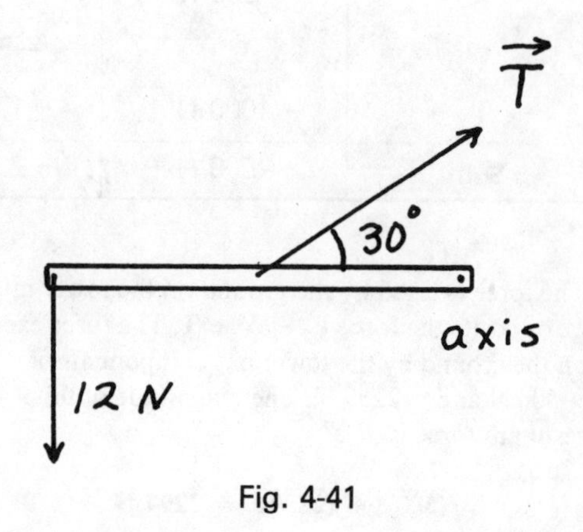

Fig. 4-41

27. Choose the axis of rotation at the center of gravity. Then there will be only two torques.

$$\tau_1 = (2\text{ N})(12\text{ m} - x)\,; \quad \tau_2 = -(6\text{ N})(x)$$

Since $\tau_1 + \tau_2 = 0$,

$$2\text{ N}\,(12\text{ m} - x) = 6\text{ N}\,(x)$$
$$x = 3\text{ m}$$

28a. The wrong value has been used for the lever arm of the force **W**. The lever arm is not the distance from the axis to the point of application of a force. The lever arm is the perpendicular distance from the axis to the *line of action* of the force. Look at the drawing.

b. You can probably avoid this error by drawing in a dotted line to represent the line of action of each force and then making sure that the lever arm is, in fact, perpendicular to this dotted line.

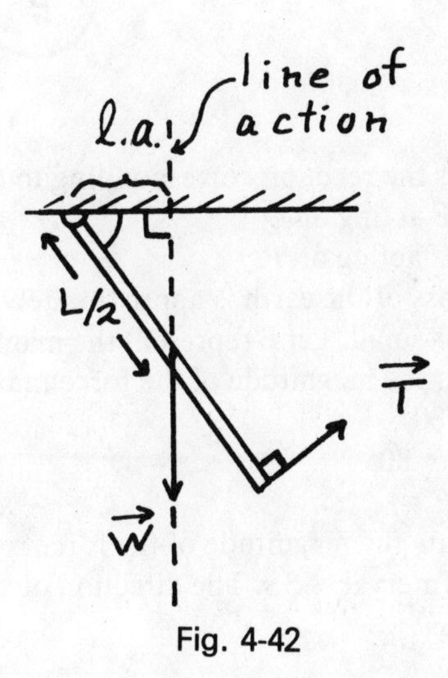

Fig. 4-42

c. $T(L) = 50\text{ N}\overbrace{\left((L/2)\cos 60°\right)}^{\text{lever arm}}$
$T = 50\text{ N}(1/2)(1/2) = 12.5\text{ N}$

REVIEW TEST FOR CHAPTERS 3 AND 4

1. According to Newton's third law of motion, forces always occur in pairs often called "action" and "reaction." In Fig. T-7 the arrow labeled **F** represents the force that object P exerts on object Q.

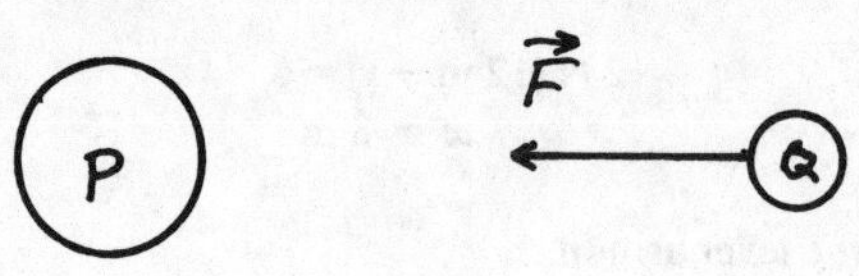

Fig. T-7

What is the reaction corresponding to this force?

a. $-\mathbf{F}$ acting on Q
b. $-\mathbf{F}$ acting on P
c. $+\mathbf{F}$ acting on P
d. $+\mathbf{F}$ acting on Q

2. The mass of the earth is approximately 6×10^{24} kg; that of the moon is 7.4×10^{22} kg, approximately 80 times as small. Let F represent the magnitude of the gravitational force that the earth exerts on the moon. What is the magnitude of the force that the moon exerts on the earth?

a. $F/80$
b. $F/\sqrt{80}$
c. $80F$
d. F
e. $\sqrt{80}\,F$

3. Calculate the magnitude of the force required to change the velocity of a 4-kg mass from 2 m/s to 8 m/s in a time interval of 3 s. The direction of **v** does not change.

4. A small object whose mass is 2 kg is subject to two forces, $\mathbf{F}_1$ and $\mathbf{F}_2$, as shown in Fig. T-8. What is the magnitude of the acceleration produced?

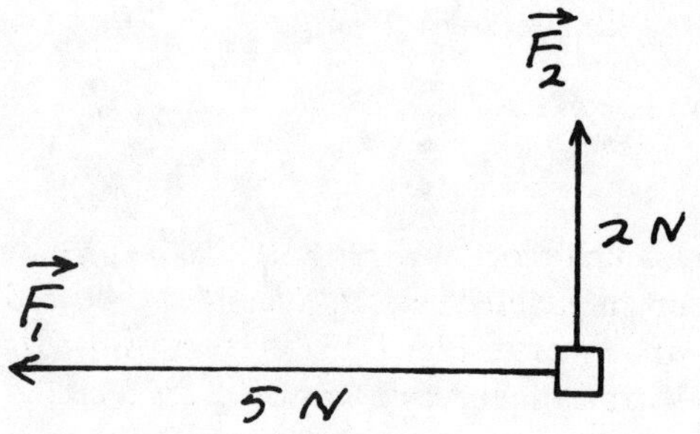

Fig. T-8

5. Two masses are connected by a light string running over a pair of pulleys as shown in Fig. T-9. Assuming that there is no friction, what is the upward acceleration of the smaller mass?

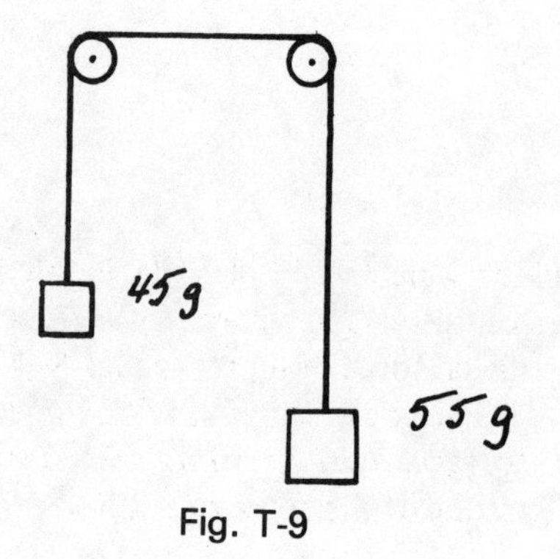

Fig. T-9

6. An object is observed to be moving in a straight line at a constant speed. Which of the following statements does *not* apply to this motion?
   a. The sum of all forces acting on the object is zero.
   b. The object is in a state of equilibrium.
   c. $\mathbf{v}$ = constant.
   d. Net **F** is a constant, non-zero force in the direction of motion.
   e. $\mathbf{a}$ = zero.

7. A small mass is in equilibrium under the action of three forces. Two of the forces are shown in Fig. T-10. Which of the arrows in Fig. T-11 represents the third force?

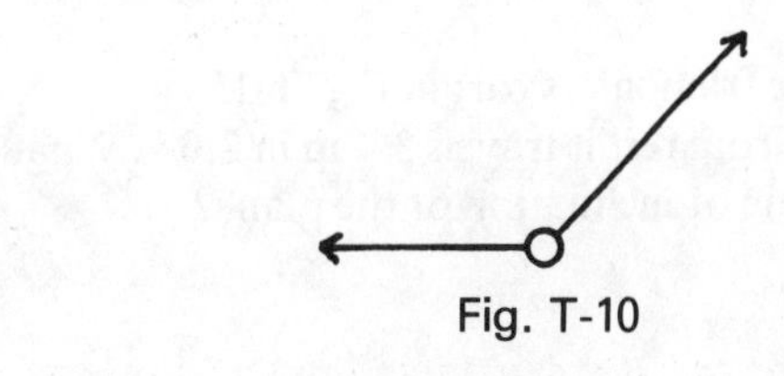

Fig. T-10

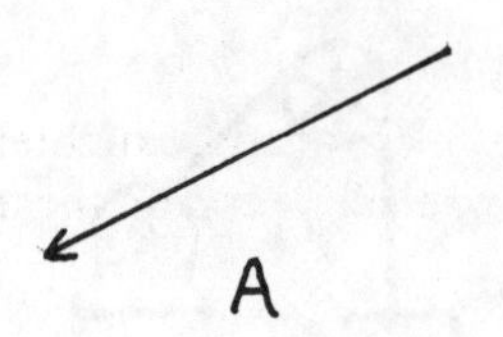

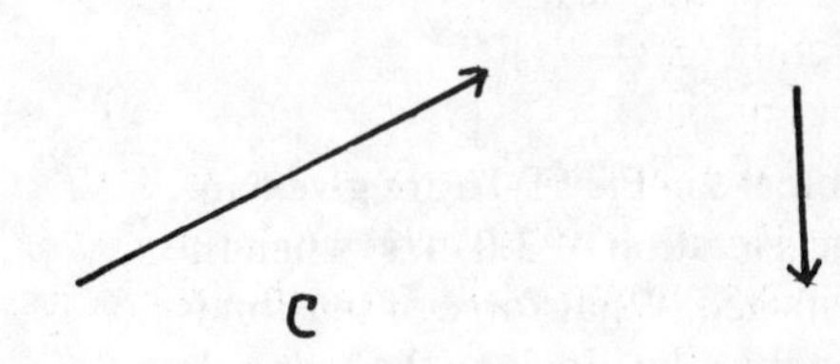

Fig. T-11

8. A 10-N weight is supported by two strings as shown in Fig. T-12. The tension T_1 in the string on the left is given by T_2 times
   a. cos 37°/sin 30°
   b. cos 30°/cos 37°
   c. sin 37°/cos 30°
   d. cos 37°/cos 30°
   e. sin 30°/sin 37°

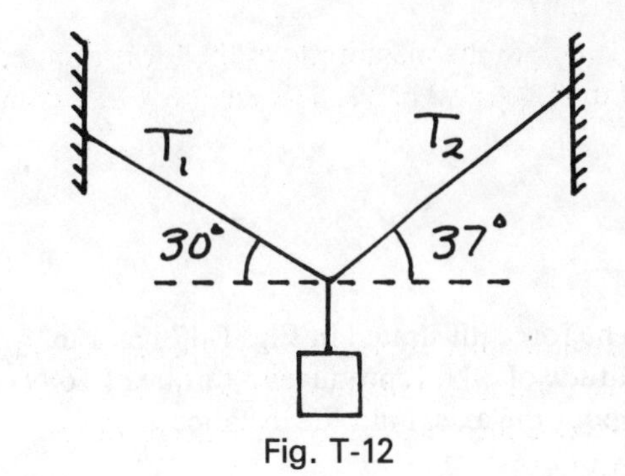

Fig. T-12

9. A small block weighing 10 N is being pulled up an inclined surface at a constant speed of 4 m/s (see Fig. T-13). Assuming no friction, what is the force of tension in the cord?

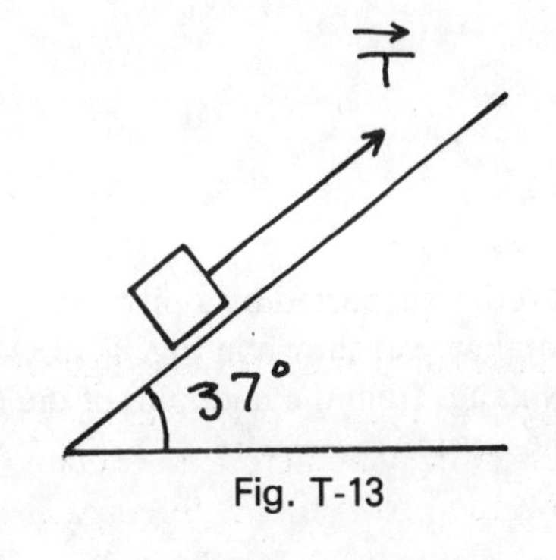

Fig. T-13

10. A block weighing 10 N is placed on a horizontal surface where it is subject to a horizontal force of magnitude 2 N (see Fig. T-14). The coefficients of friction between the two surfaces are 0.3 and 0.4. What is the magnitude of the force of friction F?

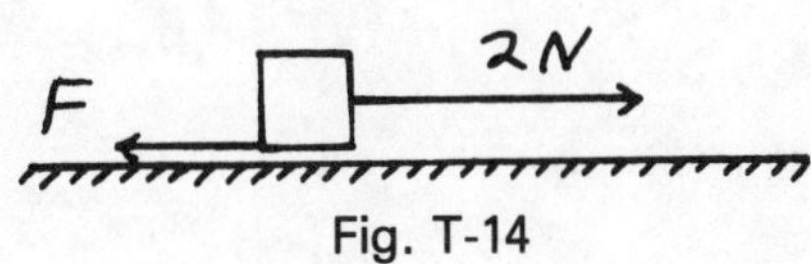

Fig. T-14

11. An elevator weighing 7000 N has an upward acceleration of 2.0 m/s². What is the tension in the supporting cable?

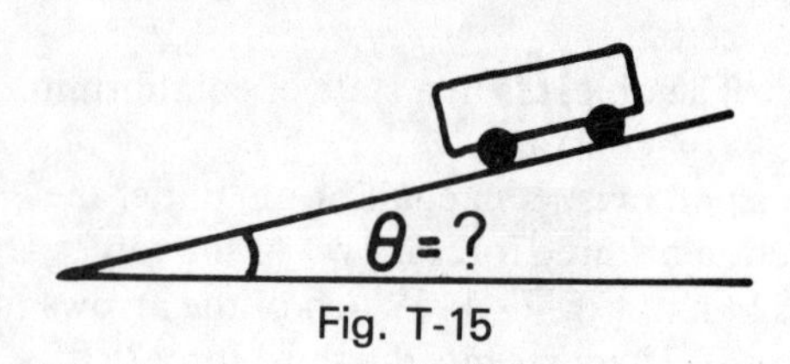

Fig. T-15

12. When the frictionless cart in Fig. T-15 is released from rest it travels 3.4 m in 2.0 s. What is the angle of inclination of the plane?

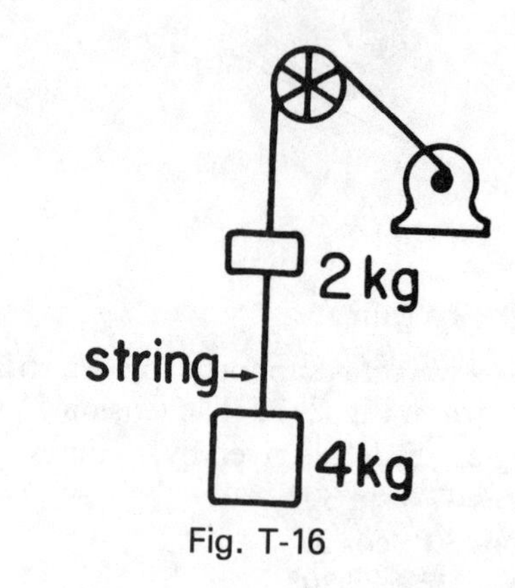

Fig. T-16

13. The two blocks in Fig. T-16 are given an upward acceleration of 3.0 m/s² when the motor is started. What force of tension is exerted on the 4-kg block by the string that connects it to the smaller block?

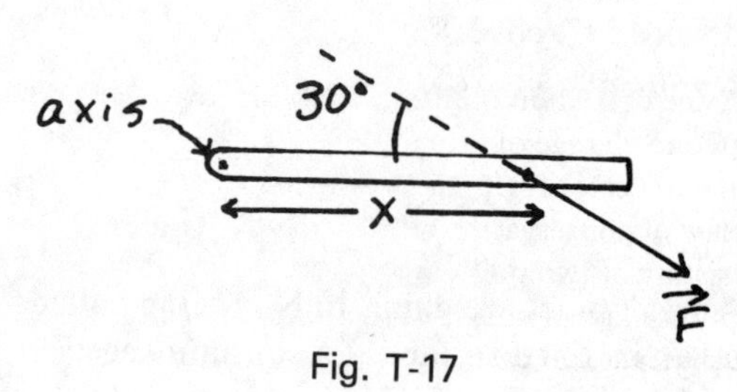

Fig. T-17

14. The force illustrated in Fig. T-17 has a magnitude of 8 N. It produces a torque of 16 N·m about the axis. Find the distance x.

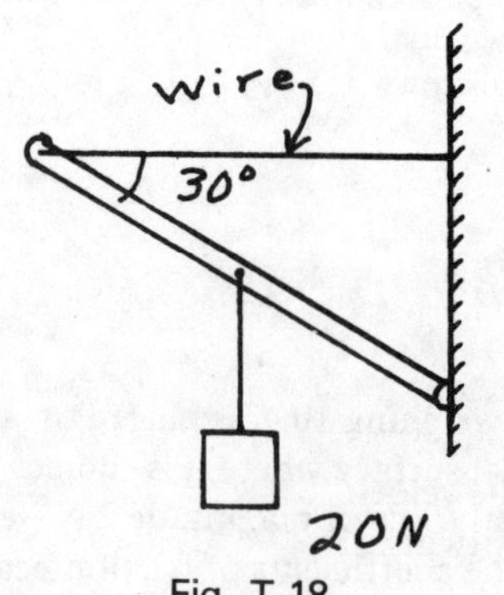

Fig. T-18

15. A light rod is supported by a pivot and a horizontal wire as shown in Fig. T-18. A weight of 20 N hangs from the midpoint of the rod. Find the tension in the wire.

CHAPTER 5
Conservation of Momentum

GOALS To gain an understanding of the conditions under which the momentum of a system is constant and to use this knowledge in problem solving.

OBJECTIVES After completing this chapter the student should be able to do the following:

1. Write the definition of linear momentum in words and in symbols.
2. Write Newton's second law in terms of momentum.
3. Give the definition of impulse and show how it is related to change of momentum.
4. Write, in vector form, the mathematical expression of the law of conservation of momentum for a system consisting of two point masses.
5. Apply the law of conservation of momentum in a one-dimensional case to determine an unknown mass or an unknown velocity.
6. Apply the law of conservation of momentum to a collision in two dimensions to determine an unknown mass or an unknown velocity.
7. State the technical term for a set of three mutually perpendicular axes relative to which positions in space can be measured.
8. Distinguish between inertial and non-inertial frames of reference.
9. Give an example of an inertial, or nearly inertial, frame of reference.
10. Calculate the apparent weight of an object in an accelerated frame of reference. Calculate the acceleration from the apparent weight of a known object.
11. Find the "inertial force" acting on a body in an accelerated frame of reference.

SUMMARY

Two of the great unifying principles of physics and, in fact, of science as a whole, are the laws of conservation of momentum and energy. Momentum, **p**, is a vector quantity equal to the product of the mass of a particle and its velocity, in symbols $\mathbf{p} = m\mathbf{v}$. Newton's second law of motion may be written either in terms of mass and velocity or in terms of momentum.

$$\text{net } \mathbf{F} = \lim_{\Delta t \to 0} m \frac{\Delta \mathbf{v}}{\Delta t} \quad \text{or} \quad \text{net } \mathbf{F} = \lim_{\Delta t \to 0} \frac{\Delta \mathbf{p}}{\Delta t}$$

The first formulation, in terms of velocity, is valid only if the mass does not change with time. The second formulation is more general since it remains valid even if the mass is a function of time, as in the launching of a rocket, for example.

The law of conservation of linear momentum states that the total linear momentum of an *isolated* system of bodies remains constant. A system of bodies is isolated in this sense if there is no net *external* force acting on the system. Internal forces of any kind, even the large forces that are present during a collision of two bodies within the system, cannot change the total linear momentum of the system. Even frictional forces between colliding bodies cannot change the total momentum of a system, although they will cause a loss of *mechanical energy* as we shall see in the following chapter. Thus, for an isolated system,

$$\Sigma \mathbf{p} = \text{a constant}$$

Measurements of position and velocity are necessarily made with respect to a coordinate system, usually a set of three mutually perpendicular axes. Such a *frame of reference* is said to be *inertial* if Newton's laws of motion are valid in that frame. In a rotating coordinate system, Newton's laws of motion in their usual formulation are not valid, thus a rotating frame of reference is *non-inertial*.

Although the mass of an object does not depend on its location in space—being the same on the moon as on the earth, for example—the *weight* of an object depends on the local value for g, the acceleration during free fall, through the relationship, $W = mg$. In an accelerated frame of reference—an orbiting satellite, for example—the *apparent weight* of an object will be much diffeent from its weight mg.

QUESTIONS AND PROBLEMS

Secs. 5-1—5-2

1. Write the definition of momentum (a vector) and the SI units in which it is expressed.

   Momentum = ________________________.

   The units of momentum are ________________________.

2. The two vectors $\mathbf{p}_1$ and $\mathbf{p}_2$ in Fig. 5-1 represent the momentum of a particle at two different times.

   The magnitudes are: $p_1 = 3\ \text{kg} \cdot \text{m/s}$;
   $p_2 = 4\ \text{kg} \cdot \text{m/s}$.

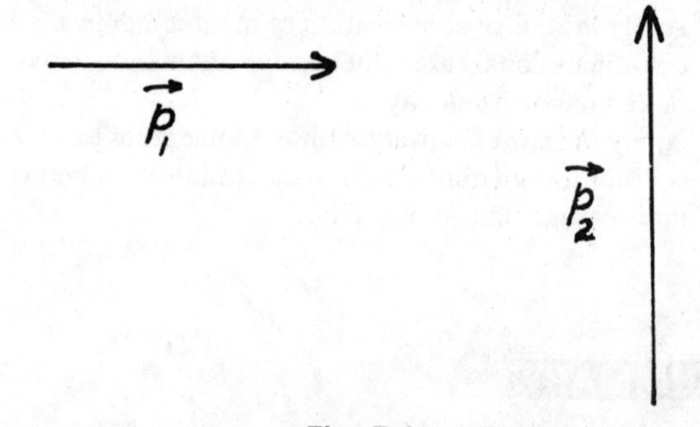

Fig. 5-1

   a. Calculate the change in momentum $\Delta\mathbf{p}$ that has taken place: $\Delta\mathbf{p} = \mathbf{p}_2 - \mathbf{p}_1$. (Remember that a vector is specified by giving both its magnitude and direction.)

   b. Supposing that the change in momentum has taken place in 0.5 s, calculate the force that caused the change in momentum.

3. A communications satellite is in a circular orbit about the earth. Two positions of the satellite are indicated by points A and B in Fig. 5-2.
   a. Draw vectors $\mathbf{p_A}$ and $\mathbf{p_B}$ to represent the momentum of the satellite at A and at B. (Check the solutions section for the right answer before proceeding. Note that the magnitudes $|\mathbf{p_A}|$ and $|\mathbf{p_B}|$ are equal.)
   b. Make a vector diagram showing how the change in momentum is found.

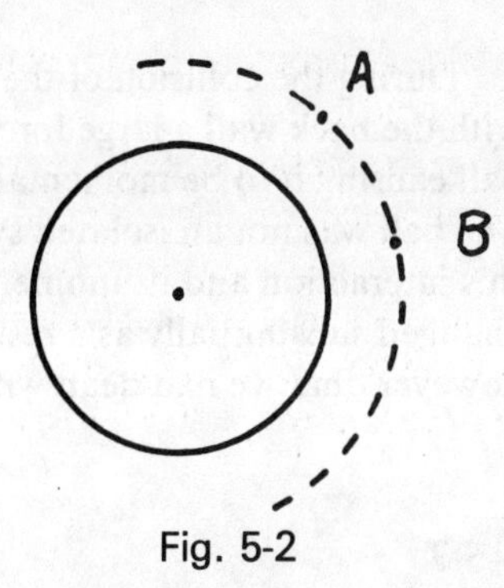

Fig. 5-2

   c. Is the satellite subject to a force? Draw a vector to represent this force. What is its direction?

   d. Is the satellite in equilibrium? Explain.

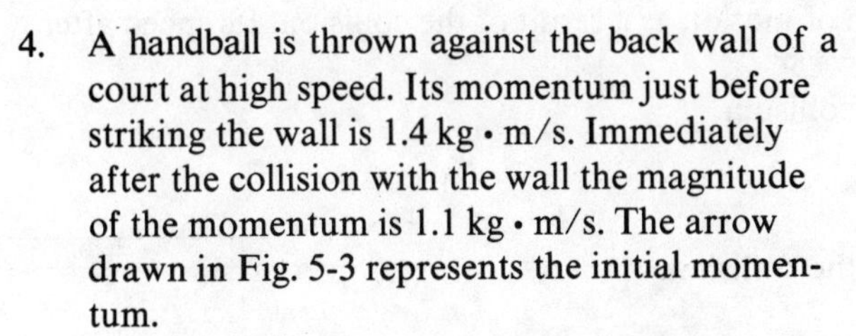

4. A handball is thrown against the back wall of a court at high speed. Its momentum just before striking the wall is 1.4 kg • m/s. Immediately after the collision with the wall the magnitude of the momentum is 1.1 kg • m/s. The arrow drawn in Fig. 5-3 represents the initial momentum.
   a. Draw a second arrow to represent $\mathbf{p}_2$, the momentum after the collision. (Assume that the ball is moving horizontally after the collision.)
   b. Draw to the same scale the vector $\mathbf{p}_2 - \mathbf{p}_1$. What is the magnitude of the change in momentum?

   c. Supposing that the ball is in contact with the wall for 0.001 s, calculate the average force that acted on the ball during the collision.

   d. The mass of the ball is 0.1 kg. Calculate its speed and its kinetic energy before the collision. $KE = \frac{1}{2}mv^2$.

$$v_1 = p_1/m = \_\_\_\_\_\_\_\_\_\_.$$

   e. Calculate the speed and the kinetic energy after the collision. You will find that some energy has been "lost" during the collision (actually converted into heat). Such a collision is said to be inelastic.

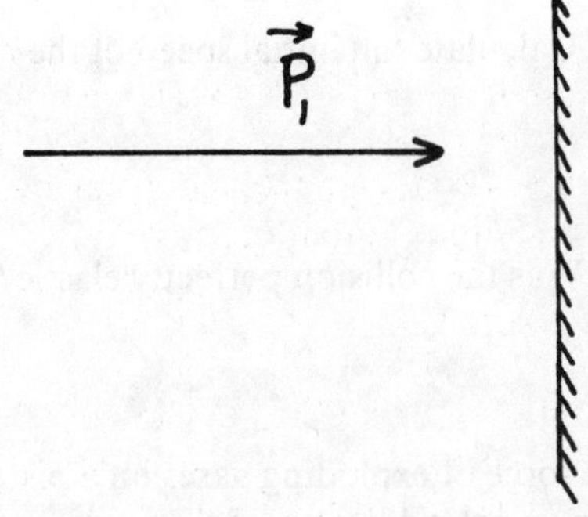

Fig. 5-3

   f. How much of the original kinetic energy was lost?

During the collision of the handball with the back wall a large force acted on the ball causing it to be momentarily flattened. The ball was not an isolated system during this interaction and its momentum was changed substantially as a result. Suppose, however, that we had dealt with a larger system, the ball plus the object with which it collided. During the collision the ball would have exerted a force **F** on the wall and the latter would have exerted an equal and opposite force $-\mathbf{F}$ on the ball. The *total* momentum of the "system" would not have been changed by these internal forces.

Secs. 5-3—5-7

5. A small projectile whose mass is 50 g strikes a 450-g block of wood and becomes imbedded in it. The velocity of the block plus projectile after the collision is 2 m/s. What was the velocity of the projectile before the collision?

When the colliding objects stick together and subsequently move as one, the collision is said to be "perfectly inelastic." The opposite extreme is perfect rebound with no loss of kinetic energy; it is called a "perfectly elastic" collision.

6. Suppose that the small projectile in the preceding problem is a rubber sphere that rebounds from the block. The masses are the same as before and the block, initially at rest, has a speed of 2 m/s after the collision. The rubber projectile reverses its direction of motion as a result of the collision. Its speed after the collision is 8 m/s.
   a. Calculate the momentum of the block after the collision.

   $m_2 v_2 =$ ____________ .

   b. Calculate the momentum of the projectile after the collision.

   $m_1 v_1 =$ ____________ .

   c. Calculate the total momentum of the system (projectile plus block) after the collision.

   $p_{tot} =$ ____________ .

   d. What was the total momentum of the system before the collision?

   $p_{tot} =$ ____________ .

   e. Calculate the initial speed of the rubber sphere.

   $v_1 =$ ____________ .

   f. Was the collision perfectly elastic?

   ____________ .

7. The force of exploding gases on a piston in an automobile engine is 2400 N. The quantity of gas involved is 0.8 g and the duration of the explosion is 0.001 s. Calculate the average speed of the molecules of the gas. (Hint: Study Ex. 5-2 in the text.)

8. a. Why is the earth not, strictly speaking, an inertial frame of reference?

b. Describe a frame of reference which is a better approximation to an inertial frame than the rotating earth.

9. An automobile passenger having a mass of 55 kg is thrown forward against her seat belt in a head-on collision with another car. If the average deceleration during the collision is 10*g* (i.e., 98 m/s²), what is the magnitude of the inertial force that pushes the passenger forward?

10. Imagine a frame of reference that is fixed to the platform of a merry-go-round. The origin is located at the center and the *x*-axis lies along one of the radial struts that support the platform. The merry-go-round is revolving at a constant angular speed of 0.2 rev/s.

a. Is this frame of reference an inertial frame? Explain your answer.

b. Suppose that you are sitting on the floor of the merry-go-round. Then you are "at rest" in this frame of reference. Is it true that $\Sigma F_x = 0$ in this frame of reference?

c. Are Newton's laws of motion valid in this frame of reference? Explain.

11. The acceleration of free fall on the surface of Mars is $g' = 3.7\ m/s^2$.

a. What is the *true weight* of an object on the Martian surface if the mass of the object is 20 kg?

b. Suppose that the 20-kg object is placed on a spring scale on the floor of a landing module. The landing module blasts off from the Martian surface with an acceleration of $2g'$. What is the *apparent weight* of the object?

12. A student, whose true weight is 650 N, is on an elevator that is descending from the 20th floor. During the brief period of acceleration, he observes that his weight appears to be reduced to 400 N.

a. What is the acceleration of the elevator?

b. When the elevator brakes to a stop at the ground floor it has a maximum deceleration of $3.0\ m/s^2$. What is the apparent weight of the student at maximum deceleration?

13. Because of the rotation of the earth, an object dropped from a great height does not strike the ground at a point directly below its point of release. For an object that is dropped from a tall building in the Northern Hemisphere (the latitude of Chicago for example), is the actual impact point to the north, south, east, or west of the expected point of impact?

Solutions

1. momentum = (mass)(velocity)
   or $\mathbf{p} = m\mathbf{v}$

   The units of momentum are kg•m/s or N•s.

2. $\Delta\mathbf{p} = \mathbf{p}_2 + (-\mathbf{p}_1)$

   a. The magnitude of $\Delta\mathbf{p}$ is 5 kg • m/s. Its direction is 37° from $\mathbf{p}_2$ as shown in Fig. 5-4.
   b. net $F = \Delta p/\Delta t$ = (5 kg • m/s)/(0.5 s) = 10 N

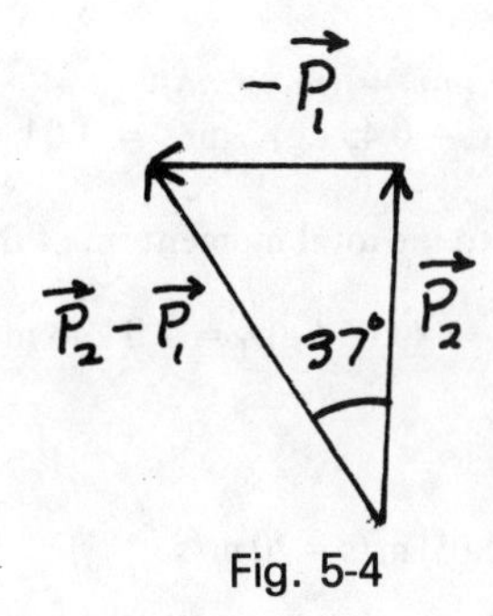

Fig. 5-4

3a.

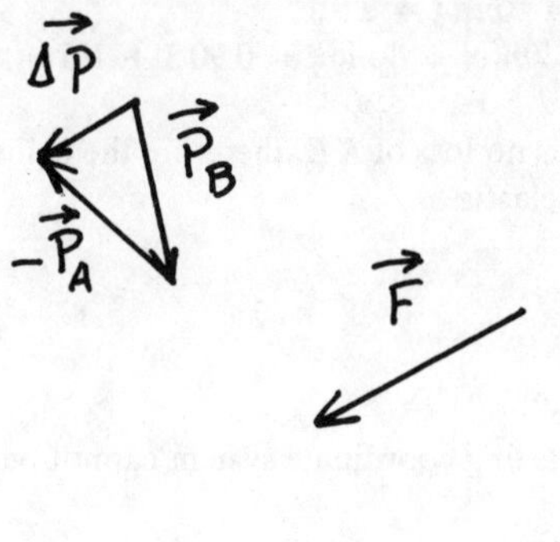

Fig. 5-5

$\Delta\mathbf{p}$ and $\mathbf{F}$ are directed toward the center.

b. $\Delta\mathbf{p} = \mathbf{p}_B - \mathbf{p}_A$
c. Yes. A force is necessary to produce a change in momentum. The force is in the direction of $\Delta\mathbf{p}$. Its magnitude is $\Delta p/\Delta t$.
d. The satellite is acted upon by an unbalanced force; net $\mathbf{F} = \Delta\mathbf{p}/\Delta t$. It is certainly *not* in equilibrium.

4a.

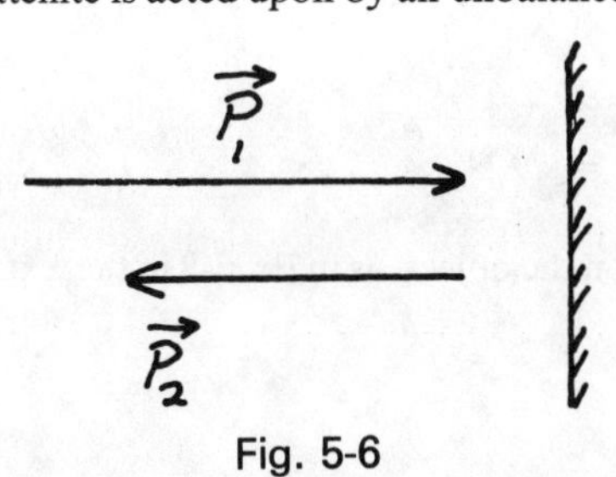

Fig. 5-6

b. $\Delta\mathbf{p} = \mathbf{p}_2 - \mathbf{p}_1$
The magnitude of $\Delta\mathbf{p}$ is 1.1 kg • m/s + 1.4 kg • m/s = 2.5 kg • m/s.

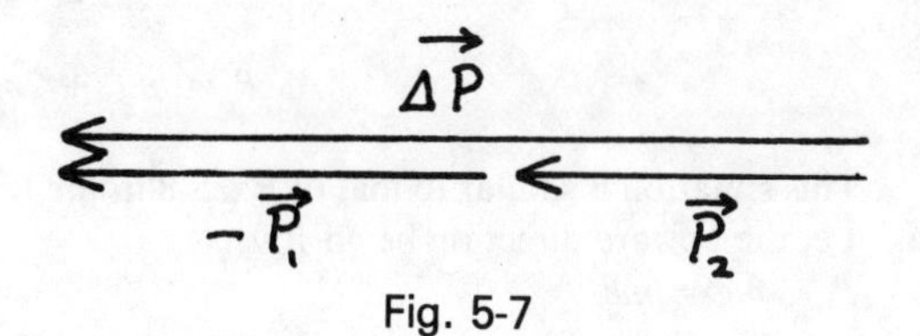

Fig. 5-7

c. $F_{(av)} = \frac{\Delta p}{\Delta t} = \frac{2.5 \text{ kg} \cdot \text{m/s}}{10^{-3} \text{ s}} = 2.5 \times 10^3 \text{ N}$

d. $v_0 = p_1/m = (1.4 \text{ kg} \cdot \text{m/s})(0.1 \text{ kg}) = 14 \text{ m/s}$
$KE = \frac{1}{2} mv^2 = 9.8 \text{ J}$

e. $v = p_2/m = (1.1 \text{ kg} \cdot \text{m/s})/(0.1 \text{ kg}) = 11 \text{ m/s}$
$KE = \frac{1}{2} mv^2 = 6.05 \text{ J}$

f. energy loss $= (9.8 - 6.05)\text{J} \cong 3.7 \text{ J}$

5. First find the total momentum of the system *after* the collision.

$$p_{tot} = m_1v_1 + m_2v_2 = (m_1 + m_2)v$$
$$= (0.05 \text{ kg} + 0.45 \text{ kg})\ 2 \text{ m/s} = 1.0 \text{ kg} \cdot \text{m/s}$$

This is equal to the total momentum of the system before the impact.

$$p_{before} = m_1v_1 = (0.05 \text{ kg})v_1 = 1.0 \text{ kg} \cdot \text{m/s}$$

Thus

$$v_1 = (1.0)/(0.05) \text{ m/s} = 20 \text{ m/s}$$

6a. $m_2v_2 = (0.45 \text{ kg})(2 \text{ m/s}) = 0.9 \text{ kg}\cdot\text{m/s}$
b. $m_1v_1 = (0.05 \text{ kg})(-8 \text{ m/s}) = -0.4 \text{ kg}\cdot\text{m/s}$
c. $p_{tot} = (0.9 - 0.4)\text{kg}\cdot\text{m/s} = 0.5 \text{ kg}\cdot\text{m/s}$
d. The total momentum did not change during the collision.
$p_{after} = p_{before} = 0.5 \text{ kg}\cdot\text{m/s}$
e. $m_1v_1 = (0.05 \text{ kg})v_1 = 0.5 \text{ kg}\cdot\text{m/s};\ v_1 = 10 \text{ m/s}$
f. $KE(\text{before}) = \frac{1}{2}m_1v_0^2 = 2.5 \text{ J}$
$KE(\text{after}) = \frac{1}{2}m_1v_1^2 + \frac{1}{2}m_2v_2^2 = 0.90 \text{ J} + 1.6 \text{ J}$
$= 2.5 \text{ J}$
Yes. There was no loss of KE, therefore the collision was perfectly elastic.

7. (See Ex. 5-2 in the text for method of solution.)
$v_{av} = 1.5 \times 10^3 \text{ m/s}$

8a. The earth rotates about its own axis, completing one revolution every 24 h. A rotating coordinate system cannot be an inertial frame of reference.
b. A frame of reference that is at rest with respect to the fixed stars is a very good approximation to an inertial frame.

9. The inertial force arises from the acceleration of the frame of reference, in this case, the car. The magnitude of the force, which exists only in the accelerated frame, is

$$ma = (55 \text{ kg})(98 \text{ m/s}^2) = 5.4 \times 10^2 \text{ N}$$

10a. No. A rotating frame of reference is never an inertial frame. Fictitious forces are always present in rotating frames.
b,c. No. Newton's laws of motion are not valid in a non-inertial frame of reference. Specifically, an object that is subject to no forces will be seen to accelerate in the rotating frame of reference.

11a. The *true* weight of the 20-kg object on Mars is

$$w' = mg' = (20 \text{ kg})(3.9 \text{ m/s}^2) = 79 \text{ N}$$

b. Let P represent the magnitude of the upward force exerted by the scale on the object, as in Ex. 5-8 in the text. Newton's second law as applied to the 20-kg object reads

$$P - mg' = ma$$

Since the acceleration a is upward with a magnitude of $2g'$,

$$P = mg' + 2mg' = 3mg' = 3W' = 2.4 \times 10^2 \text{ N}$$

12. This situation is similar to that of Ex. 5-8 in the text except that the acceleration is downward.
a. Let the upward direction be positive.
$P - mg = ma$

$$a = \frac{P - mg}{m} = \frac{400\ \text{N} - 650\ \text{N}}{(650\ \text{N})/9.8\ \text{m/s}^2} = -3.8\ \text{m/s}^2$$

b. $P - mg = ma; \quad P = mg + ma$
$P = (66.3\ \text{kg})(9.8\ \text{m/s}^2 + 3.0\ \text{m/s}^2) = 8.5 \times 10^2\ \text{N}$

13. The earth rotates toward the east at an angular speed of 1 rev/day. Because the object at the top of the building is farther from the earth than is the bottom of the building, the speed of the object toward the east is greater than is the speed of the ground. The horizontal component of the object's velocity remains constant as it falls. Thus the falling object moves farther toward the east than the earth's surface. It will strike the ground slightly to the east of the "expected" point of impact.

CHAPTER

6 Conservation of Energy

GOALS To learn to recognize the conditions under which the total energy of a system will remain constant and to use this knowledge in problem solving.

OBJECTIVES After completing this chapter the student should be able to do the following:

1. Write the definition of work in terms of force and displacement.
2. Calculate the work done by a constant force when the force and displacement vector make an arbitrary angle with each other.
3. Calculate the work done by a nonconstant force from the area under the force-displacement graph.
4. Name the two types of mechanical energy, associating one with the relative position of parts of a system and the other with the relative motion.
5. Write the formulas for kinetic energy and gravitational potential energy using standard symbols.
6. Write a general statement of the law of conservation of energy.
7. Write the law of conservation of energy for a mechanical system in mathematical terms using standard symbols.
8. Apply the energy principle to solve simple problems in mechanics.
9. Write the definition of power in words and in symbols.
10. Use this definition to calculate power in simple cases.
11. Give the technical term for a collision in which some of the kinetic energy is transformed into another form, usually heat.
12. Give the technical term for a collision in which no kinetic energy is lost.
13. Solve one-dimensional problems involving a collision by application of energy and momentum principles.

After completing the "For Further Study" sections of this chapter, the student should be able to do the following:

14. Calculate the work input, work output, and efficiency of certain simple machines (pulley systems, inclined planes).
15. Write the definition of coefficient of restitution.
16. Find the final velocities of colliding objects for which the coefficient of restitution is known.

SUMMARY

The term *work*, which has many meanings in everyday life, must be used in the context of mechanics in a very restricted and well-defined sense. The work done by a force on an object is the product of the displacement and the component of the force in the direction of the displacement. This relationship may be written $W = \mathbf{F} \cdot \mathbf{s}$ or $W = Fs \cos \theta$ where θ is the angle between the force vector **F** and the displacement vector **s**. A force that acts in a direction perpendicular to the motion of an object can do no work since the cosine of 90° is zero. If the force is not constant, the work can be found from the area under the force-displacement curve.

When work is done on a body, that body is said to gain *energy*. The energy of a body appears in two forms, *potential energy* and *kinetic energy*. Potential energy is related to the relative position of a body with respect to other bodies. Kinetic energy is that form of energy associated with the motion of the body. Gravitational potential energy is the weight of an object times its height above a reference level, $PE = Wh = mgh$. Kinetic energy is equal to one half the product of the mass and the square of the velocity, $KE = \frac{1}{2}mv^2$. The total mechanical energy of a body is the sum of its potential energy and its kinetic energy. Other forms of energy, which will be discussed in later chapters, are electrical, thermal, and chemical.

Taking all forms of energy into account, the law of conservation of energy states that the *total* energy in any isolated system remains constant, even if transformations of one form of energy into another are taking place. If a system consists of two objects approaching each other on a collision course, the kinetic energy of the system will generally be less after the collision than before because some of the kinetic energy will have been transformed into heat. The *total* energy of the system, however, cannot be changed by a collision or by any other interaction between parts of the system.

Collisions between two objects may, under special circumstances, take place in such a way that the kinetic energy of the system does not change. These are called ***perfectly elastic*** collisions. On an atomic scale, collisions between atoms and the constituents of atoms are often perfectly elastic.

Power is the rate at which work is done. This is written

$$P = \lim_{\Delta t \to 0} \frac{\Delta W}{\Delta t}$$

If the work is done at a constant rate, the power may be written simply, $P = \Delta W/\Delta t$.

QUESTIONS AND PROBLEMS

Secs. 6-1—6-2

1. The barge shown in Fig. 6-1 is being pulled along a canal by a small electric engine running on a parallel track. If the force exerted by the cable is 5 kN, how much work is done in moving the barge 10 km?

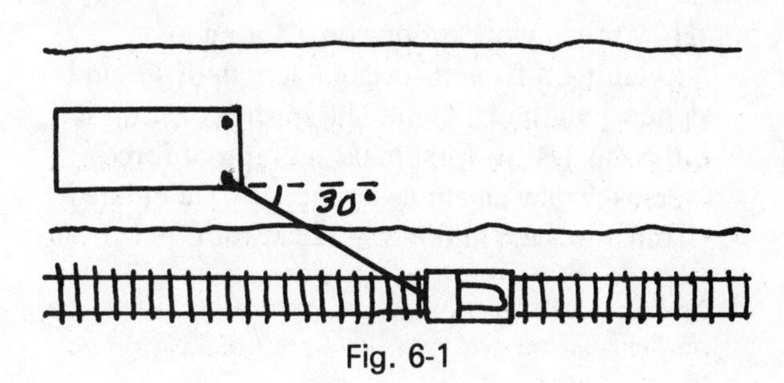

Fig. 6-1

2. A pendulum bob is swinging in an arc of radius 1.5 m. The force of tension **T** in the supporting cord is 2.5 N. During a displacement of 2 cm along the arc how much work is done by the force **T** shown in Fig. 6-2?

   Is work being done on the pendulum bob by some other force? Explain.

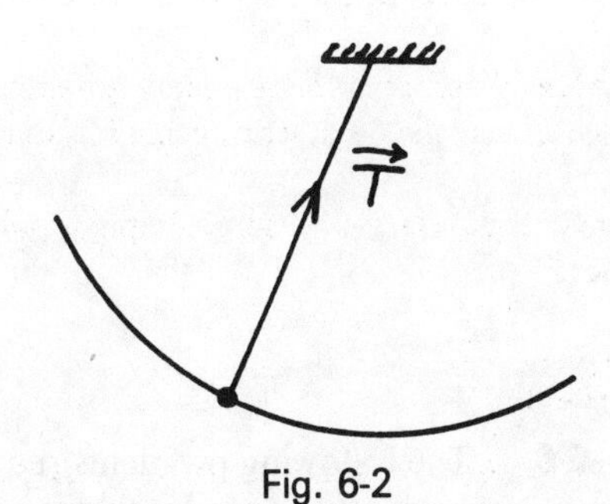

Fig. 6-2

3. The moon is held in its approximately circular orbit about the earth by a gravitational force of 2×10^{20} N. How much work is done by this force on the moon?

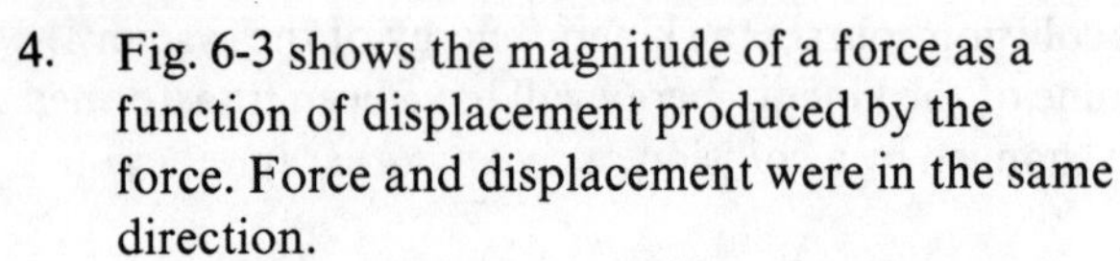

4. Fig. 6-3 shows the magnitude of a force as a function of displacement produced by the force. Force and displacement were in the same direction.
   a. What work was done by this force during the displacement from 0 to 2 m?

   b. What work was done by this force during the displacement from 2 to 4 m?

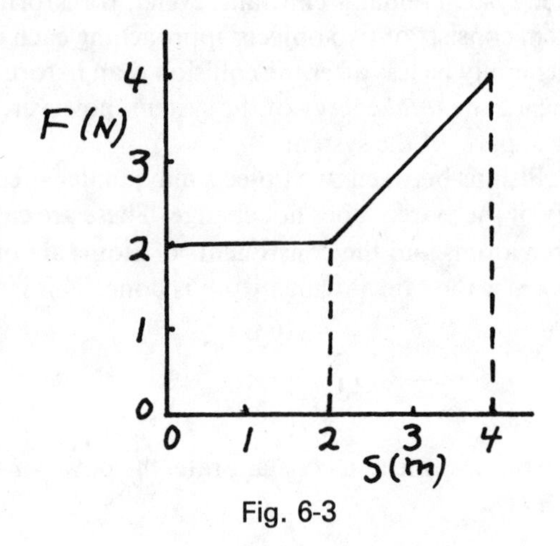

Fig. 6-3

5. How much work is done on a spring in stretching it from its original length of 12 cm to a new length of 17 cm? The spring constant is 10 N/m. (Hint: First make a graph of force versus displacement using Fig. 6-4. Calculate F from $F = k\Delta s$, taking $s = 12$ cm for $F = 0$.)

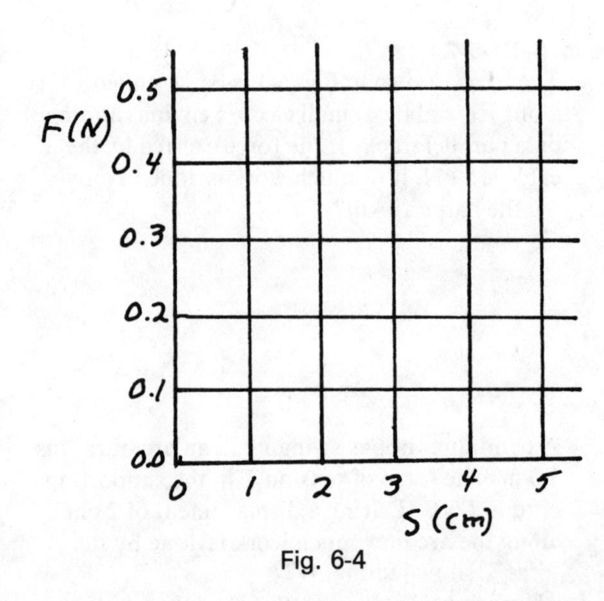

Fig. 6-4

Secs. 6-3—6-6 The following problems require a knowledge of potential energy, kinetic energy, and energy conservation. First, write the formulas that you will need.

KE = _______________. PE(gravitational) = _______________.

6. Figure 6-5 shows a frictionless cart that starts from rest and rolls down a hill. In each case the starting point is 50 m above the bottom of the hill. In case (A) the cart arrives at the bottom of the hill with a kinetic energy of 5000 N • m.

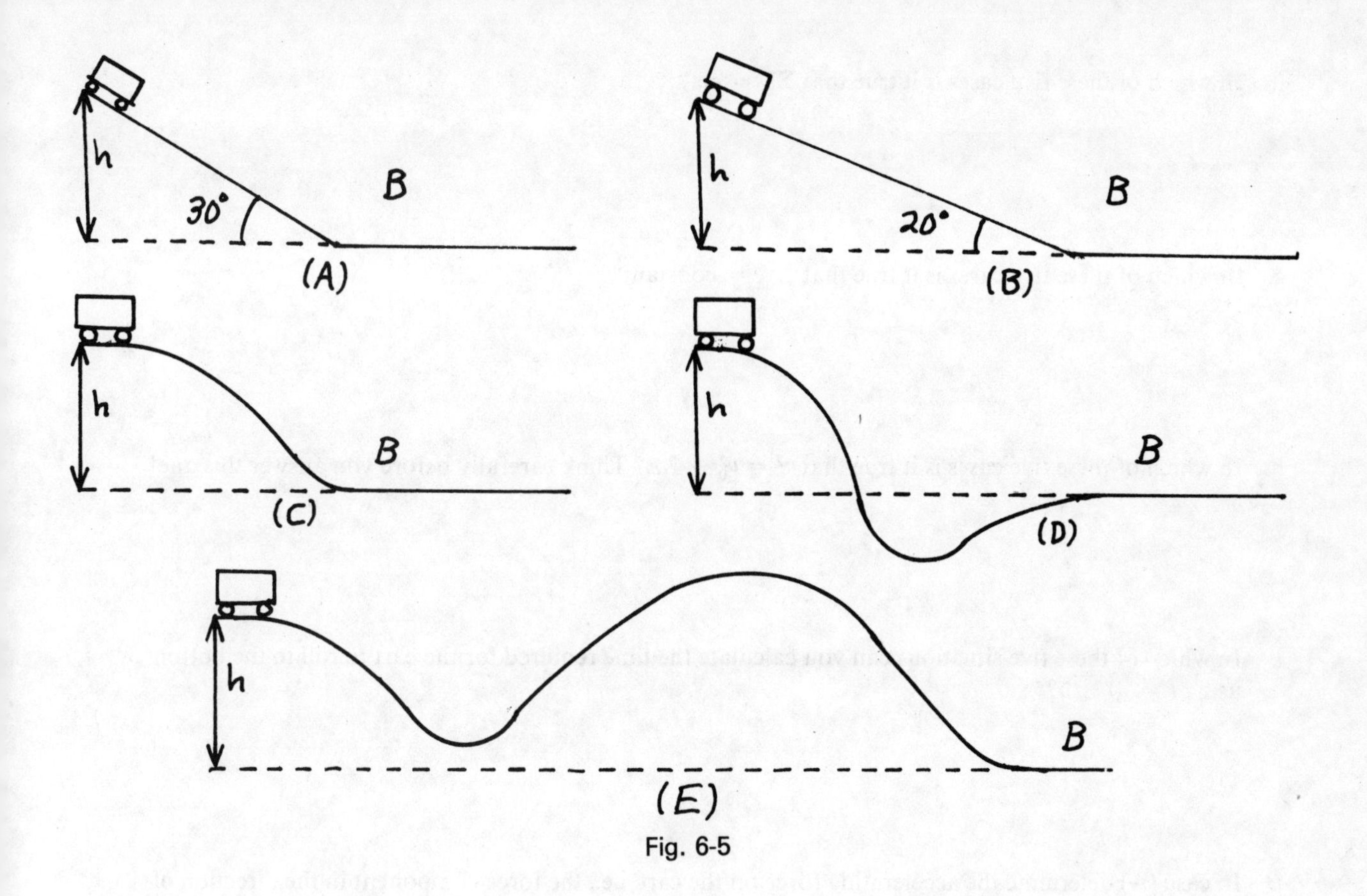

Fig. 6-5

a. What was the potential energy of the cart when it was at the top of the hill, just before starting to roll? (*PE* is to be measured relative to the zero level indicated in the drawings.)

$PE =$ ______________.

b. What is the *KE* of the cart when it arrives at the "bottom" (point B) in the other four cases? [Consider case (E) carefully!]

(B) $KE =$ ______________. (C) $KE =$ ______________.

(D) $KE =$ ______________. (E) $KE =$ ______________.

c. What is the weight of the cart? (Check your answers as you proceed.)

d. What is the mass of the cart?

e. What is the velocity (magnitude only) of the cart when it reaches the bottom of the hill at point *B*?

f. In which of these five cases is it true that $\Sigma \mathbf{F} = m\mathbf{a}$?

g. In which of these five cases is it true that E_{tot} = constant?

h. In which of these five cases is it true that $v^2 = v_0^2 + 2as$? Think carefully before you answer this one!

i. In which of these five situations can you calculate the time required for the cart to roll to the bottom using $t = \sqrt{2s/a}$?

j. In case (A) determine the accelerating force on the cart, i.e., the force component in the direction of motion.

k. Use the net F found above in part (j) to calculate the acceleration of the cart. Why is this acceleration constant?

l. Use the appropriate equation of uniformly accelerated motion to calculate the speed of the cart when it reaches the bottom of the hill.

m. You have found in part (c) that the weight of the cart is 100 N. Now suppose that the weight is changed to 50 N. The height of the hill, the slope, etc., all remain the same. Which of the following quantities are changed and in what way?

PE?

KE?

Final speed?

Initial acceleration?

Time to reach bottom?

7. A heavy trunk is to be loaded onto a truck by pushing it up a plank as shown in Fig. 6-6. The trunk weighs 1000 N (about 200 lb) and the angle of the inclined plane is 20°. The workers find that it is necessary to push with a force of 620 N to keep the crate moving. The height through which the crate is lifted is 1.5 m.

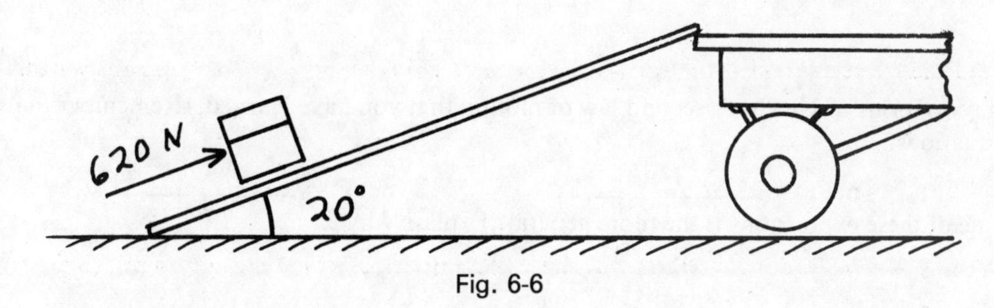

Fig. 6-6

a. How much work is done in pushing the crate up the plank?

$W(\text{in}) =$ ______________ .

b. What is the increase in potential energy (work output)?

$W(\text{out}) =$ ______________ .

c. The inclined plane may be regarded as a machine whose efficiency is the ratio of $W(\text{out})$ to $W(\text{in})$. What is the efficiency of this inclined plane?

Eff = ______________ .

d. Why is the output of this machine less than the input? Explain the energy "loss" that takes place.

e. Calculate the work done against the frictional force and then determine the force of friction from the relationship $W(\text{against } F_k) = F_k s$.

$F_k =$ ______________ .

f. There is another way in which the force of friction can be found. Make a diagram of the forces acting on the crate and resolve these forces into components parallel and perpendicular to the inclined plane. Apply the condition of equilibrium to determine the frictional force.

$F_k =$ ______________ .

These two methods of calculating F_k must give the same result within the accuracy of the calculation.

g. Since the normal force between crate and ramp is easily found, it is possible to determine the coefficient of friction. Do it.

$\mu_k =$ ______________ .

h. Let us suppose that the crate is pushed up the ramp at a steady speed of 0.2 m/s. What is the input power?

8. A ball weighing 12 N *slides* down a frictionless chute and then strikes a stationary 8-N block on a horizontal surface (see text Fig. 6-10). (We assume that all of the energy of the ball is transformed into *translational* kinetic energy. If it rolled down the plane rather than slid, rotational KE would have to be considered.) The coefficient of kinetic friction between the block (plus ball) and the horizontal surface is 0.2. If the ball and the block stick together, how far do they slide before coming to rest? (The method of solution is given in Ex. 6-11 in the text. Use $h = 2.5$ m.)

9. Write the two forms of Newton's second law of motion that you have learned. (Remember that these are vector equations.)

net **F** = ____________. net **F** = ____________.

Which one of these expressions is the more general? Explain why.

10. Write the two conservation laws. Which of these is a vectorial relationship? Which one deals with scalar quantities? (Secs. 5-3 and 6-5.)

11. In what type of collision is the total mechanical energy conserved? (Sec. 6-9.)

12. In what type(s) of collision is the total momentum conserved?

13. Is some of the sun's energy used up in causing the earth to move in a nearly circular orbit?

14. Is the earth in equilibrium as it travels around the sun in a nearly circular orbit?

15. An astronaut, working outside his spaceship, throws a heavy wrench to his partner a few meters away. The mass of the astronaut is 75 kg and the mass of the wrench is 5 kg. If the astronaut was originally at rest with respect to the spaceship and if the velocity of the wrench was 6 m/s as it left his hand, what is the velocity imparted to the astronaut? All velocities are measured with respect to the spaceship, a local frame of reference.

16. Fig. 6-7 shows two small objects about to collide at point P.

a. Assuming that the objects are perfectly elastic, which of the drawings in Fig. 6-8 represents possible outcomes of the collision? The arrows represent momenta of the objects.

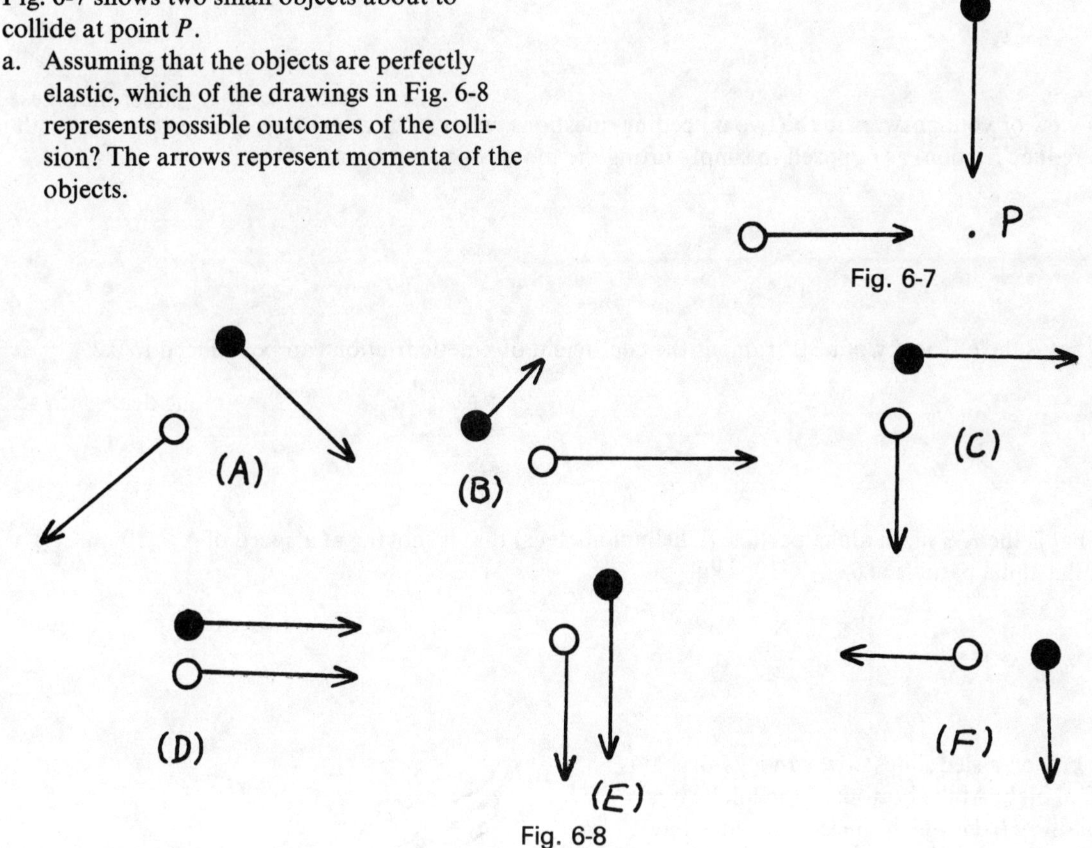

Fig. 6-7

Fig. 6-8

b. If the colliding objects are not perfectly elastic, which of the drawings, A through F, represent possible outcomes of the collision?

17a. A 400-V electric generator having a mass of 1200 kg is to be loaded onto a flatbed truck. If the bed of the truck is 1.5 m above the floor on which the generator is resting, how much work is required to load the generator by pushing it up a 30° ramp assumed to be frictionless? (Refer to Fig. 6-6.)

b. How much work would be required to load the generator without the use of the ramp?

c. In view of your answers to the two preceding questions, what is the advantage of using a ramp (with rollers to reduce friction) as opposed to simply lifting the motor onto the truck?

d. What is the efficiency of a 30° ramp if the coefficient of kinetic friction can be reduced to 0.2?

18. What is the KE of an alpha particle (a helium nucleus) that is moving at a speed of 4×10^4 m/s? The mass of the alpha particle is 6.64×10^{-27} kg.

19. A boy on a sled slides down an icy hill that is 20 m high. At the bottom of the hill he hits a roadway from which the ice and snow have been removed. He is brought to rest after sliding 25 m across the roadway. Assuming that the friction is negligible on the slope, what is the coefficient of friction between the runners and the pavement?

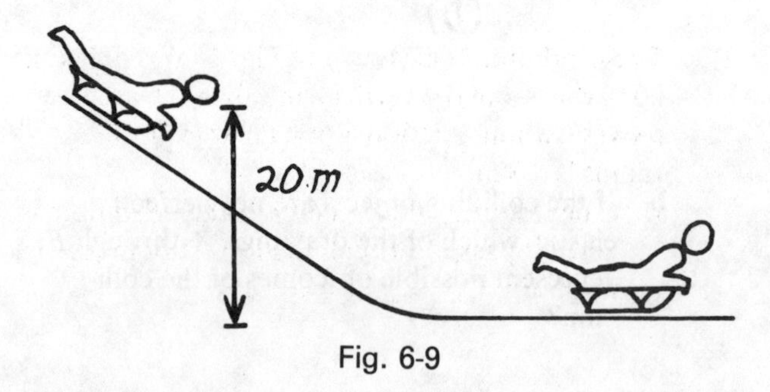

Fig. 6-9

20. A stone of mass 0.2 kg is thrown from the edge of a cliff, 60 m high, at an angle of 37° (see Fig. 6-10). The initial speed of the stone is 20 m/s. Air resistance may be neglected.

a. Using the ground as reference level, what is the total energy of the stone when it reaches the highest point of its path?

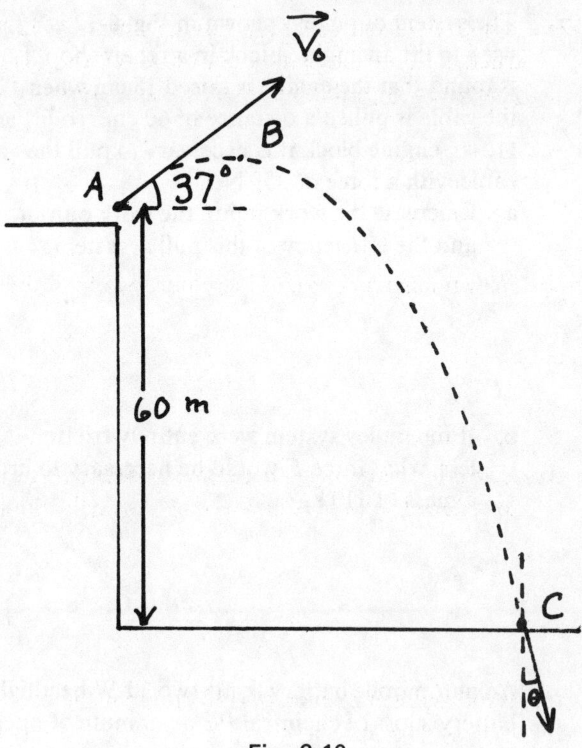

b. What is the total energy of the stone just before it strikes the ground?

c. At what speed is the stone moving when it reaches the highest point of its path?

d. What is the magnitude of the stone's velocity just before it strikes the ground?

Fig. 6-10

f. How high is the stone above the ground at the top of its trajectory?

e. At what angle with respect to the vertical is the stone moving just before it strikes the ground?

21. The pendulum bob shown in Fig. 6-11 is pulled 60° away from the vertical and then released. It moves without friction along an arc whose radius is 0.8 m.

a. Calculate v_{max}, the velocity of the pendulum when it passes through the lowest point of its path.

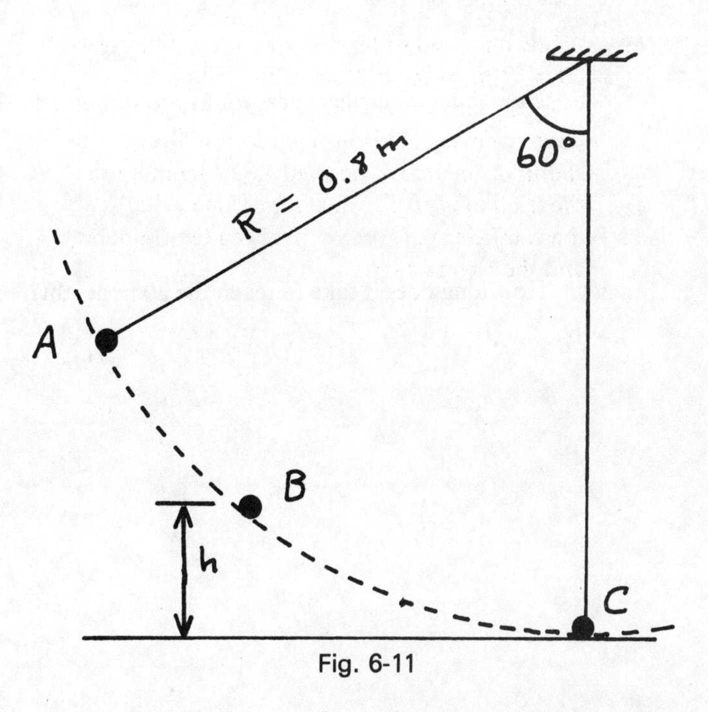

b. What is the height of the pendulum when it has the velocity $(1/2)v_{max}$?

Fig. 6-11

22. The system of pulleys shown in Fig. 6-12 is used to lift an engine block in a repair shop. It is found that the engine is raised 15 cm when the cable is pulled a distance of 60 cm. To lift a 110-kg engine block it is necessary to pull the cable with a force of 359 N.
    a. Calculate the work input, the work output, and the efficiency of this pulley system.

    b. If the pulley system were entirely frictionless, what force F would be necessary to lift a mass of 110 kg?

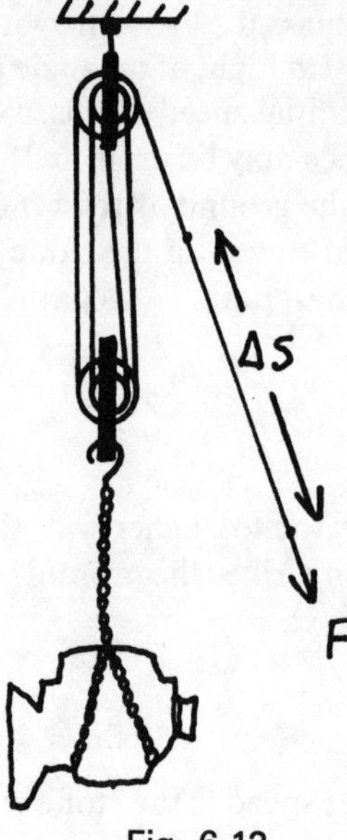

Fig. 6-12

23. An automobile battery lights two 30-W headlights and runs a 0.2-hp motor at the same time. What is the battery's loss of chemical PE per minute of operation?

24. A 5-hp electric motor is used to operate a small elevator at a construction site. The motor, which is 60% efficient, lifts a load of 15 kN (including the platform) to a height of 20 m.
    a. How much work is done?

    b. How much electrical energy must be supplied to the motor?

    c. How long does it take to reach the 20 m height?

Solutions

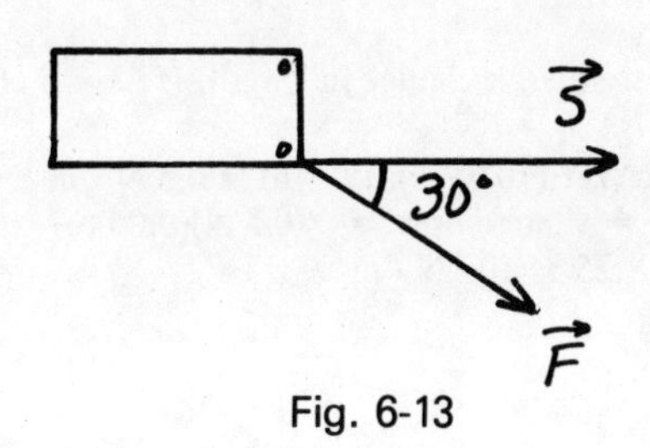

Fig. 6-13

1. Work = (force)(displacement)(cos 30°) = $Fs \cos 30°$
   $W = (5 \times 10^3\ \text{N})(0.866)(10 \times 10^3\ \text{m}) = 4.3 \times 10^7\ \text{N}\cdot\text{m}$

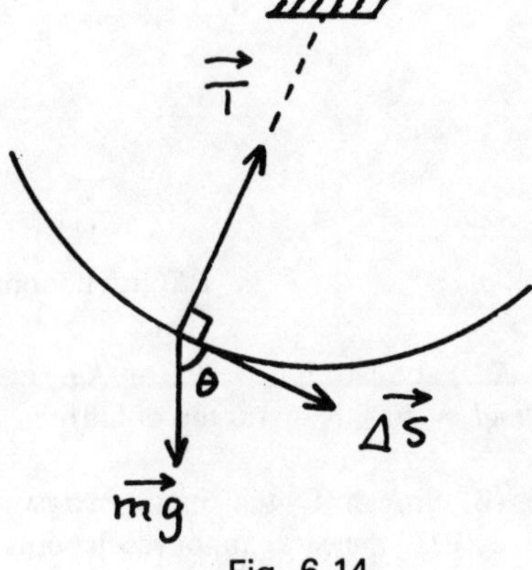

Fig. 6-14

2. The displacement is along the arc. The force **T** is radial, exactly perpendicular to the displacement. Thus the component of **T** in the direction of **s** is zero.

$$T \cos \theta = T \cos 90° = 0$$

It follows that no work is done by this force.
   The pendulum bob is also subject to the vertical force *mg*, which does have a component in the direction of the displacement. Thus the gravitational force does work on the pendulum during the downward swing causing its *KE* to increase.

3. The gravitational force that holds the moon (or any earth satellite) in orbit is directed toward the center of the earth. This force is exactly perpendicular to the displacement if the orbit is circular. Thus *no work is done.*

4. Work = area under force-displacement curve.
   a. For the displacement 0 to 2 m,

   $W = (2\ \text{N})(2\ \text{m}) = 4\ \text{N}\cdot\text{m}$

   For the displacement 2 to 4 m,

   $W = (2\ \text{m})\frac{1}{2}(2\ \text{N} + 4\ \text{N}) = 6\ \text{N}\cdot\text{m}$

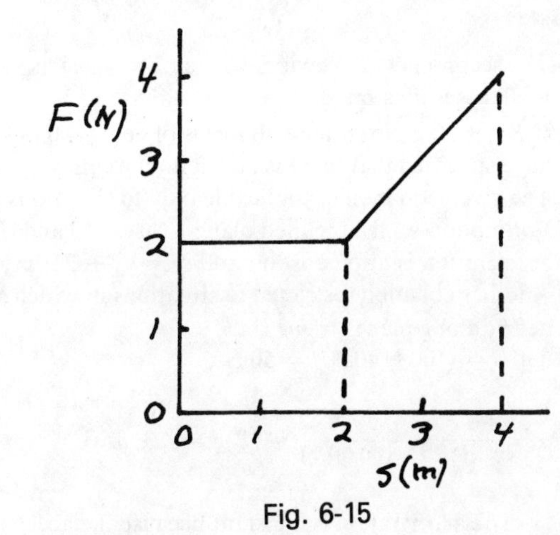

Fig. 6-15

5. Displacement is change in length. If $\Delta s = 5$ cm,

$F = k\Delta s = (10\text{ N/m})(0.05\text{ m}) = 0.5\text{ N}\cdot\text{m}$
Work = shaded area = $\frac{1}{2}(0.5\text{ N})(0.05\text{ m})$
$W = 1.25 \times 10^{-2}\text{ N}\cdot\text{m}$

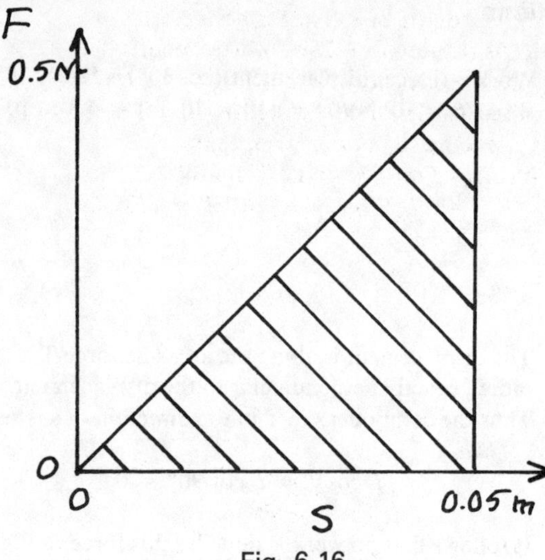

Fig. 6-16

$KE = \frac{1}{2}mv^2$ $\qquad PE(\text{gravitational}) = mgh$

6a. $(PE + KE)$ at top of hill = $(PE + KE)$ at bottom
$PE = mgh = 5000\text{ N}\cdot\text{m}$ (at top of hill)

b. In cases (B) through (D) the kinetic energy at the point B is 5000 N·m. Conservation of energy holds regardless of the path. In case (E) the cart cannot reach point B because of the intervening hill.

c. $PE = mgh$. Thus the weight is

$$mg = \frac{PE}{h} = \frac{5000\text{ N}\cdot\text{m}}{50\text{ m}} = 100\text{ N}$$

d. $$\text{mass} = \frac{mg}{g} = \frac{100\text{ N}}{9.8\text{ m/s}^2} = 10.2\text{ kg}$$

e. $$KE = \tfrac{1}{2}mv^2;\quad v = \sqrt{\frac{2KE}{m}} = \sqrt{\frac{2(5000\text{ N}\cdot\text{m})}{10.2\text{ kg}}} = 31.3\text{ m/s}$$

f. The second law of Newton, written $\Sigma\mathbf{F} = m\mathbf{a}$, is valid for *any* situation in which mass is constant. It certainly applies to all cases illustrated.

g. If E_{tot} is taken to include all forms of energy, then it applies to any isolated system including those illustrated. Since we have specified that there is no friction, we may apply the more restricted form, $PE + KE$ = constant.

h. The given equation is applicable only to situations in which the acceleration **a** is constant in magnitude and direction. For motion on the inclined plane—cases (A) and (B)—the accelerating force is constant; $F_x = mg\sin\theta$. Thus the acceleration is also constant and $v^2 = v_0^2 + 2as$ is valid.

i. Another equation restricted to situations in which $\mathbf{a}$ = constant. Not valid in cases (C) through (E).

j. net F (along plane) = $mg\sin\theta$
net F = (100 N)(0.50) = 50 N

k. $$a = \frac{\text{net } F}{m} = \frac{50\text{ N}}{(100/g)\text{ N}} = (\tfrac{1}{2})g = 4.9\text{ m/s}^2$$

The acceleration is constant because the accelerating force, $mg\sin\theta$, is constant. The latter is constant because m, g, and $\sin\theta$ are all constants. [Note, however, that $\sin\theta$ is not constant in cases (C) through (E).]

l. $v^2 = v_0^2 + 2as;\ v_0 = 0$
$s = h/\sin\theta = 2h = 100$ m
$v^2 = 2(4.9\text{ m/s}^2)100\text{ m} = 980\text{ m/s}^2$
$v = 31.3$ m/s

m. PE is reduced to 1/2 its former value.
KE is reduced to 1/2 its former value.
Final speed is not changed.
Initial acceleration is not changed.
Time to reach bottom is not changed.

7a. $W(\text{in}) = Fs = (620\text{ N})(1.5/\sin 20°)$
$= 620(1.5)/(0.342) \cong 2.72 \times 10^3\text{ J}$
$[s = 1.5/0.342 = 4.39\text{ m}]$

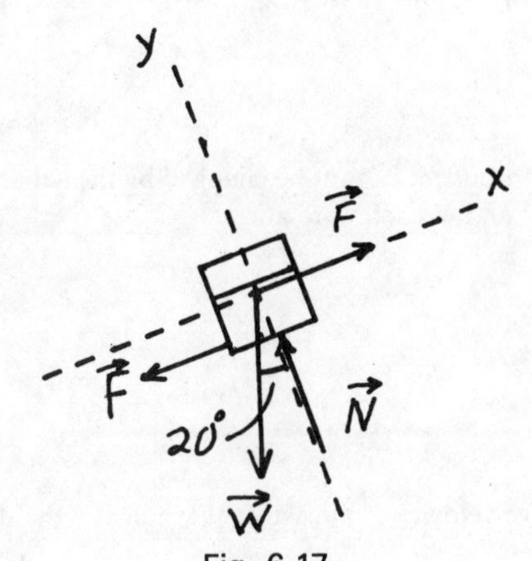

Fig. 6-17

b. $W(\text{out}) = mgh = (1000\text{ N})(1.5\text{ m}) = 1.5 \times 10^3\text{ J}$
c. Efficiency $= W(\text{out})/W(\text{in}) = 1.5/2.7 = 55\%$
d. There is evidently friction between the crate and the ramp. Work done against this force is converted to heat.
e. $W(\text{against } F_k) = W(\text{in}) - W(\text{out}) = 1.22 \times 10^3\text{ N} \cdot \text{m}$
$F_k = W/s = 278\text{ N} \cong 2.8 \times 10^2\text{ N}$

f. The component of mg parallel to the plane is $mg \sin 20° = 342$ N. It is in the negative x-direction.
$\Sigma F_x = F - F_k - 342\text{ N}$; but $F = 620$ N
$F_k = (620 - 342)\text{ N} = 278\text{ N} \cong 2.8 \times 10^2\text{ N}$
g. $N = mg \cos 20° = 940$ N
$\mu_k = F_k/N = 278/940 \cong 0.30$
h. Power is the rate at which the work is done.

$$P = \frac{\Delta W}{\Delta t}$$

Since the force and velocity are both constant, we may calculate P for any convenient time interval. Let Δt be the time to move the entire length of the plane.

$\Delta t = s/v = 4.39\text{ m}/0.2\text{ m/s} \cong 22\text{ s}$

Thus

$$P(\text{in}) = \frac{\Delta W}{\Delta t} = \frac{2.7 \times 10^3\text{ J}}{22\text{ s}} \cong 1.2 \times 10^2\text{ watt}$$

8a. $v = 7.0$ m/s
b. $V = 4.2$ m/s
c. $s = 4.5$ m

9. $\text{net } \mathbf{F} = \dfrac{\Delta \mathbf{p}}{\Delta t}$ $\quad \text{net } \mathbf{F} = m\dfrac{\Delta \mathbf{v}}{\Delta t}$

The first equation says that net **F** is equal to the rate of change of momentum with respect to time. The second says that net **F** is equal to the mass times the rate of change of velocity. But momentum is the product of mass and velocity. Thus

$\dfrac{\Delta \mathbf{p}}{\Delta t} = \dfrac{\Delta(m\mathbf{v})}{\Delta t}$ *If* the mass is constant, $\dfrac{\Delta(m\mathbf{v})}{\Delta t} = m\dfrac{\Delta \mathbf{v}}{\Delta t}$

Thus the second equation above is a special case of the first and is valid only if the mass is constant. The first equation,

$$\text{net } \mathbf{F} = \frac{\Delta \mathbf{p}}{\Delta t}$$

remains true, however, even if the mass is changing with time. For example, during the first few minutes of a rocket launch the mass of the rocket decreases rapidly as the fuel is burned. The expression in terms of momentum is valid for this situation; the other is not.

10. If $\mathbf{p}$ is the total momentum of an isolated system, then $\mathbf{p}$ = const. This is a vectorial relationship. If E is the total energy of an isolated system, all forms of energy being included, then E = const. This is a scalar relationship. All forms of energy are scalar quantities.
11. A collision in which total mechanical energy ($KE + PE$) is conserved is called *elastic* or, for emphasis, *perfectly elastic*. This is an idealized case that is difficult to achieve in practice.
12. The total momentum of a system is conserved in *any* collision process whether it is elastic or not. The internal forces, especially friction, which cause a colliding system to lose mechanical energy, *do not* cause the system to lose momentum.
13. The gravitational attraction of the sun upon the earth is a vector $\mathbf{F}$ directed toward the sun. The displacement $\mathbf{s}$ is a vector tangent to the circular orbit. These two vectors are at right angles. Thus work is

$$W = F(\cos 90°)s = 0$$

14. Since net $\mathbf{F} \neq 0$, the earth is not in equilibrium.
15. The system, astronaut plus wrench, has a total momentum that is constant; it cannot be changed by the act of throwing the wrench. Before the wrench was thrown, both astronaut and wrench were at rest.

$$\text{Momentum before} = \text{momentum after}$$
$$0 = (5\text{ kg})(6\text{ m/s}) + (75\text{ kg})v$$

$$v = -\frac{30}{75}\text{ m/s} = -0.4\text{ m/s}$$

Since the velocity of the wrench was taken to be positive, the negative velocity of the astronaut indicates that he is moving in a direction opposite to that of the wrench.

16a. The momentum of the two objects before the collision is represented by an arrow directed downward and toward the right. Among the six choices shown for the momenta after the collision, only case (C) is clearly downward and toward the right. Case (B) can be excluded because it has an upward component. Case (F) is not possible because it has a leftward component. None of the other cases give a final momentum equal to the initial momentum.

b. Conservation of momentum is valid even if the mechanical energy is not conserved. The answer is still case (C).

17a. If the ramp is frictionless, the generator can be kept moving along the ramp at a constant speed by a force of $mg \sin 30° = 5.88 \times 10^3$ N. The distance moved along the ramp will be twice the height. Thus $s = 3.0$ m. Since $\mathbf{F}$ and $\mathbf{s}$ are in the same direction, we get

$$W = Fs \cos\theta = (5.88 \times 10^3\text{ N})(3.0\text{ m})$$
$$= 1.76 \times 10^4\text{ J}$$

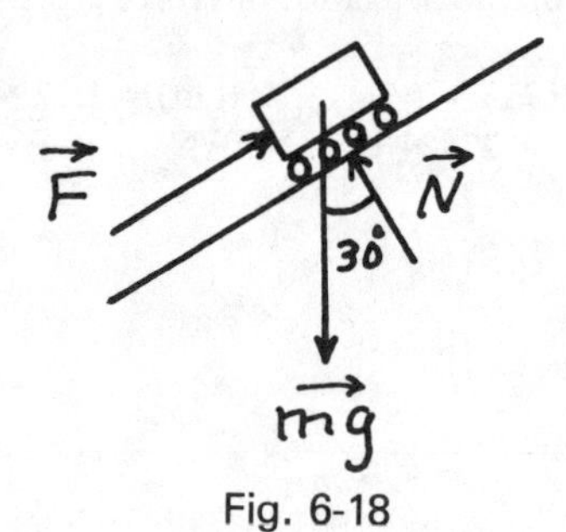

Fig. 6-18

b. Without the ramp, the generator would be lifted straight up.

$$s = 1.5\text{ m};\ F = W = (1200\text{ kg})(9.8\text{ m/s})$$
$$= 11.76 \times 10^3\text{ N}$$

The work is

$$W = Fs\cos\theta = (11.76 \times 10^3\text{ N})(1.5\text{ m}) = 1.76 \times 10^4\text{ J}$$

c. The work is the same in the two cases, as expected. The advantage of the inclined ramp is that the job can be done with a smaller force, in this case one half the force required without the ramp.

d. First find the normal force, then the frictional force.

$$N = mg\cos 30° = (11.76 \times 10^3\text{ N})(0.866) = 1.02 \times 10^4\text{ N}$$
$$F_k = \mu N = (0.2)(1.02 \times 10^4\text{ N}) = 2.04 \times 10^3\text{ N}$$

The force required to push the object along the ramp is $5.88 \times 10^3\text{ N} + 2.04 \times 10^3\text{ N} = 7.92 \times 10^3$ N. The efficiency is the work output, that is the increase in potential energy, divided by the work input.

$$\text{Efficiency} = \frac{mgh}{Fs} = \frac{1.76 \times 10^4\text{ J}}{(7.92 \times 10^3)(3.0\text{ m})} = 0.74 = 74\%$$

18. $$KE = \tfrac{1}{2}mv^2 = \tfrac{1}{2}(6.64 \times 10^{-27}\text{ kg})(4 \times 10^4\text{ m/s})^2$$
$$= 5.3 \times 10^{-18}\text{ J}$$

19. The *KE* acquired by the boy and sled is equal to their potential energy at the top of the hill.

$KE = mgh$

This *KE* is dissipated by the frictional force as the sled slides along the road through a distance *s*.

$F_k = KE$

The frictional force can be found from

$F_k = \mu_k N = \mu_k mg$

Thus

$(\mu_k mg)s = mgh \; ; \; \mu_k = \frac{h}{s}$

$\mu_k = \frac{20\text{ m}}{25\text{ m}} = 0.8$

20a. In the absence of friction, this is a conservative system; the total mechanical energy is the same at every point of the path. This energy is most easily found at point A, just as the stone is thrown.

$$E_{tot} = KE + PE = \tfrac{1}{2}mv_0^2 + mgh$$
$$E_{tot} = \tfrac{1}{2}(0.20\text{ kg})(20\text{ m/s})^2 + (0.20\text{ kg})(9.8\text{ m/s}^2)(60\text{ m})$$
$$E_{tot} = 40\text{ J} + 117.6\text{ J} = 158\text{ J}$$

b. $E_{tot} = 158\text{ J}$

c. At the highest point of its trajectory, the stone is moving horizontally. Since the horizontal component of the stone's velocity does not change, it is simply the horizontal component of the initial velocity.

$$v_x = v_0 \cos 37° = (20\text{ m/s})(0.8) = 16\text{ m/s}$$

d. Just before striking the ground, the stone's energy is entirely in the form of *KE*.

$$\tfrac{1}{2}mv^2 = 158\text{ J}$$

$$v = \sqrt{\frac{2(158\text{ J})}{(0.2\text{ kg})}} = 39.7\text{ m/s}$$

This velocity can also be found by adding vectorially the horizontal and vertical components. The vertical component is obtained from the equation for uniformly accelerated motion. Taking the downward direction to be positive,

$$v_y^2 = v_{y0}^2 + 2as$$
$$v_y^2 = [(-20\text{ m/s})(0.6)]^2 + 2(9.8\text{ m/s}^2)(60\text{ m}) = 1320\text{ m}^2/\text{s}^2$$
$$v_y = 36.3\text{ m/s}$$

Thus

$$v = \sqrt{v_x^2 + v_y^2} = \sqrt{(16\text{ m/s})^2 + (36.3\text{ m/s})^2} = 39.7\text{ m/s}$$

e. $\sin\theta = \frac{v_x}{v} = \frac{16\text{ m/s}}{39.7\text{ m/s}} = 0.40$

$\theta \cong 24°$

f. The vertical distance from A to B is found from one of the equations of uniformly accelerated motion. Let the upward direction be positive.

$$v_y^2 = v_{y0}^2 + 2as$$

$$s = \frac{v_y^2 - v_{y0}^2}{2g} = \frac{0 - (12 \text{ m/s})^2}{2(-9.8 \text{ m/s}^2)} = 7.3 \text{ m}$$

The height above the ground is 7.3 m + 60 m = 67.3 m.

21a. This is a conservative system. The total energy at the lowest point, entirely kinetic, is equal to the total energy at the highest point, entirely potential.

$$\tfrac{1}{2} mv^2 = mgh \; ; \; v = \sqrt{2gh}$$

The vertical distance from point A to point C is

$$h = 0.8 \text{ m} - (0.8 \text{ m})(\cos 60°) = 0.4 \text{ m}$$

Thus

$$v = \sqrt{2(9.8 \text{ m/s}^2)(0.4 \text{ m})} = 2.8 \text{ m/s}$$

b. At point B the pendulum has both *KE* and *PE*. Its total energy is the same as at A. Thus,

$$(KE)_B + (PE)_B = (PE)_A$$

Since the velocity at B is one half that at C we find,

$$(KE)_B = \tfrac{1}{2}m(v_C/2)^2 = \tfrac{1}{4}(\tfrac{1}{2}mv_C^2) = \tfrac{1}{4}(KE)_C = \tfrac{1}{4}(PE)_A$$

Thus,

$$\tfrac{1}{4}(PE)_A + (PE)_B = (PE)_A$$
$$(PE)_B = (PE)_A - \tfrac{1}{4}(PE)_A = \tfrac{3}{4}(PE)_A$$
$$mgh = (3/4) \text{ mg } (0.4 \text{ m})$$
$$h = 0.3 \text{ m}$$

22a. $W_{in} = (359 \text{ N})(0.6 \text{ m}) = 215 \text{ N}\cdot\text{m}$

$W_{out} = (110 \text{ kg})(9.8 \text{ m/s}^2)(0.15 \text{ m}) = 162 \text{ N}\cdot\text{m}$

$$\text{Efficiency} = \frac{W_{out}}{W_{in}} = \frac{162 \text{ N}\cdot\text{m}}{215 \text{ N}\cdot\text{m}} = 75\%$$

b. In the absence of friction $W_{in} = W_{out}$

$$F(0.6 \text{ m}) = 162 \text{ N}\cdot\text{m}$$
$$F = 270 \text{ N}$$

23. First find the total power expended.

$$P = 2(30 \text{ W}) + (0.2 \text{ hp})\left(\frac{0.746 \times 10^3 \text{ W}}{1 \text{ hp}}\right)$$

$$P = 60 \text{ W} + 149 \text{ W} = 209 \text{ W}$$

Since P = energy/time,

energy $= Pt = (209 \text{ J/s})(60 \text{ s}) = 1.25 \times 10^4 \text{ J}$

24a. $W_{out} = mgh = (15 \times 10^3 \text{ N})(20 \text{ m}) = 3.0 \times 10^5 \text{ J}$

b. $\text{Efficiency} = \dfrac{W_{out}}{W_{in}}$; $W_{in} = \dfrac{W_{out}}{\text{Eff}}$

$$W_{in} = \frac{3.0 \times 10^5 \text{ J}}{0.60} = 5.0 \times 10^5 \text{ J}$$

c. $$t = \frac{W(\text{by motor})}{P(\text{of motor})} = \frac{(5.0 \times 10^5 \text{ J})}{(5 \text{ hp})}\left(\frac{1 \text{ hp}}{746 \text{ J/s}}\right) = 134 \text{ s}$$

REVIEW TEST FOR CHAPTERS 5 AND 6

1. During impact with the back wall of a handball court, a ball undergoes a change in momentum whose magnitude is 3 kg • m/s. If the ball was in contact with the wall for 2×10^{-3} s, what was the average force exerted on the ball?

2. An isolated system consists of two particles, m_1 and m_2, moving at velocities $\mathbf{v}_1$ and $\mathbf{v}_2$ as shown in Fig. T-19. There are no external forces acting on the system. The masses are $m_1 = 1.5$ kg, $m_2 = 4$ kg. The velocities are: $v_1 = 2$ m/s, $v_2 = 1$ m/s.
Find the magnitude of the total momentum of this system.

Fig. T-19

3. What is the total KE of the system described in Prob. 2 above?

4. A 2-kg ball is dropped from a height of 3 m above the laboratory floor. Its initial speed is zero. What is the total energy of the ball when it is two-thirds of the way to the floor, i.e., at a height of 1 m?

5. The two arrows on the diagram to the right represent the momentum of a ball before ($\mathbf{p}_1$) and after ($\mathbf{p}_2$) striking the floor. Which arrow below represents the change in momentum of the ball due to the collision?

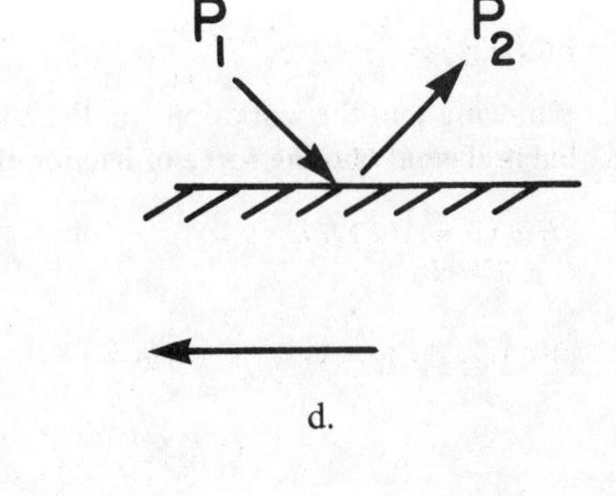

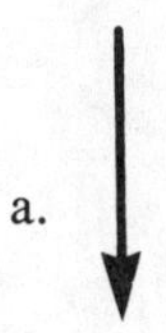
a.

b.

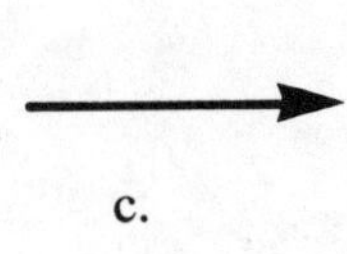
c.

d.

Fig. T-20

6. A force whose magnitude is 5.0 N acts on an object whose mass is 4.0 kg. During an interval of 6.0 s the object is displaced a distance of 3.0 m. The work done by the force is 7.5 J. What is the angle between the force and the displacement?

7. An earth resources satellite is in a circular orbit about the earth. It is held in its circular path by a gravitational force of 1250 N directed toward the center of the earth. During a displacement of 20 m, how much work is done on the satellite by the gravitational force?

8. A small object of unknown weight is pushed up an inclined surface at constant speed by a horizontal force of 20 N as shown below. How much work is done in moving the object a distance of 10 m along the inclined surface?

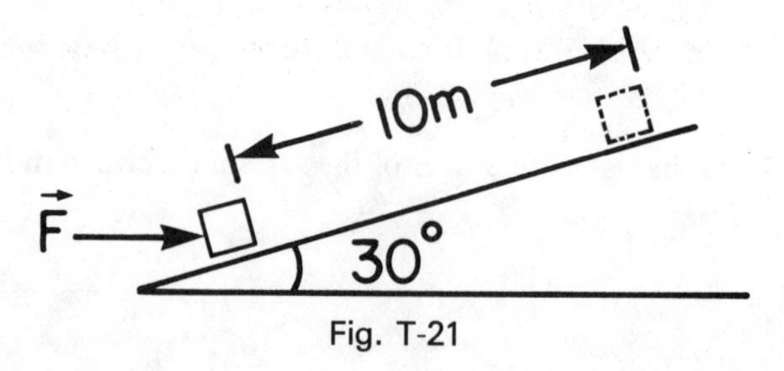

Fig. T-21

9. Assuming that the object in the preceding problem had a mass of 2 kg, what was its increase in potential energy as it was pushed 10 m along the inclined surface?

10. Knowing that the work done on the 2-kg object in Prob. 8 was 75.2 J greater than the increase in potential energy, what was the force of friction that acted on the sliding object?

11. A pendulum is released from point A where $\theta = 60°$ and $v_0 = 0.0$ m/s. It reaches its maximum speed of 2.0 m/s as it passes through point C, the lowest point of its path (see Fig. T-22). At what value of θ will the pendulum bob have a speed of 1.0 m/s?

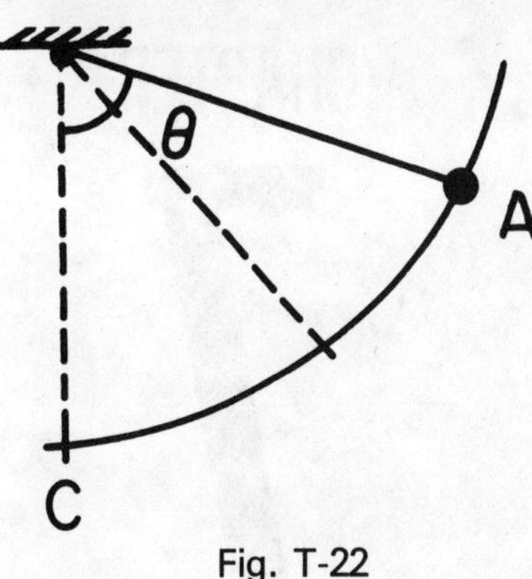

Fig. T-22

12. The coal car shown in Fig. T-23 is to be pulled out of a mine shaft at a steady speed of 2.0 m/s. The loaded car weighs 5×10^4 N and the ramp makes an angle of 30° with respect to the horizontal. If all frictional forces are neglected, what is the power rating of the smallest motor that will do the job?

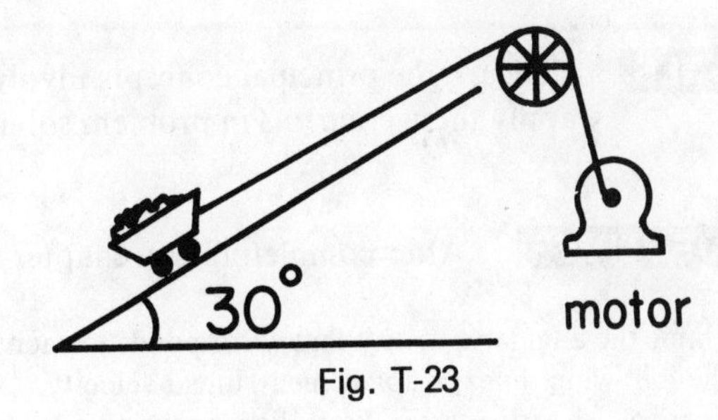

Fig. T-23

CHAPTER 7

Rotation

GOALS To learn the principal concepts involved in rotary motion about a fixed axis and to be able to apply these concepts in problem solving.

OBJECTIVES After completing this chapter the student should be able to do the following:

1. Name the angular quantity that corresponds to each of the following: linear displacement, linear velocity, linear acceleration, mass, force, linear momentum.
2. Write the correct symbol or symbols corresponding to each angular quantity named above.
3. Write the equations relating the quantities s, v, and a to θ, ω, and α.
4. Give the definitions of angular velocity and angular acceleration.
5. State the relationship between period and frequency, period and angular velocity, and frequency and angular velocity.
6. Calculate the centripetal acceleration of a point mass in uniform circular motion from its radius and any of the following: linear speed, angular speed, period, or frequency.
7. Describe the direction of centripetal acceleration and centripetal force.
8. Write the rotational form of Newton's second law of motion using commonly accepted symbols.
9. State the law of conservation of angular momentum.
10. Calculate the moment of inertia of point masses about a given axis of rotation.
11. Solve problems in rotational motion involving centripetal force, angular momentum, torque, and energy.

SUMMARY

In earlier chapters we dealt with motion along a straight-line path and motion along a parabolic path. A third type of motion, one that plays an important role in any technological society, is motion along a circular path, angular motion. The equations that govern angular motion have exactly the same form as those previously developed for linear motion. This close analogy between the two types of motion makes it possible to write down any equation in angular motion by simply replacing linear quantities, acceleration a, speed v, and displacement s by the corresponding angular quantities, acceleration α, speed ω, and displacement θ. If α is constant, the four equations for angular motion are

$$\omega = \omega_0 + \alpha t \qquad \omega^2 = \omega_0^2 + 2\alpha\theta$$

$$\theta = \frac{\omega_0 + \omega}{2} t \qquad \theta = \omega_0 t + \tfrac{1}{2}\alpha t^2$$

Angular velocity ω and angular acceleration α are defined in an analogous way to the corresponding linear quantities.

$$\omega = \lim_{\Delta t \to 0} \frac{\Delta \theta}{\Delta t} \qquad \alpha = \lim_{\Delta t \to 0} \frac{\Delta \omega}{\Delta t}$$

A linear quantity is obtained from the corresponding angular quantity by multiplication by the radius r of the circular path. Thus $s = r\theta, v = r\omega$, and $a_t = r\alpha$. The subscript t is added to distinguish the *tangential* acceleration a_t from the *centripetal* acceleration $a_c = \omega^2 r$, which *has no counterpart in linear motion.* For the most general circular motion, the total linear acceleration **a** has the magnitude $\sqrt{a_t^2 + a_c^2}$.

If the angular acceleration α is zero, we have the special case of uniform circular motion with $a_t = 0$, and $a_c = \omega v = \omega^2 r = v^2/r$.

A body moving in a circular path at a constant speed does *not* have a constant *linear velocity* **v**. In fact its velocity vector **v** changes direction continuously thus giving rise to an acceleration $\mathbf{a}_c$ directed toward the center of the circle. The force that causes this acceleration is likewise directed toward the center of the path, its magnitude being $F_c = ma_c$. $\mathbf{F}_c$, like **v**, is constant in magnitude only.

The centripetal force, which is absolutely necessary to keep the object in a circular path, may be supplied in various ways. On an unbanked highway curve, the centripetal force is supplied by friction between the tires and the road. If the curve is banked, gravity may supply all or part of $\mathbf{F}_c$.

Further analogies exist between rotating bodies and objects moving in a straight line. That which *causes* angular acceleration of a body is *torque* τ. It is the rotational analogue of force. That which *resists* angular acceleration is *moment of inertia* I, the rotational analogue of mass.

Newton's second law of motion, written in its rotational form, is net $\tau = I\alpha$ (or $\Sigma\tau = I\alpha$). The moment of inertia of a body depends on its mass, its shape, and the location of its axis of rotation.

Rotating bodies possess KE of rotation equal to $\frac{1}{2}I\omega^2$. Work W is required to impart energy to a rotating object, $W = \tau\theta$. The angular momentum $I\omega$ is analogous to linear momentum mv; both are vector quantities. The angular momentum of a body or system of bodies remains constant in the absence of *external torques*. *Internal torques* that one part of a system exerts on another part of the system cannot change the total angular momentum of the system.

QUESTIONS AND PROBLEMS

Secs. 7-1—7-5

1. What are the angular sizes (in radians) of the moon and the sun? The diameter of the moon is 3.5×10^6 m and its distance from the earth is 3.9×10^8 m. The sun has a diameter of 1.4×10^9 m and is 1.5×10^{11} m from the earth.

2. Express the following in radians:

a. 180°

b. 30°

c. 90°

d. 40°

e. 0.30 rev

3. While balancing the wheel of an automobile, an electric motor is used to bring the wheel to a speed of 720 rev/min. By applying the brakes, the wheel is brought to rest in 3 s.

a. Calculate the angular acceleration produced by the brakes.

b. Calculate the number of revolutions that takes place while the wheel is brought to rest.

4. Calculate the linear velocity of a point on the surface of the earth at the equator. The radius of the earth is 6.4×10^6 m.

5. A small mass m is moving in a circular path of radius $r = 10$ cm. In Fig. 7-1, two positions of the mass are shown at A and B. The angular displacement $\Delta\theta$ is 0.1 rad and the time required for this displacement Δt is 0.2 s.

a. Calculate the angular speed ω and the linear (tangential) speed v.

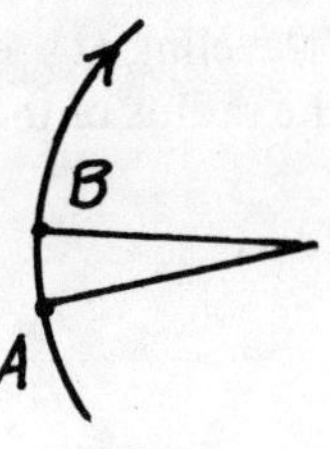

Fig. 7-1

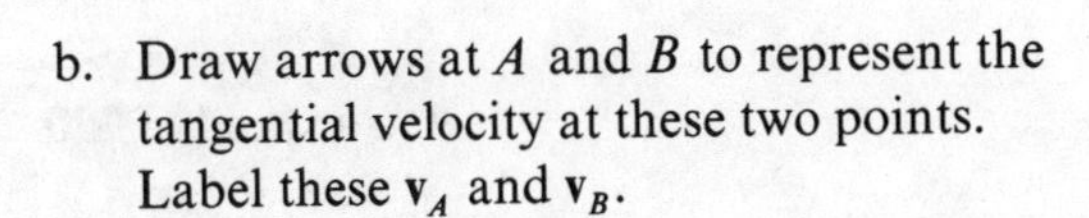

b. Draw arrows at A and B to represent the tangential velocity at these two points. Label these $\mathbf{v}_A$ and $\mathbf{v}_B$.

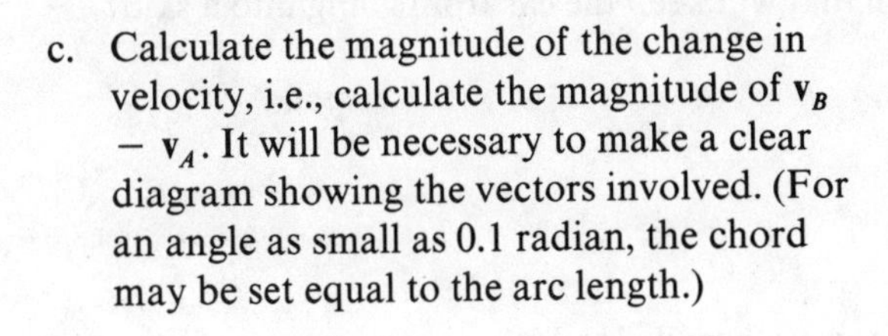

c. Calculate the magnitude of the change in velocity, i.e., calculate the magnitude of $\mathbf{v}_B - \mathbf{v}_A$. It will be necessary to make a clear diagram showing the vectors involved. (For an angle as small as 0.1 radian, the chord may be set equal to the arc length.)

d. After verifying the correctness of your results, calculate the magnitude of the acceleration from the definition

$$a = \lim_{\Delta t \to 0} \frac{\Delta v}{\Delta t} \cong \frac{\Delta v}{\Delta t}$$

e. Draw an arrow on Fig. 7-1 to indicate the direction of this acceleration.

f. Now calculate the acceleration from

$$a_c = \omega^2 r$$

and compare the result you obtained in d above.

6. Calculate the angular velocity of the earth and the centripetal acceleration of a point on the equator.

7. In Ex. 7-5 in the text, the mass of the block is to be doubled and the angular speed increased to 200 rev/min, the radius being constant.

a. What is the new tension in the rope?

b. Is the block in equilibrium?

8. A racing car traveling at a steady speed of 144 km/h completes a quarter turn (90°) in 6.28 s.
   a. What is the radius of the turn, assuming the path to be a segment of a circle?

   b. What is the centripetal acceleration of the car?

   c. What force produces this acceleration?

   d. What is the minimum coefficient of static friction that will keep the car from going into a skid? (Assume that the curve is not banked.)

   e. Would your answer to part d be different if the car were heavily loaded?

9. Fig. 7-2 shows an automobile on the banked turn of a test track. Make a force diagram showing the forces acting on the automobile. The speed of the car is constant (i.e., $|\mathbf{v}|$ = const.) and you may assume that it is the speed for which the track was designed. (This is the speed for which no frictional force is needed to keep the car in its circular path.)

Fig. 7-2

   a. Using R for the radius of the turn, write the expression for the centripetal acceleration.

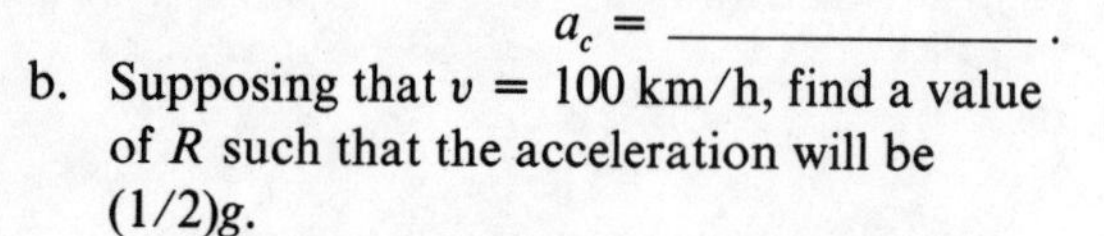

   a_c = ____________ .

   b. Supposing that v = 100 km/h, find a value of R such that the acceleration will be (1/2)g.

   R = ____________ .

   c. What is the centripetal force required to keep a 1800 kg automobile in a circular path of radius 157 m if the speed is 100 km/h?

   F_c = ____________ .

d. Make a force diagram showing all forces acting on the automobile as it travels along a banked curve. Indicate the forces by means of arrows on Fig. 7-3. Assume that there is no frictional force between the car and the road.

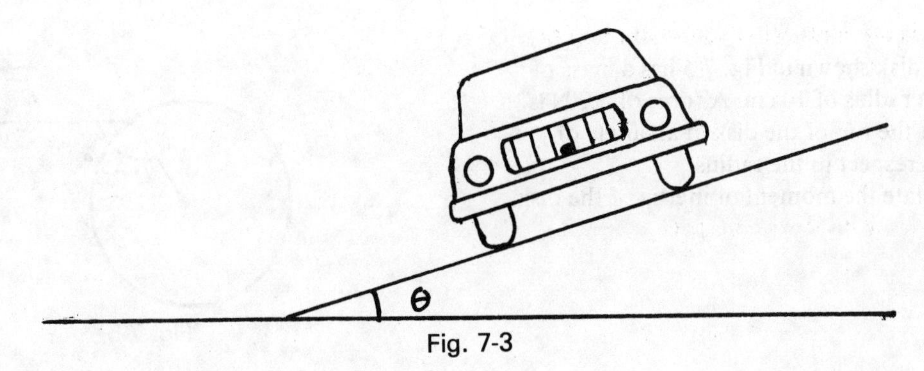

Fig. 7-3

e. Since the weight and the centripetal force are both known, the angle θ at which the road must be banked can be calculated. What is it?

Secs. 7-6—7-9

10. A light rod pivoted at one end carries a mass of 3 kg. A force of 2 N is applied to the rod as shown in Fig. 7-4. The force is applied at a point 10 cm from the axis.

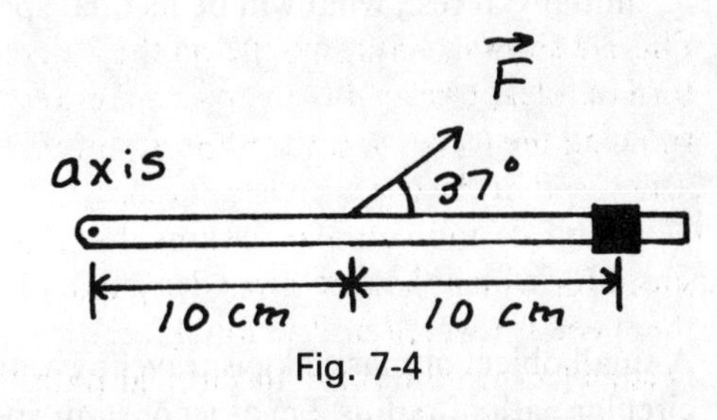

Fig. 7-4

a. Calculate the torque associated with this force.

$\tau =$ ______________.

b. Calculate the moment of inertia I of the mass about the axis of rotation.

$I =$ ______________.

c. Write Newton's second law for rotational motion.

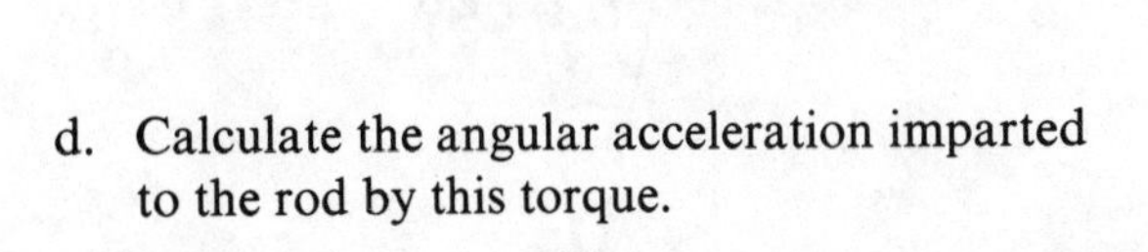

d. Calculate the angular acceleration imparted to the rod by this torque.

$\alpha =$ ______________.

11. The flywheel in a piece of machinery has the form of a solid uniform disk with a radius of 25 cm and a mass of 12 kg. If a constant torque of 0.3 N • m is applied, what time will be required to bring the flywheel to a speed of 2000 rev/min?

12. The solid disk shown in Fig. 7-5 has a mass of 4 kg and a radius of 10 cm. A force of 0.2 N is applied at the rim of the disk at an angle of 120° with respect to the radius.
   a. Calculate the moment of inertia of the disk.

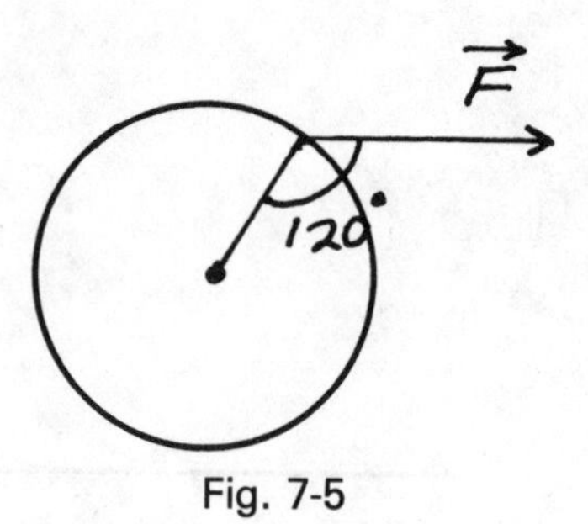

Fig. 7-5

   b. What torque is acting on the disk?

   c. Calculate the angular acceleration.

   d. If the force acts for 0.2 s and the disk was initially at rest, what will be its final speed?

   e. Find the angular displacement of the disk.

   f. Calculate the work done on the disk during the 0.2 s interval.

   g. Calculate the KE of the rotating disk at the end of 0.2 s. Compare with the result of (e) above.

13. A small object of mass 3 kg is traveling along a circular path of radius 2 m at a constant speed. Fig. 7-6 shows the velocity of the object at two times 4 s apart.
   a. Calculate the distance traveled along the arc Δs and the linear speed v.

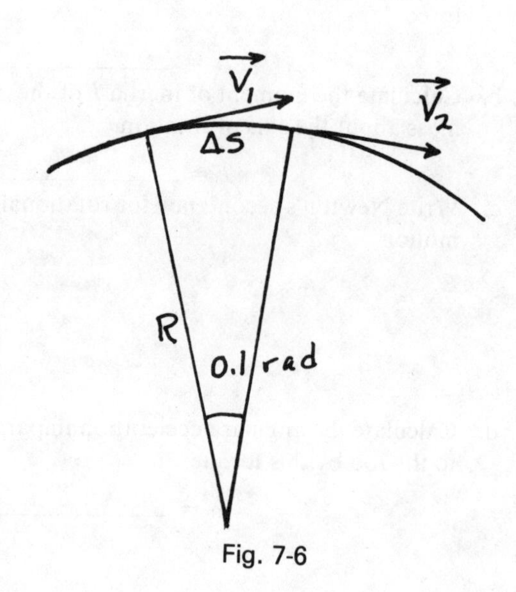

Fig. 7-6

b. Make a careful drawing to show how the change in velocity $\Delta \mathbf{v}$ may be found. What is the direction of $\Delta \mathbf{v}$? Verify the correctness of this result before proceeding.

c. Determine the magnitude of $\Delta \mathbf{v}$. (Hint: Because the angle is small—about 5.7° —the chord is approximately equal to the arc length.)

d. Calculate the magnitude of the average acceleration over this 4 s interval. State the direction of $\mathbf{a}_{av}$.

e. Since the magnitude of **v** is constant, the centripetal acceleration may be calculated from an exact formula. Carry out this calculation and compare it with the result you obtained in (d) above.

14. A student is standing on a frictionless turntable as illustrated in text Fig. 7-18. She has a 3-kg mass in each hand. With her arms fully extended—each mass 80 cm from the axis of rotation—she is given an initial speed ω_1 of 0.5 rev/s. Calculate the new angular speed after the student has brought the masses to a distance of 10 cm from the axis. Assume that the moment of inertia of the student and platform is a constant 1.2 kg • m².

Solutions

1. The angular sizes of the moon and sun are found as follows:

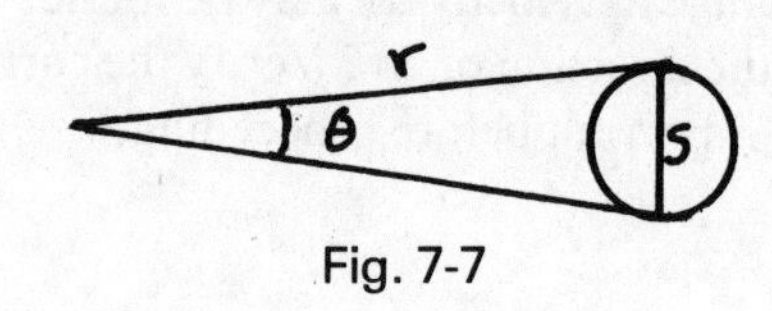

Fig. 7-7

a. The moon:

$$\theta = \frac{s}{r} = \frac{3.5 \times 10^6 \text{ m}}{3.9 \times 10^8 \text{ m}} = 9 \times 10^{-3} \text{ rad}$$

b. The sun:

$$\theta = \frac{s}{r} = \frac{1.4 \times 10^9 \text{ m}}{1.5 \times 10^{11} \text{ m}} = 9 \times 10^{-3} \text{ rad}$$

2. a. $180° = \pi$ rad
b. $30° = \pi/6$ rad $= 0.52$ rad
c. $90° = \pi/2$ rad
d. $40° = 0.70$ rad
e. $0.30 \text{ rev} = (0.30 \text{ rev})(2\pi \text{ rad/rev}) = 1.89 \text{ rad}$

3a. $\alpha = \dfrac{\Delta\omega}{\Delta t} = \dfrac{0-720 \text{ rev/min}}{3 \text{ s}} = -\dfrac{12 \text{ rev/s}}{3 \text{ s}}$

$\alpha = -4 \text{ rev/s}^2 = -8\pi \text{ rad/s}^2 = -25.2 \text{ rad/s}^2$

b. $\omega_{av} = \dfrac{\Delta\theta}{\Delta t}$; $\omega_{av} = \dfrac{\omega + \omega_0}{2} = 6 \text{ rev/s}$

$\Delta\theta = \omega_{av}\,\Delta t = (6 \text{ rev/s})(3 \text{ s}) = 18 \text{ rev}$

4. $v = r\omega = (6.4 \times 10^6 \text{ m})(2\pi \text{ rad}/24 \text{ h})(1 \text{ h}/3600 \text{ s}) = 465 \text{ m/s}$

5a. $\omega = \dfrac{\Delta\theta}{\Delta t}$; $v = r\omega$

$$\omega = \frac{0.1 \text{ rad}}{0.2 \text{ s}} = 0.5 \text{ rad/s}$$

$v = (0.1 \text{ m})(0.5 \text{ rad/s}) = 0.05 \text{ m/s}$

b. See Fig. 7-8.

c. $\Delta\theta \cong \dfrac{\Delta v}{v_A}$; $\quad \Delta v = \Delta\theta v_A$

$\Delta v = (0.1 \text{ rad})(0.05 \text{ m/s}) = 5 \times 10^{-3} \text{ m/s}$

d. $a \cong \dfrac{\Delta v}{\Delta t} = \dfrac{5 \times 10^{-3} \text{ m/s}}{0.2 \text{ s}} = 2.5 \times 10^{-2} \text{ m/s}^2$

e. The direction of **a** is that of Δ**v**.

f. $a_c = \omega^2 r = (0.5 \text{ rad/s})^2(0.1 \text{ m}) = 2.5 \times 10^{-2} \text{ m/s}^2$
The results are the same, as they must be.

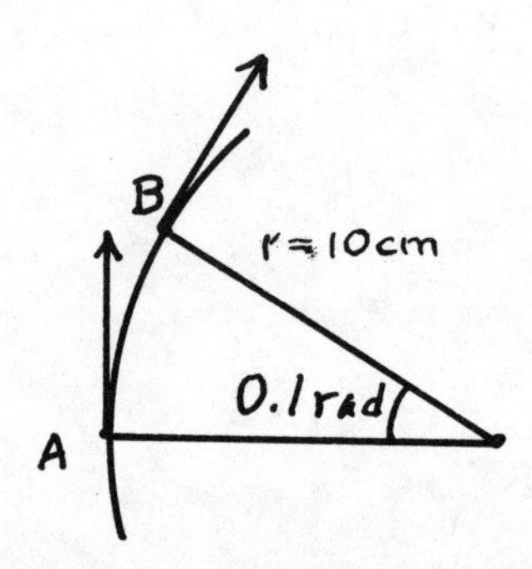

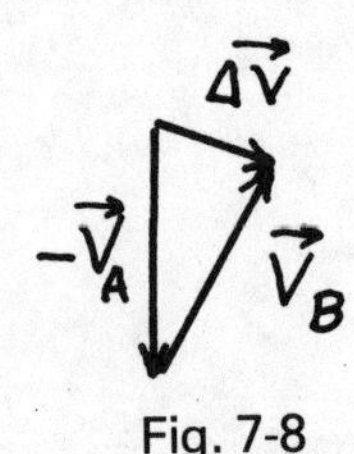

Fig. 7-8

6. $\omega = (2\pi \text{ rad}/24 \text{ h})(1 \text{ h}/3600 \text{ s}) = 7.27 \times 10^{-5} \text{ rad/s}$
$a_c = r\omega^2 = (6.4 \times 10^6 \text{ m})(7.27 \times 10^{-5} \text{ rad/s})^2 = 3.4 \times 10^{-2} \text{ m/s}^2$

7. $a_c = \omega^2 r = [(200 \text{ rev/min})(1 \text{ min}/60 \text{ s})(2\pi \text{ rad/rev})]^2(2 \text{ m})$
$= 876 \text{ m/s}^2$

a. $T = ma_c = (20 \text{ kg})(876 \text{ m/s}^2) = 1.75 \times 10^4 \text{ N}$
b. No. The block is being accelerated.

8a. Let s be the distance traveled.
$s = vt = (1.44 \times 10^5 \text{ m/h})(1 \text{ h}/3600 \text{ s})(6.28 \text{ s}) = 251 \text{ m}$
$s = (1/4)C = (1/4)(2\pi R) = (\pi/2)R$
$R = s(2/\pi) = (251 \text{ m})(2/\pi) = 160 \text{ m}$

b. $a_c = v^2/R = (40 \text{ m/s})^2/160 \text{ m} = 10 \text{ m/s}^2$

c. The frictional force F_s that the pavement exerts on the tires. It is directed toward the center of the path.

d. $F_s = \mu_s N \quad [\mu_s]_{\min} = \dfrac{F_s}{N}$

$$[\mu_s]_{\min} = \frac{ma_c}{mg} = \frac{10 \text{ m/s}^2}{9.8 \text{ m/s}^2} = 1.02$$

e. No. The masses of the car and contents cancel out.

9a. $a_c = R\omega^2 = v^2/R$

b. $R = v^2/a_c = \dfrac{[(10^5 \text{ m/h})(1 \text{ h}/3600 \text{ s})]^2}{(4.9 \text{ m/s}^2)} = 157 \text{ m}$

c. $F_c = ma_c = (1800 \text{ kg})(4.9 \text{ m/s}^2) = 8820 \text{ N}$

d. $\mathbf{N} + \mathbf{W} = \mathbf{F}_c$

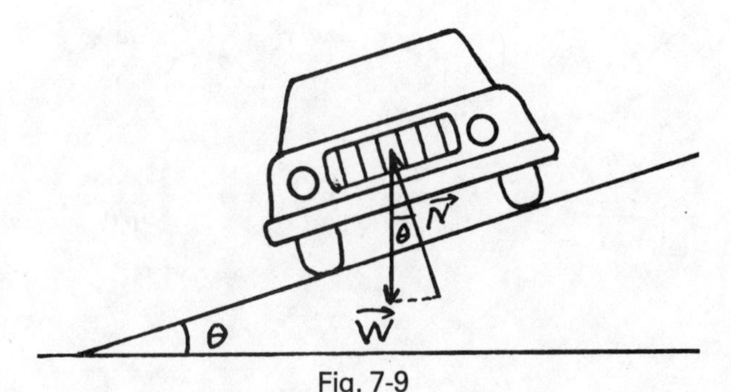

Fig. 7-9

e. $\tan\theta = \dfrac{F_c}{W} = \dfrac{8820 \text{ N}}{17{,}640 \text{ N}} = 0.50 \quad \theta = 26.5$

10a. τ = force × moment arm

$= (2 \text{ N})(0.1 \text{ m})\sin 37° = 0.2(0.6) = 0.12 \text{ N} \cdot \text{m}$

b. $I = mr^2 = 3 \text{ kg}(0.2 \text{ m})^2$

$I = 0.12 \text{ kg} \cdot \text{m}^2$

c. net $\tau = I\alpha$

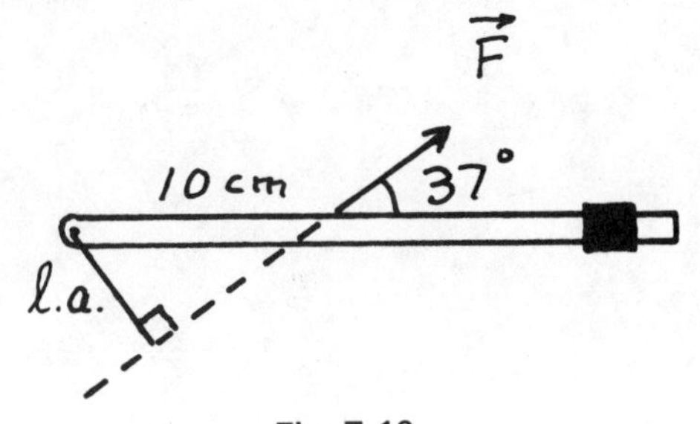

Fig. 7-10

d. $\alpha = \dfrac{\tau}{I} = \dfrac{0.12 \text{ N} \cdot \text{m}}{0.12 \text{ kg} \cdot \text{m}^2} = 1.0 \text{ rad/s}^2$

11. For a solid disk about its axis of symmetry,

$I = \frac{1}{2}MR^2 = \frac{1}{2}(12 \text{ kg})(0.25 \text{ m})^2 = 0.375 \text{ kg} \cdot \text{m}^2$

$\alpha = \dfrac{\tau}{I} = \dfrac{0.3 \text{ N} \cdot \text{m}}{0.375 \text{ kg} \cdot \text{m}^2} = 0.80 \text{ rad/s}^2$

Since the angular acceleration is constant,

$\omega = \omega_0 + \alpha t$. Also $\omega_0 = 0$

$t = \dfrac{\omega - \omega_0}{\alpha} = \dfrac{2000 \text{ rev/min } (2\pi \text{ rad})(1 \text{ min})}{0.80 \text{ rad/s}^2 \text{ } (1 \text{ rev})(60 \text{ s})}$

$t = 2.6 \times 10^2 \text{ s}$

12a. $I = \frac{1}{2}MR^2 = \frac{1}{2}(4 \text{ kg})(0.1 \text{ m})^2 = 0.02 \text{ kg} \cdot \text{m}^2$

b. First find the moment arm.

$m.a. = (0.1 \text{ m})(\cos 30°)$

$= (0.1 \text{ m})(0.866) = 8.66 \times 10^{-2} \text{ m}$

$\tau = F(m.a.) = (0.2 \text{ N})(0.866) = 8.66 \times 10^{-2} \text{ m}$

$= 1.73 \times 10^{-2} \text{ N} \cdot \text{m}$

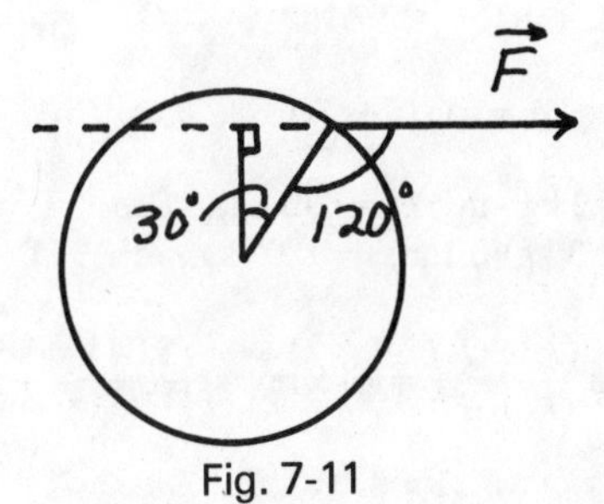

Fig. 7-11

c. $\alpha = \dfrac{\tau}{I} = \dfrac{1.73 \times 10^{-2} \text{ N} \cdot \text{m}}{2 \times 10^{-2} \text{ kg} \cdot \text{m}^2} = 0.866 \text{ rad/s}^2$

d. $\omega = \omega_0 + \alpha t; \quad \omega_0 = 0$

$\omega = (0.866 \text{ rad/s}^2)(0.2 \text{ s}) = 0.173 \text{ rad/s}$

e. $\theta = \omega_0 t + \frac{1}{2}\alpha t^2; \omega_0 = 0$

$= \frac{1}{2}(0.866 \text{ rad/s}^2)(0.2 \text{ s})^2 = 1.73 \times 10^{-2} \text{ rad}$

f. Work $= \tau\theta = (1.73 \times 10^{-2} \text{ N} \cdot \text{m})(1.73 \times 10^{-2} \text{ rad})$

$= 3 \times 10^{-4} \text{ N} \cdot \text{m}$

g. $KE = \frac{1}{2}I\omega^2 = \frac{1}{2}(0.02 \text{ kg} \cdot \text{m}^2)(0.173 \text{ rad/s})^2$

$= 3 \times 10^{-4} \text{ N} \cdot \text{m}$

is equal to the *KE* that it gains. This must be true if the disk rotates without friction as we have assumed.

13a. $\Delta s = r\,\Delta\theta = (2 \text{ m})(0.1 \text{ rad}) = 0.2 \text{ m}$

$v = \dfrac{\Delta s}{\Delta t} = \dfrac{0.2 \text{ m}}{4 \text{ s}} = 0.05 \text{ m/s}$

b. $\Delta\mathbf{v} = \mathbf{v}_2 - \mathbf{v}_1 = \mathbf{v}_2 + (-\mathbf{v}_1)$

$\Delta\mathbf{v}$ is directed toward the center of the path.

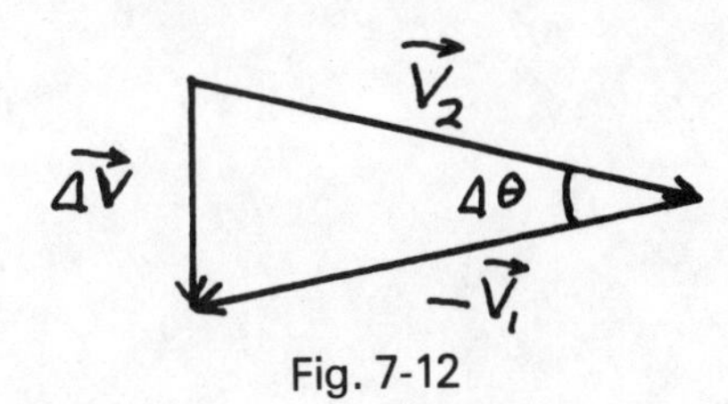

Fig. 7-12

c. $\Delta v \cong v_1\,\Delta\theta = (0.05 \text{ m/s})(0.1 \text{ rad})$

$= 5 \times 10^{-3} \text{ m/s}$

d. $a_{av} = \frac{\Delta v}{\Delta t} = \frac{5 \times 10^{-3}\ \text{m/s}}{4\ \text{s}} = 1.25 \times 10^{-3}\ \text{m/s}^2$

$\mathbf{a}_{av}$ is directed toward the center of the circular path.

e. $a_c = \frac{v^2}{r} = \frac{(5 \times 10^{-2}\ \text{m/s})^2}{2\ \text{m}} = 1.25 \times 10^{-3}\ \text{m/s}^2$

The two methods of calculation of centripetal acceleration lead to the same result, as expected.

14. Assuming that friction may be neglected, the angular momentum of the system will not change.

Before After

$\omega_1(I_s + I_1) = \omega_2(I_s + I_2)$

$I_s = 1.2\ \text{kg} \cdot \text{m}^2;\ I_1 = 2(3\ \text{kg})(0.8\ \text{m})^2 = 3.84\ \text{kg} \cdot \text{m}^2$
$I_2 = 2(3\ \text{kg})(0.1\ \text{m})^2 = 0.06\ \text{kg} \cdot \text{m}^2$

$$\omega_2 = \omega_1\frac{(I_s + I_1)}{(I_s + I_2)} = (0.5\ \text{rev/s})\frac{(5.04\ \text{kg} \cdot \text{m}^2)}{(1.26\ \text{kg} \cdot \text{m}^2)} = 2.0\ \text{rev/s}$$

CHAPTER 8

Gravitation and Planetary Motions

GOALS To learn the exact mathematical form of the law of universal gravitation and how it determines the motions of the planets.

OBJECTIVES After completing this chapter the student should be able to do the following:

1. Write the mathematical statement of Newton's law of universal gravitation using the standard symbols, F, G, m, and r.
2. Calculate the mass of the earth when its radius and the gravitational constant are given.
3. Calculate the magnitude of the gravitational acceleration due to a body other than the earth given G, the mass of the attracting object, and the distance between centers.
4. Write a mathematical definition of the gravitational field, g.
5. Calculate the magnitude of the gravitational field at a distance r from a body of mass m.
6. Describe the direction of the gravitational field in the vicinity of a body.
7. Calculate the magnitude and direction of the resultant gravitational field at a point in space due to the presence of two massive objects.
8. Write a correct statement of Kepler's law of areas (Kepler's second law).
9. State the condition that must be satisfied by the force for the law of areas to be valid.
10. Write a statement of Kepler's first law, the law of orbits.
11. State the condition that the force law must satisfy in order for the orbiting body to follow a closed, elliptical path.
12. Write a mathematical statement of Kepler's third law, the law of periods.
13. Use the law of periods to calculate the period of a planet given its mean distance from the sun plus the period and distance of another planet.

SUMMARY

One of Newton's outstanding contributions to eighteenth century science was the establishment of a simple yet elegant mathematical model that represented the solar system to an unprecedented degree of precision. Newton was able to show that any two objects were attracted toward each other with a force proportional to the product of their masses and inversely proportional to the square of the distance between their centers. Mathematically,

$$F = G\frac{m_1 m_2}{r^2}$$

This *law of universal gravitation* gave a firm mathematical basis for understanding the three laws of planetary motion discovered empirically by Johannes Kepler some 50 years earlier.

Kepler made very precise calculations of the motions of the planets, which led him to state three laws: (1) The orbit of any planet around the sun is an ellipse with the sun at one focus. (2) The line joining any planet to the sun sweeps out equal areas in equal times. (3) For any two planets in the solar system, the squares of the periods of revolution are in the same proportion as the cubes of their average distances from the sun. Newton was able to show that Kepler's empirical laws are a necessary consequence of the law of universal gravitation.

Although the gravitational force is large enough to hold the planets in their orbits about the sun, it is very small for objects of human size, such as those used in an earthbound laboratory. Nevertheless, in 1798 Henry Cavendish developed an extremely sensitive apparatus that allowed him to measure the gravitational force between two lead balls a few centimeters apart. His measurements give the first determination of the universal constant of gravitational attraction ($G = 6.7 \times 10^{-11}\ \text{N} \cdot \text{m}^2/\text{kg}^2$).

A gravitational field is said to exist at a point in space if a test body placed there experiences a gravitational force. The direction of the field is that of the gravitational force, and the magnitude of the field is defined as the magnitude of the gravitational force divided by the mass of the test body. Thus defined, a gravitational field is independent of the test bodies used to observe or measure it. A gravitational field—like force, velocity, and acceleration—is a vector quantity.

QUESTIONS AND PROBLEMS

1. Write the mathematical expression for the gravitational force between a spherical object of mass m_1 and a point mass m_2, which is illustrated in Fig. 8-1.

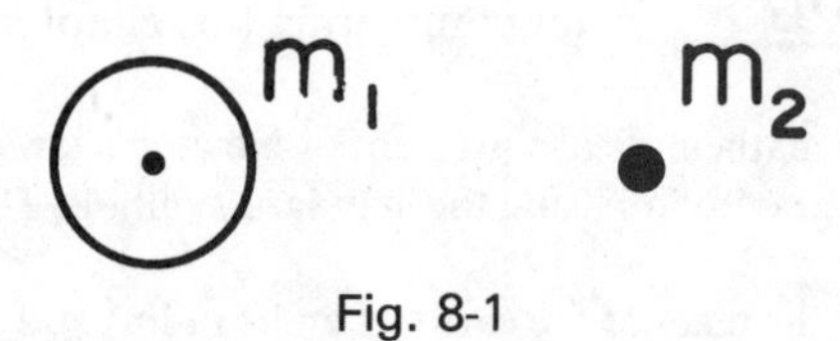

Fig. 8-1

2a. What exactly is the significance of the quantity r that appears in the law of universal gravitation?

b. Does r represent the radius of either body?

3. An interplanetary space probe experiences a gravitational attraction toward the sun of 120 N when the probe is at a distance of 120 million kilometers. What will the gravitational force become when the interplanetary probe is only 40 million kilometers from the sun? (Hint: First solve algebraically for the unknown force. Substitute numerical values only in the last step.)

4. The "weight" of an astronaut hovering in an orbit about the moon is the gravitational force exerted on the astronaut by the moon. On the moon's surface an astronaut had a weight of 140 N. Calculate the weight of this astronaut when he was in an orbit 1740 km above the moon's surface. (The moon's radius is 1740 km.) (Hint: In this problem m_1, m_2, and G are all constants.)

5. A helium atom consists of two electrons orbiting about a positively charged nucleus. Compute the gravitational force of attraction between one of these electrons and the nucleus using the following data: Mass of nucleus, 6.6×10^{-27} kg; mass of electron, 9.1×10^{-31} kg; radius of electron orbit, 5.3×10^{-11} m.

6. During an exploration of the lunar landing site, an astronaut drops a small stone from a height of 3.2 m and discovers that it requires 2.0 s to reach the surface. With this information plus the known radius of the moon (1740 km) and the value of G, the astronaut calculates the mass of the moon.
   a. Use the astronaut's data to determine the acceleration of gravity on the moon's surface.

   b. Now show how the astronaut calculates the mass of the moon.

7a. The space probe in Prob. 3 experienced a gravitational force of 120 N at a certain point in space. If the mass of the probe is 600 kg, what is the magnitude of the gravitational field at this point? (Don't forget to specify the units!)

b. What is the direction of the vector **g**?

c. What force would be experienced by a 10,000-kg asteroid at this point?

8. A point P is 0.02 m from a 20 kg mass and 0.04 m from a 100 kg mass as shown in Fig. 8-2.

a. Calculate the magnitude of the gravitational field at P due to the larger mass alone.

b. Calculate the magnitude of the gravitational field at P due to the small mass alone.

c. Draw arrows to represent the directions of the two gravitational fields just found.

d. Now calculate the magnitude of the resultant gravitational field at P due to both masses acting together.

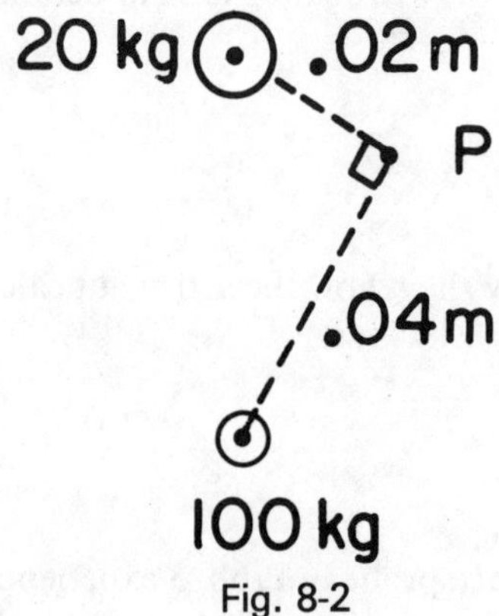

Fig. 8-2

9a. The elliptical path of a planet about the sun is shown in Fig. 8-3. The time required for the planet to move from A to B is the same as that required to move from A' to B'. What relationship exists between the two shaded areas on the drawing?

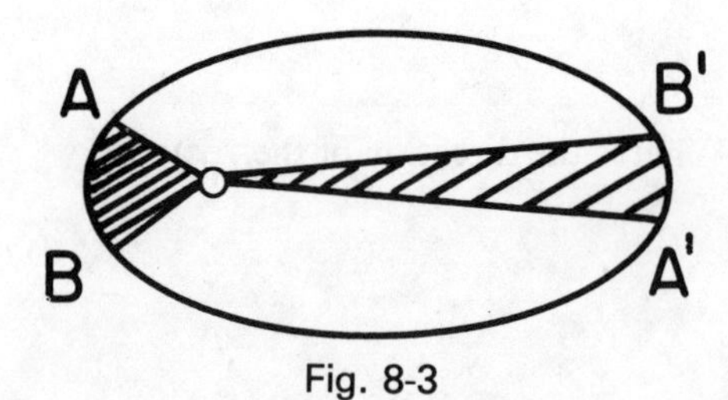

Fig. 8-3

b. Kepler's law of areas is valid for the gravitational force which is directed toward the attracting body (the sun in this case) and is proportional to $1/r^2$. Would the law of areas be valid for a central force proportional to $1/r^3$?

10. A certain small planet has a perihelion distance r_1 that is 1/3 its aphelion distance r_2. If the linear speed of the planet at perihelion is 40 km/s, what is its speed at aphelion? [Hint: Let Δt be a small time interval. The area swept out by the planet at perihelion may be written $A_1 = \frac{1}{2}(v_1 \Delta t)r_1$.]

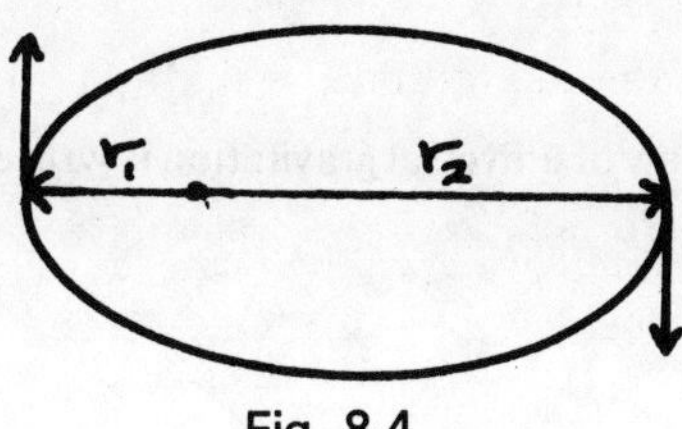

Fig. 8-4

11. The average distances of the earth and Saturn from the sun are 1.47×10^6 km and 14.0×10^6 km, respectively. Knowing that the period of revolution of the earth is one year, calculate the period of revolution of Saturn in earth years.

12. Kepler's first law states that the orbit of any planet around the sun is an ellipse with the sun at one focus. What special form of a central force leads to elliptical orbits?

13. Write a mathematical statement of Kepler's law of periods.

14. The period of revolution of the earth about the sun is 365 days and its mean distance from the sun is 1.50×10^{11} m. The period of revolution of the planet Saturn is 1.08×10^4 days. Use this information to calculate the mean radius of Saturn's orbit.

Solutions

1. Newton's law of universal gravitation is written as

$$F = G\frac{m_1m_2}{r^2}$$

The direction of the force on m_2 is toward the center of m_1 and vice versa. The gravitational force is always an attractive force.

2a. The quantity r is the distance from the center of mass of one object to the center of mass of the other.

b. No, r does not represent the radius of either body. If a very small object is on the surface of a very large spherical object, such as the earth, the distance r will be nearly equal to the radius of the large object. This special situation is illustrated in Fig. 8-5.

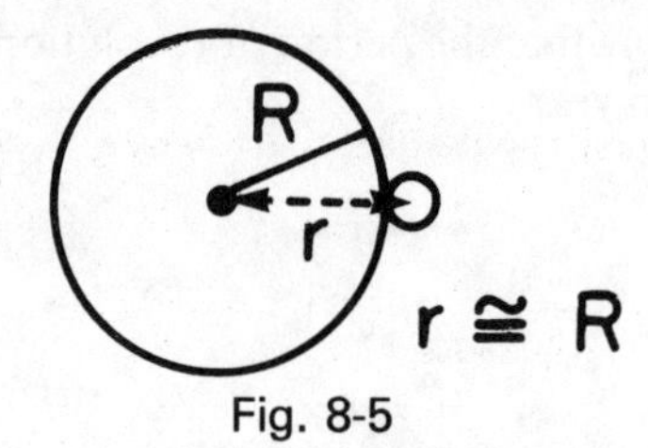

Fig. 8-5

3. We will distinguish the two cases by subscripts 1 and 2.
Case 1: distance $= d_1 = 12 \times 10^7$ km
force $= F_1 = 1.2 \times 10^2$ N
Case 2: distance $= d_2 = 4 \times 10^7$ km
force $= F_2 = ?$

$$F_1 = G\frac{m_1m_2}{d_1^2} \qquad F_2 = G\frac{m_1m_2}{d_2^2}$$

Eliminating the constant, $G\ m_1m_2$,

$$\frac{F_2}{F_1} = \frac{d_1^2}{d_2^2}; \quad F_2 = F_1\left(\frac{d_1}{d_2}\right)^2$$

$$F_2 = (1.2 \times 10^2 \text{ N})\left(\frac{12 \times 10^7 \text{ km}}{4 \times 10^7 \text{ km}}\right)^2 = 1.08 \times 10^3 \text{ N}$$

4. When the astronaut is on the surface, the separation is $d_1 = 1740$ km. When the astronaut is in orbit at an altitude of 1740 km, the separation becomes 1740 km + 1740 km = 3480 km. Using the relationship found in the preceding problem,

$$F_2 = F_1\frac{d_1^2}{d_2^2} = 140 \text{ N}\left(\frac{1740}{3480}\right)^2; \quad F_2 = 35 \text{ N}$$

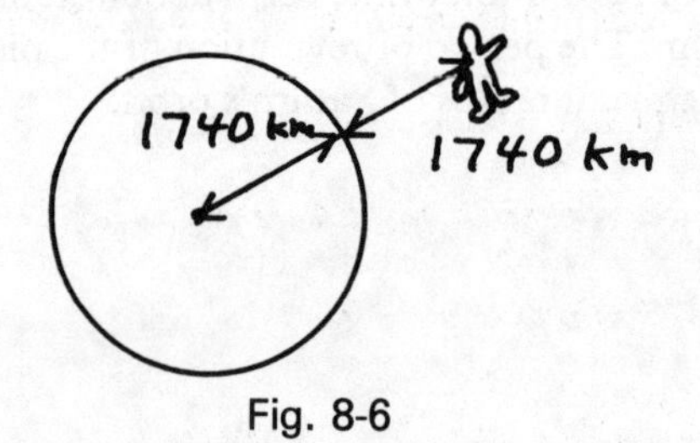

Fig. 8-6

5. The gravitational force can be computed from the universal law of gravitation.

$$F = G\frac{m_1m_2}{d^2}$$

$$F = \left(6.7 \times 10^{-11}\frac{\text{N}\cdot\text{m}^2}{\text{kg}^2}\right)\frac{(6.6 \times 10^{-27} \text{ kg})(9.1 \times 10^{-31} \text{ kg})}{(5.3 \times 10^{-11} \text{ m})^2}$$

$$F = 1.4 \times 10^{-46} \text{ N}$$

6a. Since the initial velocity of the stone was zero, the distance is given by $s = \frac{1}{2}gt^2$. Solving for g,

$$g = \frac{2s}{t^2} = \frac{2(3.2\ \text{m})}{(2.0\ \text{s})^2} = 1.6\ \text{m/s}^2$$

b. The equation for the mass of the attracting body, the moon, can be found by equating the two expressions for the gravitational force.

$$F = mg \quad \text{and} \quad F = G\frac{Mm}{r^2}$$

Thus $mg = G\dfrac{Mm}{r^2}$; $M = gr^2/G = \dfrac{(1.6\ \text{m/s}^2)(1.74 \times 10^6\ \text{m})^2}{6.7 \times 10^{-11}\ \text{N·m}^2/\text{kg}^2} = 7.2 \times 10^{22}\ \text{kg}$

7a. The gravitational field is given by $\mathbf{g} = \dfrac{\mathbf{F}_{\text{grav}}}{m}$

Thus $g = \dfrac{120\ \text{N}}{600\ \text{kg}} = 0.02\ \text{N/kg}$

b. The gravitational field is directed toward the object which causes it, in this case, the sun.

c. $F = mg = (10{,}000\ \text{kg})(0.2\ \text{N/kg}) = 2000\ \text{N}$

8a. $g_1 = \dfrac{Gm_1}{r_1^2} = \dfrac{(6.7 \times 10^{-11}\ \text{N·m}^2/\text{kg}^2)(20\ \text{kg})}{(0.02\ \text{m})^2} = 3.35 \times 10^{-6}\ \text{N/kg}$

b. $g_2 = \dfrac{Gm_2}{r_2^2} = \dfrac{(6.7 \times 10^{-11}\ \text{N·m}^2/\text{kg}^2)(100\ \text{kg})}{(0.04\ \text{m})^2} = 4.19 \times 10^{-6}\ \text{N/kg}$

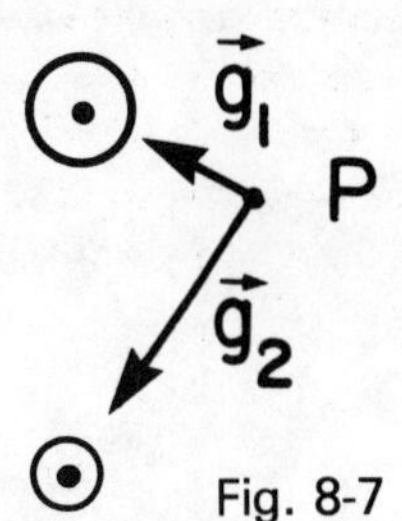

Fig. 8-7

c. Each gravitational field is directed toward the object that produced it.

d. Since the two vectors being added are at right angles, the Pythagorean theorem may be used.

$$g_R = [(3.35 \times 10^{-6})^2 + (4.19 \times 10^{-6})^2]^{1/2} = 5.36 \times 10^{-6}\ \text{N/kg}$$

9a. The line joining the planet and the sun sweeps out equal areas in equal times. Thus the two shaded areas are equal.

b. Yes. The law of areas is valid for *any* central force.

10. The radius vector from the planet to its sun sweeps out equal areas in equal times. At perihelion and at aphelion the velocity vector is perpendicular to the radius. Thus $\Delta A_1 = \frac{1}{2}s_1r_1$ and $\Delta A_2 = \frac{1}{2}s_2r_2$. If Δt is a small time interval,

$$s_1 = v_1\Delta t \quad \text{and} \quad s_2 = v_2\Delta t$$

According to the law of areas, $\Delta A_1 = \Delta A_2$. Thus

$$\tfrac{1}{2}(v_1\Delta t)r_1 = \tfrac{1}{2}(v_2\Delta t)r_2$$

$$v_2 = v_1\left(\frac{r_1}{r_2}\right) = 40\ \text{kg/s}\ (1/3) = 13.3\ \text{km/s}$$

11. Use the subscript 1 for earth and 2 for Saturn. Kepler's third law may be written

$$\frac{T_1^2}{T_2^2} = \frac{r_1^3}{r_2^3} \quad \text{or} \quad \frac{T_1}{T_2} = \left(\frac{r_1}{r_2}\right)^{3/2}$$

Thus $T_2 = T_2\left(\dfrac{r_2}{r_1}\right)^{3/2} = (1\ \text{y})\left(\dfrac{14.0}{1.47}\right)^{3/2} = 29.4\ \text{y}$

12. Elliptical orbits result if the central force is an inverse-square force, that is, a force proportional to $1/r^2$.
13. Kepler's law of periods may be written

$$\frac{T_1^2}{T_2^2} = \frac{r_1^3}{r_2^3}$$

14. Let the subscript 1 refer to the earth and 2 to Saturn. Then the law of periods leads to

$$r_2^3 = r_1^3 \frac{T_2^2}{T_1^2}$$

$$r_2 = r_1[T_2^2/T_1^2]^{1/3} = 1.50 \times 10^{11}\ \text{m}[(1.08 \times 10^4 \text{d})^2/(365\ \text{d})^2]^{1/3} = 1.43 \times 10^{12}\ \text{m}$$

CHAPTER 9 Elasticity and Vibration

GOALS To study the vibratory motion that is characteristic of elastic systems.

OBJECTIVES After completing this chapter the student should be able to do the following:

1. Write the equations expressing the relationship between stress and strain for each of the three types of distortion of an elastic solid.
2. Give the symbol for the elastic constant, the modulus, in each of the three cases.
3. Solve for any one of the quantities ΔF, A, ΔL, L, or E when the others are given.
4. Define simple harmonic motion (SHM) in terms of restoring force and displacement.
5. Give the correct name for that type of one-dimensional motion in which the acceleration is proportional to the displacement and in the opposite direction.
6. State the relationship that exists between uniform circular motion and SHM.
7. Write in standard notation the equation relating the velocity of a point in uniform circular motion to the radius and period of the motion.
8. Calculate the maximum velocity of a vibrating object given the amplitude and period of the vibration.
9. Calculate the maximum acceleration of a point in SHM from its maximum velocity and its amplitude.
10. Use prior knowledge of work done by a variable force (Sec. 6-2) to obtain the formula for the potential energy stored in a spring.
11. Write the formula for the period of vibration of a mass subject to an elastic restoring force.
12. Calculate the period and frequency of vibration of an elastic system from a knowledge of the displacement produced by a known force.
13. Calculate the velocity, acceleration, and energy of a mass moving in SHM at any point of its vibration.
14. Draw a complete force diagram for the simple pendulum and show that the force constant for small angles is given by

$$k \cong mg/L$$

15. Combine this result with the formula obtained for the period of vibration to derive the formula for the period of a pendulum.

SUMMARY

The length of a wire increases when a force is applied to it. If the wire is not stretched so much as to be permanently deformed, the elongation will be proportional to the applied force, the length, and the reciprocal of the area. The relationship, called Hooke's law, may be written

$$\frac{\Delta F}{A} = E\,\frac{\Delta L}{L}$$

where the constant E is the *stretch modulus*, a property of the material of which the wire is made. The ratio of force to cross-sectional area $\Delta F/A$ is the stretching *stress* and the ratio of elongation to length $(\Delta L/L)$ is the stretching *strain*.

When a force is applied to a solid in such a way as to change the shape of the object rather than change its length, the elastic modulus involved is the *shear modulus n*. Hooke's law for this case is written as

$$\frac{\Delta F}{A} = n\frac{\Delta x}{h}$$

where h is the dimension of the object perpendicular to the applied force.

When the external pressure on a solid or fluid is increased its volume will decrease. The relationship between volume stress ΔP and volume strain $\Delta V/V$ may be written as

$$P = -B\frac{\Delta V}{V}$$

where B, the bulk modulus, is a constant, characteristic of the solid or fluid. In the case of a fluid, the bulk modulus is the only one that is non-zero.

Simple harmonic motion (SHM) is a special type of vibratory motion in which the restoring force is directly proportional to the displacement. Symbolically $F = ks$ where k is called the *force constant*. The force is always in the opposite direction to the displacement. The motion that results has the useful property of being the projection of a point moving around a circle at a constant speed. The use of this *reference circle* greatly facilitates the solution of problems in SHM. The position, velocity, and acceleration of a vibrating object can all be found as projections of the position, velocity, and acceleration of the corresponding point in the reference circle.

From the fact that mechanical energy, $KE + PE$, is conserved, it can be shown that the *period* of a vibrating mass, the time for a complete oscillation, is $T = 2\pi\sqrt{m/k}$.

A pendulum vibrating through small angles is an example of SHM. Its period is given by $T = 2\pi\sqrt{L/g}$ where L is the length of the pendulum.

QUESTIONS AND PROBLEMS

Sec. 9-1 Solve the following problems using the elastic constants listed in Table 9-1 in the text.

1. Two wires of the same material and same length are subject to equal forces. If the radius of the first wire is twice that of the second, what is the ratio of the elongations produced? (Given $r_1 = 2r_2$, find $\Delta L_1/\Delta L_2$.)

2. What stress applied to an aluminum wire will cause its length to increase by 0.05%?

3. Figure 9-1 shows the results of an experiment on an annealed copper wire. The load was increased until the wire broke (point D).

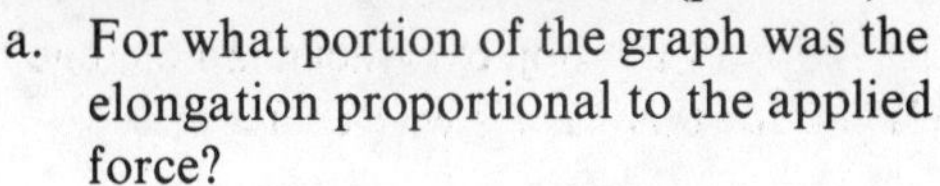

a. For what portion of the graph was the elongation proportional to the applied force?

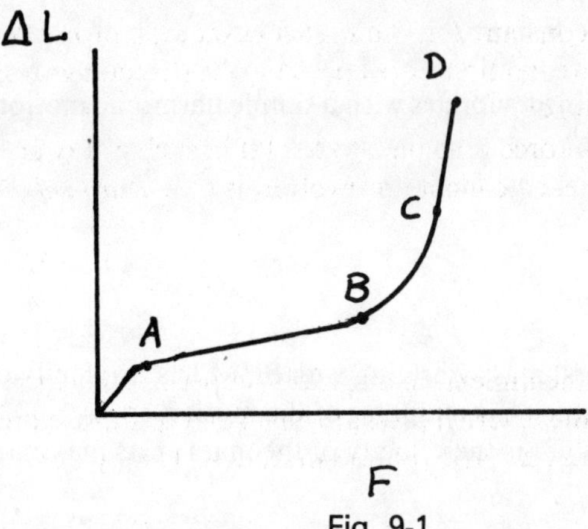

Fig. 9-1

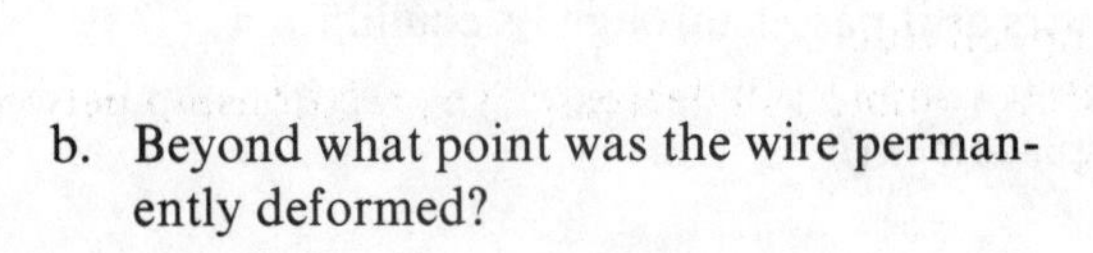

b. Beyond what point was the wire permanently deformed?

c. What is a possible explanation of the behavior between the origin and point A?

4. The cylinder of a bicycle pump has a diameter of 4 cm and a length of 40 cm at the top of its stroke. (See Fig. 9-2.) The piston is now pushed down a distance of 4 cm. What force is required to hold the piston at this position? (Assume that no air escapes from the pump.)

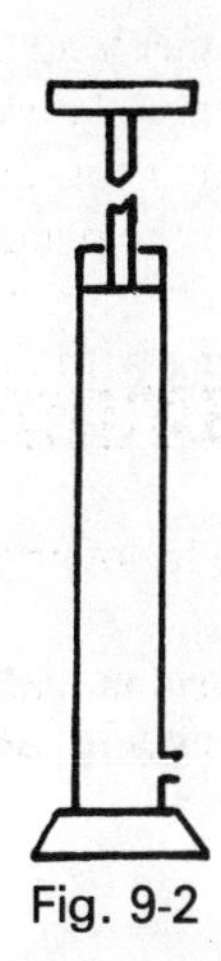
Fig. 9-2

5. Two iron wires have the same mass but one wire is twice as long as the other, $L_1 = 2L_2$. When the same tensile force is applied to each wire, different elongations are produced. Find the ratio of the elongations $\Delta L_1/\Delta L_2$. (Hint: Since the wires have the same mass, they also have the same volume, $L_1A_1 = L_2A_2$.)

6. A sample of human bone, a femur about 12 cm long, is subjected to a force of tension of 1500 N. The measured elongation is 0.36 mm. Use the approximate stretch modulus of bone, 2.1×10^{10} N/m^2, to calculate an average cross-sectional area of this bone.

Secs. 9-2—9-6

7. An object vibrates with a simple harmonic motion of amplitude 5 cm. What is the diameter of the reference circle?

8. A small mass vibrating with SHM has a velocity of 0.5 m/s as it passes through its equilibrium position—the midpoint of the motion.
   a. What is the velocity of the mass at its maximum displacement?

   b. What is the velocity of the point in the reference circle?

   c. If the amplitude of the vibration is 5 cm, what is its period?

9. For a point moving with SHM, the acceleration is proportional to the displacement. Complete this equation

$$a = (\qquad)\, s$$

inserting the correct constant of proportionality.

10. For a point moving with SHM, is the magnitude of the velocity proportional to the displacement? Explain your answer.

* * * * * * *

The solution of problems in SHM may be facilitated by the use of the *reference circle.* Sec. 9-3 of the text explains that the motion of a particle that is vibrating in SHM is closely related to the motion of a point moving around a circle at a constant speed. To be more specific, SHM is the *projection* of uniform circular motion. We will examine the implications of this statement in the following problems.

11. We have a particle vibrating in SHM along the vertical line between the points S and S'. If the distance from S to S' is 20 cm, then the amplitude of the SHM is $A =$ ______________ . (Write in the answer.) Let M be the midpoint of the motion. A circle of radius A has been drawn with its center on the perpendicular bisector of SS'. This is the *reference circle.* Let P be a point moving counterclockwise around the circle at a constant speed. From P we drop a perpendicular to the vertical line SS'. The point P' is called the *projection* of P. As P moves around the circle at a constant speed, its projection P' moves with simple harmonic motion between the end points S and S'.

    Suppose, for example, that P makes one revolution of the reference circle every two seconds. Answer the following questions by referring to Fig. 9-3. Check your answers as you go and correct any mistakes.

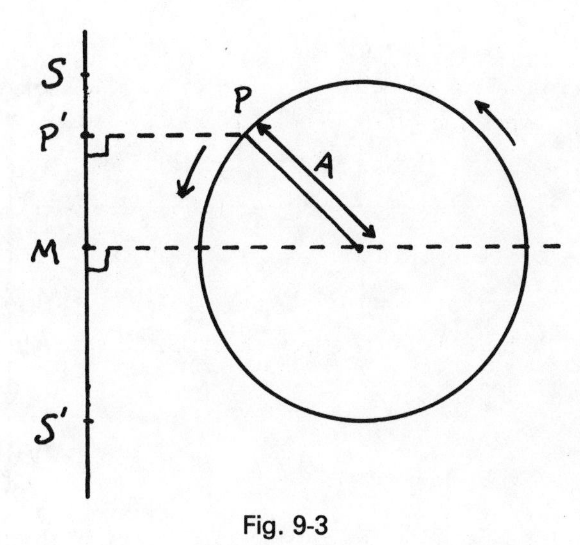

Fig. 9-3

12. The angular speed of the point in the reference circle is ______________ rev/s or ______________ rad/s.
13. The linear speed of the point in the reference circle is ______________ m/s.
14. When P', the vibrating point, moves from S to S' the point in the reference circle goes halfway around the circle. Thus the time required for P' to go from S to S' is ______________ seconds. When P' moves from S' to M its companion point P moves ______________ (what fraction?) of the distance around the circle. Thus the time required for P' to move from S' to M is ______________ seconds.

15. We have just shown that it takes 1/4th of the period ($T/4 = 0.5$ s) for the vibrating point to go from M to S'. Now let's find out how long it takes P' to move from M to Q, which is halfway between M and S'. Figure 9-4 shows the two positions of P' and the two corresponding positions of P. Even without making any measurements we can see that P does not move 1/8 the distance around the circle. One eighth of 360° is 45° and the angle through which P has moved is obviously less than this. Through what angle does P move? (Hint: θ is an angle in a right triangle whose hypotenuse is A.)

θ = ______________.

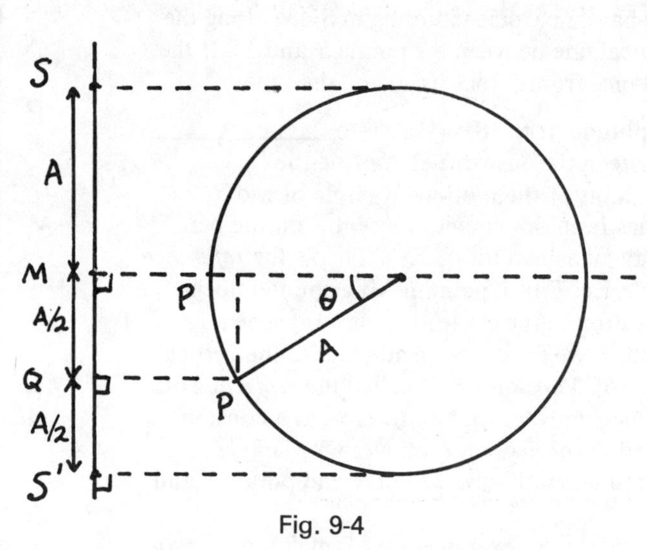

Fig. 9-4

16. Having found the angular displacement of P, it is a simple matter to find the linear distance (the length of the arc) through which P has moved and the time required for P to move through this distance. The arc length through which P has moved is ______________ cm. The time required to move through the angle θ is ______________ s.

17. How long does it take the vibrating particle to move a distance of 5 cm *starting from S*?

t = ______________.

* * * * * * *

In the preceding problems we have made use of the fact that the *position* of a particle vibrating with SHM is the projection of a point moving in uniform circular motion. It is also true that the *velocity* of a particle in SHM is the projection of the velocity of a point moving in a circular path at constant speed. This idea is illustrated in Fig. 9-5. A point P is moving around a circle at a constant speed of 10 m/s. An arrow has been drawn to represent the velocity $\mathbf{v}$ of the point at a particular instant. The projection of P on a horizontal line is indicated by P'. The velocity of the vibrating object is exactly the projection of the vector $\mathbf{v}$ on the horizontal line. The enlarged drawing (Fig. 9-6) shows how the projection may be found by dropping perpendiculars from each end of the arrow. It is evident from the geometry that the projection of $\mathbf{v}$ is just the horizontal component of $\mathbf{v}$.

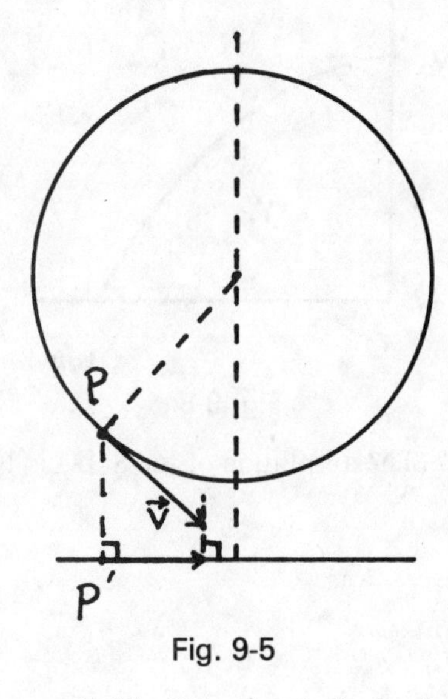

Fig. 9-5

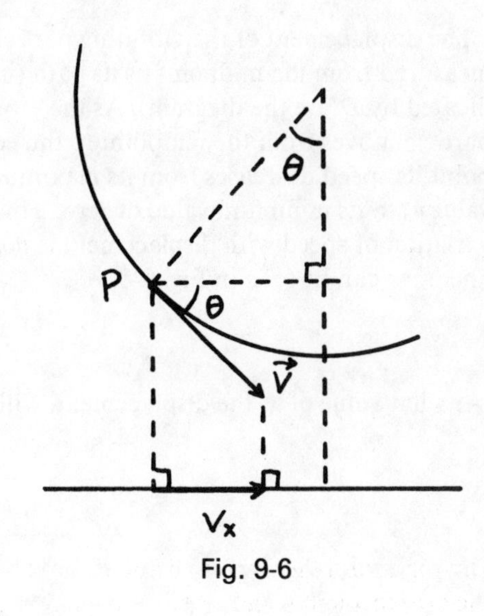

Fig. 9-6

18. If the angle θ is 60°, what is the horizontal component of $\mathbf{v}$?

v_x = ______________.

19. On the reference circle in Fig. 9-7, five different positions of the moving point are shown. Place a small arrow at each position to indicate the velocity $\mathbf{v}$ of the moving point as it passes through *A, B, C, D,* and *E*. Remember that each arrow must have the same length since P moves around the circle at a constant speed.

    At which of the five positions does the horizontal component of $\mathbf{v}$ have a maximum value? At which of the five positions does the horizontal component of $\mathbf{v}$ have a minimum value?

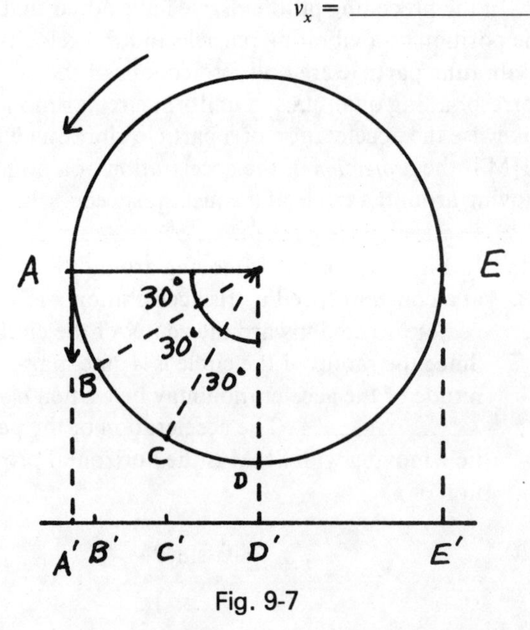

Fig. 9-7

20. The particle in SHM vibrates back and forth between A' and E'. Its velocity varies from a minimum value of ______________ to a maximum value of ______________. (The speed of a point on the rim of the reference circle is a constant 10 m/s.)

21. A particle in SHM has its maximum speed as it passes through the ______________. It has a speed of zero at each ______________ of the vibration.

22. Referring again to Fig. 9-7, what is the speed of the vibrating particle as it passes through point C'?

$v_x =$ ______________.

The displacement of the vibrating particle is measured from the midpoint of its path (indicated by D' on the diagram). As the vibrating particle moves from the midpoint to the endpoint its speed decreases from its maximum value v_0 to its minimum value of zero. This variation of speed with displacement is *not* linear, as can be seen in Fig. 9-8.

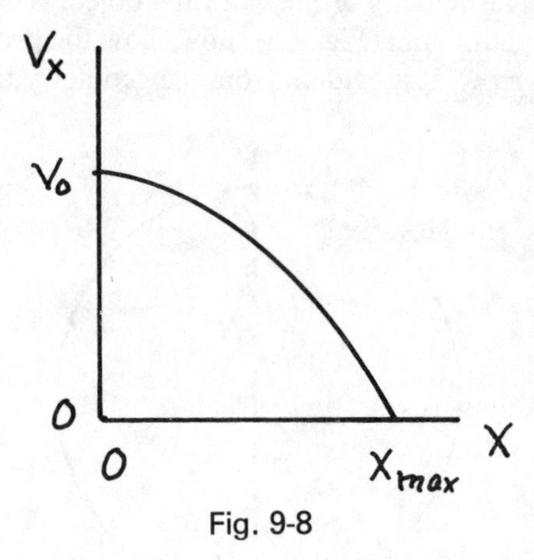

Fig. 9-8

23. At what value of x, the displacement, will the speed be $(0.707)v_0$? (The amplitude of the SHM is 10 cm.)

* * * * * * *

In the preceding problems, we have noted that the position of a vibrating particle and the velocity of a vibrating particle are both projections of the corresponding quantities in uniform circular motion. Likewise the acceleration of a particle vibrating with SHM is the *projection* of the acceleration of a point moving around a circle at a constant speed.

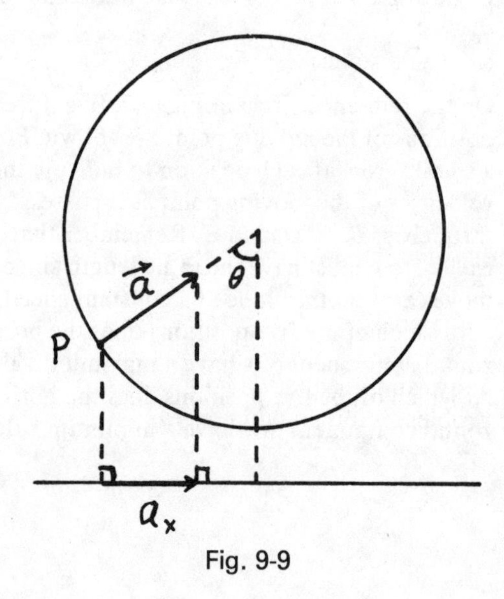

Fig. 9-9

24a. In Fig. 9-9, point P is moving around the circle at a constant speed v_0. Its acceleration **a** is a vector directed toward the center of the circle. Since the radius of the circle is A, the magnitude of the acceleration may be written $a =$ ______________. The acceleration of the particle moving with SHM is the horizontal projection of **a**.

$$a_x = a \sin \theta$$

b. At what points in its motion does the *vibrating* particle have its maximum acceleration?

25. At what point in its motion does the vibrating particle have its minimum acceleration?

26. For the particle in SHM a_{max} = ______________ and a_{min} = ______________.

27. At what value of the displacement is $a = (1/2)a_{max}$?

x = .

28. Suppose that it takes 0.5 s for the vibrating point to go from the midpoint to its maximum displacement of 10 cm. How long will it take for the particle to go from the midpoint to a position where $a = (0.866)a_{max}$?

t = ______________.

29. A mass is vibrating with simple harmonic motion between points A and D, illustrated below. The mass has a velocity of 10 cm/s at B (the midpoint) and a velocity of 0 cm/s at D.

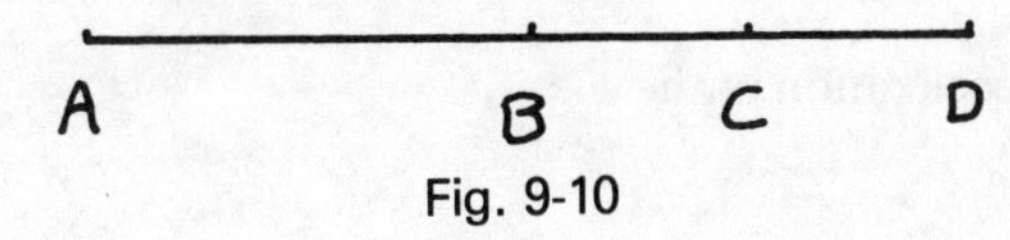

Fig. 9-10

a. What is the velocity at point C, halfway between B and D? (Hint: Use the reference circle.)

b. The distance A to B, the amplitude, is 8 cm. How long does it take for the mass to go from C to D?

30. In some earlier chapters, especially Chapter 2, you have used the equation

$$s = v_0t + \frac{1}{2}at^2$$

Why can this equation *not* be used to find the time to travel a distance s in Prob. 29b? For what type of problem is this equation valid?

31. A mass of 1 kg is suspended from a light spring as shown in Fig. 9-11. When a second 1-kg mass is added, the spring elongates 9.8 cm. The second mass is removed and the system is set into vibration with that single 1-kg mass, the amplitude being 5 cm.

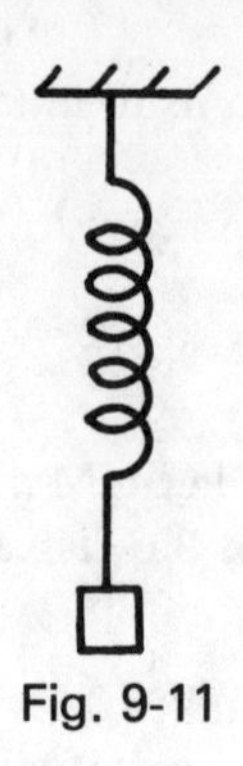
Fig. 9-11

a. Calculate the frequency of vibration.

b. Calculate the maximum velocity of the vibrating mass.

c. Calculate the maximum acceleration of the vibrating mass.

d. At what value of the displacement will the velocity be exactly one half its maximum value?

32. The pendulum of a clock vibrates with a frequency of 0.5 Hz when located on earth. What would the frequency of vibration of this same pendulum be on the surface of the moon, where g has a value 1/6 that of g on the earth.

33. A 400-g mass suspended from a spring vibrates with a frequency of 1.2 Hz. What would be the frequency of vibration of this same mass-spring system on the surface of the moon?

34. In light of your answers to Probs. 32 and 33, which of these two vibrating systems could be used for keeping time in space travel.

35. Suppose that the mass of the pendulum bob and the mass attached to the spring are both doubled. What are the new vibration frequencies of these two systems on the earth and on the moon?

Solutions

1. The elongations are inversely proportional to the cross-sectional area.

$\Delta L \propto 1/A$

$$\Delta L_1/\Delta L_2 = A_2/A_1 = \pi r_2^2/\pi r_1^2 = \frac{r_2^2}{(2r_2)^2}$$

Thus

$\Delta L_1/\Delta L_2 = 1/4$

2. stress = (stretch modulus)(strain)

$$\frac{\Delta F}{A} = (7.0 \times 10^{11}\ \text{N/m}^2)(5 \times 10^{-4}) = 3.5 \times 10^7\ \text{N/m}^2$$

3a. The linear region is A to B.

b. Point B.

c. There were perhaps some bends or kinks in the wire, which allowed it to stretch easily until the kinks disappeared.

4. The change in pressure is related to the change in volume.

$$\Delta P = -B\,\Delta V/V$$
$$\Delta P = -(1.01 \times 10^5\ \text{N/m}^2)(-0.1) \cong 1 \times 10^4\ \text{N/m}^2$$
$$F/A = \Delta P;\ F = \Delta PA = (10^4\ \text{N/m}^2)(4\pi \times 10^{-4}\ \text{m}^2) = 12.6\ \text{N}$$

5. $$\Delta L_1 = \frac{\Delta F}{E A_1} L_1;\ \Delta L_2 = \frac{\Delta F}{E A_2} L_2$$

Dividing the first equation by the second,

$$\frac{\Delta L_1}{\Delta L_2} = \frac{L_1 A_2}{L_2 A_1}$$

Since the volumes are equal,

$$L_1 A_1 = L_2 A_2 \quad \text{or} \quad \frac{A_2}{A_1} = \frac{L_1}{L_2}.$$

Thus

$$\frac{\Delta L_1}{\Delta L_2} = \left(\frac{L_1}{L_2}\right)^2 = \left(\frac{2L_2}{L_2}\right)^2 = 4$$

6. $\frac{\Delta F}{A} = E\frac{\Delta L}{L}$ Solving for A gives

$$A = \frac{\Delta F\, L}{E\,\Delta L} = \frac{(1.5 \times 10^3\ \text{N})(0.12\ \text{m})}{(2.1 \times 10^{10}\ \text{N/m}^2)(3.6 \times 10^{-4}\ \text{m})}$$
$$= 2.4 \times 10^{-5}\ \text{m}^2 = 0.24\ \text{cm}^2$$

7. Diam. = $2A$ = 10 cm

8. $v_0 = 0.5$ m/s

a. At maximum displacement $v = 0$.

b. The point in the reference circle has a velocity of 0.5 m/s.

c. $$T = \frac{2\pi A}{v_0} = \frac{0.1\ \pi m}{0.5\ \text{m/s}} = 0.63\ \text{s}$$

9. $a = (k/m)s$

Since **a** and **s** have opposite directions, this relationship is often written $a = -(k/m)s$.

10. No. The velocity is *not* a linear function of the displacement. The correct relationship is $y = A \sin(2\pi t/T)$ if the motion is timed from the moment the vibrating particle passes the midpoint.

* * * * * * *

11. The SHM described has an amplitude of $A = 10$ cm.
12. 0.5 rev/s or π rad/s
13. The circumference of the reference circle is 20π cm or 62.8 cm. The time required to go around once is 2 s. Thus the speed in the reference circle is

$$v = \frac{C}{T} = \frac{62.8 \text{ cm}}{2 \text{ s}} = 31.4 \text{ cm/s}$$

14. 1; 1/4; 0.5
15. $\theta = 30°$ or 1/12 revolution or $\pi/6$ rad.

16. $\text{arc length} = (1/12)C = \dfrac{62.8 \text{ cm}}{12} = 5.24 \text{ cm}$

$30° = \pi/6$ rad

Since $\omega = \dfrac{\theta}{t}$; $\quad t = \dfrac{\theta}{\omega} = \dfrac{\pi/6}{\pi \text{ rad/s}} = 1/6 \text{ s}$

17. First identify the corresponding points in the reference circle. These are S' and U'. Next draw the two radii and mark the angle θ. The side adjacent to θ is 5 cm long. The hypotenuse is 10 cm long. Thus $\theta = 60°$ or $\pi/3$ rad.

$$t = \theta/\omega = \frac{\pi/3}{\pi \text{ rad/s}} = 1/3 \text{ s}$$

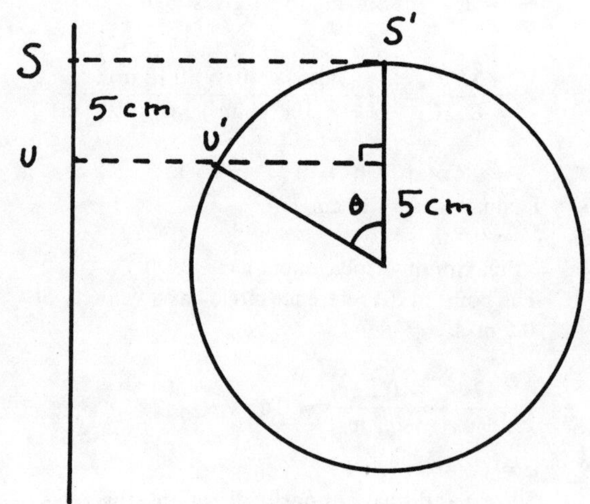

Fig. 9-12

18. $v = 10$ m/s
$v_x = v \cos 60° = (10 \text{ m/s})(0.5) = 5 \text{ m/s}$

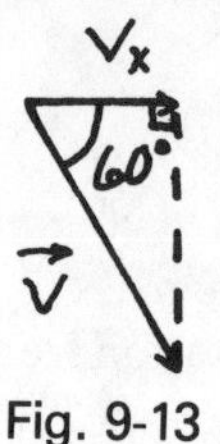

Fig. 9-13

19. v_x is a maximum at D;
v_x is a minimum at A; and also at E.

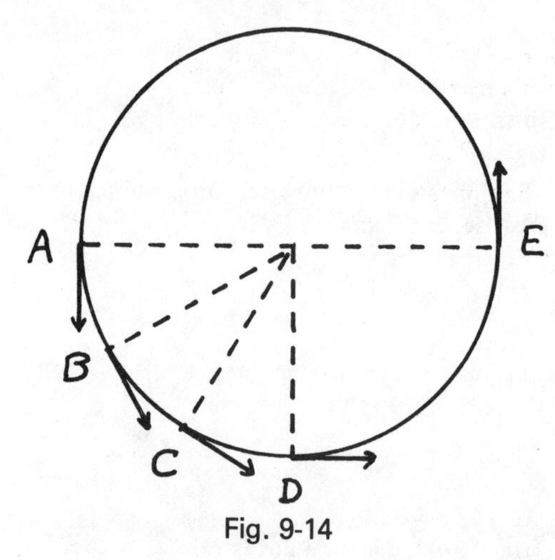

Fig. 9-14

20. zero; 10 m/s
21. midpoint or equilibrium position; endpoint
22. $v_x = v \cos 30° = 0.866\,(10 \text{ m/s})$
$v_x = 8.66$ m/s

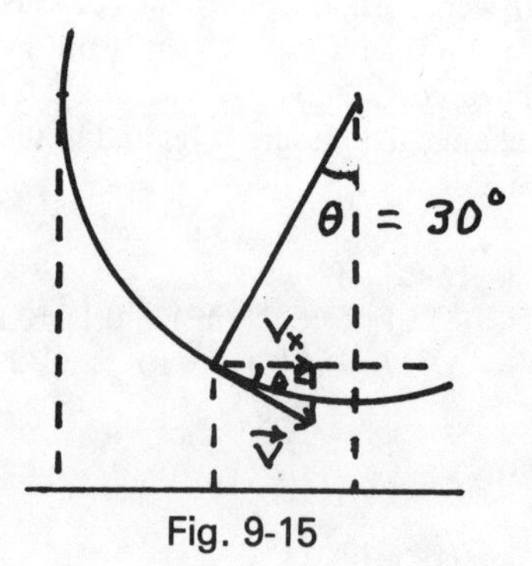

Fig. 9-15

23. In this case, θ is an unknown angle.

$$\cos\theta = \frac{v_x}{v} = \frac{(0.707)v_0}{v_0} = 0.707$$

$$\theta = 45°$$

$$\sin 45° = \frac{x}{A}; \; x = A \sin 45°$$

$$x = (0.707)\, 10 \text{ cm} = 7.07 \text{ cm}$$

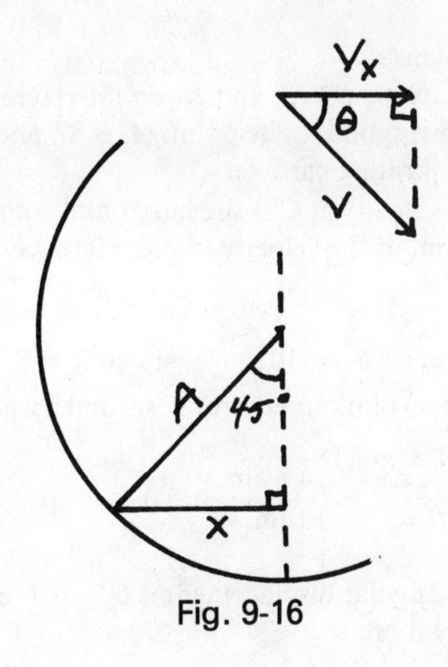

Fig. 9-16

24a. $a = v^2/R$

b. a is a maximum at the endpoints.

25. a is a minimum (zero) at the midpoint.

26. $a_{max} = R\omega^2$; $a_{min} = 0$

27. We note that the triangle having sides a and a_x is similar to the larger triangle having sides A and x. Thus

$$\frac{a_x}{a} = \frac{x}{A}$$

$$x = A(a_x/a) = A(1/2) = 5 \text{ cm}$$

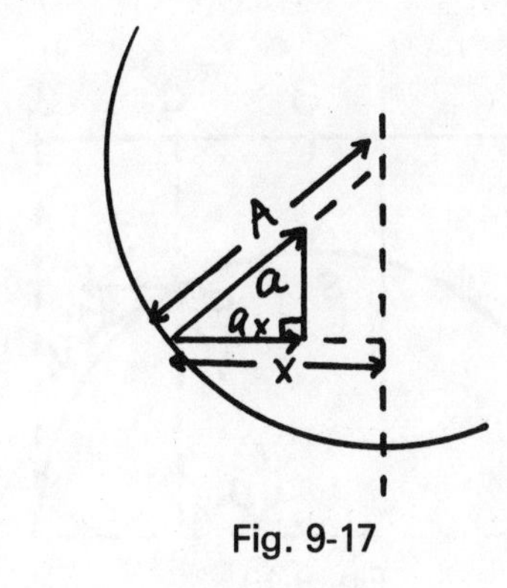

Fig. 9-17

28. First find the angular displacement, θ.

$$\sin\theta = \frac{a_x}{a} = \frac{(0.866)a_{max}}{a_{max}}$$

$$\theta = 60°$$

In 0.5 s the vibrating point goes from M' to S'. The corresponding angular displacement is $\theta = 90°$. Thus

$$\frac{t}{60°} = \frac{0.5 \text{ s}}{90°} \qquad t = (2/3)(0.5 \text{ s}) = 0.33 \text{ s}$$

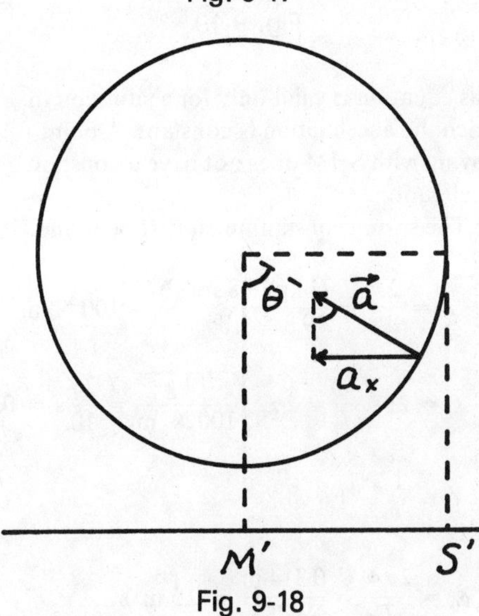

Fig. 9-18

29. $v_0 = 10$ cm/s
The points A', B', C', and D' on the reference circle correspond to the points A, B, C, and D for the vibrating particle.

a. The velocity at C is the horizontal component of the velocity in the reference circle.

$$v = v_0 \sin\theta = (10 \text{ cm/s})(\sin 60°) = 8.7 \text{ cm/s}$$

b. One revolution requires T seconds, where

$$T = \frac{2\pi A}{v_0} = \frac{16\,\pi \text{ cm}}{10 \text{ cm/s}} = 5.0 \text{ s}$$

The angular displacement is 60° or 1/6 of a revolution.

$$t = \frac{1}{6}T = 0.84 \text{ s}$$

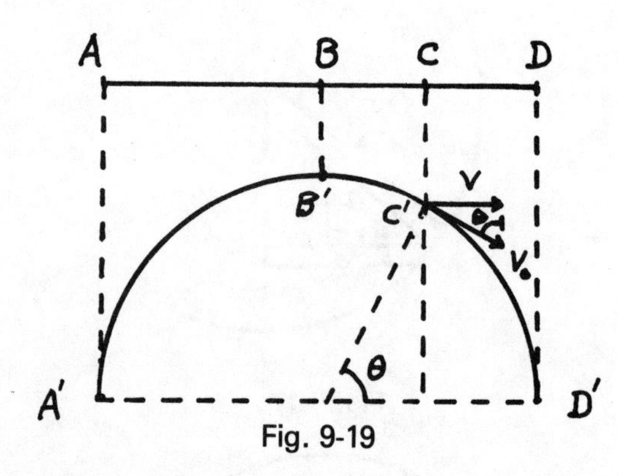

Fig. 9-19

30. This equation is valid only for a situation in which the acceleration is constant. A point moving with SHM does not have a constant acceleration.

31. a. The spring constant must first be found.

$$k = \frac{\Delta F}{\Delta x} = \frac{(1 \text{ kg})(9.8 \text{ m/s}^2)}{9.8 \times 10^{-2} \text{ m}} = 100 \text{ N/m}$$

$$T = 2\pi\sqrt{\frac{m}{k}} = 2\pi\sqrt{\frac{1 \text{ kg}}{100 \text{ N/m}}} = \frac{2\pi}{10} \text{s} = 0.63 \text{ s}$$

$$f = \frac{1}{T} = 1.6 \text{ vib/s}$$

b. $v_0 = \dfrac{2\pi A}{T} = \dfrac{0.314 \text{ m}}{0.628 \text{ s}} = 0.5 \text{ m/s}$

c. $a = r\omega^2 = A\left(\dfrac{2\pi}{T}\right)^2 = 4\pi^2 A/T^2$

$$a = \frac{4\pi^2 (5 \times 10^{-2} \text{ m})}{(6.28 \times 10^{-1})^2 \text{s}^2} = 5 \text{ m/s}^2$$

The maximum acceleration can also be obtained from the force constant.

$$F = kx = kA = ma$$

$$a = \left(\frac{k}{m}\right)A = \frac{100 \text{ N/m}}{1 \text{ kg}}(5 \times 10^{-2} \text{ m}) = 5 \text{ m/s}^2$$

d. Let x be the displacement satisfying this condition.

vertical component of $\mathbf{v}_0 = \frac{1}{2}v_0$

In Fig. 9-20, the vertical component of $\mathbf{v}_0$ is $v_0 \cos\theta$, thus $v_0 \cos\theta = 1/2 v_0$; $\cos\theta = 0.5$; $\theta = 60°$
In the larger triangle,

$$x = A \sin\theta = A \sin 60° = 5 \text{ cm}(0.866) = 4.33 \text{ cm}$$

This displacement may be either above or below the midpoint.

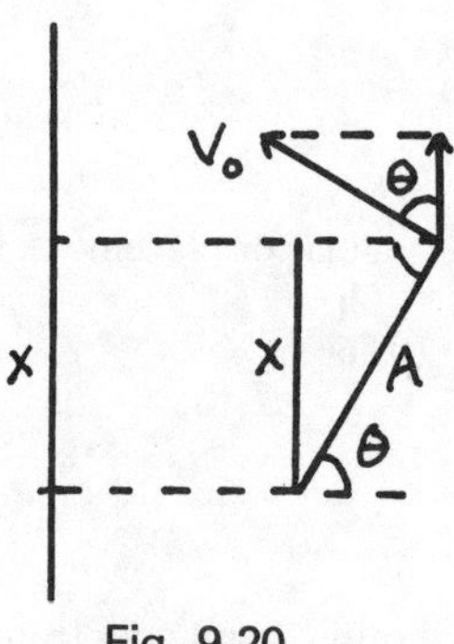

Fig. 9-20

32. On the earth $T = 2\pi\sqrt{\dfrac{L}{g}}$; $f = \dfrac{1}{2\pi}\sqrt{\dfrac{g}{L}}$

On the moon

$$f' = \frac{1}{2\pi}\sqrt{\frac{g/6}{L}} = \frac{1}{\sqrt{6}}\frac{1}{2\pi}\sqrt{\frac{g}{L}} = f/\sqrt{6} = (0.5/\sqrt{6}) \text{ Hz}$$

$$= 0.2 \text{ Hz}$$

33. $T = 2\pi \sqrt{\frac{m}{k}}$ $\quad f = \frac{1}{2\pi}\sqrt{\frac{k}{m}}$

The frequency is not changed because it is independent of g.

34. The mass-spring system. Its frequency is independent of the local acceleration of gravity.

35. The frequency of the pendulum is not changed when the mass is doubled.

$f_{earth} = 0.5 \text{ Hz} \quad f_{moon} = 0.2 \text{ Hz}$

The frequency of the mass-spring system is changed.

$f' = f/\sqrt{2} = 1.2 \text{ Hz}/\sqrt{2} = 0.85 \text{ Hz}$

This is the frequency on earth and on the moon.

REVIEW TEST FOR CHAPTERS 7–9

1. A point moving in a circular path undergoes an angular displacement of 0.2 rad. Its linear displacement is 6 cm. What is the radius of the circular path?

2. Two vectors, **A** and **B**, are shown. Which arrow in Fig. T-24 represents the difference **B** − **A**?

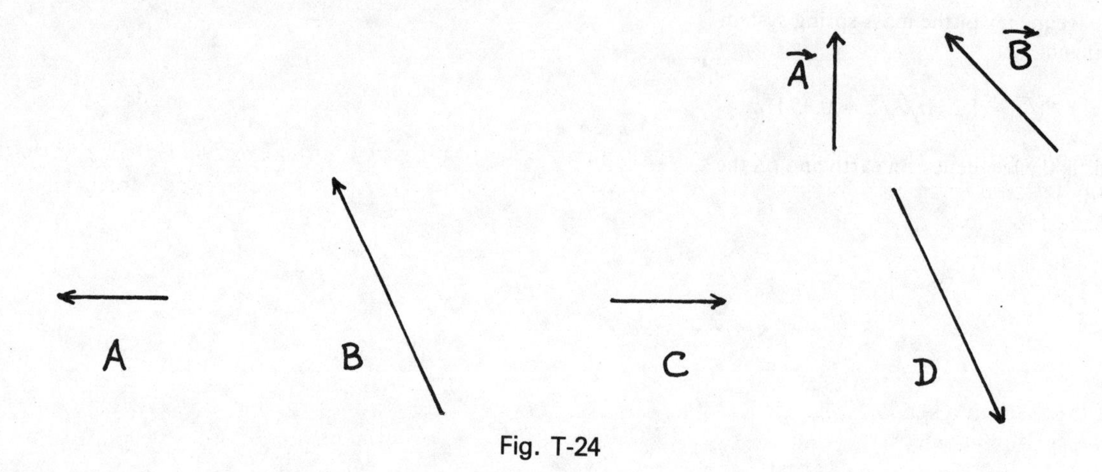

Fig. T-24

3. A small object is moving along a circular path at a constant speed in a counterclockwise direction. Its acceleration may be represented in Fig. T-25 by
   a. arrow **A**
   b. arrow **B**
   c. arrow **C**
   d. arrow **D**
   e. an arrow of zero length

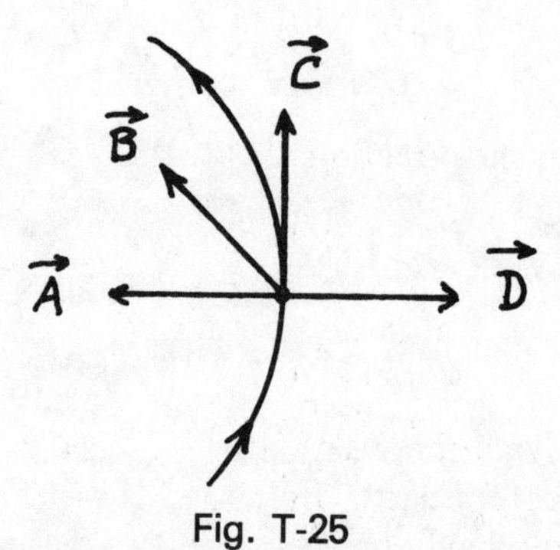

Fig. T-25

4. A small mass is moving along a circular path at a constant speed. Which of the following relationships is *not* applicable to this motion?
   a. $s = v_0t + 1/2\ at^2$
   b. $v = (2\pi R)/T$
   c. net $\mathbf{F} = m\mathbf{a}$
   d. $\omega = v/R$
   e. $a = v^2/R$

5. The moon travels about the earth in a circular path, completing one revolution of its orbit each 27.3 days. Given that the radius of the moon's path is 3.84×10^8 m, calculate the centripetal acceleration of the moon toward the earth.

6. A large flywheel whose moment of inertia is 1.8×10^3 kg·m^2 stores 10^8 joules in the form of KE. What is its angular velocity?

7. During a skating exhibition, a skater is observed to be rotating on the ice with her arms fully extended at an angular speed of $\omega_1 = 0.5$ rev/s. By bringing her arms in close to her body she increases her angular speed to $\omega_2 = 1.5$ rev/s. Calculate the ratio of her moment of inertia I_1 with arms extended to her moment of inertia I_2 with her arms in close.

8. The navigator of a spaceship determines that the distance to a nearby planet is 8.0×10^4 km and that the gravitational force on her ship is 5.0×10^3 N. What will this gravitational force become when the spaceship moves to a distance of 2.0×10^4 km from the planet?

9a. If the mass of a spaceship is 2.5×10^4 kg and it experiences a gravitational force of 5.0×10^3 N due to a nearby planet, what is the magnitude of the gravitational field at the location of the spaceship?

b. What is the direction of this gravitational field?

10. A geosynchronous satellite is placed in orbit above the equator at a distance of $r_1 = 4.23 \times 10^7$ m from the center of the earth. Its period of revolution is the same as the rotational period of the earth, 24 h. What is the period in hours of a satellite placed in a circular orbit at one half this distance ($r_2 = 2.115 \times 10^7$ m)?

11. A steel wire 3.00 m long and 1.00 mm in radius is stretched to a length of 3.005 m when a force of 1000 N is applied. What is the stretch modulus of the steel used?

12. In order for an object to move in simple harmonic motion when it is displaced from its equilibrium position, the object must
    a. experience a gravitational force.
    b. move in a circular path at a constant speed.
    c. experience a restoring force proportional to its displacement.
    d. be suspended like a pendulum.
    e. have a momentum at all times greater than zero.

13. A particle is vibrating with simple harmonic motion. Its amplitude is 8 cm and its maximum speed is 10 cm/s. What is the speed of the particle when its displacement is 8 cm?

14. What is the speed of the above particle when its displacement is only 4 cm, one half the amplitude?

15. A mass suspended from a spring is set into vibration with an amplitude of 6 cm. Twenty vibrations take place in 30 s. What is the speed of the vibrating mass as it passes through the equilibrium position?

16. A simple pendulum has a period of 2 s in a laboratory on earth. What period will the pendulum have on the moon where the gravitational acceleration is 1/6 that on the earth?

CHAPTER 10 Wave Motion

GOALS To study the physical characteristics common to all types of wave motion.

OBJECTIVES After completing this chapter the student should be able to do the following:

1. Name three examples of wave motion and state what disturbance is propagated in each case.
2. Use dimensional analysis to choose the correct physical relationship from a group of "possible" relationships.
3. Write the expression for the velocity of a compression wave in an elastic solid.
4. Given a list of common wave motions, state which are longitudinal, which are transverse, and which allow polarization.
5. Complete the sentence: The motion or disturbance of a single particle of the medium is termed a(n) ____________.
6. Complete the sentence: The interrelated set of vibrations of many particles in a medium is called a(n) ____________.
7. Calculate any one of the quantities v_w, E, or d when the other two are given.
8. State the physical significance of the symbols B, E, d, and v_w.
9. State the type of disturbance involved in the propagation of a sound wave.
10. Given a list of characteristics of a periodic wave, state the technical term and symbol associated with each.
11. Calculate any one of the quantities v_w, T, f, or λ when a sufficient number of the others are given.
12. Write a statement of the superposition principle as applied to the resultant of two disturbances at a particular point in space.
13. Calculate the beat frequency for two waves of known frequencies.
14. Calculate the frequency of a sound wave from a moving source as heard by a stationary observer.
15. Carry out a similar calculation for the case of stationary source and moving observer.

SUMMARY

Wave motion is one of the two important methods by which energy is transported from one point to another. The energy that the earth receives from the sun has been carried some 93 million miles through space by electromagnetic waves. The energy of a sonic boom, often great enough to break windows several kilometers from the aircraft that produced it, is carried through the atmosphere by sound waves. A wave may be defined as the *propagation of a disturbance*. In the case of radiation from the sun, the "disturbance" being propagated is a fluctuation in the electric and magnetic field. The "disturbance" propagated by a sound wave is a change in the air pressure or, equivalently, a fluctuation in the air density.

Compression waves may be propagated through solids, liquids, and gases. The speed of propagation in a solid depends on the elastic modulus and the density, the exact relationship being $v_w = \sqrt{E/d}$ where E is the stretch modulus of the material and d is its density.

Longitudinal waves are those for which the disturbance, or displacement, is parallel to the direction of propagation. *Transverse* waves have displacements that are perpendicular to the direction of propagation. Transverse waves may be polarized; longitudinal waves cannot.

The motion of a single particle of the medium is called a *vibration*. The interrelated set of vibrations of many particles is a *wave*. A wave may be propagated over long distances; however, the individual particles involved in the propagation vibrate over relatively small distances. The distance that a wave travels during one period is the *wavelength*. The wave velocity is equal to the product of the wavelength and the frequency.

The superposition theorem states that the resultant disturbance at a point is the algebraic sum of the disturbances due to each wave motion present.

The frequency of a wave motion perceived by an observer will differ from the frequency emitted by the source if there is relative motion between source and observer. This phenomenon is called the *Doppler effect*. If source and observer are moving toward each other, no matter which is moving, the apparent frequency will be greater than the frequency emitted.

QUESTIONS AND PROBLEMS

Secs. 10-1—10-3

1. A simple pendulum has a mass m, a length L, and is situated in a gravitational field g. Let us use the technique of dimensional analysis (Sec. 10-3) to discover how the period must depend on these quantities. We begin with the equation

$$T = km^x L^y g^z$$

where k is a dimensionless constant and x, y, and z are unknown exponents. Find three equations involving x, y, and z; solve them and obtain an equation for the period.

2. A railroad switchman with his ear to the rails hears the sound of a derailment. Two seconds later the sound of the accident reaches him by propagation through the air. At what distance did the derailment occur? (Hint: Let v_1 and v_2 be the velocities of sound in steel and air, respectively. The distance to the accident may be written $d = v_1t_1 = v_2t_2$ where t_1 and t_2 are the two times of transmission. Now write Δt for the difference in the two times. $\Delta t = t_2 - t_1$. Obtain an equation for the distance d in terms of the two known velocities and the time interval Δt, also known.)

   Check your equation before proceeding with the numerical calculation. Use 5.06×10^3 m/s and 340 m/s for the velocities in steel and air, respectively.

3. A metal rod 80 cm long is tapped lightly at one end. A small microphone detects the compressional pulse at the other end of the rod 0.24 ms later. The stretch modulus of the metal is 10×10^{10} N/m^2.
   a. Calculate the velocity of the compressional pulse in the rod.

   b. Use the result of (a) with the known modulus of elasticity to calculate the density of the metal.

4. The sound of a whistle at A is detected by a microphone placed at B. If the microphone is rotated about the line AB, will the intensity of the detected signal change? Explain. How will the detected signal change as the whistle is rotated about AB?

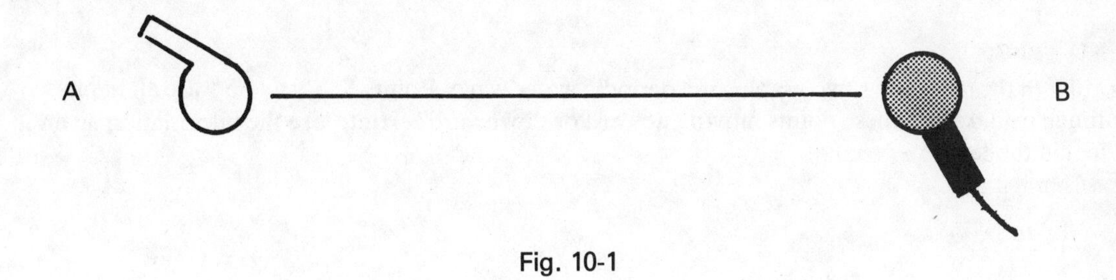

Fig. 10-1

5. A compressional wave having a frequency of 440 Hz is propagated in a copper rod at a speed of 3.33×10^3 m/s.
   a. Calculate the period of this wave.

   b. Calculate the wavelength in copper of a 440 Hz sound wave.

   c. Calculate the wavelength in air of a 440 Hz sound wave. Use $v_w = 310$ m/s.

6. Two waves of equal frequency, wavelength, and velocity, and of amplitudes 2 and 4 (units unspecified), are traveling in the same direction through a medium.
   a. What is the maximum amplitude that the resultant wave could have?

   b. What condition would give rise to the resultant wave having this maximum amplitude?

   c. What is the minimum amplitude that the resultant wave could have?

7. Fig. 10-7 in the text shows two graphs of a periodic water wave. Points *S, S′*, and *S″* are all in phase. At the time $t = 3.0$ s, are these points moving upward or downward? (Hint: Use the information given in Fig. 10-7(b) in the text.)

8. Examine text Fig. 10-7, both (a) and (b), to determine which way (toward the right or the left) the wave is traveling. Explain your choice.

9. What is the wave velocity of the water wave pictured in text Fig. 10-7?

10. Calculate the speed of sound in m/s and in ft/s in pure water. (Consult Table 9-1 in the text for needed information.)

11. Calculate the distance between rarefactions in a sound wave whose velocity is 330 m/s and whose frequency is 220 Hz.

12. When the E-string of an old piano is sounded simultaneously with a tuning fork of frequency 660 Hz, a beat frequency of 2 Hz is heard. What are the two possible frequencies of the E-string?

13. The ultrasonic generator of a cleaning bath vibrates at a frequency of 40 kHz. Calculate the wavelength in air of this ultrasonic wave. (Use $v_w = 340$ m/s.)

14. A siren on a police car emits a sound whose most intense component has a frequency of 2000 Hz. The car is approaching a stationary observer at a speed of 100 km/h.
    a. What frequency will the observer hear as the police car approaches her?

    b. What frequency will the observer hear after the police car has passed her and is moving away at 100 km/h?

15. Hydrogen atoms in the laboratory emit an electromagnetic wave whose frequency is 4.57×10^{14} Hz. Light from a certain star, known to have hydrogen in its atmosphere, contains an emission line at 4.53×10^{14} Hz.
    a. What is the relative velocity between this star and the earth?

    b. Is the star approaching or receding?

Solutions

1. $T = km^x L^y g^z$

The first step is to write the corresponding dimensional equation in which each physical quantity, T, m, L, and g, is replaced by its dimensional expression.

$T \rightarrow [T] \quad m \rightarrow [M] \quad L \rightarrow [L]$
$g \rightarrow [LT^{-2}] \quad k$ is dimensionless
$[T] = [M]^x[L]^y[LT^{-2}]^z$
$[T]^1[M]^0[L]^0 = [M]^x[L]^{y+z}[T]^{-2z}$

Since the exponents must be the same on both sides of the equation,

$x = 1, \; y + z = 0, \; -2z = 1$

Solving these equations simultaneously,

$x = 1, \qquad z = -\tfrac{1}{2}, \qquad y = +\tfrac{1}{2}$

Thus the equation for the period must be

$$T = k\sqrt{\frac{L}{g}}$$

Dimensional analysis does not determine the value of the dimensionless constant k. It was shown in Sec. 9-5 that $k = 2\pi$.

2. Since $d = v_1 t_1$ and $d = v_2 t_2$, the difference in the two times of transmission may be written

$$\Delta t = \frac{d}{v_2} - \frac{d}{v_1} = d\left(\frac{1}{v_2} - \frac{1}{v_1}\right)$$

or

$$d = \Delta t \bigg/ \left(\frac{1}{v_2} - \frac{1}{v_1}\right) = \frac{\Delta t v_1 v_2}{v_1 - v_2}$$

Substituting the given values

$$d = \frac{(2.0\text{ s})(5.06 \times 10^3)(340)\text{ m}^2/\text{s}^2}{(5.06 \times 10^3 - 0.34 \times 10^3)\text{ m/s}}$$

$$d = 729\text{ m}$$

3a. $$v = \frac{s}{t} = \frac{0.80\text{ m}}{0.24 \times 10^{-3}\text{ s}} = 3.3 \times 10^3\text{ m/s}$$

b. The wave velocity is given by

$$v_w = \sqrt{E/d}$$

Thus

$$d = \frac{E}{v_w^2} = \frac{10 \times 10^{10}\text{ N/m}^2}{(3.3 \times 10^3)^2\text{ m}^2/\text{s}^2}$$

$$d = 9.2 \times 10^3\text{ kg/m}^3$$

4. Sound waves are longitudinal waves. There is no asymmetry about the line of propagation. Thus the detected signal will not change if the microphone is rotated about the line AB, the line of propagation. Rotation of the source about the line of symmetry has no effect for the same reason.

5. $v_w = f\lambda$ or $v_w = \lambda/T$

a. $T = \frac{1}{f} = \frac{1}{440\ s^{-1}} = 2.27 \times 10^{-3}$ s

b. $\lambda = \frac{v_w}{f} = \frac{3.33 \times 10^3 \text{ m/s}}{440\ s^{-1}} = 7.57$ m

c. $v = \frac{v_w \text{ (in air)}}{f} = \frac{310 \text{ m/s}}{440\ s^{-1}} = 0.704$ m

6a. The maximum amplitude is 6 units.

b. It will be obtained when the two waves have the same phase.

c. The minimum amplitude is 2 units. It will occur when the two waves have a phase difference of 180° or π rad.

7. The motion of the point S' is shown in Fig. 10-7(b) in the text. At $t = 3$ s, point S' has a displacement of zero. One second later ($t = 4$ s), the displacement, read from graph (b), is +50 cm. Thus point S' is moving upward. Since points S and S'' are in phase with S', they are also moving upward at $t = 3$ s.

8. If the wave shown in graph (a) moves toward the right, point S' will move upward as it should. Thus the wave is certainly traveling to the right.

9. $v_w = \frac{\lambda}{T} = \frac{24 \text{ m}}{4 \text{ s}} = 6$ m/s

The wavelength is obtained from text Fig. 10-7(a) and the period is obtained from text Fig. 10-7(b).

10. The velocity of a compression wave in a liquid is given by

$$v_w = \sqrt{\frac{B}{d}}$$

Thus

$$v_w = \sqrt{\frac{0.20 \times 10^{10} \text{ N/m}^2}{1.0 \times 10^3 \text{ kg/m}^3}} = 1.41 \times 10^3 \text{ m/s}$$

$$(1.41 \times 10^3 \text{ m/s})\left(\frac{1 \text{ ft}}{0.305 \text{ m}}\right) = 4.62 \times 10^3 \text{ ft/s}$$

11. The distance between rarefactions is equal to the wavelength.

$$\lambda = \frac{v_w}{f} = \frac{330 \text{ m/s}}{220\ s^{-1}} = 1.50 \text{ m}$$

12. 658 Hz and 662 Hz

13. $\lambda = \frac{v_w}{f} = \frac{340 \text{ m/s}}{40 \times 10^3\ s^{-1}} = 8.5 \times 10^{-3}$ m

14. Since the source is moving, we use the equation

$$f' = \frac{f}{1 \pm v/v_w}$$

a. Since the observed frequency must be greater than the emitted frequency when the source is approaching, the minus sign must be used to make f' greater than f.

$$f' = \frac{2000\ \text{Hz}}{1 - (1 \times 10^5\ \text{m/s})(1\ \text{h}/3600\ \text{s})/(340\ \text{m/s})} = 2.2 \times 10^3\ \text{Hz}$$

b. The source is still moving but it is now receding from the listener. Thus the plus sign must be used.

$$f' = \frac{2000\ \text{Hz}}{1 + (1 \times 10^5\ \text{m/s})(1\ \text{h}/3600\ \text{s})/(340\ \text{m/s})} = 1.85 \times 10^3\ \text{Hz}$$

15. The relationship between frequency shift Δf and relative velocity v is

$$\frac{\Delta f}{f} = \frac{v}{v_w}$$

a. Thus the relative velocity is

$$v = v_w \frac{\Delta f}{f} = (3 \times 10^8\ \text{m/s}) \frac{(0.04 \times 10^{14}\ \text{s}^{-1})}{(4.5 \times 10^{14}\ \text{s}^{-1})}$$

$$v = 2.7 \times 10^6\ \text{m/s}$$

b. Since the observed frequency is less than the emitted frequency, the source must be receding.

CHAPTER 11

Interference and Standing Waves

GOALS To study the various effects that arise when two or more wave motions are superimposed, especially standing waves of various types, and to learn some physical characteristics of musical sounds.

OBJECTIVES After completing this chapter the student should be able to do the following:

1. Express the distance between nodes in a standing wave in terms of wavelength.
2. State the physical significance of the symbols that appear in the formula

$$v_w = \sqrt{\frac{F}{m/L}}$$

3. Describe at least three different modes of vibration of a stretched string fixed at each end.
4. For each mode of vibration listed above describe the number and location of the nodes and the relationship between wavelength and length of string.
5. Choose tension and mass per unit length such that a stretched wire will vibrate with a specified frequency.
6. Describe the number and location of nodes and antinodes for at least three simple modes of vibration of the air in a pipe open at one or both ends.
7. Describe the conditions that permit the building up of large vibrations by means of repeated applications of small impulses.
8. Name the three psychological sensations that characterize musical sounds and the physical attribute associated with each of them.
9. State the ratio of the frequencies of two sounds that are one octave apart.
10. Name the technical term for the rate of flow of energy in a wave, power per unit cross-sectional area.
11. Calculate the difference in intensity levels between two sounds whose intensity ratio is known.
12. Name the three major divisions of the human ear.
13. Write the technical term for the membrane that separates the outer ear from the middle ear.
14. Name the two principal components of a musical instrument.

SUMMARY

Interference is the general term for the superposition of two or more wave motions in the same region of space. Since the amplitudes of the individual waves add *algebraically*, it is possible for the resultant amplitude at a point to be zero—*destructive interference*—or very large—*constructive interference*.

Of special interest is the superposition of two waves having the same frequency and amplitude but moving in opposite directions, say the positive x direction and the negative x direction. The resultant disturbance along the

x axis moves neither to the right nor to the left; it is called a *standing* or *stationary* wave. There will be regions at which no motion takes place—the motional nodes—and other regions at which the amplitude of vibration is a maximum—the motional antinodes. The distance between adjacent nodes is one-half the wavelength.

Standing waves can be produced in a stretched string (or wire) provided the frequency of the two oppositely traveling waves is one of the *natural* frequencies of the string. Since the two fixed ends of the string are necessarily motion nodes, the greatest wavelength allowed is $2L$, twice the length of the string. The corresponding frequency, $f = v_w/(2L)$ is the lowest frequency or *fundamental* frequency of the stretched string. The other allowed frequencies, called *overtones*, are multiples of the fundamental frequency.

Stationary waves may also exist in air columns confined to pipes. The closed end of a pipe is a motion node; the open end is an antinode. The fundamental frequency of a pipe closed at one end and open at the other is $f = v_w/(4L)$. The other allowed frequencies for the closed pipe are *odd* multiples of the fundamental frequency; even multiples are not possible.

A system may be set into vibration by a succession of small impulses. If the frequency of the impulses corresponds to one of the natural frequencies of the system, very large vibrations may be built up, a phenomenon called *resonance*.

Musical sounds are characterized by *pitch, loudness,* and *quality*. The physical attributes that lead to these three psychological sensations are *frequency, intensity,* and *overtone structure.*

The range of frequencies to which the human ear is sensitive varies from one individual to another. For a young person with normal hearing the audible range extends from about 20 Hz to 20 kHz. The term *ultrasonic* refers to compression waves of frequencies greater than 20 kHz.

The *intensity level* of a sound wave is measured in *bels* or, more commonly, in *decibels* (1 dB = 1/10 bel). The difference in the intensity levels of two sounds (in dBs) of intensities I_1 and I_2 is given by $\Delta\beta = 10 \log(I_2/I_1)$. Two intensities differing by a factor of 100 have intensity levels that differ by 20 dB.

The human ear is an extremely sensitive detector of small pressure fluctuations (sound waves) especially so in the mid-frequency range from about 1 kHz to 10 kHz. Under ideal conditions the human ear can detect pressure fluctuations of only 1 part in 5 billion. The ear has three distinct regions, the *outer ear*, the *middle ear*, and the *inner ear*. The eardrum, or tympanum, separates the outer ear from the middle ear. Three small bones in the middle ear transmit movements of the eardrum to the oval window and thence to the inner ear.

Musical instruments consist of a *generator* of sound, such as a vibrating air column or string, and a *resonator*, which selects and accentuates certain of the overtones. The distinctive quality of a particular musical instrument depends on the relative intensities of the various overtones.

QUESTIONS AND PROBLEMS

1. When two identical waves are traveling through a medium in opposite directions the resulting wave pattern is called a(n) ______________.
2. When a stationary wave is set up in a string there are narrow regions in which the displacement remains zero. These regions are called ______________.
3. What is the name for those regions in which the amplitude of the disturbance is a maximum?

In the remainder of this chapter we will use the term "node" to mean "motional node" and "antinode" to mean "motional antinode."

4. A string is tightly stretched between fixed points P and Q. Stationary waves are set up on the string. What is the minimum number of nodes? Show their location in Fig. 11-1.

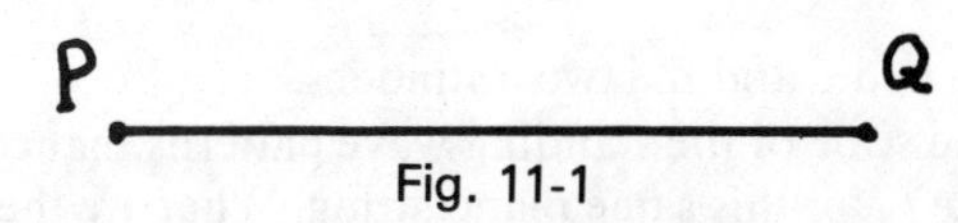

Fig. 11-1

5. Fig. 11-1 in the text shows how two identical waves traveling in opposite directions give rise to standing waves. At the top of the drawing the two waves are easily distinguished. Each wave has a wavelength of 2.5 cm. At the bottom of the figure a time exposure of the resulting stationary wave is shown. What is the distance between adjacent nodes? How many centimeters? What fraction of a wavelength?

6. Write the equation for the velocity of propagation of a transverse wave along a wire. Identify the symbols F, m, and L that appear in your equation.

7. In Example 11-1 in the text the time for a transverse pulse to travel a certain distance is calculated. The result is 0.4 s. What would this time become if the tension in the rope were increased from 400 N to 1600 N? No detailed calculation is required.

8. A guitar string 0.5 m long has a mass of 0.0125 kg and is under tension of 4×10^3 N. What is the frequency of the fundamental and the first overtone. (A similar problem is solved in Example 11-2 in the text.)

9. Fig. 11-2 shows a picture of a piano wire vibrating in its first overtone. The frequency of vibration is $f_2 = 880$ Hz.

Fig. 11-2

Mark the positions of the three nodes and the two antinodes.
Now draw a picture to the same scale of the standing wave patterns that correspond to the fundamental tone f_1 and the second overtone f_3 for this same piano string. What are these frequencies?

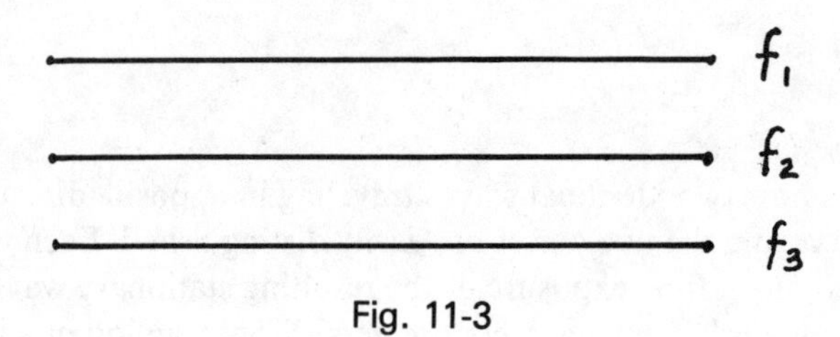

Fig. 11-3

f_1 = ____________. f_2 = ____________. f_3 = ____________.

10. If the total length of the wire is L, what is the wavelength for each of these three cases in Prob. 9? (Express λ in terms of L.)

fundamental λ_1 = ____________.

1st overtone λ_2 = ____________.

2nd overtone λ_3 = ____________.

11. Suppose that the piano wire in Prob. 9 has a length of 1 m and a mass of 0.5 g. What tension in this wire will permit it to produce a tone of 440 Hz when vibrating in its fundamental node? (Hint: First determine the wave velocity. A similar problem is solved in Example 11-3 in the text.)

12. A transverse wave is traveling

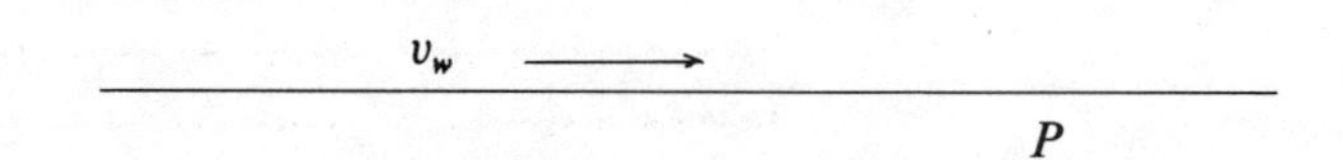

to the right along this string as indicated by the arrow marked v_w. A small section of the string is indicated at P. Put a double arrow ↔ near P to indicate the way in which this part of the string will move as the wave passes it.

13. Fig. 11-4 shows a sound wave produced by a loud speaker near one end of a hollow tube. The sound wave

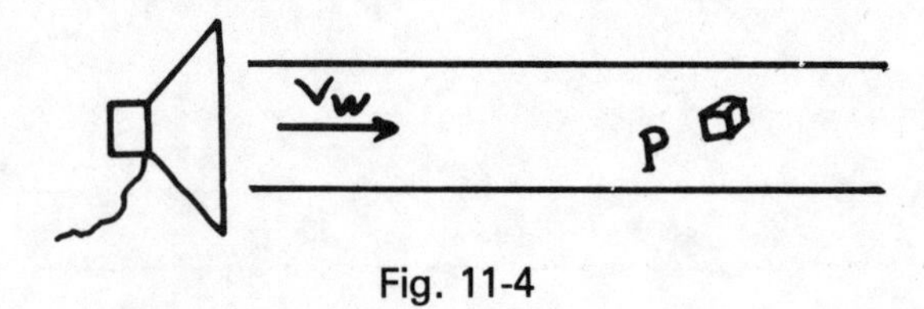

Fig. 11-4

is traveling to the right as indicated by the arrow marked v_w. A small volume of gas is indicated at P. Put a double arrow ↔ near P to indicate the motion imparted to the gas molecules as a result of the passage of the sound wave. This type of wave motion is called ____________ in contrast to the transverse waves generated in a string.

14. When standing waves are set up in a pipe closed at one end, the air molecules are free to move at the open end but are not free to move (longitudinally) at the closed end. Thus the closed end must be

_______________ and there must be _______________ at or near the open end.

As we have seen, the waves propagated in a gas or liquid are longitudinal in that the molecules, in transmitting the wave, move back and forth along a line parallel to the direction of propagation. Such a compression wave may be represented graphically by plotting the displacement—a pressure fluctuation in this case—as a function of distance.

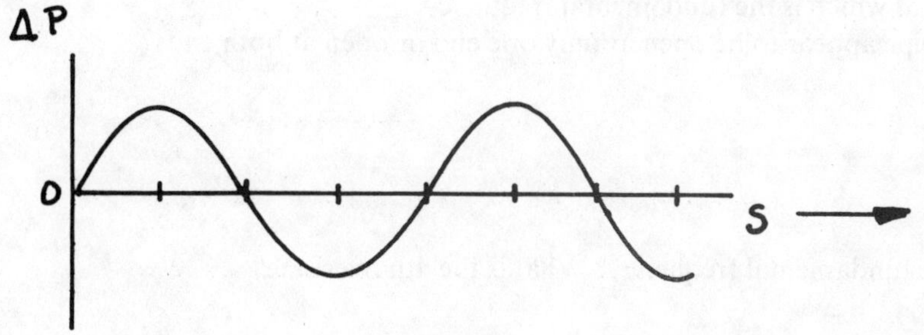

ΔP

O

S

Fig. 11-5

Standing waves may be pictured in like fashion. For example, a possible mode of vibration in a pipe closed at one end is illustrated in Fig. 11-6.

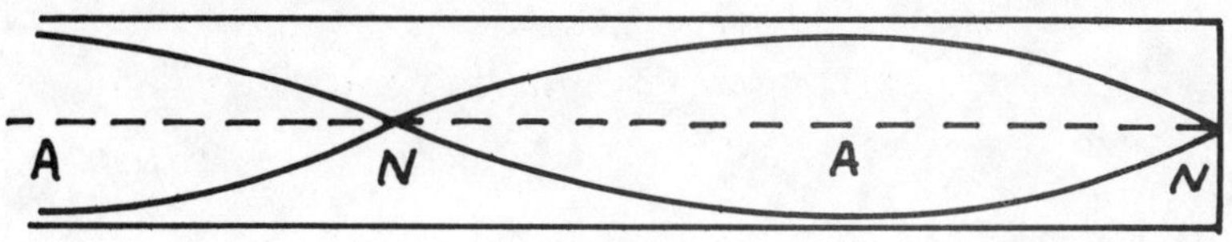

A N A N

Fig. 11-6

15. Make a sketch of a standing wave of lowest possible frequency in a pipe closed at one end.

L

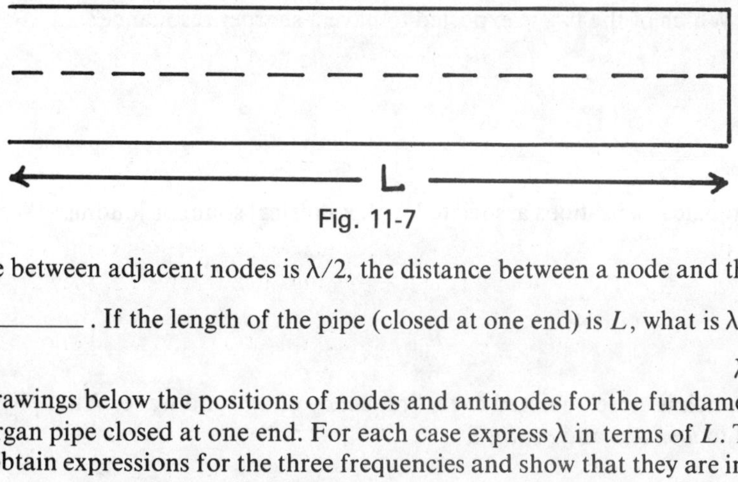

Fig. 11-7

Since the distance between adjacent nodes is $\lambda/2$, the distance between a node and the nearest antinode must be _______________. If the length of the pipe (closed at one end) is L, what is λ in terms of L?

$\lambda =$ _______________.

16. Indicate on the drawings below the positions of nodes and antinodes for the fundamental and first two overtones in an organ pipe closed at one end. For each case express λ in terms of L. Then, using the relation $f = v/\lambda$, obtain expressions for the three frequencies and show that they are in the ratio 1:3:5.

Fundamental  $\lambda_1 =$ _______________.

1st overtone 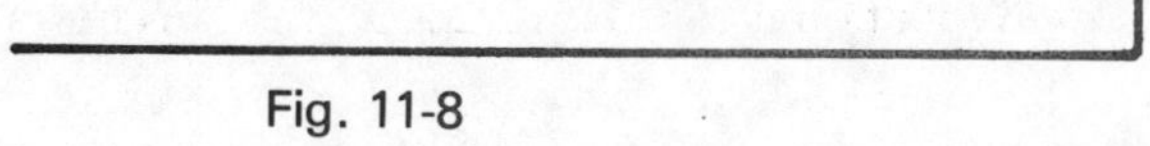 $\lambda_2 =$ _______________.

2nd overtone $\lambda_3 =$ _______________.

Fig. 11-8

17. The frequency of the 1st overtone of a pipe closed at one end is 600 Hz. What is the frequency of the 2nd overtone?

18. Among the natural frequencies of an air column in a pipe are the values 440 Hz, 660 Hz, 880 Hz, and 1100 Hz, none of which is the fundamental frequency.
    a. Does this pipe appear to be open at only one end or open at both ends?

    b. What is the fundamental frequency? What is the 4th overtone?

    c. Assuming a wave velocity of 340 m/s, how long is the air column?

19. The building up of a large vibration by repeated applications of small impulses whose frequency equals one of the natural frequencies of the resonating body is called ______________ .

20. Two guitar strings are plucked at the same time. One of them continues to vibrate for a short time after the other has ceased. Which of the two is expected to have a sharper resonance?

21. One of the psychological sensations associated with a musical sound is loudness. What are the other two?

22. The sensation of pitch is closely related to the frequency of a sound wave. What physical attributes of sound are related to *loudness* and *quality*?

23. Which of the following frequencies are one octave apart? 220 Hz, 300 Hz, 400 Hz, 440 Hz, 500 Hz, 600 Hz, 660 Hz.

24. The power output of a certain high-fidelity amplifier is 20 dB less at 22 kHz than at 10 kHz. What is the ratio of the output power at these two frequencies?

25. List the three principal parts of the human ear. Which parts are separated by the tympanum?

Solutions

1. standing wave
2. motional nodes
3. motional antinodes
4. There must be at least two nodes, one at each end.
5. 1.25 cm; ½ wavelength
6. $$v_w = \sqrt{\frac{F}{m/L}}$$

   F = tension; m = mass of string or wire; L = length of wire.
7. $v_w \propto \sqrt{F}$. If F is multiplied by 4, v_w is multiplied by 2. Thus the time to travel the distance is reduced from 0.4 s to 0.2 s.
8. First find the wave velocity

   $$v_w = \sqrt{\frac{F}{m/L}} = \sqrt{\frac{(4 \times 10^3 \text{ N})(0.5 \text{ m})}{(1.25 \times 10^{-2} \text{ kg})}} = 4.0 \times 10^2 \text{ m/s}$$

   The frequency of the fundamental is

   $$f_1 = \frac{v_w}{\lambda_1} = \frac{v_w}{2L} = \frac{3.1 \times 10^2 \text{ m/s}}{2(0.5 \text{ m})} = 3.1 \times 10^2 \text{ Hz}$$

   The frequency of the first overtone is just twice that of the fundamental.

   $$f_2 = \frac{v_w}{\lambda_2} = \frac{v_w}{L} = 6.2 \times 10^2 \text{ Hz}$$

9. fundamental: The three frequencies of vibration are:

   $\lambda_1 = 2L$ $\qquad f_1 = \frac{v}{\lambda_1} = \frac{v}{2L}$

   1st overtone:

   $\lambda_2 = L$ $\qquad f_2 = \frac{v}{\lambda_2} = 2\frac{v}{2L}$

   2nd overtone: $\qquad f_3 = \frac{v}{\lambda_3} = 3\frac{v}{2L}$

   $\lambda_3 = \frac{2L}{3}$

   Since f_2 = 880 Hz, f_1 = 440 Hz and f_3 = 1320 Hz.

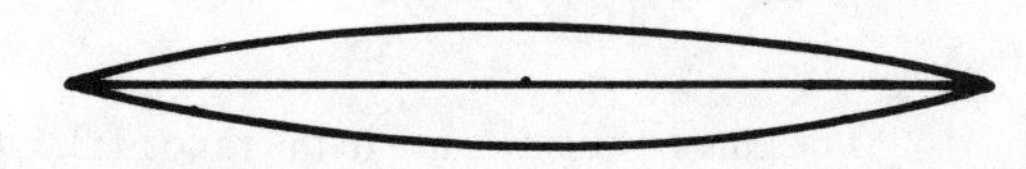

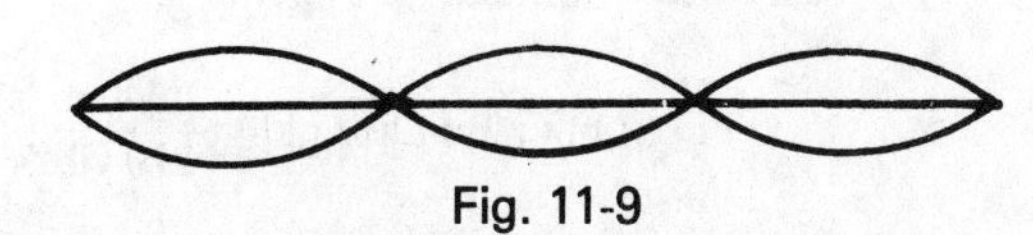

Fig. 11-9

10. $\lambda_1 = 2L$; $\lambda_2 = L$; $\lambda_3 = 2L/3$
11. $v = f_1\lambda_1 = (440 \text{ vib/s})(2)(1 \text{ m}) = 800 \text{ m/s}$

$$v_w = \sqrt{\frac{F}{m/L}}; \qquad F = v_w^2(m/L)$$

$$F = (8.8 \times 10^2 \text{ m/s})^2(0.5 \times 10^{-3} \text{ kg/m}) = 3.9 \times 10^2 \text{ N}$$

12.

Fig. 11-10

13.

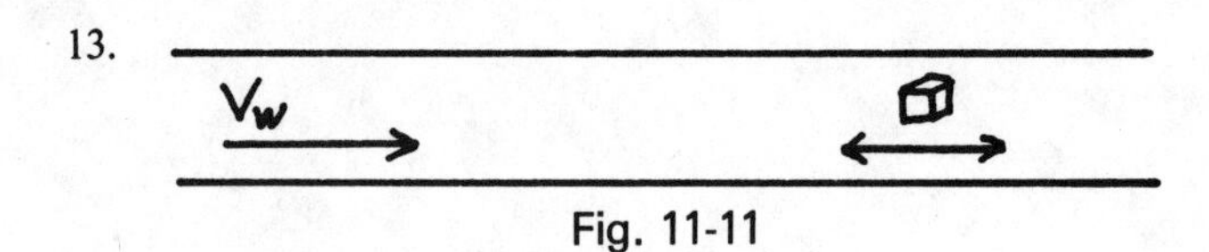

Fig. 11-11

longitudinal

14. a node; an antinode
15.

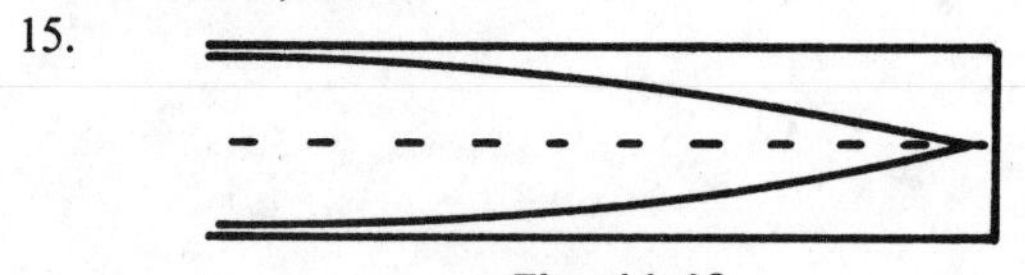

Fig. 11-12

one quarter of a wavelength
$\lambda = 4L$

16.

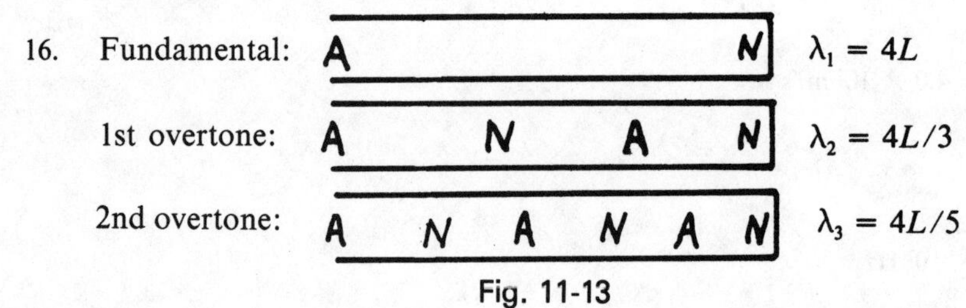

Fig. 11-13

$$f_1 = \frac{v}{\lambda_1} = \frac{v}{4L}; f_2 = \frac{v}{\lambda_2} = \frac{v}{4L} \times 3 = 3f_1$$

$$f_3 = \frac{v}{\lambda_3} = \frac{v}{4L} \times 5 = 5f_1$$

Thus the frequencies are in the ratios 1:3:5. Only the odd harmonics are present.

17. $f_1 = 200$ Hz; $f_3 = 1000$ Hz

18a. The given frequencies are ... 440, 660, 880, 1100, ... Hz. They are in the ratios ... 2:3:4:5 ...
Since the even harmonics are present, this cannot be a pipe closed at one end; it is therefore a pipe open at both ends.

b. The fundamental frequency is 220 Hz. The 4th overtone is 1100 Hz.

c. Consider the fundamental mode of vibration.
We have $\lambda_1 = 2L$ and $f_1 = v/\lambda_1$.

$$L = \lambda_1/2 = 2(v/f_1) = \frac{2(340 \text{ m/s})}{220 \text{ vib/s}} = 0.77 \text{ m}$$

19. resonance
20. The string that vibrates longer has a sharper resonance.
21. pitch; quality
22. intensity; overtone structure
23. 220 Hz and 440 Hz; 300 Hz and 600 Hz
24. $\Delta\beta = 10 \log (P_2/P_1)$

    $20 = 10 \log (P_2/P_1)$

    Thus

    $P_2/P_1 = 10^2$

25. The three parts are the outer ear, the middle ear, and the inner ear. The tympanum (eardrum) separates the outer ear from the middle ear.

CHAPTER 12

Fluids

GOALS To study the physical characteristics of fluids at rest and in motion.

OBJECTIVES After completing this chapter the student should be able to do the following:

1. State the relationship between the cross-sectional area of an object, the force exerted on it, and the resulting pressure.
2. State the proper units for expressing pressure in the various systems of units used in this course.
3. Calculate the pressure exerted on an object when the force and cross-sectional area are given.
4. Calculate the difference in pressure between two points in a fluid that is at rest.
5. Write a correct statement of Pascal's principle.
6. Solve problems that involve applying a force on a fluid at one point and having the fluid exert a force on an object at some other point.
7. State the relationship between gauge pressure and absolute pressure.
8. State Archimedes' principle and describe the physical cause of the buoyant force supplied by a fluid.
9. Solve for the volume, density, or mass of an object immersed in a fluid when given sufficient information to apply Archimedes' principle.
10. Calculate the specific gravity of an object or of a liquid in which it is immersed from a knowledge of weight and buoyant force.
11. State a correct definition of surface tension.
12. Calculate any of the quantities, h, γ, or r when the others are given.
13. State the conditions that must exist in a fluid for Bernoulli's equation to correctly describe the motion of the fluid.
14. Calculate the volume flowing through a tube in a given time from a knowledge of the speed and cross section.
15. Use the conservation of volume to calculate the velocity in a fluid flowing through a tube when the cross-sectional area is known.
16. Apply Bernoulli's equation to a variety of situations that are similar to the examples in the text.

SUMMARY

A *fluid* is any substance that has no rigidity; liquids and gases are fluids. A *liquid* is a fluid with a well-defined surface and a volume that changes only slightly with external pressure. A *gas* is a fluid that can occupy any volume.

Pressure is force per unit area. Its SI units are the *pascal* (*Pa*) or the combination N/m^2. The pressure at any point in a liquid at rest is the pressure on the surface of the liquid plus an additional pressure proportional to the vertical distance from the surface to the point in question. The equation is $P = P_0 + dgh$ where d is the density (assumed to be constant) and h is the vertical distance from the surface.

Pascal's principle states that a change of pressure exerted at any point in a confined fluid is transmitted undiminished to all points in the fluid. This is the physical principle underlying the hydraulic press, a fluid analogue of the lever, which permits a virtually unlimited multiplication of force.

The general term for a pressure-measuring device is *manometer*. If the pressure being measured is that of the atmosphere, the device is called a *barometer*.

Archimedes' principle states that the apparent loss of weight of a body, wholly or partially immersed in a fluid, is equal to the weight of the fluid displaced by the body. The apparent loss of weight may be ascribed to an upward force, the *buoyant force* exerted by the fluid. The solution of problems involving Archimedes' principle is often simplified by using an alternate definition of specific gravity,

$$\text{sp. gr.} = \frac{\text{weight of a substance}}{\text{weight of equal volume of } H_2O}$$

A liquid has a definite surface because the molecules have cohesive forces. As a result of these intermolecular forces, a liquid left to itself will assume a shape that has the least surface area for a given volume. The *coefficient of surface tension* of a liquid is the *PE* per unit surface area or, equivalently, the contractile force per unit length. Symbolically, $\gamma = F/L$.

A fluid is said to be in *streamline flow* when the molecules move from point to point without *any* rotational motion or turbulence. For any fluid in steady, streamline flow through a tube there is conservation of mass; the mass of fluid crossing any point during a time Δt is equal to the mass of fluid crossing any other point during that time interval. If the fluid is incompressible, there is also conservation of volume; the volume of fluid crossing any point during Δt is equal to the volume crossing any other point during the same Δt. Symbolically, $\Delta V_1/\Delta t_1 = \Delta V_2/\Delta t_2$ or $v_1A_1 = v_2A_2$.

Bernoulli's principle states that if an incompressible fluid is in streamline flow, then the quantity $(\frac{1}{2}dv^2 + hdg + P)$ is constant at every point in the fluid.

Viscosity may be described as the internal molecular friction in a fluid. Viscosity limits the flow of a liquid through a pipe and the velocity of an object moving through a fluid. At low speeds, where the flow is streamline, the viscous drag is proportional to speed. At higher speeds, as the flow becomes turbulent, the viscous drag increases as the square or cube of the speed.

QUESTIONS AND PROBLEMS

1. Which of the following units could not be used for expressing pressure: newton/m², joule/in.², lb/ft², kg/s², atmospheres, ft-lb/s², kg/m-s², g-cm/s²?

2. Given that a 20-N force is exerted on a piston that is 16 cm in diameter, what is the pressure on the piston?

3. Why is it incorrect to apply the formula $P = hdg$ to calculate the pressure difference between two points in the earth's atmosphere? (See text Ex. 12-3.)

4. Given a tank of arbitrary shape filled with a liquid of density d, and with a pressure P_0 at a point M,
   a. What is the pressure at a point M' that is a vertical distance h above M?

   b. What is the pressure at a point M'' that is a vertical distance h below M?

5. A U-tube initially is partially filled with water. Another liquid, which does not mix with water, is poured in side 1 until the two levels at D and C differ by an amount d (see Fig. 12-1). The top of the water on side 1 is a height ℓ below the level of side 2. What is the ratio of the density of the unknown liquid to that of the water?

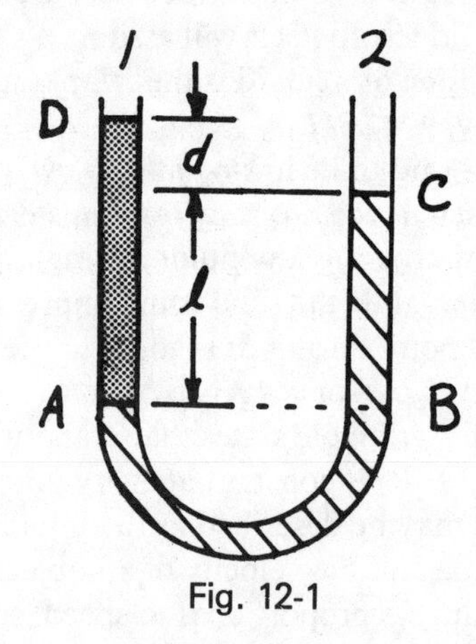

Fig. 12-1

6. Find the pressure in N/m^2 at a point 200 m below the surface of the ocean. Assume the ocean is a still body of water and the density of sea water is $1.03\ g/cm^3$.

7. Fig. 12-3 in the text is a schematic diagram of a hydraulic lift. Calculate the force F required to balance an automobile weighing 16,000 N if the diameters of the two pistons are 30 cm and 5 cm.

8. Using the data of Prob. 7 above, calculate the distance through which the smaller piston must move to lift the car 2 cm.

9. The small cylinder in Fig. 12-2 has a diameter of 2 cm and is connected to a large cylinder with a diameter of 12 cm. Each cylinder contains a frictionless piston. A force of 400 N is applied to the smaller cylinder. What is the total force exerted by the fluid on the larger piston?

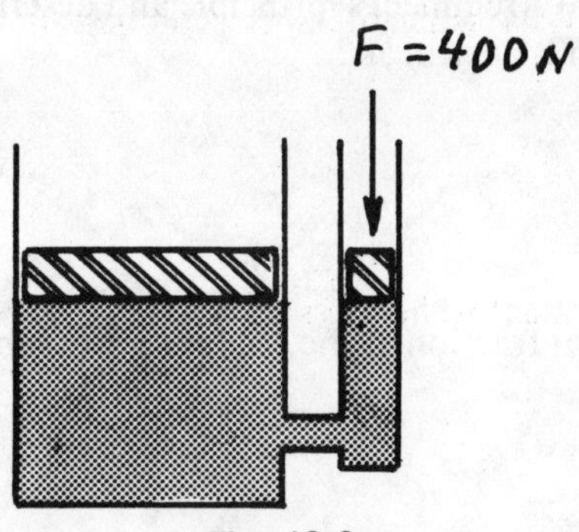

Fig. 12-2

10a. The manufacturer of a certain steel-belted radial tire recommends an operating pressure of 2.6×10^5 N/m^2 (38 lb/in^2). Is this an absolute pressure or gauge pressure?

b. Suppose that your tire is completely flat as a result of a blowout. If you put a standard tire gauge on the valve, what pressure reading would you expect?

c. A tire has been partially inflated so that the tire gauge reads 1.38×10^5 N/m^2 (20 lb/in^2). What is the absolute pressure inside the tire?

11. The vertical pipe in Fig. 12-3 is connected to a vacuum pump. As the air is pumped out, the water rises to a height h and remains there. What is this maximum height?

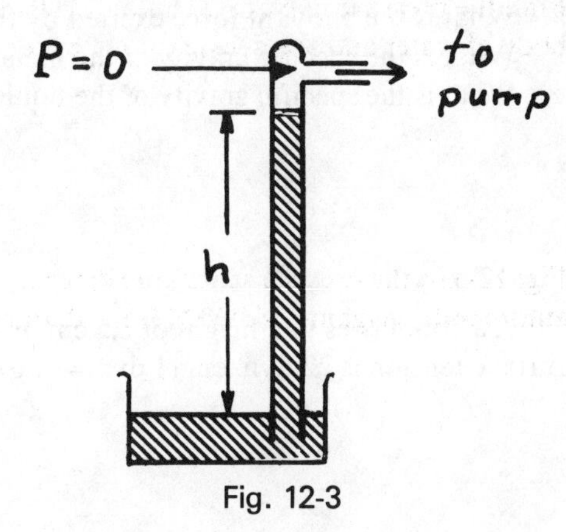

Fig. 12-3

12. State Archimedes' principle and describe the physical reason why an object immersed in a fluid weighs less than it does in air.

13. What fraction of the volume of an iceberg is above water: $d_{ice} = 0.92\ g/cm^3$ or $d_{sea\ water} = 1.03\ g/cm^3$?

14. A rock of irregular shape has an average density of 8.0 g/cm³ and a weight (in air) of 90 N. When the rock is suspended in water as shown in Fig. 12-4, what is the tension in the supporting wire? (Hint: Make a diagram showing all forces acting *on* the rock. Use the relationship sp. gr. = (wt. of rock)/(wt. of H_2O displaced).)

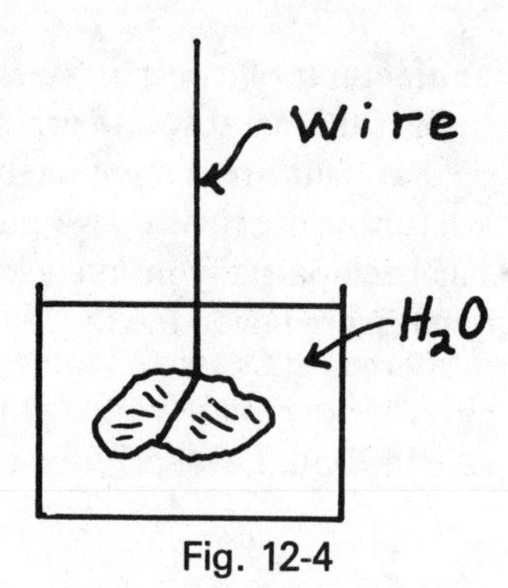

Fig. 12-4

15. A solid piece of an unknown metal weighs 5.5 N in air, 4.0 N in water and 4.5 N in an unknown liquid.
 a. What is the buoyant force exerted by the water on the object?
 b. What is the specific gravity of the metal object?
 c. What is the specific gravity of the liquid?

16. Ethyl alcohol rises to a height of 8.5 cm in a clean glass tube. What is the inside diameter of the tube? The surface tension is 22 dyn/cm (1 dyn = 1 g·cm/s²) and the density of ethyl alcohol is 0.79 g/cm³.

17. State the conditions that must exist in a fluid for Bernoulli's equation to apply.

18. The ship in Fig. 12-5 has a mast that is rotated by a motor. In what direction will the ship move if the wind is from the north and the mast rotates

a. from north to east, or clockwise as shown by arrow *a*?

b. from north to west, or counterclockwise as shown by arrow *b*?

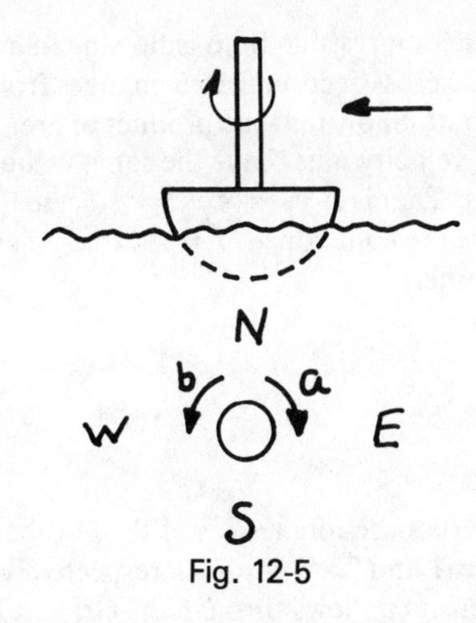

Fig. 12-5

19. The cylindrical tank in Fig. 12-6 has an inlet pipe of internal diameter 1.0 cm and an outlet pipe of internal diameter 2.0 cm. The tank and both pipes are completely filled with an incompressible liquid, water for example.

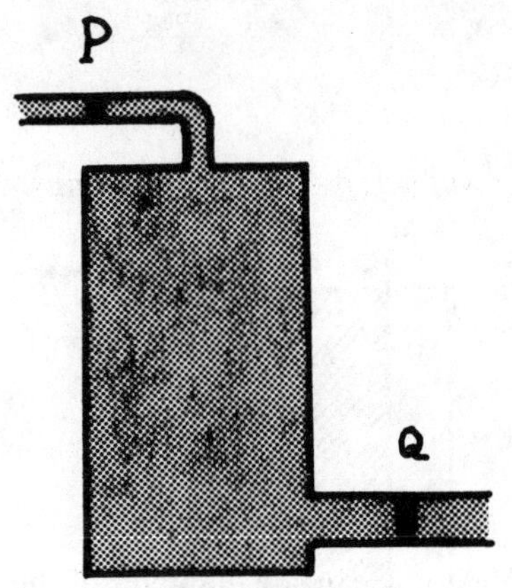

Fig. 12-6

a. During the time of 5 s, the mass of water passing through the cross section at *P* was 50 grams. What mass of water passed through the cross section *Q* in this interval of 5 s?

m (at Q) = ____________.

b. During the time interval of 5 s the volume of water entering the tank was $V = m/d = 50 \text{ g}/(1 \text{ g/cm}^3) = 50 \text{ cm}^3$. What was the volume of water leaving the tank during that same time interval?

V (at outlet) = ____________.

c. The velocity of flow in the input pipe can be found as follows: During a time interval t, the water in the pipe moves a distance s given by $s = vt$ where v is the velocity at which the fluid is moving. The volume of fluid that passes a cross section P in time t is

$$V = \text{area} \times s = \frac{\pi d^2}{4} s$$

Thus

$$V = \frac{\pi d^2}{4} vt \quad or \quad \frac{V}{t} = \frac{\pi d^2}{4} v$$

Assuming that 50 cm^3 of water crosses P each 5 s, calculate the velocity of flow in the smaller pipe.

v_1 = ____________.

d. Now calculate the velocity of flow in the output pipe.

v_2 = ____________.

20. An incompressible fluid is flowing in a tube whose cross-sectional area changes from point to point. Show that the product of area and flow velocity must have the same value at each point. That is $A_1v_1 = A_2v_2 = A_3v_3$ etc. (Hint: Write the equation $V/t = Av$ for each part of the tube.)

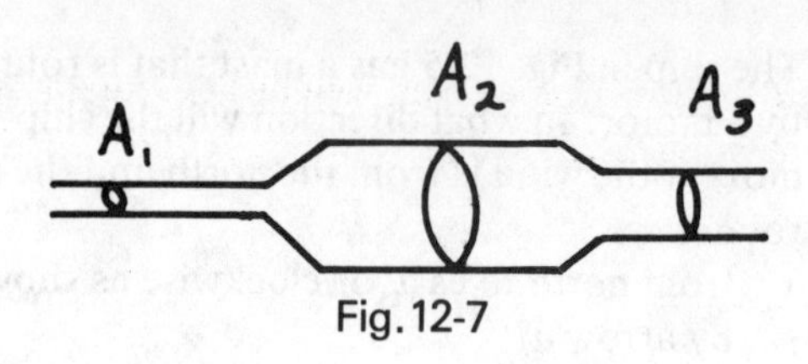

Fig. 12-7

21. The cross-sectional areas of the U-tube at points 1 and 2 are A and a, respectively. A gas of density d flows through the straight tube at a speed v and also fills the upper portion of the U-tube. A liquid of density d' is in the lower portions such that there is a difference h between its level in the two arms of the tube. Find v. (Remember that $vA = v'a$ where v' is the speed of the fluid at point 2. Assume that the density of the fluid through the straight tube is much less than that of the liquid in the U-tube.)

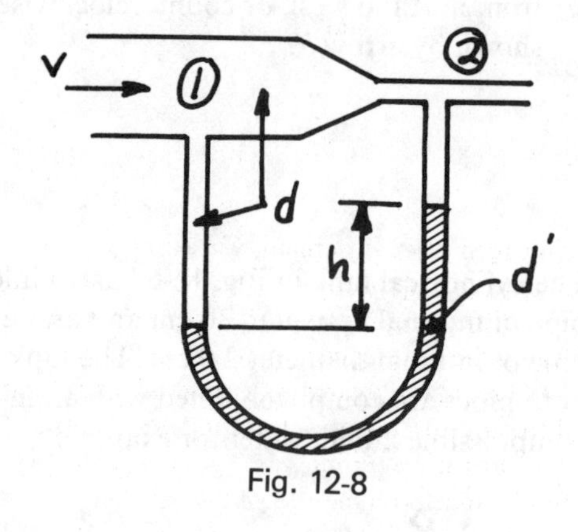

Fig. 12-8

22. The tank in Fig. 12-9 has an inside diameter of 2.0 m. It is filled with water to a depth of 4.0 m.
  a. Calculate the total pressure at point P on the bottom of the tank. (Do not forget atmospheric pressure.)

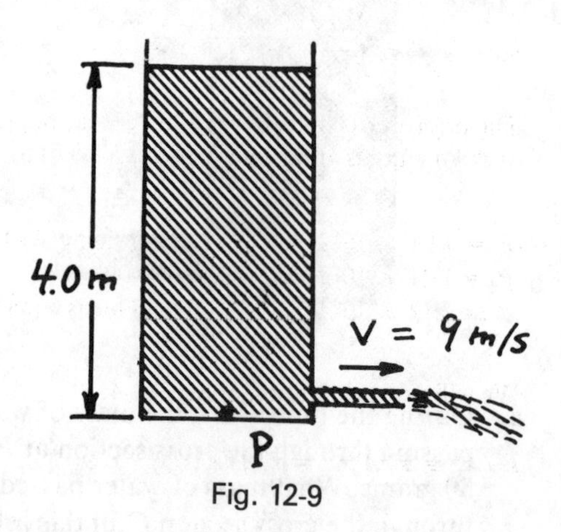

Fig. 12-9

  b. The outlet pipe has an inside diameter of 1.0 cm. Water is flowing out of the pipe with a velocity of approximately 9 m/s. How long will it take for the water level in the tank to be lowered by 2 cm?

23. A hollow iron casting weighs 60 N in air and 40 N in water. What is the volume of the cavity in the casting? $d_{iron} = 7.8\ g/cm^3$.

Solutions

1. Since pressure equals force divided by area, the following are proper units: newton/m², lb/ft², kg/m • s². Atmospheres are proper units by definition.

2. $P = F/A = 20\ N/\pi(0.08\ m)^2 = 995\ N/m^2$

3. This formula was derived by assuming that the density of the fluid under consideration is a constant. This is not true for the atmosphere.

4a. The difference in pressure between M and M' is dgh. Since M' is above M, $P_M > P_{M'}$, therefore $P_{M'} = P_M - dgh$.

b. Again the difference in pressure is dgh. Since M is above M'', $P_{M''} > P_M$. $P_{M''} = P_0 + dgh$.
The fact that the tank is of irregular shape and M, and M', and M'' are not directly above one another makes no difference. In a motionless fluid, all points at the same height are at the same pressure.

5. The pressure at point A is the same as the pressure at B. This is so because A and B are at the same height in the same fluid. Since both mouths of the tube are open to the atmosphere, C and D are at atmospheric pressure P_0.
As in Prob. 4, $P_C = P_B - \ell g d_w$ and $P_0 = P_A - (\ell + d)gd$, where d_w = density of water; d_u = density of unknown. Since $P_D = P_C = P_0$,

$$P_A - (\ell + d)gd_u = P_B - \ell g d_w, \text{ but } P_A = P_B$$

so $$\frac{d_u}{d_w} = \frac{\ell}{\ell + d}$$

6. The pressure 200 m below the surface is greater than the pressure at the surface by an amount (200 m)gd. The pressure at the surface is atmospheric pressure, $1.01 \times 10^5\ N/m^2$.

$$P = 1.01 \times 10^5\ N/m^2 + (200\ m)(9.8\ m/s^2)(1.03 \times 10^3\ kg/m^3)$$
$$P = 1.01 \times 10^5\ N/m^2 + 2.02 \times 10^6\ N/m^2$$
$$P \cong 2.12 \times 10^6\ N/m^2 \text{ at 200 m (This is equal to 21 atmospheres!)}$$

7. We will use the notation of Sec. 12-4:

$$F'/A' = F/A;\ F = F'A/A'$$

$$F = (16 \times 10^3\ N)\frac{(\pi/4)(5\ cm)^2}{(\pi/4)(30\ cm)^2} = \frac{16 \times 10^3}{36} N = 4.4 \times 10^2\ N$$

8. Assuming 100% efficiency, work input = work output. Thus,

$$Fs = F's';\ s = s'F'/F$$

$$s = (2\ cm)\frac{16 \times 10^3\ N}{4.4 \times 10^2\ N} = 72\ cm$$

9. The pressure just below the small piston is the same as the pressure just below the large piston. Thus

$$\frac{F_1}{A_1} = \frac{F_2}{A_2};\ F_1 = F_2A_1/A_2$$

$$F_1 = 400\ \text{N}\frac{(\pi/4)(12\ \text{cm})^2}{(\pi/4)(2\ \text{cm})^2} = 1.44 \times 10^4\ \text{N}$$

10a. The manufacturer's recommended pressure is always *gauge* pressure.
b. The tire gauge should read zero when the tire is completely flat. Tire gauges read gauge pressure.
c. The absolute pressure is the gauge pressure plus atmospheric pressure.
Thus P(absolute) = $1.38 \times 10^5\ \text{N/m}^2 + 1.01 \times 10^5\ \text{N/m}^2 = 2.39 \times 10^5\ \text{N/m}^2$.

11. The pressure at the surface in the open vessel is 1 atmosphere or $1.01 \times 10^5\ \text{N/m}^2$. Thus,

$$dgh = 1\ \text{atm};\ h = \frac{1\ \text{atm}}{dg}$$

$$h = \frac{1.01 \times 10^5\ \text{N/m}^2}{(10^3\ \text{kg/m})(9.8\ \text{m/s}^2)} = 10.3\ \text{m}$$

In British units $\quad h = 10.3\ \text{m}\ \frac{1\ \text{ft}}{0.305\ \text{m}} \cong 34\ \text{ft}$

12. A body immersed in a fluid experiences an apparent loss of weight equal to the weight of fluid displaced. Reread Sec. 12-6 for discussion of this point.

13. Primed quantities apply to sea water, unprimed to ice. Two forces act on the iceberg, its weight and a buoyant force. Since the iceberg is in equilibrium, these forces must be equal.

B = buoyant force = weight of water displaced
$B = d'V'g \qquad W$ = weight of ice = dVg

so that $V'd' = Vd$

$$\frac{V'}{V} = \frac{d}{d'} = \frac{0.92}{1.03} = 89\%$$

Since V'/V is the fraction of the iceberg beneath the water, 11% of the iceberg is above water.

14. The rock is subject to three forces as shown in Fig. 12-10. The specific gravity of the rock is

$$\text{sp. gr.} = \frac{\text{wt. of rock}}{\text{wt. of water displaced}} = \frac{W}{B}$$

Thus $\quad$ B(buoyant force) $= \frac{W}{\text{sp. gr.}} = \frac{(90\ \text{N})}{8.0} = 11.3\ \text{N}$

Since the rock is in equilibrium,

$$T + B - W = 0$$
$$T = W - B = 90\ \text{N} - 11\ \text{N} = 79\ \text{N}$$

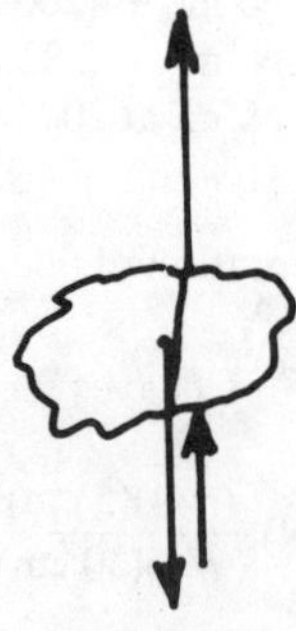

Fig. 12-10

15a. The buoyant force exerted by the water is 5.5 N − 4.0 N = 1.5 N.

b. $$\text{sp. gr.} = \frac{\text{weight of object}}{\text{weight of equal volume of water}} = \frac{5.5\text{ N}}{1.5\text{ N}} = 3.7$$

c. $$\text{sp. gr.} = \frac{\text{weight of liquid displaced}}{\text{weight of equal volume of water}} = \frac{1.0\text{ N}}{1.5\text{ N}} = 0.67$$

16. Rewriting text Eq. 12-4,

$$r = \frac{2\gamma}{hdg} = \frac{2(22\text{ dyn/cm})}{(8.5\text{ cm})(0.79\text{ g/cm}^3)(980\text{ cm/s}^2)}\ \frac{1\text{ g}\cdot\text{cm/s}}{1\text{ dyn}}$$

$$= 6.7 \times 10^{-3}\text{ m} = 6.7\text{ mm}$$

The diameter of the tube is 13.4 mm.

17. The fluid must: (a) be incompressible, so that the mass entering a volume equals the mass leaving; (b) have no rotational motion or turbulence in the motion of its molecules in the fluid. If this is not true, eddies or whirlpools will absorb energy whereas Bernoulli's equation was derived on the assumption that energy is conserved.

18. As with the rotating baseball, the mast drags air along with it, thereby causing the speed of air on one side of the mast to be greater than on the other side.

a. $v_a = W - u$
$v_b = W + u$
By Bernoulli's equation since $v_b > v_a$ $P_a > P_b$ $F_a \rightarrow \bigcirc \leftarrow F_b$
since F is proportional to P, $F_a > F_b$
so the ship experiences a net force toward the east.

b. Exact opposite; net F is toward the west.

19. The pressure just below the small piston is the same as the pressure just below the large piston. Thus

$$\frac{F_1}{A_1} = \frac{F_2}{A_2} \quad F_1 = F_2A_1/A_2$$

$$F_1 = 100\text{ lb}\frac{(\pi/4)(6\text{ in.})^2}{(\pi/4)(1\text{ in.})^2} = 3600\text{ lb}$$

a. These masses have to be the same. If 50 g of water enter the tank in 5 s, 50 g must also leave the tank.

$$m(\text{at } Q) = 50\text{ g}$$

b. Water is incompressible. This means that the density is the same everywhere.

$$V_2 = \frac{m_2}{d_2} = \frac{m_1}{d_1} = V_1$$

The same volume leaves the tank as enters it.

c. $$v_1 = \frac{V(4)}{t\,\pi(d_1^2)} = \frac{(50\text{ cm}^3)4}{(5\text{ s})\pi(1\text{ cm}^2)} = \frac{40}{\pi}\frac{\text{cm}}{\text{s}} = 12.7\text{ cm/s}$$

d. $$v_1 = \frac{V(4)}{t\,\pi(d_2^2)} = \frac{(50\text{ cm}^3)4}{(5\text{ s})\pi(4\text{ cm}^2)} = \frac{10}{\pi}\frac{\text{cm}}{\text{s}} = 3.2\text{ cm/s}$$

20. At the first cross section,

$A_1 v_1 = V/t$

At the second cross section,

$A_2 v_2 = V/t$

Since the liquid is incompressible, V/t has the same value everywhere. Thus

$A_1 v_1 = A_2 v_2 = A_3 v_3$, etc.

21. Bernoulli's equation at points 1 and 2 gives $P_1 + \frac{1}{2}dv^2 + dgy_1 = P_2 + \frac{1}{2}dv'^2 + dgy_2$ where y_1 and y_2 are the heights of these points relative to an arbitrary point. Let's choose A as our reference point, which is the level of the fluid d'. In the U-tube the fluid is static so that

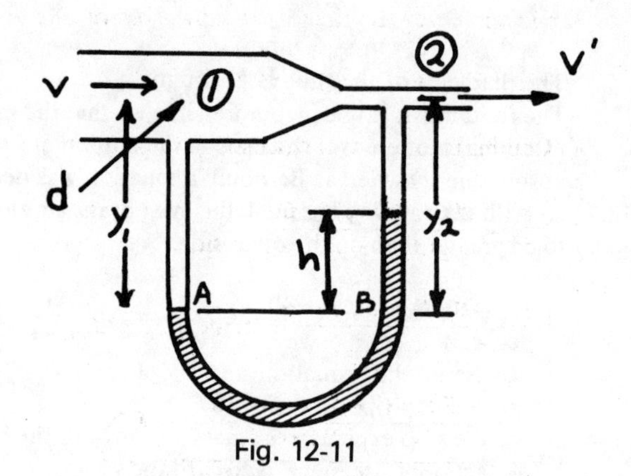

Fig. 12-11

$P_B - P_2 \cong d'gh$ and $P_A \cong P_1$

These are good approximations if d is much less than d'.

Since A and B are at the same height, $P_A = P_B$ and $P_1 - P_2 = d'gh$.
Since points 1 and 2 are at the same height, $y_1 = y_2$.
Thus Bernoulli's equation simplifies to

$\frac{1}{2}dv^2 = \frac{1}{2}dv' - d'gh$

The velocities at 1 and 2 are related by $vA = v'a$.
Eliminating v',

$\frac{1}{2}dv^2 = \frac{1}{2}dv^2(A/a)^2 - d'gh$

Solving for v^2,

$$v^2 = \frac{2d'gh}{d[(A/a)^2 - 1]} \quad \text{or} \quad v = \sqrt{\frac{2d'gh}{d[(A/a)^2 - 1]}}$$

22a. $P = dgh + P_0$; $P_0 = 1$ atm
$P = 10^3\ \text{kg/m}^3\ (9.8\ \text{m/s}^2)(4.0\ \text{m}) + (1.01 \times 10^5\ \text{N/m}^2)$
$P = 0.39 \times 10^5\ \text{N/m}^2 + 1.10 \times 10^5\ \text{N/m}^2 = 1.40 \times 10^5\ \text{N/m}^2$

b. One method of solution is to use the result of the previous problem, $A_1 v_1 = A_2 v_2$. Let "1" refer to the tank and "2" to the outlet pipe.

$$v_1 = v_2 A_2/A_1 = 9\ \text{m/s}\ \frac{(\pi/4)(1\ \text{cm})^2}{(\pi/4)(200\ \text{cm})^2}$$

$v_1 = 9/(4 \times 10^4) = 2.25 \times 10^{-4}$ m/s
$t = s/v = 0.02\ \text{m}/(2.25 \times 10^{-4}\ \text{m/s})$
$t = 89\ \text{s} \cong 1\frac{1}{2}$ min

23. d_W = density of water = $1.0\ g/cm^3$
$d_{iron} = 7.8\ g/cm^3$ d_{air} = density of air
V_{iron} = volume occupied by the iron
V_{air} = volume occupied by the cavity. The weight of water displaced by the cavity must equal the change in the apparent weight of the casting.
$d_w g V = 20\ N \quad V = V_{air} + V_{iron}$

$$V_{air} + V_{iron} = \frac{20\ N}{g d_w}$$

The weight of the casting in air equals the weight of the iron plus the weight of the air in the cavity, ignoring the buoyant force supplied by the surrounding air. So that

$$g(d_{iron} V_{iron} + d_{air} V_{air}) = 60\ N$$

Combining these two equations, we get

$$V_{air} = \frac{20(d_{iron}/d_W) - 60}{g(d_{iron} - d_{air})}$$

$$\cong \frac{20(d_{iron}/d_W) - 60}{g d_{iron}} \quad \text{since } d_{iron} \gg d_{air}$$

$$V_{air} = \frac{\dfrac{(20\ N)(7.8 \times 10^3\ kg/m^3)}{(1 \times 10^3\ kg/m^3)} - 60\ N}{(9.8\ m/s^2)(7.8 \times 10^3\ kg/m^3)}$$

$$= 1.26 \times 10^{-3}\ m^3$$

REVIEW TEST FOR CHAPTERS 10–12

1. A compression wave has a speed of 8.0×10^3 m/s in a steel rail. When heated, the stretch modulus of the steel is reduced to 64% of its original value. What is the new value of the wave velocity?

2. All of the following statements concerning certain transverse waves are correct. Which one of these statements would lead you to conclude that these transverse waves are polarized?
   a. They obey the principle of conservation of energy.

   b. They travel at a finite speed.

   c. They exhibit a preferred plane of vibration.

   d. They obey a superposition principle.

   e. They have a well-defined frequency.

3. The velocity of sound in air is about 340 m/s. What is the wavelength of a sound wave whose frequency is 500 Hz?

4. Two loudspeakers are made to produce pure tones whose wavelengths are 0.385 m and 0.386 m. What beat frequency will be heard? (Use 340 m/s as the speed of sound.)

5. Red light emitted from the element hydrogen in the laboratory (source at rest) has a frequency of 4.571×10^{14} Hz. The same light emitted by hydrogen gas in a distant galaxy appears, to an astronomer on earth, to have a frequency of 4.561×10^{14} Hz. What is the apparent velocity of the galaxy? Is the galaxy receding or approaching?

6. Let f represent the fundamental frequency of a guitar string. What does the frequency become when the tension in the string is doubled?
   a. $2f$
   b. $f + 2$ Hz
   c. $\sqrt{2}f$
   d. $f/2$
   e. $4f$
7. The air column within a closed pipe is vibrating in its first overtone. The length of the pipe is 30 cm and the speed of sound in air is 340 m/s. Calculate the frequency of the tone produced.

8. An opera singer hits a certain high note and a nearby goblet shatters. This is an example of the phenomenon of ______________ .
   a. conservation of energy
   b. beats
   c. resonance
   d. bulk compressibility.
9. The fundamental frequency of an open-ended pipe is 1600 Hz. What does the frequency become when one end of the pipe is closed?
   a. 3200 Hz
   b. 800 Hz
   c. 40 Hz
   d. 1200 Hz
   e. 2400 Hz
10. A certain high-fidelity amplifier has an output at 20 kHz which is 3 dB less than its output at 2 kHz. If the amplifier has an average output power of 100 watts at 2 kHz, what is its power output at 20 kHz?

11. "Change of pressure exerted at any point in a confined fluid is transmitted undiminished in all directions to all points in the fluid" is a statement of
   a. Pascal's principle
   b. Archimedes' principle
   c. the superposition principle
   d. Bernoulli's principle
   e. the Doppler effect
12. A glass U-tube open to the atmosphere at both ends contains mercury (d = 13.6 g/cm³) plus a clear liquid of unknown density. The levels of the liquids are shown in Fig. T-26. Find the density of the unknown liquid. (The lower liquid is the mercury.)

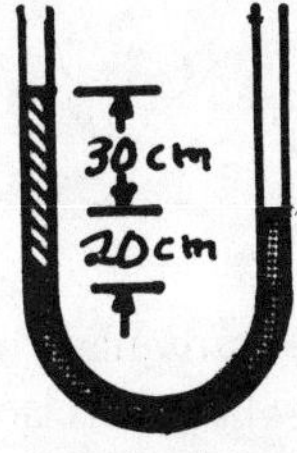

Fig. T-26

13. A moon rock of irregular shape is weighed in air and in an inert liquid with a specific gravity of 0.80. The observed weights are 2.2 N in air and 1.8 N in the liquid. What is the specific gravity of the moon rock?

14. An incompressible fluid is flowing smoothly through this horizontal glass tube. The velocity of the fluid shown in Fig. T-27 is 0.6 m/s at A and 1.2 m/s at B. The density of the fluid is 8.0×10^2 kg/m³. What is the difference in pressure between points A and B? Which pressure is higher?

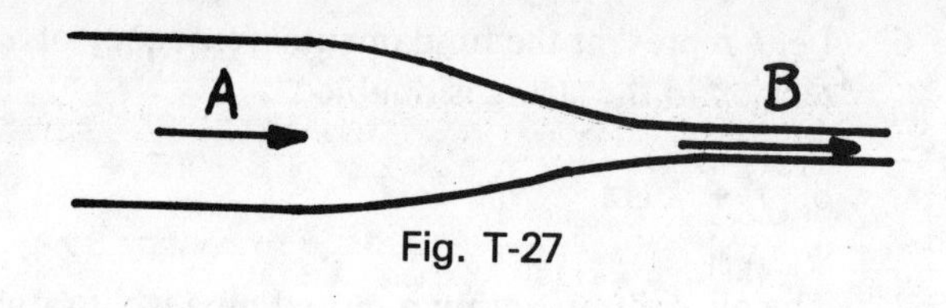

Fig. T-27

15. The diameter of the larger portion of the glass tube in Prob. 14 is D. What is the smaller diameter?
    a. $D/2$
    b. $D/\sqrt{2}$
    c. $D/4$
    d. $D/2\sqrt{2}$

16. Figure T-28 shows a metal can filled with water to a depth of 25 cm with a small hole 10 cm from the bottom. At what horizontal distance R will the stream of water strike a surface on which the can is placed? (Hint: First find the horizontal velocity of the stream as it leaves the hole. The trajectory of the water is that of any small mass having the same initial velocity.)

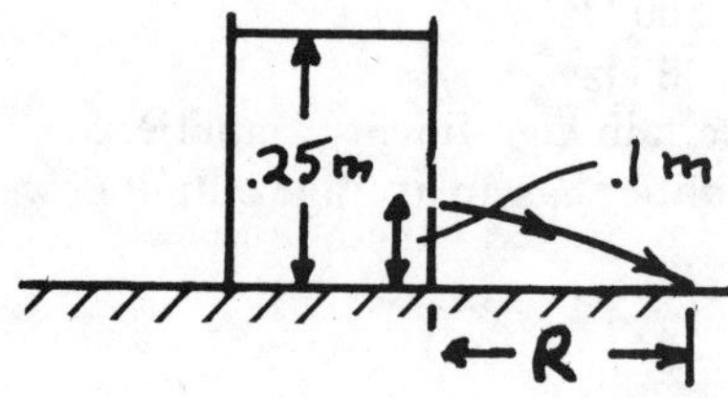

Fig. T-28

CHAPTER 13

Temperature and Thermal Expansion

GOALS To learn how the two most common temperature scales are related to fixed reference points, and to study the thermal expansion of solids, liquids, and gases.

OBJECTIVES After completing this chapter the student should be able to do the following:

1. Name two temperature scales in common use. State the values assigned by each scale to the fixed points, the melting point of ice and the boiling point of water.
2. State correctly the relative size of a Celsius degree as compared to a Fahrenheit degree and vice versa.
3. Find the Celsius temperature corresponding to a Fahrenheit temperature and vice versa.
4. Write the equation relating the change in length of a rod to the change in temperature and the original length.
5. Calculate any one of the quantities α, $\Delta \ell$, ΔT, or ℓ when the others are given.
6. State the physical significance of the symbols that appear in the equation for expansion of a liquid, $\Delta V = V\beta\Delta T$.
7. Calculate the amount by which a mercury column rises in a thermometer for a given change in temperature.
8. Calculate the temperature indicated by a constant-pressure gas thermometer when the volume is given at 0°, 100°, and at the unknown temperature.

SUMMARY

The *temperature* of an object is a measure of how hot or cold it is. The concept of temperature becomes quantitative when it is measured with a *thermometer*. A *temperature scale* is established by assigning numbers to well-defined and reproducible temperatures.

The two temperature scales in most general use are the *Celsius* scale (improperly called centigrade) and the *Fahrenheit* scale. Both of these use, as fixed points, the freezing point (0°C, 32°F) and the boiling point (100°C, 212°F) of pure water. A temperature *change* of 10° on the Celsius scale is equal to a temperature *change* of 18° on the Fahrenheit scale. On the other hand, a temperature *reading* of 10° on the Celsius scale corresponds to a temperature *reading* of 50° on the Fahrenheit scale. Temperature intervals ΔT must be clearly distinguished from temperatures T.

A property of a substance that depends uniquely on temperature is called a *thermometric property*. Examples are pressure of a confined gas, resistivity of a solid, and color of a molten metal.

A characteristic of almost all solids is that they expand upon heating. Any linear dimension—length, width, circumference, etc., whether interior or exterior—increases with temperature. Furthermore, the relationship is strictly linear. The basic equation for the expansion of a solid is $\Delta \ell = \ell\alpha\Delta T$ where ℓ is the length and α is the *coefficient of linear expansion*. The bimetallic strip in a thermostatic relay exploits the different coefficients of expansion of dissimilar metals.

When a liquid or a gas is heated, its volume increases. The change in volume is given by $\Delta V = V\beta\Delta T$ where V is the volume, β is the coefficient of volume expansion, and ΔT is the temperature change. The mercury column in a mercury-in-glass thermometer rises with increasing temperature because the volume coefficient of expansion of mercury is greater than the volume coefficient of expansion of glass. The volume coefficient of expansion is nearly the same for all gases. It is considerably larger than the coefficient for liquids.

QUESTIONS AND PROBLEMS

1. In a calorimetry experiment the temperature of a water bath is increased by 27°F. What is this temperature change on the Celsius scale?

2. The temperature of the anti-freeze in an automobile radiator is 27°F. To what temperature does this correspond on the Celsius scale?

3. A temperature *difference* on the Celsius scale may be indicated by the symbol ______________, whereas a *temperature* on that scale is written ______________.
4. What is the Celsius temperature corresponding to 72°F?

5. The temperature of liquid nitrogen (used in the storage of whole blood) is −196°C. Express this temperature on the Fahrenheit scale.

6. Fig. 13-1 shows two commonly used temperature scales. Fill in the missing numbers.

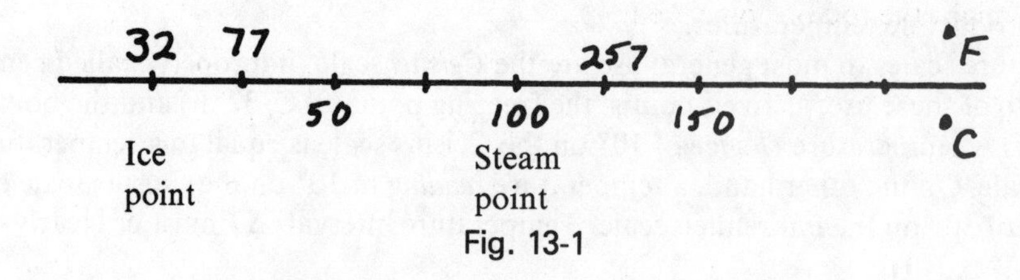

Fig. 13-1

7. Write the basic equation for the linear expansion of a solid.

8. The two rods illustrated in Fig. 13-2 are made of steel. They are 50 cm long and have diameters of 0.2 cm and 0.4 cm. Calculate the elongation of the smaller rod for a temperature change of 18°F. $\alpha = 11 \times 10^{-6}$ $(°C)^{-1}$

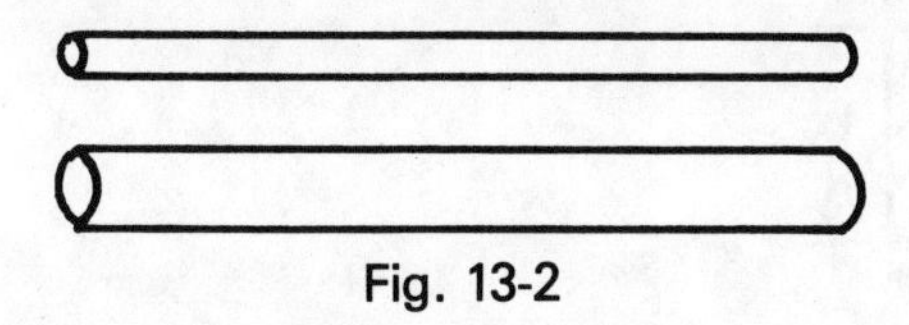

Fig. 13-2

9. What is the elongation of the larger rod for the same temperature change?

10. The temperature coefficient of linear expansion of concrete is 12×10^{-6} $(°C)^{-1}$. The sections of a reinforced concrete bridge are 30 m long. How large an expansion joint should be allowed if a temperature change of 117°F is anticipated?

11. Fig. 13-3 shows a bimetallic strip consisting of a thin strip of brass (on top) welded to a similar piece of iron. The left end of the assembly is held rigidly. Make a sketch to show how the bimetallic strip will bend when it is heated. (Consult the text for the coefficients of expansion.)

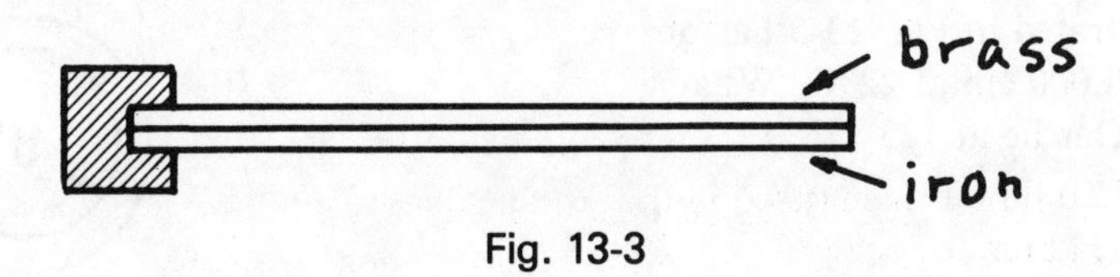

Fig. 13-3

12. A solid rod of an unknown material expands from 12.035 cm to 12.053 cm when it is heated from 20°C to 470°C. Determine the coefficient of linear expansion of this material. Of which materials listed in Table 13-1 in the text might the rod be composed?

13a. Fig. 13-4 shows a U-shaped piece of aluminum. Some of its linear dimensions are indicated by the letters a, b, h, and ℓ.

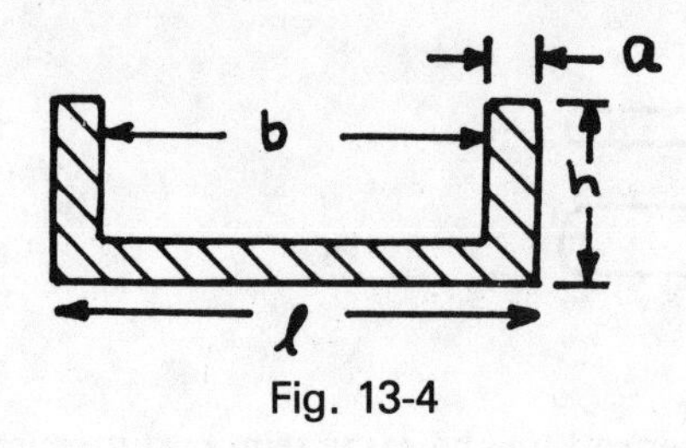

Fig. 13-4

If the piece is heated uniformly, which of these dimensions will increase?

b. If the relative change in ℓ is $\Delta\ell/\ell = 26 \times 10^{-4}$, what is the relative change in the dimension a? What is it for b?

$$\frac{\Delta a}{a} = \_\_\_\_\_\_\_\_\_\_. \qquad \frac{\Delta b}{b} = \_\_\_\_\_\_\_\_\_\_.$$

c. What temperature change would produce a relative change in h of 26×10^{-4}°C?

14. The brass ring illustrated in Fig. 13-5 has an inside diameter of 8.000 cm at 22°C. What will its inside diameter be at 125°C?

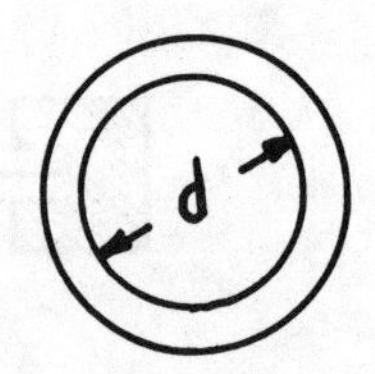

Fig. 13-5

15. The length of a steel block increases by 0.1% when a temperature change ΔT takes place. By what percent will the volume change under the same conditions?

16. A Pyrex flask is brimful of carbon tetrachloride at 0°C. It cannot hold another drop. The volume of carbon tetrachloride at 0°C is 314.5 cm³. What volume of liquid overflows when the temperature of the flask and contents is increased by 170°C?

17. What is the relative or apparent volume coefficient of expansion for methyl alcohol in glass?

18. Fig. 13-6 shows the relationship between the volume of a certain liquid and its temperature. What is the liquid that has this peculiar behavior? Label the three points indicated on the abscissa. In what portion of the graph is the volume coefficient of expansion negative?

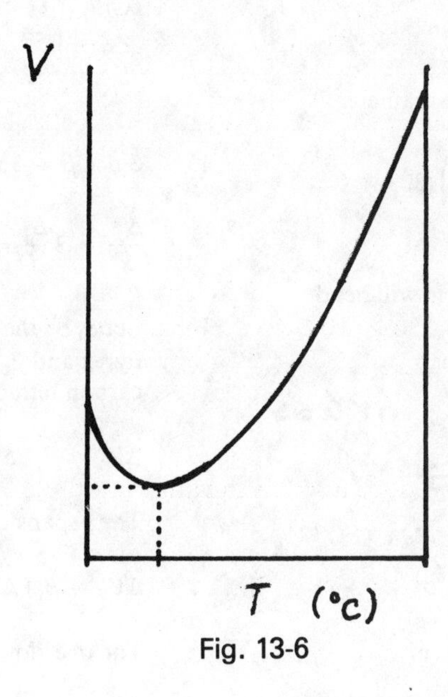

Fig. 13-6

Solutions

1. A temperature change of 9°F corresponds to 5°C thus 27°F corresponds to 15°C.
2. 27°F is five degrees below the freezing point. 32°F − 27°F = 5°F. Calculate the corresponding interval on the Celsius scale.

$$\frac{\Delta T}{5°\text{F}} = \frac{5°\text{C}}{9°\text{F}}; \ \Delta T = \frac{25}{9}\ °\text{C} = 2.8°\text{C}$$

Thus 27°F corresponds to −2.8°C.

3. ΔT, °C; T, °C
4. 72°F − 32°F = 40°F

$$\frac{\Delta T}{40°\text{F}} = \frac{5°\text{C}}{9°\text{F}}; \ \Delta T = 22.2°\text{C}$$

Thus 72°F corresponds to ~22°C

5. $$\frac{\Delta T}{196°\text{C}} = \frac{9°\text{F}}{5°\text{C}}; \ \Delta T = 353°\text{F}$$

The corresponding temperature is 32°F − 353°F = −321°F.

6.

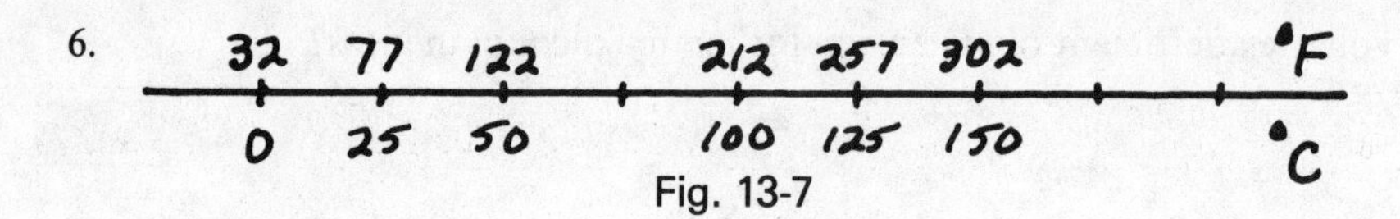

Fig. 13-7

7. $\Delta L = \alpha L \Delta T$

8. $\Delta L = (1 \times 10^{-6})(°C)^{-1}(50 \text{ cm})(18°F)\left(\frac{5°C}{9°F}\right)$

$= 5.5 \times 10^{-3}$ cm

9. The same. ΔL does not depend on the diameter.

10. $\Delta L = (12 \times 10^{-6})(°C)^{-1}(117°F)\left(\frac{5°C}{9°F}\right)(30 \text{ m})$
$= 2.3 \times 10^{-2}$ m $= 2.3$ cm

11. Brass expands more than iron. The strip will bend downward, as shown below.

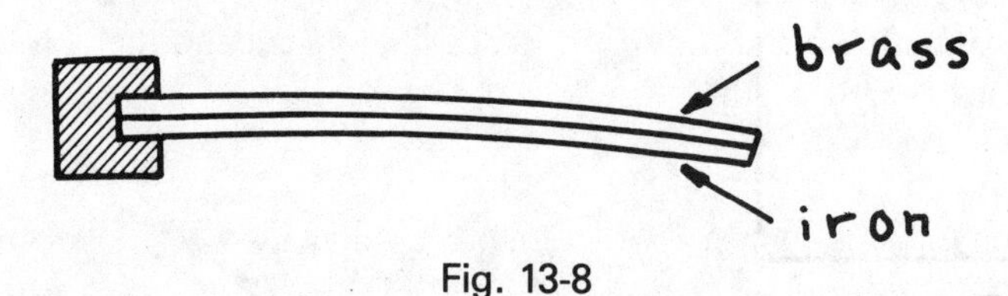

Fig. 13-8

12. $L = 12.053$ cm $- 12.035$ cm $= 0.018$ cm
$T = 470°C - 20°C = 450°C$

$$= \frac{\Delta L}{L\Delta T} = \frac{1.8 \times 10^{-2} \text{ cm}}{(12.05 \text{ cm})(450°C)} = 3.3 \times 10^{-6} (°C)^{-1}$$

The rod appears to be made of Pyrex glass.

13a. All of the dimensions will increase, including the inside dimension b.

b. $\frac{\Delta a}{a} = \frac{\Delta b}{b} = \frac{\Delta \ell}{\ell} = 26 \times 10^{-4}$

The relative change in any linear dimension depends only on α and ΔT.

c. For aluminum, $\alpha = 26 \times 10^{-6} (°C)^{-1}$.

$$\Delta T = \frac{\Delta \ell / \ell}{\alpha} = \frac{26 \times 10^{-4}}{26 \times 10^{-6} (°C)^{-1}} = 100°C$$

$\Delta T = 100°C$

14. The inside diameter increases in the same way as any other dimension.

$\Delta T = 125°C - 22°C = 103°C$
$\Delta d = \alpha d \Delta T = (19 \times 10^{-6})(°C)^{-1}(8.00 \text{ cm})(103°C)$
$= 1.57 \times 10^{-3}$ cm
$d' = d + \Delta d = 8.000$ cm $+ 0.0016$ cm
$= 8.002$ cm

15. Since $\beta = 3\alpha$,

$$\frac{\Delta V}{V} = 3\frac{\Delta \ell}{\ell} = 3 \times 0.1\% = 0.3\%$$

16. Let β_1 be the volume coefficient of expansion for Pyrex and β_2 the volume coefficient of expansion for carbon tetrachloride. The expansion of the glass is

$$\Delta V_1 = \beta_1 V \Delta T$$

The expansion of the liquid is

$$\Delta V_2 = \beta_2 V \Delta T$$

The overflow volume is

$\Delta V_2 - \Delta V_1 = (\beta_2 - \beta_1) V \Delta T$
$= (581 - 3.3) \times 10^{-6}(°C)^{-1}(314.5 \text{ cm}^3)(170°C)$
$= 30.9 \text{ cm}^3$

17. The apparent volume coefficient of expansion is

$(1134 - 26) \times 10^{-6}(°C)^{-1} = 1108 \times 10^{-6}(°C)^{-1}$

18. This liquid is water. The coefficient of expansion is negative between 0°C and 4°C.

CHAPTER 14 Heat and Heat Transfer

GOALS To study heat as a form of energy, and to examine some of the effects produced by heat and the means by which heat is transferred from one place to another.

OBJECTIVES After completing this chapter the student should be able to do the following:

1. State the definition of the calorie and the British thermal unit.
2. State the numerical relationship between the calorie and joule.
3. Write a correct statement of the first law of thermodynamics.
4. Apply the first law of thermodynamics to situations involving a conversion of mechanical energy to heat energy and vice versa.
5. Write a definition of specific heat capacity and give the equation relating quantity of heat, mass, specific heat, and temperature change.
6. Solve problems involving the application of the specific heat equation and the first law of thermodynamics.
7. Give the names of the three states or phases in which matter may exist and name the latent heats associated with two changes of phase.
8. Solve calorimetry problems involving one or two changes of phase. (The unknown quantity may be any of the following: mass, final temperature, heat of fusion, specific heat capacity.)
9. Give a definition of heat of combustion.
10. Convert a heat of combustion from MJ/kg to kcal/kg and vice versa.
11. Name and describe briefly the three processes by which heat energy may be transferred from one place to another.
12. Name the type of radiation involved in the transfer of energy from the sun to the earth and state the velocity at which this radiant energy is transmitted.

SUMMARY

Experiments by Black, Rumford, and Davy in the latter half of the eighteenth century clearly established that heat is not a fluidlike substance, nor even a weightless fluid. It is, rather, a form of energy and is associated with the motions of the molecules of which a body is composed.

The first law of thermodynamics states that all the heat energy added to a system can be accounted for as mechanical work, an increase in internal energy, or both. The numerical relationship between heat and mechanical energy is 1 cal = 4.186 J, where a *calorie* (cal) is the amount of heat required to raise the temperature of one gram of water one Celsius degree.

The *specific heat capacity c* of a substance is the number of calories required to raise the temperature of one gram of the substance one Celsius degree. The specific heat capacity of water, the standard substance, is 4186

J/kg·°C or 1 cal/g·°C. The amount of heat required to raise the temperature of a body of mass m is $\Delta Q = mc\Delta T$ where ΔT is the temperature change.

The amount of heat required to melt one gram of a substance—to change it from a solid to a liquid without changing its temperature—is the *latent heat of fusion*, or simply *heat of fusion*. The amount of heat required to vaporize one gram of a substance—to change it from a liquid to a vapor without changing its temperature—is the *latent heat of vaporization*. The two latent heats of water are 80 cal/g (fusion) and 540 cal/g (vaporization).

The *heat of combustion* of a fuel is the thermal energy released per unit mass when the fuel is combined with enough oxygen to ensure complete combustion. The heat of combustion of gasoline is 11,500 kcal/kg or 48 MJ/kg.

Heat energy can be transferred from one location to another in any of three ways—*convection, conduction,* and *radiation*. Convection involves the movement or flow of a fluid as a whole. It cannot take place in a solid. Convection currents are produced in a room as heated air rises toward the ceiling, displacing the more dense, cooler air that flows downward toward the floor.

Conduction takes place through molecular interactions. Faster-moving molecules transfer some of their KE to neighboring molecules, causing these in turn to move more rapidly. Heat can be transferred by conduction from one body to another only if the two are in physical contact.

The third method of heat transfer, *radiation*, involves the emission of electromagnetic waves by one object and their absorption by another. The energy transfer via radiation does not require that the bodies involved be close together or that a medium be present in the space between them. It is by this means, electromagnetic radiation, that the earth receives energy from the sun. Electromagnetic waves travel at a speed of 3×10^8 m/s through empty space. All hot bodies emit electromagnetic radiation to some extent. The amount of energy radiated depends very strongly on the temperature of the body.

QUESTIONS AND PROBLEMS

1. Write the names of the forms of energy that you have encountered thus far in this course.

2. Complete the following definition: A calorie is the amount of _____________ required to raise the temperature of 1 g of _____________ 1°C.
3. Whenever mechanical energy (KE or PE) is transformed into heat energy, the rate of exchange is always _____________ J for each calorie.
4. A 20 N lead weight is dropped on a concrete platform from a height of 4 m. The collision is perfectly inelastic. How much heat is generated?

5. How many calories are required to raise the temperature of 1 g of silver 1°C? What is the name of this quantity of heat?

6. How much heat is required to raise the temperature of 300 g of silver by 10°C?

7. What quantity of water (in grams) would be raised 10°C by this same quantity of heat?

8. What is the water equivalent of 300 g of silver?

9. Write an equation defining specific heat capacity, identifying each term that it contains.

10. A metal object having a mass of 350 g is heated from 70°F to 250°F. In the process, the metal absorbs 1.07 kcal heat energy. Calculate the specific heat capacity of the metal.

11. What quantity of heat is required to change 60 g of ice at 0°C to water at 20°C?

12. It has been found experimentally that approximately 80 cal of heat is required to melt 1 g of ice, the temperature remaining constant at 0°C. This quantity of heat is called the ________.

13. The heat of vaporization of water is ________ (include units).

14. What happens when 15 g of steam at 100°C is added to 25 g of ice at 0°C? (Hint: Study Ex. 14-5 in the text.)

15. Suppose that the heat energy contained in 40 liters (10.6 gal) of gasoline is converted into mechanical work with an efficiency of 25%.
   a. How many joules of energy is obtained?

   b. To what height can a 1500 kg (1.5 tonne) automobile be lifted? (The density of gasoline is about $0.675\ g/cm^3 = 675\ kg/m^3$.)

16. Write the names of the three methods of heat transfer.

17. By what means does the sun's energy reach the earth?

18. A metal railing and a wooden door are at the same temperature yet one feels cold to the touch and the other does not. Which one feels cold? Explain why.

19. How long does it take for energy from the sun to reach the earth?

20. What is the wavelength in angstroms of the most intense radiation from an electric heater whose temperature is 700°C? (See Fig. 14-4 in the text.)

Solutions

1. kinetic energy, potential energy, heat energy
2. heat energy; water
3. 4.186 J for each calorie

4. The potential energy is converted into kinetic energy during free fall. When the collision takes place the KE is converted into heat energy.

$$\Delta Q = mgh = Wh = (20\text{ N})(4\text{ m}) = 80\text{ J}$$

$$\Delta Q = (80\text{ J})\left(\frac{1\text{ cal}}{4.186\text{ J}}\right) = 19.1\text{ cal}$$

5. From Table 14-1 in the text, the specific heat capacity of silver is 0.056 cal/g·°C.
6. $\Delta Q = mc\Delta T = (300\text{ g})(0.056\text{ cal/g°C})(10\text{°C}) = 168\text{ cal}$
7. 16.8 g of water
8. The water equivalent of 300 g of silver is 16.8 g.
9. $c = \Delta Q/m\Delta T$ where ΔQ is the quantity of heat required, m is the mass of the substance heated, and ΔT is the temperature rise.

10. $$c = \frac{\Delta Q}{m\Delta T} = \frac{1.07 \times 10^3\text{ cal}}{(350\text{ g})(180\text{°F})}\left(\frac{9\text{°F}}{5\text{°C}}\right) = 3.06 \times 10^{-2}\text{ cal/g} \cdot \text{°C}$$

11. $$\Delta Q = (60\text{ g})(80\text{ cal/g}) + (60\text{ g})(1\text{ cal/g} \cdot \text{°C})(20\text{°C})$$
$$= 4800\text{ cal} + 1200\text{ cal} = 6000\text{ cal}$$
12. Heat of fusion
13. 540 cal/g
14. The heat that would be released if all the steam were to condense is (15 g)(540 cal/g) = 8100 cal. The amount of heat required to melt the ice is (25 g)(80 cal/g) = 2000 cal. The amount of heat required to raise the temperature of the ice water to 100°C is (25 g)(1 cal/g · °C)(100°C) = 2500 cal. We see that the 8100 cal available from the condensation of the steam is more than enough to melt all the ice and bring the temperature up to 100°C (2000 cal + 2500 cal = 4500 cal). We conclude that only part of the steam will condense and that the final temperature will be 100°C. Let m be the mass of steam that condenses.

$$m(540\text{ cal/g}) = 4500\text{ cal}$$
$$m = 8.33\text{ g}$$

15a. First calculate the mass of gasoline.

$$m = Vd = (40\text{ lit})(1\text{ m}^3/10^3\text{ lit})(675\text{ kg/m}^3) = 27\text{ kg}$$

$$\Delta Q = (27\text{ kg})(48 \times 10^6\text{ J/kg}) = 1.30 \times 10^9\text{ J}$$

b. $PE = mgh \quad h = PE/mg$

$$h = \frac{1.23 \times 10^9\text{ N} \cdot \text{m}}{(1.500 \times 10^3\text{ kg})(9.8\text{ m/s}^2)} = 8.4 \times 10^4\text{ m}$$

16. conduction, convection, radiation
17. radiation, specifically electromagnetic radiation, which travels through space at a speed of 3×10^8 m/s
18. The metal, being a good conductor, absorbs heat more readily thus cooling the skin, which is in contact with the metal surface.

19. $$t = \frac{\text{distance}}{\text{velocity}} = \frac{14.9 \times 10^9\text{ m}}{3 \times 10^8\text{ m/s}} \cong 50\text{ s}$$

20. The intensity distribution corresponding to 700°C has a peak at 10^{14} Hz. The corresponding wavelength is

$$\lambda = c/f = \frac{3 \times 10^8\text{ m/s}}{10^{14}\text{ s}^{-1}} = 3 \times 10^{-6}\text{ m}$$

CHAPTER

15 Thermal Behavior of Gases

GOALS To study the properties of ideal gases and some characteristics of a vapor, especially water vapor.

OBJECTIVES After completing this chapter the student should be able to do the following:

1. Name the four quantities which, taken together, determine the "state" of a sample of gas.
2. Write the mathematical relationship that exists between the pressure and volume of a given mass of gas when the temperature is kept constant.
3. Choose from several graphs the one that correctly represents P versus V for a gas at constant temperature.
4. Construct a graph of pressure versus temperature for a fixed mass of gas in a container of constant volume.
5. Determine graphically the intercept of the pressure-temperature graph with the temperature axis and give the name assigned to the temperature thus obtained.
6. Given the equation for the straight line on a pressure-temperature graph,

$$P_T = P_0(1 + 0.00366T_C)$$

calculate the temperature corresponding to $P_T = 0$.
7. State the relationship between absolute temperature T and Celsius temperature T_C.
8. Write the general gas law using standard symbols.
9. Calculate the value of the universal gas constant from the set of conditions corresponding to standard temperature and pressure; i.e., $P = 1$ atm, $n = 1$ mole, $T = 273$ K, and $V = 22.4$ liters.
10. Calculate any one of the quantities, P, V, n, T, or R when the others are given.
11. Calculate the relative humidity of air from the density of water vapor and the maximum value of that density if the air were saturated.
12. Calculate the mass of water required to bring the relative humidity to a stated value.
13. Use the phase diagram for H_2O to determine the temperature at which a phase transition takes place when the pressure is given.
14. Use the phase diagram to describe the changes that take place as the temperature is increased, the pressure being held constant.

SUMMARY

An *ideal gas* is a useful mathematical model whose properties may be calculated exactly. The molecules of an ideal gas have a negligible volume and exert no cohesive forces on one another. Many real gases at ordinary temperatures and pressures behave very nearly like ideal gases.

If a fixed mass of an ideal gas is kept at a constant temperature, the product of its pressure and volume is constant. Symbolically, $PV =$ constant. Taking into account the number of moles n of a gas, the quantity PV/n is a constant if the temperature is fixed.

When a confined gas is heated, its pressure increases linearly with temperature. The relationship may be written $P_T = P_0(1 + bT_C)$ where P_T is the pressure at temperature T_C, P_0 is the pressure at 0°C, and b is the *pressure coefficient* for the gas. The value of the pressure coefficient for an ideal gas is 0.00366. *Absolute zero* is the temperature at which an ideal gas would have a pressure of zero. This temperature, on the Celsius scale, is −273 C.

Absolute temperature is the Celsius temperature T_C plus 273° and is denoted by T. Temperatures on this scale are measured in kelvins (K). Absolute zero may be written either −273°C or 0 K.

The relationship between pressure, volume, absolute temperature, and number of moles is given by the *general gas law*

$$\frac{PV}{nT} = R$$

where R is the *universal gas constant.*

Dalton's law of *partial pressures* states that in a mixture of gases the total pressure is the sum of the partial pressures that each gas would exert if it alone occupied the entire volume.

A volume above the surface of a liquid is said to be *saturated* with vapor when the rate at which molecules leave the liquid and enter the vapor phase is equal to the rate at which molecules of the vapor return to the liquid phase. The partial pressure of a vapor at saturation is a function of temperature only.

The ratio of the density of water vapor in the air at a given temperature to the density that would exist at saturation is the *relative humidity.*

The phase diagram for a substance shows the pressures and temperatures at which the various phases (solid, liquid, and vapor) may exist and the conditions of pressure and temperature at which transitions between phases take place. In addition to the well-known transition from solid to liquid (melting) and from liquid to vapor (evaporation), there may also be a transition from solid to vapor (sublimation). The one temperature and pressure at which all three phases may exist in contact with one another is called the *triple point.*

QUESTIONS AND PROBLEMS

1. In the table below, four measurable quantities pertaining to a gas are to be listed in the column on the left. The standard symbols used to represent them are to be given in the second column and typical units in the third column. Complete this table.

Measurable Quantity	Symbol	Units
(a)		atm, N/m^2
(b) volume		
(c)		°C
(d)		kg

2. A fixed mass of gas (0.012 kg) is in a container whose volume may be varied. If the temperature is kept constant, what relationship exists between pressure P and volume V?

3. In the situation described above the pressure is increased from 0.5 atm to 2.5 atm. The original volume was 2.0 liters. Calculate the final volume.

4. At the beginning of its stroke, the cylinder of an air compressor contains 300 cm³ of air at 30°C and 0.8 atm. This gas is slowly compressed, keeping the temperature constant, until the volume is 50 cm³ and the pressure is 4.5 atm. Did any air leak out during the process of compression? Explain.

5. A compressed-air storage tank whose volume is 56 liters contains 1.20 kg of air at a pressure of 18 atm. How much air would have to be forced into the tank to increase the pressure to 25 atm?

6. A steel cylinder contains 250 moles of argon gas at a pressure of 12.5 atm. The volume of the cylinder is 493 liters. An identical cylinder contains carbon monoxide gas at the same temperature and pressure (T = 300 K). How many moles of CO are contained in the second cylinder?

7. Let m_1 be the mass of argon gas and m_2 the mass of carbon monoxide in the preceding problem. Calculate the ratio m_1/m_2.

8. Write the expression for the general gas law in terms of P, V, n, R, and T.

9. Use the data of Prob. 6 to calculate the gas constant R. (1 liter = 0.0224 m³.)

10. Which one of the graphs in Fig. 15-1 could be used to represent pressure as a function of volume for a fixed mass of gas at a constant temperature?

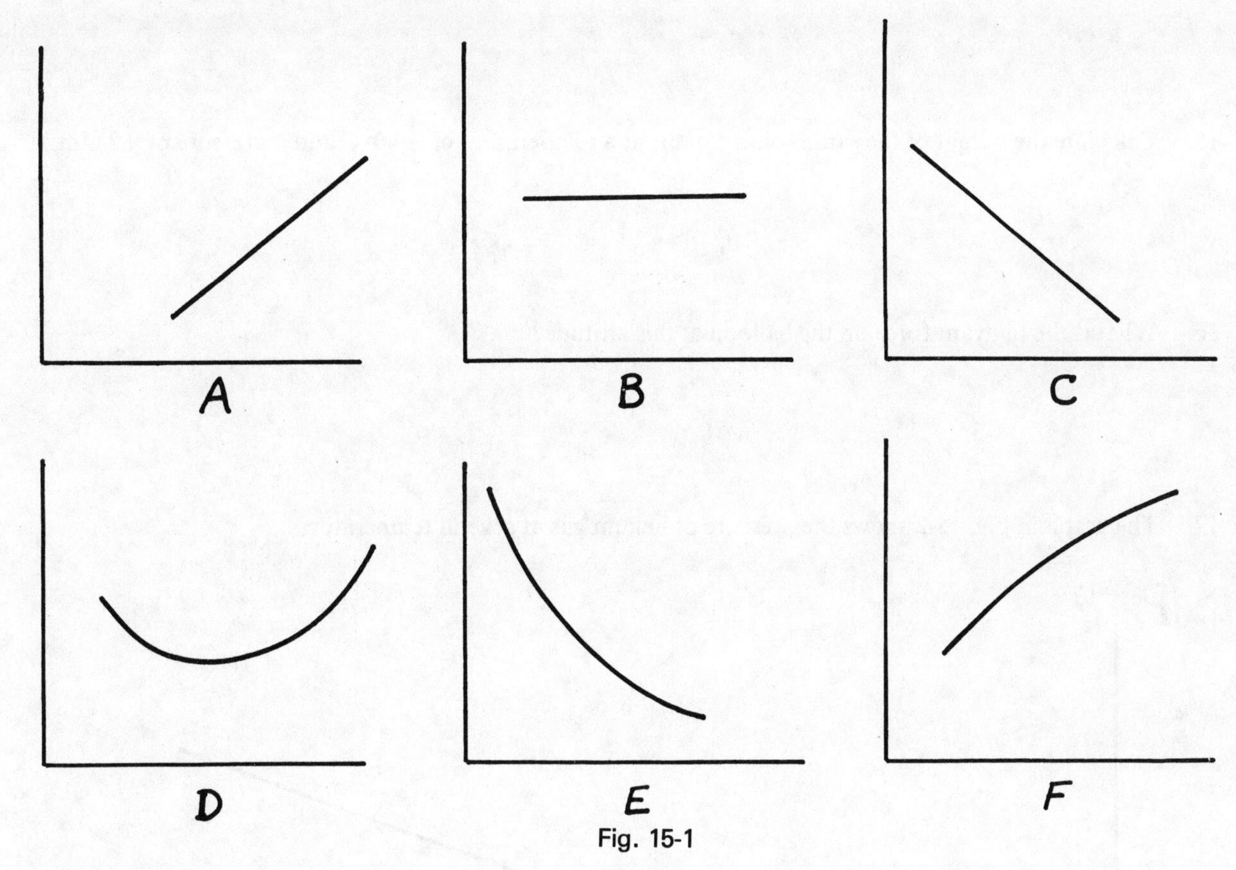

Fig. 15-1

11. Which of these graphs could represent the pressure of a confined gas (constant volume) as a function of temperature?

12. Which of these graphs could represent the volume of a gas as a function of temperature?

13. A balloon containing 0.34 m^3 of hydrogen at 20°C and 1 atm pressure rises to an altitude at which the temperature is −40°C and the pressure is 0.2 atm. Assuming that the balloon bag is free to expand, what is its new volume? (Hint: Use the ratio method as described in Sec. 15-7 in the text.)

14. Calculate the weight (in newtons) of this quantity of hydrogen gas.

15. Calculate the weight of the same volume of air at a temperature of $-40°C$ and a pressure of 0.2 atm.

16. What is the buoyant force on the balloon at this altitude?

17. The graph in Fig. 15-2 shows the pressure of helium gas at several temperatures.

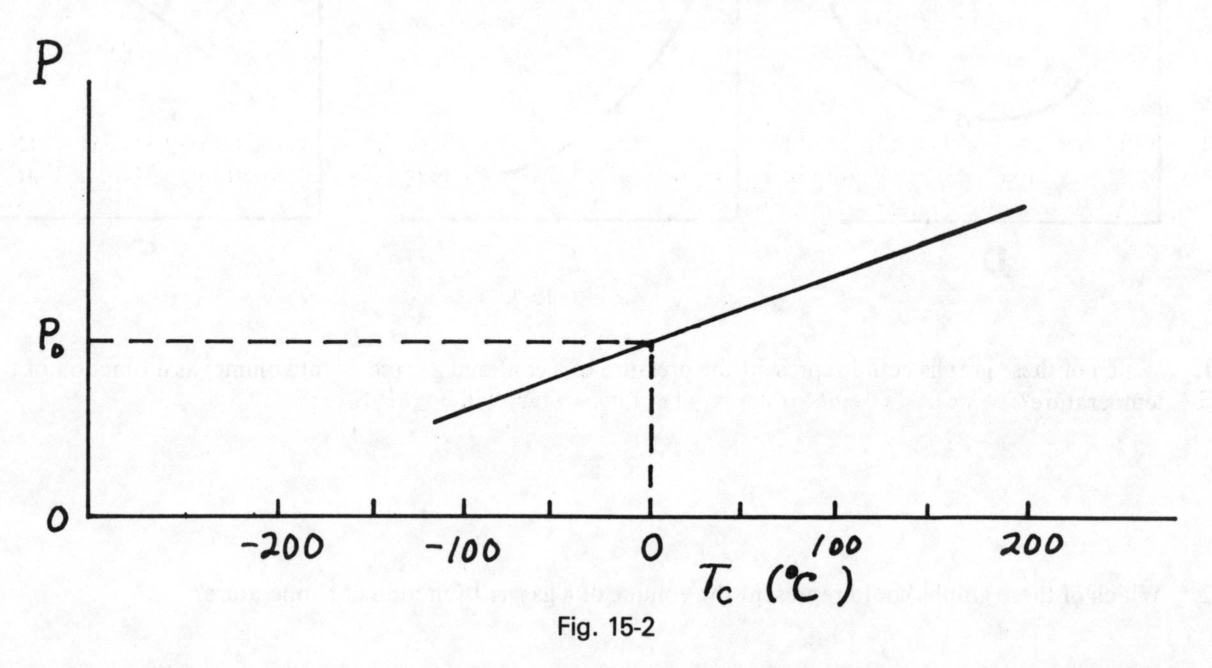

Fig. 15-2

a. What relationship do you find between P and T_C over the temperature range $-100°C$ to $+100°C$?

b. Make the *assumption* that this linear relationship holds down to zero pressure and find the temperature that corresponds to $P = 0$. What is the name of the temperature found in this way?

18. Suppose that the density of water vapor in the air within a room is 7.0 g/m^3. If the air were saturated with water vapor at that temperature, the density of water vapor would be 17.30 g/m^3. What is the relative humidity under these conditions?

19. Suppose that the temperature in a basement room is 20°C and the relative humidity is 80%. Use the information in Table 15-2 in the text to calculate the water vapor density and the partial pressure of water vapor under these conditions.

20. Supposing that the basement room in the preceding problem has a volume of 60 m^3, what is the mass of H_2O present?

21. Solid and liquid H_2O are in equilibrium at a pressure of 0.5 atm. Use the phase diagram for H_2O (Fig. 15-5 in the text) to estimate the equilibrium temperature. What is the name of this temperature? (*Assume* that the pressure scale is linear up to 760 torr.)

22. Assuming the same pressure (0.5 atm), at what temperature will boiling take place?

23. In what phases can H_2O exist if the pressure is less than 4.6 torr?

24. In what phases can H_2O exist if the temperature is above 20°C?

Solutions

1.

Measurable Quantity		Symbol	Units
(a)	pressure	P	atm, N/m^2
(b)	volume	V	m^3, liter
(c)	temperature	T_C, T	°C, K
(d)	mass	m	kg

2. PV = constant

3. $P_1V_1 = P_2V_2$; $V_2 = V_1P_1/P_2$

$$V_2 = 2.0 \text{ liters} \left(\frac{0.5}{2.5}\right) = 0.4 \text{ liters}$$

4. $P_1V_1 = (300 \text{ cm}^3)(0.8 \text{ atm}) = 240 \text{ cm}^3 \text{ atm}$
$P_2V_2 = (50 \text{ cm}^3)(4.5 \text{ atm}) = 225 \text{ cm}^3 \text{ atm}$

Since $PV = nRT$ and T was held constant, n must have decreased. Some of the gas leaked out.

5. $\dfrac{P_1V_1}{m_1} = \dfrac{P_2V_2}{m_2}$; $m_2 = m_1 \dfrac{P_2V_2}{P_1V_1}$

$$= (1.20 \text{ kg}) \frac{(25 \text{ atm})(56 \text{ liters})}{(18 \text{ atm})(56 \text{ liters})} = 1.67 \text{ kg}$$

Thus 0.47 kg of air must be added.

6. The number of moles in the cylinder depends only on the pressure, the volume, and the temperature, none of which has changed.

$$n = \frac{PV}{RT}$$

Thus the second cylinder contains 250 moles of CO gas.

7. Argon: $m_1 = (40 \text{ g/mole})(250 \text{ moles})$
CO: $m_2 = (28 \text{ g/mole})(250 \text{ moles})$
$m_1/m_2 = 1.43$

8. $PV = nRT$

9. $R = \dfrac{PV}{nT} = \dfrac{(12.5 \text{ atm})(493 \text{ liters})}{(250 \text{ moles})(300 \text{ K})}$

$$= 0.082 \frac{\text{liters} \cdot \text{atm}}{\text{mole} \cdot \text{K}}$$

10. P versus V at const. T: graph E
11. P versus T at const. V: graph A
12. V versus T at const. P: graph A

13. $\frac{P_1V_1}{T_1} = \frac{P_2V_2}{T_2}$; $V_2 = V_1\frac{P_1T_2}{T_1P_2}$

$$V_2 = 340 \text{ liters}^3 \left(\frac{1 \text{ atm}}{0.2 \text{ atm}}\right)\left(\frac{233 \text{ K}}{293 \text{ K}}\right) = 1350 \text{ liters} = 1.35 \text{ m}^3$$

14. $n = \frac{PV}{RT} = \frac{(1 \text{ atm})(340 \text{ liters})}{(8.21 \times 10^{-2} \text{ liters} \cdot \text{atm/mole} \cdot \text{K})(293 \text{ K})} = 14.1 \text{ moles of } H_2$

$m = (14.1 \text{ moles})(2.02 \text{ g/mole}) = 28.4 \text{ g}$
$W = mg = (2.84 \times 10^{-2} \text{ kg})(9.8 \text{ m/s}^2) = 0.278 \text{ N}$

15. Although air is a mixture of gases, the same method of calculation can be used.

$$n = \frac{PV}{RT} = \frac{(0.2 \text{ atm})(1350 \text{ liters})}{(8.21 \times 10^{-2} \text{ liters} \cdot \text{atm/mole} \cdot \text{K})(233 \text{ K})} = 14.1 \text{ ``moles'' of air}$$

(Note however that this calculation is not necessary!)

$W = mg = (14.1 \text{ moles})(29 \text{ g/mole})(10^{-3} \text{ kg/g})(9.8 \text{ m/s}^2) = 4.00 \text{ N}$

16. The buoyant force is the weight of the displaced air, in this case 4.0 N.
17a. In the range $-100°$C to $+100°$C (and beyond) pressure is a linear function of temperature.
b. A linear extrapolation to the horizontal axis gives an intercept of about $-250°$C. If the graph had been constructed more carefully, an intercept of $-273°$C would have been found. This temperature is absolute zero on the Celsius scale.
18. Relative humidity = 7.0/17.3 = 40.5%
19. Water vapor density = $(0.80)(17.30 \text{ g/m}^3) = 13.8 \text{ g/m}^3$
Partial pressure of H_2O = (0.80)(17.55 torr) = 14.0 torr
20. mass of $H_2O = (13.8 \text{ g/m}^3)(60 \text{ m}^3) = 0.83 \text{ kg}$
21. A horizontal line may be drawn on the graph halfway between the horizontal axis and the dashed line at P = 760 torr. This horizontal line crosses the curve representing the solid-liquid phase transition at approximately 5°C. This is the *melting point* of ice at 0.5 atm.
22. The horizontal line at P = 0.5 atm intersects the liquid-vapor transition curve at about 53°C. Thus the boiling point of water at 0.5 atm is about 53°C.
23. Below 4.6 torr only the solid and vapor phases are possible.
24. At temperatures greater than 20°C only the liquid and vapor phases can exist.

REVIEW TEST FOR CHAPTERS 13–15

1. The steel washer shown in Fig. T-29 has an outside diameter d of 2.0 cm and an inside diameter of 1.0 cm. When the washer is heated uniformly to 400°C the outside diameter becomes 2.01 cm, an increase of 0.5%. What will the inside diameter become at 400°C?
   a. 1.01 cm
   b. 0.99 cm
   c. 1.005 cm
   d. 0.995 cm
   e. 1.02 cm

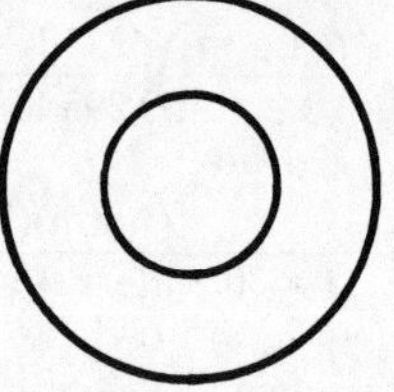

Fig. T-29

2. During a calorimetry experiment a metal has its temperature increased by 77°F. What is the corresponding temperature change on the Celsius scale?

3. What Celsius temperature corresponds to 77°F?

4. The coefficient of linear expansion for platinum is $\alpha = 9.0 \times 10^{-8}\ (°C)^{-1}$. A cylinder of platinum has a volume of 10 cm^3 at 20°C. By how much does the volume change when the cylinder is heated to 120°C?

5. A certain quantity of hydrogen gas occupies a volume of 600 cm^3 at 0°C. The gas is heated to 2.5°C, and the volume is allowed to increase just enough to keep the pressure constant at 1 atm. What is the volume of the gas at the higher temperature?

6. When 100 cal of heat is added to a system it is found that the system does work in the amount of 83.7 J. By how many calories does the internal energy of the system increase?

7. An electric heater adds 836 J of heat energy to 10 g of water originally at 10°C. What is the resulting temperature of the water?

8. An unknown amount of ice at 0°C is mixed with 10 g of steam at 100°C. The result is water at 100°C. Find the mass of the ice.

9. It is found that 640 cal of heat raises the temperature of 100 g of a substance from 6°C to 14°C. What is the specific heat capacity of the substance?

10. Which one of the three methods of heat transfer can take place when the bodies involved are separated by empty space?

11. A fixed mass of an ideal gas is contained within a cylinder whose volume can be varied. The temperature is kept constant. Which graph below represents the relationship between pressure and volume in this gas?

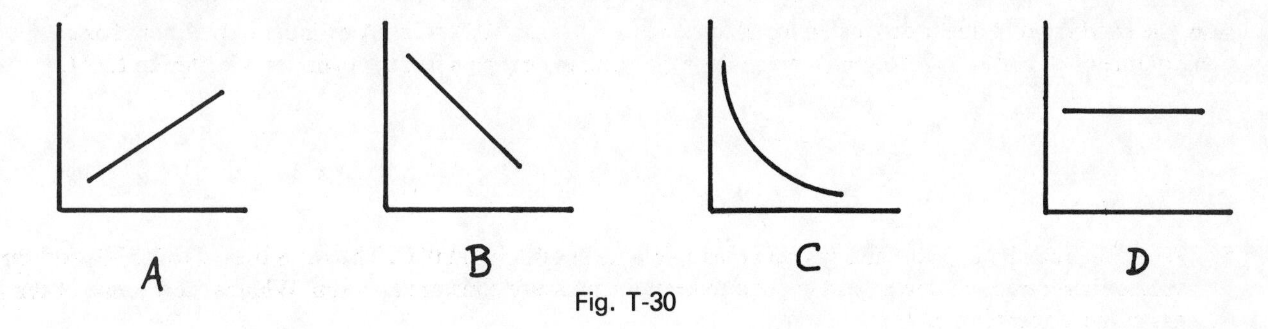

Fig. T-30

12. Two sealed flasks have equal volumes and are at the same pressure and temperature. The first flask contains 14×10^{21} molecules of N_2 gas. The second flask contains pure He. (The molecular weights are 28 and 4, respectively.) How many molecules of He are in the second flask?

13. The density of water vapor in a certain room is 10 g/m³. If the air were completely saturated at that temperature (about 20°C), the density of water vapor would be 17 g/m³. What is the relative humidity in the room?

14. Temperatures and pressures at which liquid and vapor are in equilibrium are specified by the
 a. vapor pressure curve
 b. triple point
 c. sublimation curve
 d. critical point

CHAPTER

16 The Kinetic Theory of Gases

GOALS To study the kinetic theory of gases and some of its predictions regarding the behavior of matter in the gaseous state.

OBJECTIVES After completing this chapter the student should be able to do the following:

1. Name the property of a gas or liquid that is proportional to the average translational KE per molecule due to random motions.
2. Write the equation relating the pressure and volume of a gas to the average speed of a molecule.
3. Given the equation $PV = (1/3)Nmv^2$, state the physical significance of each symbol.
4. Calculate one of the quantities P, V, m, or v when the others are known.
5. Calculate the KE of 1 mole of a gas from a knowledge of the universal gas constant R and the absolute temperature T.
6. Write the equation which defines Boltzmann's constant, k.
7. Calculate the KE per molecule of an ideal gas knowing Boltzmann's constant and the temperature.
8. Calculate the specific heat capacity at constant volume of an ideal gas.
9. Name the types of energy that a molecule may possess in addition to the KE of translation.
10. Calculate the work done by a gas when it expands (at constant pressure) against the surroundings.
11. Compare C_p and C_v, stating which is larger and why.
12. State the theoretical value of C_p/C_v for an ideal monatomic gas.
13. Name the extra type of internal energy possessed by a diatomic gas that is not present in a monatomic gas.
14. State the theoretical value of C_p/C_v for an ideal diatomic gas.

SUMMARY

In a gas or liquid, molecules are in continual random motion. Rapidly moving molecules collide with the walls of a container, giving rise to the phenomenon of pressure. An application of Newton's second law of motion to the collision process in an ideal gas leads to the result $PV = (1/3)Nmv^2$ where P and V are the pressure and volume of the gas, respectively, m is the mass, v the average speed, and N the number of molecules in the given volume.

The *absolute temperature* of a gas or liquid is proportional to the average *translational* KE per molecule due to the random molecular motions. This relationship may be written

$$\text{KE per molecule} = kT$$

where the constant k is the gas constant per molecule called Boltzmann's constant, $k = R/N_A$.

From a mathematical point of view, an ideal gas is a much simpler system than either a solid or a liquid. In fact, many bulk properties of a gas, including specific heat capacity, can be calculated directly from the mathematical model. In the case of helium, which is very close to the ideal gas of the kinetic theory, the calculated specific heat capacity is 0.745 cal/g·°C, in excellent agreement with the measured value.

The law of equipartition of energy states that the energy given to an assemblage of molecules is, on the average, equally shared among all possible modes of energy absorption. The specific heat capacity of monatomic gases, helium, neon, etc., is 2.98 cal/mole·°C. The specific heat capacity of a diatomic gas, such as H_2, O_2, or N_2, is about 4.9 cal/mole·°C. More heat energy is required to raise the temperature of a diatomic gas because some of the energy goes into increasing the *rotational* and *vibrational* KE of the molecules whereas only the *translational* KE affects the temperature.

The amount of heat required to raise the temperature of a gas depends on the way in which the operation is carried out. In particular, the heat required is greater when the gas is allowed to expand, keeping the pressure constant, and less when the gas is confined to a fixed volume. The specific heat capacities corresponding to these two processes are C_p and C_v. Their ratio, written $\gamma = C_p/C_v$, can be calculated, with the result $\gamma = 5/3$ for helium and other monatomic gases.

QUESTIONS AND PROBLEMS

1. Complete the sentence: The absolute temperature in a gas or liquid is proportional to the ______________ due to random motions.
2. Let m and v represent the mass and the average speed of a molecule in an ideal gas. Then the product of pressure and volume is:

   $PV =$ ______________.
3. Use the above equation to calculate the average speed of a molecule of H_2 at 0°C. (Using the procedure of Ex. 16-1 in the text, take the volume to be 22.4 liters and the pressure 1 atm.)
4. How does the average speed of a hydrogen molecule compare with that of an oxygen molecule at the same pressure and temperature? Obtain a formula for the ratio of the average speeds of two different gases, v_2/v_1, under identical conditions.
5. Suppose that the absolute temperature is doubled. What is the new average speed of a hydrogen atom?
6. Calculate the KE of 1 mole of argon gas at 200 K and then at 202 K.

7a. Calculate the average KE per molecule of an ideal gas at 27°C and at 227°C. (Boltzmann's constant is 1.38×10^{-23} J/molecule•K.)

b. Use the result of part a to calculate the mean speed of a molecule of argon and a molecule of helium, both at 27°C.

8. Given that the atomic weight of argon is 40, calculate the specific heat capacity at constant volume for argon using the data of Prob. 6.

9. Argon is a monatomic gas. Do you expect the measured specific heat capacity to agree with the calculation just carried out? Explain.

10. Compare the following two processes:

(1) 0.2 mole of helium gas is heated from 0°C to 50°C with the helium confined to a fixed volume.
(2) 0.2 mole of helium gas is heated from 0°C to 50°C with the helium allowed to expand while the pressure is kept constant.

a. In which process is the increase in internal energy equal to 30 cal?

b. In which process is work done by the gas against its surroundings?

c. How much work is done?

d. Which process requires the greater amount of heat?

11a. Give the technical term for a process in which the temperature does not change.

b. What is the term for a process in which no heat is transferred into or out of a system?

12. A gas, originally confined at high pressure in a rather small volume, is allowed to expand slowly enough that the temperature remains constant. This is, by definition, an isothermal process. Is it also an adiabatic process?

13. Which term, adiabatic or isothermal, correctly describes the melting of an ice cube in a glass of water?

Solutions

1. average translational KE per molecule
2. $PV = (1/3)\, Nmv^2$

3. Solving for v^2, $v^2 = \dfrac{PV}{(1/3)Nm}$

For H_2, $Nm = 2.0 \text{ g} = 2.0 \times 10^{-3} \text{ kg}$
1 atm $= 1.01 \times 10^5 \text{ N/m}^2$

$$v^2 = \frac{(1.01 \times 10^5 \text{ N/m}^2)(3)(0.0224 \text{ m}^3)}{2.0 \times 10^{-3} \text{ kg}}$$

$v^2 = 3.40 \times 10^6 \text{ m}^2/\text{s}^2$; $v = 1.84 \times 10^3 \text{ m/s}$

4. The average speed of a hydrogen molecule, 1.84×10^3 m/s, is greater than the average speed of an oxygen molecule, found in Example 16-1 in the text, to be 0.46×10^3 m/s. The ratio $v(H_2)/v(O_2)$ is exactly 4.

$$\frac{v_2}{v_1} = \frac{\sqrt{3PV/Nm_2}}{\sqrt{3PV/Nm_1}} = \frac{\sqrt{m_1}}{\sqrt{m_2}}$$

Since the O_2 molecule is 16 times more massive than the H_2 molecule, the average speed of the O_2 molecule must be $\sqrt{16} = 4$ times less.

5. Since $T \propto$ KE

$$\frac{T_2}{T_1} = \frac{(1/2)mv_2^2}{(1/2)mv_1^2} = \frac{v_2^2}{v_1^2}$$

Thus

$$v_2 = v_1\sqrt{T_2/T_1} = v\sqrt{2}$$
$$v_2 = 1.41(1.84 \times 10^3 \text{ m/s}) = 2.60 \times 10^3 \text{ m/s}$$

6. KE = $(3/2)nRT$
KE (at 200 K) = (1.50)(1 mole)(8.32 J/mole•K)(200 K) $= 2.50 \times 10^3$ J
KE (202 K) $= 2.525 \times 10^3$ J
[Note that KE (at 202 K) $= 1.01 \times$ KE (at 200 K).]

7a. The average KE per molecule is

$$\text{KE} = (3/2)kT = (3/2)(1.38 \times 10^{-23} \text{ J/molecule}\cdot\text{K})(27 + 273)\text{K} = 6.2 \times 10^{-21} \text{ J}$$

At the higher temperature, 500 K, the result is 10.3×10^{-21} J.

b. $\text{KE} = (1/2)mv^2;\ v = \sqrt{\frac{2\text{KE}}{m}}$

The mass of an argon atom is 0.04 kg/(6.02×10^{23}) $= 6.6 \times 10^{-26}$ kg.
The mass of a helium atom is ten times less, 6.6×10^{-27} kg.
Thus the mean speed of a molecule of argon is

$$v = [2(6.2 \times 10^{-21} \text{ J})/(6.6 \times 10^{-26} \text{ kg})]^{1/2} = 4.3 \times 10^2 \text{ m/s}$$

For the case of helium, the mean speed is

$$v = [2(6.2 \times 10^{-21} \text{ J})/(6.6 \times 10^{-27} \text{ kg})]^{1/2} = 13.7 \times 10^2 \text{ m/s}$$

8. The specific heat (at 200 K) is

$$c = \frac{Q/\Delta T}{m} = \frac{(2.50 \times 10^3 \text{ J}) \times (1 \text{ cal})}{40 \text{ g } (4.18 \text{ J}) \text{ 2 K}}$$

$$c = 7.5 \text{ cal/g} \cdot {}^\circ\text{C}$$

9. Yes, because all the internal energy should be in the form of translational KE.

10a. The increase in internal energy is the same in both cases. It is equal to the heat added in case (1) where no work is done.

$$\Delta U = \Delta\text{KE} = (3/2)nR\Delta T = (3/2)(0.2 \text{ mol})(2.0 \text{ cal/mol}\cdot\text{K})(50 \text{ K}) = 30 \text{ cal}$$

b. In case (2) work is done by the gas as it expands.

c. $\Delta W = nR\Delta T = (0.2 \text{ mol})(2.0 \text{ cal/mol}\cdot\text{K})(50 \text{ K}) = 20 \text{ cal}$

d. The second process requires more heat, 50 cal versus 30 cal for case (1).

11a. an isothermal process

b. an adiabatic process

12. No, heat must be added to the gas from the surroundings to maintain a constant temperature during the expansion.

13. The ice cube and the water in contact with it remain at 0°C while the ice melts. This is an isothermal process.

CHAPTER

17

The Second Law of Thermodynamics

GOALS To study some of the implications of the second law of thermodynamics, especially with regard to heat engines, the heat pump, and the refrigerator.

OBJECTIVES After completing this chapter the student should be able to do the following:

1. Write a correct statement of the first law of thermodynamics.
2. Write the technical term that is a measure of the disorder of a system.
3. Given two systems that differ only in their degree of disorder, state which system has the greater entropy.
4. Write a statement of the second law of thermodynamics involving the concept of entropy.
5. Write a statement of the second law of thermodynamics in terms of the transfer of energy from one place to another.
6. Identify the quantities appearing in this equation:

$$\frac{Q_H - Q_C}{Q_H} = \frac{T_H - T_C}{T_H}$$

7. Calculate the Carnot efficiency of an engine operating between given temperatures.
8. Calculate the first-law efficiency of a power plant, or other system, from a knowledge of the quantity of fuel consumed and the useful output.
9. Calculate the maximum coefficient of performance of a heat pump and the actual coefficient of performance, given the necessary performance data.
10. Write a definition of absolute zero based on the Carnot engine.

SUMMARY

The first law of thermodynamics is a generalization of the law of conservation of mechanical energy introduced in Chapter 6. The first law of thermodynamics may be written: "All the heat energy added to a closed system can be accounted for as mechanical work, an increase in internal energy, or both."

There is a universal tendency in nature for systems to become more disordered as time goes on. *Entropy* is the measure of the disorder of a system. Thus, the second law of thermodynamics may be stated as follows: "The entropy of the universe never decreases; during any process the entropy either remains constant or increases."

An equivalent statement of the second law of thermodynamics is the following: "It is impossible to have a process whose only result is to transfer energy from a cold place to a warm place."

A *heat engine* is a device that takes in heat energy and produces useful work. Only a portion of the input heat can be converted to useful work, the remainder is exhausted to a low temperature reservoir. The efficiency of a heat engine is the useful work output divided by the heat input; symbolically, $e = W/Q$. This ratio is called the *first-law efficiency*.

The maximum possible efficiency of a heat engine, assuming no frictional losses of any kind, depends only on the two temperatures between which the engine works. This *Carnot efficiency* is given by $(T_H - T_C)/T_H$ where T_H is the temperature of the hot reservoir and T_C is the temperature of the cold reservoir.

The laws of thermodynamics require that a large fraction of the heat produced by burning a fuel be exhausted to some low temperature reservoir, usually the atmosphere, sometimes a river or lake. Thus a large power plant unavoidably produces thermal pollution.

A *heat pump* is a device that moves heat energy from a low-temperature reservoir to a high-temperature reservoir. Work is required to operate a heat pump since heat does not naturally flow from a low temperature to a higher temperature. The *coefficient of performance* of a heat pump is the heat moved divided by the work input.

The concept of the Carnot engine leads to a rigorous definition of absolute zero as the temperature of a reservoir to which a Carnot engine would reject no heat.

QUESTIONS AND PROBLEMS

1. Let Q represent the quantity of heat added to a closed system, W the work done, and ΔU the increase in internal energy of the system. Write an equation connecting Q, W, and ΔU.

2. A perfectly ordered deck of playing cards has each suit separately arranged in exact numerical order—A, 2, 3, . . . J, Q, K. Such a deck of cards is subjected to the following treatments: (a) the deck is cut twice, (b) the deck is shuffled twice, and (c) the deck is thrown into the air, falls on the floor, and is picked up and stacked by a person who is blindfolded. Which of these three treatments results in the least entropy? Which the most?

3. Complete the following statement: The entropy of the universe never ____________; during any process, the entropy either ____________ ____________ or ____________.

4. Complete the statement: It is impossible to have a process whose only result is to transfer energy from a ____________ place to a ____________ place.

5. The two statements above are alternative, but equivalent, statements of the ____________ ____________ ____________.

6. An engine produces 30 J of work for every 47.8 cal of heat provided to it. What is the thermal efficiency, or first-law efficiency of this engine?

7. Suppose that the combustion temperature in a diesel engine is 1500 K and the exhaust temperature is 750 K. What is the maximum possible efficiency of this engine in the absence of friction and all other losses?

8. A certain diesel engine operates at 70% of its Carnot efficiency. Its exhaust temperature is 452°C and its overall efficiency (first-law efficiency) is such that in one hour an input of 3.2×10^9 cal of heat produces 5.5×10^9 J. What is the temperature in the combustion chamber?

9. The coal-fired, steam-electric plant described in Ex. 17-2 in the text has a Carnot efficiency of 43.1% and a first-law efficiency of 40%. Supposing that this ratio, $0.40/0.431 \cong 0.93$, can be maintained, and that the electric output is kept at 680 MW, what input temperature would be required to produce a 5% savings in the amount of coal burned? The output temperature is 40°C as before.

10. A window air conditioner transports thermal energy from the inside of a room to the outside at the rate of 4.0×10^6 J/h. The temperature of the cooling coil inside the room is 15°C and the exhaust temperature is 35°C.
    a. Calculate the maximum possible COP.

    b. Calculate the second-law efficiency of the air conditioner if the actual COP is 2.1.

    c. What is the required power rating of the motor that drives the air conditioner?

Solutions

1. $Q = \Delta U + W$
2. In treatment (a), the deck is still highly ordered; most cards are still in sequence. Treatment (a) produces the lowest entropy. Treatment (c) produces more disorder than treatment (b) because many of the cards will now be face up. It has the highest entropy.
3. The entropy of the universe never *decreases*; during any process, the entropy either *remains constant* or *increases.*
4. It is impossible to have a process whose only result is to transfer energy from a *cold* place to a *warm* place.
5. Second law of thermodynamics
6. $$e = \frac{\text{work output}}{\text{heat input}} = \left(\frac{30\text{ J}}{47.8\text{ cal}}\right)\left(\frac{1\text{ cal}}{4.186\text{ J}}\right) = 0.15$$
7. The maximum possible efficiency is the Carnot efficiency

$$\text{Eff} = \frac{T_H - T_C}{T_H} = \frac{1500 - 750}{1500} = 50\%$$

8. First find the first-law efficiency of the engine:

$$\text{efficiency} = \frac{W}{Q_H} = \frac{5.5 \times 10^9 \text{ J (1 cal)}}{(3.2 \times 10^9 \text{ cal})(4.18 \text{ J})} = 0.41$$

(Carnot efficiency)(0.70) = 0.41

$$\text{Carnot efficiency} = 0.59 = \frac{T_H - 725 \text{ K}}{T_H}$$

$T_H - 0.59\, T_H = 725$ K; $0.41\, T_H = 725$ K
$T_H = 1.77 \times 10^3$ K

9. Calculate the new heat input.

$Q_H = 0.95\,(1.48 \times 10^{14} \text{ J}) = 1.41 \times 10^{14}$ J

The overall efficiency is W/Q_H

$$\frac{(680 \times 10^6 \text{ J/s})(8.64 \times 10^4 \text{ s})}{1.41 \times 10^{14} \text{ J}} = 0.417$$

The new Carnot efficiency is 0.417/0.93 = 0.448

$$0.448 = \frac{T_H - (273 + 40) \text{ K}}{T_H}$$

$T_H - 0.448\, T_H = 313$ K
$0.552\, T_H = 313$ K; $T_H = 567$ K = 294°C

A 5% savings in fuel consumption would result from an increase in the input steam temperature from 277°C to 294°C.

10a. $$\text{COP}_{max} = \frac{T_{cold}}{\Delta T} = \frac{(15 + 273) \text{ K}}{(35 - 15) \text{ K}} = 14.4$$

b. $$E = \frac{\text{COP}_{actual}}{\text{COP}_{max}} = \frac{2.1}{14.4} = 0.146$$

c. $$W_{in} = \frac{\text{heat moved}}{\text{COP}} = \frac{4.0 \times 10^6 \text{ J}}{(2.1)} = 1.9 \times 10^6 \text{ J}$$

$$\text{Power} = \frac{W}{t} = \frac{1.9 \times 10^6 \text{ J}}{3600 \text{ s}} = 529 \text{ W}$$

$$529 \text{ W} = \frac{1 \text{ hp}}{746 \text{ W}} = 0.71 \text{ hp}$$

REVIEW TEST FOR CHAPTERS 1–17

1. All mechanical quantities may be expressed in terms of three basic dimensional quantities. These are
   a. weight, mass, time
   b. length, weight, volume
   c. time, acceleration, length
   d. length, time, weight
   e. time, mass, length
2. Which of the following equations is *not* dimensionally consistent?
   a. $v = \Delta s/\Delta t$
   b. $Fs = (1/2)mv^2$
   c. $v^2 = 2as$
   d. $mv = Ft$
   e. $Fv = 2mv/t^2$
3. A solid sphere 4.0 cm in diameter has a mass of 47.5 g. What is the specific gravity of this material?
   a. 2.83
   b. 1.4
   c. 3.53
   d. 8.8
   e. 5.95
4. A small cart rolls without friction down an inclined plane. Starting from rest and moving with a constant acceleration, it covers the first 50 cm in 2 s. How much time is required to traverse the second 50 cm?
   a. 1.0 s
   b. 0.82 s
   c. 1.41 s
   d. 2.0 s
   e. 0.41 s
5. A projectile is fired with initial velocity v_0 directed 30° upward from the horizontal. When it reaches the highest point of its path, the projectile has an acceleration of
   a. 9.8 m/s², directed downward
   b. 9.8 (cos 30°) m/s², horizontal
   c. 9.8 (sin 30°) m/s², horizontal
   d. 9.8 m/s², upward
   e. zero
6. The mass of the earth is approximately 6×10^{24} kg; that of the moon is 7.4×10^{22} kg, approximately 80 times less. Let F represent the magnitude of the gravitational force that the earth exerts on the moon. What is the magnitude of the force that the moon exerts on the earth?
   a. $80F$
   b. $F/80$
   c. $F/\sqrt{80}$
   d. F
   e. $\sqrt{80}\,F$
7. What is the acceleration of the 6-kg mass in Fig. T-31? Assume that there is no friction. Express the result in terms of g, the acceleration of free fall.
   a. $(1/5)g$
   b. $(1/3)g$
   c. $(1/6)g$
   d. $(1/2)g$
   e. g

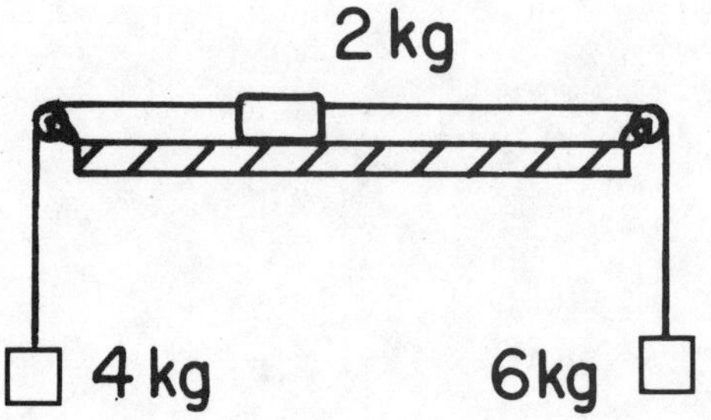

2 kg

4 kg

6kg

Fig. T-31

8. In physics the term "equilibrium" is given a precise and limited meaning. In which of the following situations is the italicized object *not* in equilibrium?
   a. A *car* is parked on a hill with its brakes securely locked.
   b. A small *meteor fragment* is speeding through interstellar space millions of light-years from any other object.
   c. An *automobile* is rounding a curve on a level road at a constant speed of 50 mi/h.
   d. A *4 kg mass* is subject to three forces such that $\mathbf{F}_1 + \mathbf{F}_2 = -\mathbf{F}_3$.
   e. A *250 lb swimmer* is sitting motionless on the end of a diving board.

9. A cart of unknown mass is released from rest and rolls, without friction, a measured distance *s*, in a time *t* (see Fig. T-32). This experiment, if carried out on the surface of the moon, would allow us to calculate
   a. the mass of the cart
   b. the weight of the cart when returned to the earth
   c. the acceleration due to gravity on the moon
   d. the radius of the moon
   e. the earth-to-moon distance

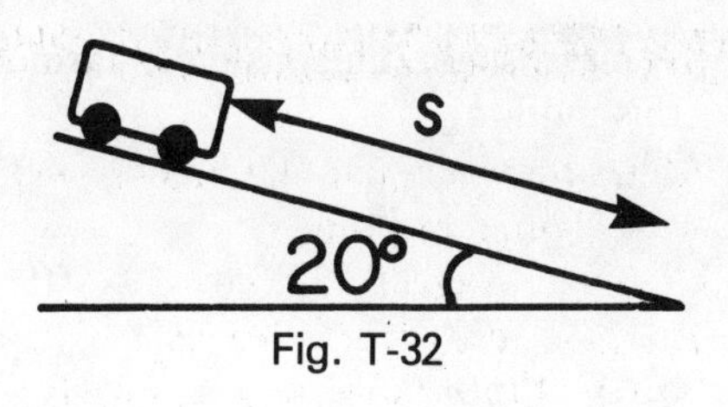

Fig. T-32

10. A rubber ball falls vertically on a concrete ramp and bounces off in a horizontal direction (see Fig. T-33). The drawing shows the momentum before the collision $\mathbf{P}_1$ and the momentum after the collision $\mathbf{P}_2$. Which arrow below represents the *change* in momentum that has taken place?

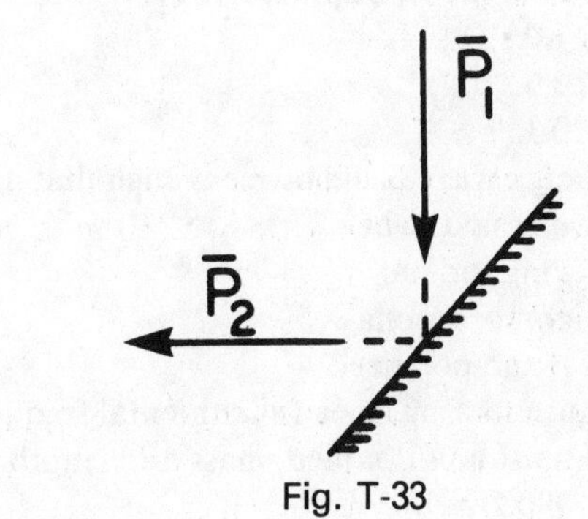

Fig. T-33

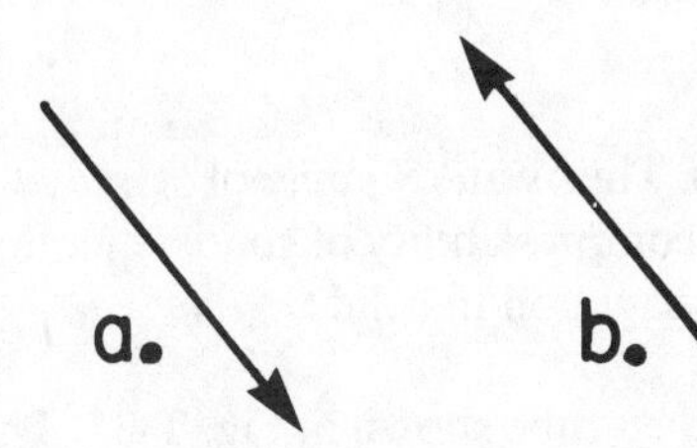

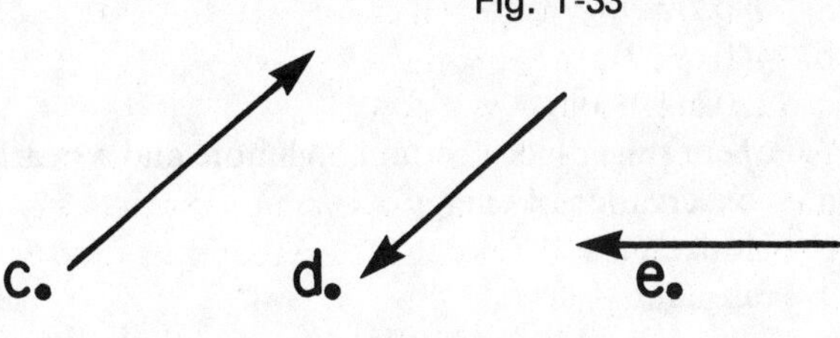

11. A satellite in circular orbit about the earth is traveling at a speed of 7.5×10^4 m/s. The force of gravitational attraction toward the earth (see Fig. T-34) is 200 N. How much work is done by this force during a displacement of 4 m?
   a. 1500 J
   b. 800 J
   c. 50 J
   d. 0.02 J
   e. zero

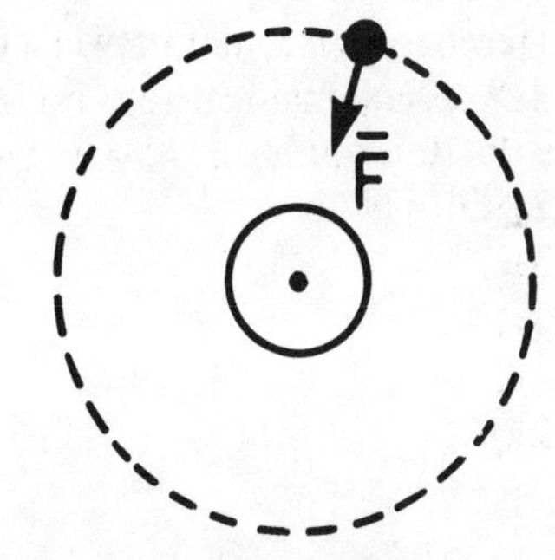

Fig. T-34

12. An object is moving around a circle at a constant angular speed of 5 rad/s. The radius of the circle is 0.05 m. What is the magnitude of the acceleration?
   a. $200\,\pi$ m/s^2
   b. $\pi/2$ m/s^2
   c. $5\,\pi^2$ m/s^2
   d. zero
   e. 1.25 m/s^2

13. A disk with moment of inertia $I = 10$ kg$\cdot$m^2 is initially at rest. A torque $\tau = 100$ Nm is applied to the disk. What is the angular acceleration α of the disk?
   a. 1000 rad/s^2
   b. 10 rad/s^2
   c. 10 rev/s^2
   d. 1000 rev/min^2
   e. 1000 rad/min^2

14. A mass is vibrating in simple harmonic motion. Which of the following relationships is *not* applicable to this motion?

a. $x = A \sin \frac{2\pi t}{T}$
b. $T \propto \sqrt{m/k}$
c. $v = v_0 + at$
d. $F = ma$
e. PE + KE = a constant

15. A small object is vibrating in simple harmonic motion with a period of 4 s and a frequency of 0.25 vib/s. How long does it take this object to move from the midpoint of its path (the equilibrium position) to a point where its displacement is $A/2$?

a. 1.0 s
b. 0.5 s
c. 0.33 s
d. 0.67 s
e. Depends on the value of A.

16. When a wave disturbance is such that all particles in the underlying medium move in the same plane, the wave is said to be

a. longitudinal
b. compressional
c. plane-polarized
d. Doppler-shifted
e. none of these

17. A guitar string has a fundamental frequency of 440 Hz. What will the fundamental frequency become if the tension is doubled, mass and length remaining the same?

a. 880 Hz
b. $440\sqrt{2}$ Hz
c. 220 Hz
d. $440/\sqrt{2}$ Hz
e. 1760 Hz

18. An opera singer hits a certain high note and a nearby glass shatters. This is an example of

a. conservation of energy
b. Hooke's law
c. resonance
d. the bulk compressibility of solids
e. wave propagation in solids

19. An incompressible fluid is flowing smoothly through a horizontal glass tube shown in Fig. T-35. The velocity of the fluid is 0.6 m/s at A and 1.2 m/s at B. The density of the fluid is 8.0×10^2 kg/m³. What is the difference in pressure between points A and B?

a. 2.4×10^2 N/m²
b. 4.3×10^2 N/m²
c. 7.2×10^2 N/m²
d. 8.6×10^2 N/m²
e. 4.2×10^3 N/m²

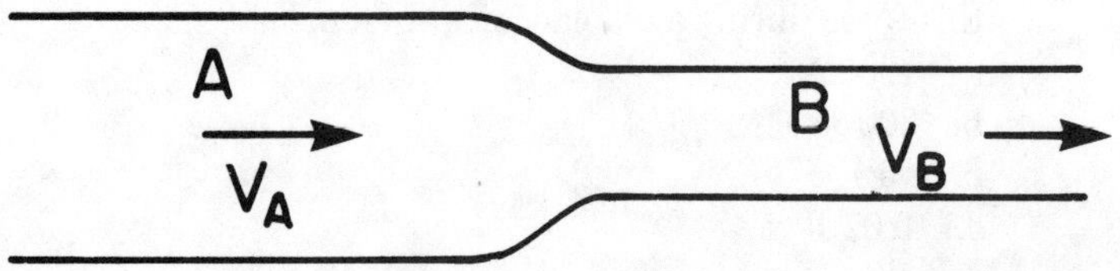

Fig. T-35

20. When a wooden rod 12 cm long is allowed to float in a water-filled, graduated cylinder, 4 cm of the rod project above the surface (see Fig. T-36). What is the specific gravity of the wood?

a. 0.25
b. 0.33
c. 0.50
d. 0.67
e. 1.50

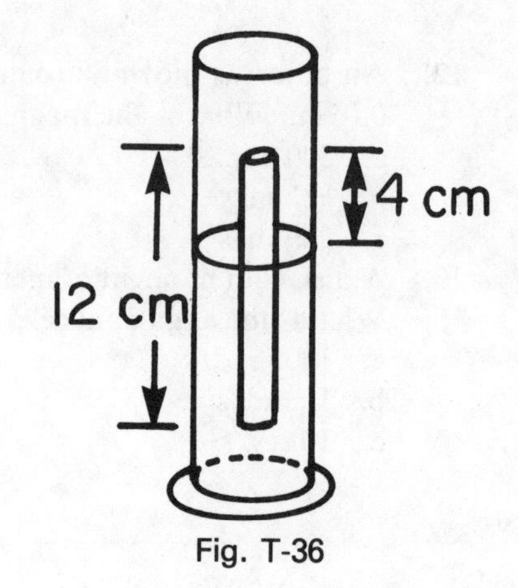

Fig. T-36

21. An aluminum plate 40 cm × 18 cm contains a rectangular hole with inside dimensions 8 cm × 3 cm (see Fig. T-37). When the plate is heated uniformly, its length (40 cm) increases by 0.002%. In what way will the dimension d change?

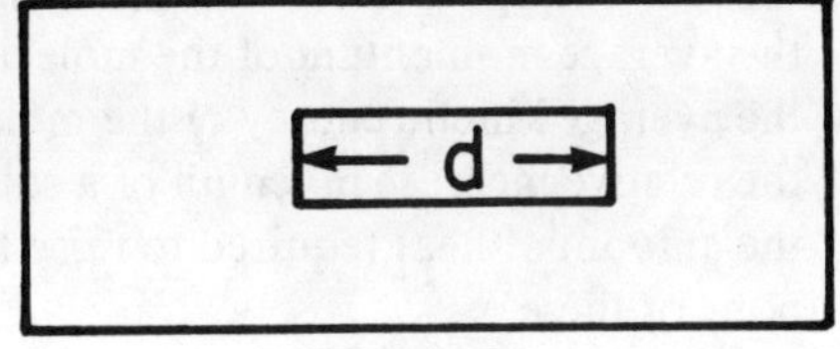

Fig. T-37

a. increase by 1.6×10^{-4} cm
b. increase by 8.0×10^{-4} cm
c. decrease by 1.6×10^{-4} cm
d. decrease by 8.0×10^{-4} cm
e. remain the same

22. It is found that 640 cal of heat raises the temperature of 100 g of a substance from 6°C to 14°C. What is the specific heat capacity of the substance?
a. 0.1 cal/g·°C
b. 0.05 cal/g·°C
c. 0.8 cal/g·°C
d. 1.05 cal/g·°C
e. 0.44 cal/g·°C

23. A fixed mass of an ideal gas is contained within a cylinder whose volume can be varied. The temperature is kept constant. Which of the following graphs is a correct description of the relationship between pressure and volume?

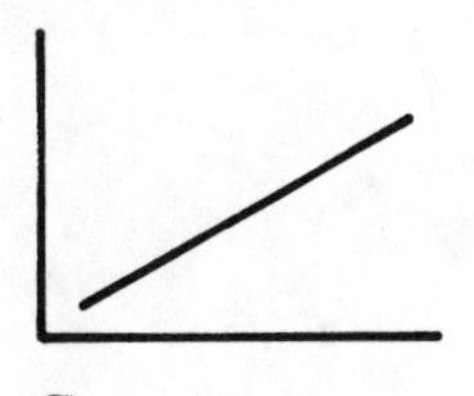
a.

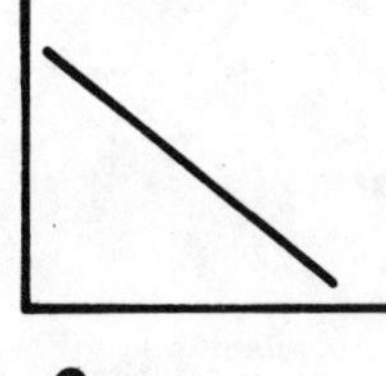
c.

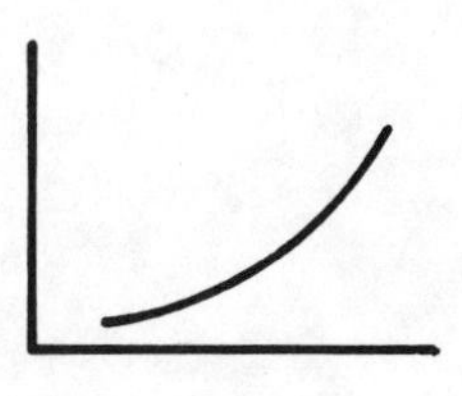
e.

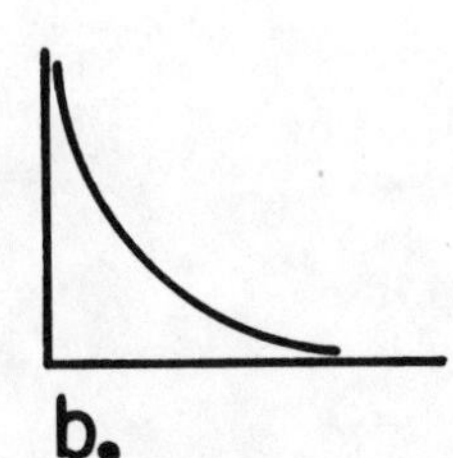
b.

d.

Fig. T-38

24. Temperatures and pressures at which liquid and vapor are in equilibrium are specified by the
a. vapor pressure curve
b. triple point
c. sublimation curve
d. critical point
e. none of these

25. The absolute temperature T is proportional to
   a. the average momentum of the molecules of a gas
   b. the average kinetic energy of the molecules of a gas
   c. the relative increase in length of a solid for a temperature increase of 1 °C
   d. the amount of heat required to raise the temperature of a solid by 1 °C
   e. none of these

26. The conclusion that no heat engine can convert its entire heat input into work is a consequence of
   a. the first law of thermodynamics
   b. the second law of thermodynamics
   c. the conservation of energy
   d. the ideal-gas law
   e. none of these

CHAPTER

18 Electric Charge

GOALS To gain an understanding of how electrically charged bodies interact with one another.

OBJECTIVES After completing this chapter the student should be able to do the following:

1. Associate the term *Coulomb force* with the force between static charges and the term *magnetic force* with the force between moving charges (or currents).
2. Name the two kinds of electric charge and state which one is normally free to "flow" in most large-scale applications.
3. State which type of charge is associated with the nucleus of an atom and which type is associated with the cloud of electrons that surround the nucleus.
4. Describe a process by which a body may be given an excess of charge.
5. Name two good conductors of electricity and two poor conductors (insulators).
6. Describe what happens to the movable leaf of an electroscope when an excess charge of either sign is placed on the knob.
7. Describe where the charge is located when a conducting body is given an excess static charge.
8. State the law of conservation of electric charge.
9. Solve simple problems involving the law of conservation of electric charge.
10. State Coulomb's law in words and in mathematical form using the commonly accepted symbols.
11. Name the SI unit of charge and the SI unit of current.
12. Write, in conventional symbols, the relationship between charge, current, and time.
13. Solve for any of the three quantities Q, I, or t when the other two are given.
14. Calculate the electron flow from the current in amperes and vice versa.
15. Calculate any of the quantities F, k, Q, or r when the others are given.
16. Solve problems involving the ratio of forces, charges, and distances.
17. Describe what happens to the molecules of silver nitrate ($AgNO_3$) when a small quantity of silver nitrate is dissolved in water.
18. Name the charge carriers in a solution of $AgNO_3$ in water.
19. Solve problems involving electrolysis in which the unknown quantity is current, time, atomic mass, or charge on the ion.

SUMMARY

Electrically charged particles, which are basic constituents of all matter, exert forces on one another. If the charges are at rest, the electric force follows a fairly simple law called Coulomb's law.

$$F = k\frac{Q_1Q_2}{r^2}$$

Like the gravitational force between two masses,

$$F = G\frac{m_1 m_2}{r^2}$$

the Coulomb force varies inversely with the square of the distance. There is, however, a fundamental difference: the electrostatic force may be either attractive or repulsive whereas the gravitational force is always attractive. In other words, electric charge may be either positive or negative; mass is always positive.

If the charges are in motion, additional forces appear that are, in general, neither attractive nor repulsive. These are the more complicated magnetic forces, which will be studied in later chapters.

Most of the electrons in the atoms of matter are tightly bound to the nucleus. In some materials, notably the metals, a few of the outer electrons of each atom are free to move through the crystal. Thus the metals are good conductors of electricity. Materials in which all electrons are tightly bound, such as glass and rubber, are very poor conductors, and are called *insulators*.

An important characteristic of electric charge is that it is *conserved*. This means that the total charge, positive plus negative, on any isolated system must remain constant. The smallest unit of charge of ordinary matter is the charge on the electron. In electrical measurements a much larger unit of charge is used, the coulomb, which is equal to the charge of 6.24×10^{18} electrons.

An electrolyte is a solution that readily conducts electricity. The conductivity in this case does not depend on the movement of free electrons, as in a metal, but the movement through the solute of positive and negative ions. Electrolytes play an important role in the plating of metals and in the operation of batteries.

QUESTIONS AND PROBLEMS

Secs. 18-1—18-4

1. The force that exists between stationary charged bodies is called the ______________ force. If the charged bodies are in motion additional forces appear; these are called ______________ forces.
2. The two kinds of electric charges are ______________ and ______________. In a metal there are free electrons that carry the current. Thus it is only the ______________ charges that are free to move in an ordinary conductor.
3. The symbol for an isotope of iron is $^{54}_{26}Fe$. The nucleus of this atom contains ______________ protons and ______________. (Give the numbers of each particle. See Chapter 1 if necessary.) The nucleus of this iron atom is surrounded by a cloud of ______________ (how many?) electrons.
4. Some of the properties of electrically charged bodies can be investigated with nothing more than a comb made of plastic or hard rubber and a few tiny bits of paper not more than a few millimeters in any dimension. Try various means of charging the comb, such as rubbing it on portions of your clothing, running it through your hair, or even rubbing it with a sheet of paper. After any of these treatments you should find that small bits of paper will be attracted to the comb, perhaps from as far as an inch away. You can compare the effectiveness of several techniques for charging the comb by observing how far the bits of paper will jump.

   Which technique for charging the comb seemed to be the most effective?

5. Assuming that the comb receives a negative charge of Q coulombs, what is the charge on the object with which it is rubbed?

6. Since the bits of paper were electrically neutral, why were they attracted to the charged comb?

7. You probably noticed that the paper, after being attracted to the comb and resting there for a few seconds, subsequently flew away again. Why?

8. A plastic pen can be charged in the same way but you will probably not be able to charge a metallic object; why not?

9. What does it mean to say that charge is conserved? To which of the above questions does this concept apply? Can you name other physical quantities that are conserved?

10. Make a list of five good conductors and five insulators.

Secs. 18-5—18-6

11. Write the law of force between two charges at rest. This is called C____________ law.

12. The current flowing through the bulb of a flashlight is about 0.1 A. How much charge passes through the filament of the bulb in 15 minutes? (The charge should be expressed in coulombs.)

Q = ____________ .

13. How many electrons pass through this filament in 0.2 s?

14. A positive charge of 12 μC ($\mu = 10^{-6}$) is 20 cm from another positive charge of 2 μC. Is the force attractive or repulsive? What is the magnitude of the force?

F = ______________.

15. When two electric charges are separated by a distance of 0.1 m the Coulomb force is F. What does the force become when the distance is increased to 0.3 m?

F' = ______________.

16. In a metallic conductor, copper for example, the charge is carried entirely by the ______________, the positive charges on the nuclei not being free to move. In an electrolyte, on the other hand, the carriers of charge are ______________ and ______________ moving in opposite directions.

17. A current of 12 A is maintained for 30 min through an electrolytic cell containing Ag^+ ions.
    a. How many atoms of silver are deposited on the negative electrode?

    ______________ atoms.

    b. What mass of silver is deposited?

    m = ______________ grams.

    c. If the silver is deposited uniformly over a surface of 20 cm², how thick is the deposit? (The density of silver is 10.5 g/cm³.)

    thickness = ______________ mm.

18. How long will it take to plate out 4 mg of copper from a solution containing Cu^{++} ions if the current available is 0.5 A?

Chapter Review

19. An uncharged metallic cup is placed on an insulating surface, illustrated in Fig. 18-1. A positively charged sphere is let down into the cup, allowed to touch the bottom, and then removed without touching the cup again.

    Describe the resulting states of charge of both the sphere and the cup. Explain what happened.

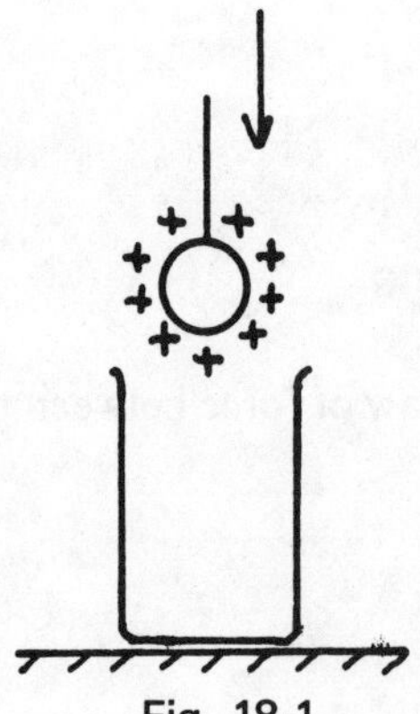

Fig. 18-1

20. Using the equipment described in Prob. 19, a slightly different experiment is performed. The positively charged sphere is lowered almost to the bottom of the cup but is not allowed to touch the cup. Then the outside of the cup is connected momentarily to ground while the sphere remains in position. Finally the sphere is pulled out of the cup without touching it.

Again describe the resulting states of charge of both the sphere and the cup. Explain any differences in the outcome of the two experiments. Give the technical term for this method of charging an object.

21. Two insulated spheres are connected by a thin wire, as shown in Fig. 18-2. When a positively charged rod is brought up close to the sphere on the right (but not touching it) that sphere acquires a charge of $-6\ \mu C$. The charge on the other sphere is

a. negative and much less than $6\ \mu C$.
b. positive and much less than $6\ \mu C$.
c. zero
d. equal to $-6\ \mu C$.
e. equal to $+6\ \mu C$.

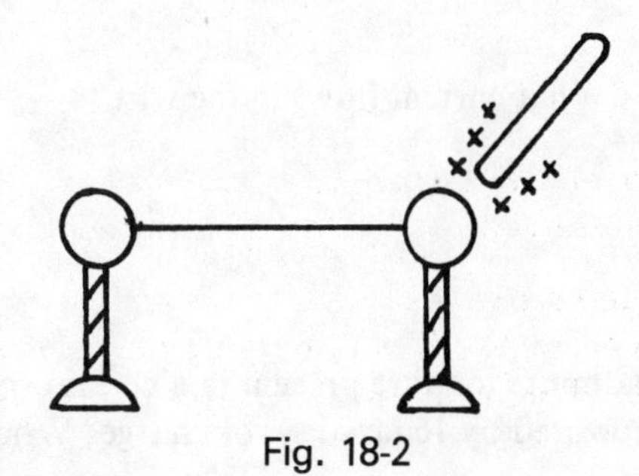

Fig. 18-2

22. A charge of $2\ \mu C$ is separated by a distance d from a second charge of $8\ \mu C$. Charge Q_1 ($2\ \mu C$) exerts a force of 4 N on charge Q_2 ($8\ \mu C$). What is the magnitude of the force that Q_2 exerts on Q_1?

a. 1 N
b. 2 N
c. 4 N
d. 8 N
e. 16 N

23. A point charge q is placed at a distance d_1 from a larger point charge Q_1. The force acting on q in this case is $F_1 = 8 \times 10^{-8}$ N. When the point charge q is placed at a distance d_2 from another point charge Q_2, the force that q experiences is $F_2 = 2 \times 10^{-8}$ N. If $Q_2 = 3\ Q_1$, what is the ratio of the distances d_2/d_1?

24.

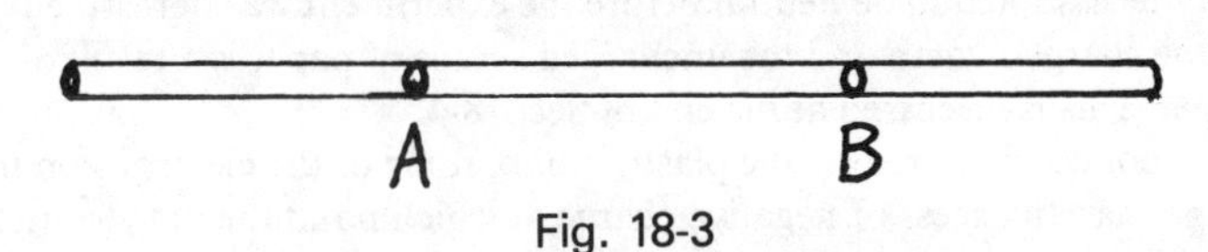

Fig. 18-3

Electrons are moving through the wire shown in Fig. 18-3 at the rate of 3.6×10^{18} electrons/s.

a. How many electrons pass point A during an interval of $0.75\ \mu s$?

b. How many electrons pass point B during the same interval?

c. What conservation law was invoked in answering (b)?

d. In what way would your answer to (b) change if the wire had a larger diameter at B?

e. What current flows in the wire?

25. Cadmium ions are present in a certain solution in the form Cd^{++}. How many moles of cadmium will be deposited by 16 faradays of charge? What is the corresponding mass of cadmium?

Solutions

1. Coulomb; magnetic
2. positive; negative; negative
3. 26 protons; 28 neutrons; 26 electrons
4. If your hair is dry it will probably allow a greater charge to be transferred to the comb than any other method.
5. A negative charge of Q transferred to the comb will leave a positive charge of Q on the object with which it was rubbed. Both objects may be assumed to be neutral before the experiment was performed.
6. The attraction between the charged comb and the uncharged scraps of paper is a result of induced charges that appear on the surface of the paper. This is discussed at the end of Sec. 18-4.
7. Although paper is a very poor conductor, as is the plastic comb, some of the electrons on the comb will slowly transfer to the paper until the paper has an excess of negative charge at which point the net electrical force becomes repulsive.
8. Because a metal is a conductor, any charge placed on it rapidly spreads over the entire surface. If you were holding the object in your fingers while rubbing it, any charge that might have been transferred to it would be conduced away by your body.
9. Charge can move from place to place but it can neither be destroyed nor created. The algebraic sum of the electric charge in any closed system remains constant.
10. All of the metals are conductors. Some insulators are wood, paper, asbestos, rubber, plastic, paint, varnish, all gases (except under high voltage), pure water, pure alcohol, etc.
11. Coulomb's law:

$$F = \frac{k\,Q\,Q'}{r^2} \quad \text{or} \quad F = \frac{Q\,Q'}{4\,\pi\,\epsilon_0\,r^2}$$

12. $Q = It = (0.1\ \text{C/s})(15 \times 60\ \text{s}) = 90\ \text{C}$

13. $Q = It = (0.1\ \text{C/s})(0.2\ \text{s})\left(\dfrac{1\ \text{electron}}{1.60 \times 10^{-19}\ \text{C}}\right) = 1.25 \times 10^{17}\ \text{electrons}$

14. The force between charges of like sign is always repulsive.

$$F = \frac{k\,Q\,Q'}{r^2} = \frac{(9 \times 10^9\ \mathrm{Nm^2/C^2})(12 \times 10^{-6}\ \mathrm{C})(2 \times 10^{-6}\ \mathrm{C})}{(0.2\ \mathrm{m})^2} = 5.4\ \mathrm{N}$$

15. The new distance is three times greater. Thus r^2 is nine times greater and $1/r^2$ is nine times less. Thus $F' = F/9$.

16. free electrons; positive and negative ions

17a. $Q = It = (12\ \mathrm{C/s})(30 \times 60\ \mathrm{s})(1\ \mathrm{elec}/1.6 \times 10^{-19}\ \mathrm{C}) = 1.35 \times 10^{23}$ electrons. Thus 1.35×10^{23} Ag atoms are deposited.

b. Mass of Ag $= (1.35 \times 10^{23}\ \text{atoms})\dfrac{108\ \mathrm{g}}{6.02 \times 10^{23}\ \text{atoms}} = 24.2\ \mathrm{g}$

c. $d = m/V$; $V = m/d$
Volume = area × thickness; $V = At$

Thus

$At = m/d$; $t = m/(d\,A)$

$$t = \frac{24.2\ \mathrm{g}}{(20\ \mathrm{cm}^2)(10.5\ \mathrm{g/cm^3})} = 0.115\ \mathrm{cm} = 1.15\ \mathrm{mm}$$

18. $Q = It$; $t = Q/I$

Find the charge first.

$$Q = (4 \times 10^{-3}\ \mathrm{g})\left(\frac{6.02 \times 10^{23}\ \text{atoms}}{63.6\ \mathrm{g}}\right)\left(\frac{1.60 \times 10^{-19}\ \mathrm{C}}{1\ \text{electron}}\right)\left(\frac{2\ \text{elec}}{\text{atom}}\right)$$

$Q = 12.1\ \mathrm{C}$ $\quad I = 0.5\ \mathrm{A} = 0.5\ \mathrm{C/s}$
$t = Q/I = (12.1\ \mathrm{C})/(0.5\ \mathrm{C/s}) = 24.2\ \mathrm{s}$

19. During contact with the inside of the cup the sphere loses all of its charge, becoming electrically neutral. The positive charge originally on the sphere is distributed on the *outer* surface of the cup. There is no further change when the sphere is removed.

20. The region inside a conductor must always be electrically neutral. Thus a negative charge must appear on the inner surface of the cup exactly equal to the positive charge Q on the sphere. Since the cup is insulated, it must remain electrically neutral, which is possible only if a positive charge equal to Q appears on its outer surface. In other words, as the positively charged sphere is let down into the cup, electrons flow from the outside of the cup to the inside, which effectively neutralizes the positive charge on the sphere.

When the outer surface of the cup is connected to ground, enough electrons flow from ground to neutralize the positive charge on the outer surface of the cup. Finally, when the inner sphere is removed, the excess electrons on the cup distribute themselves over its outer surface. The three steps of the process, called *charging by induction*, are illustrated in Fig. 18-4.

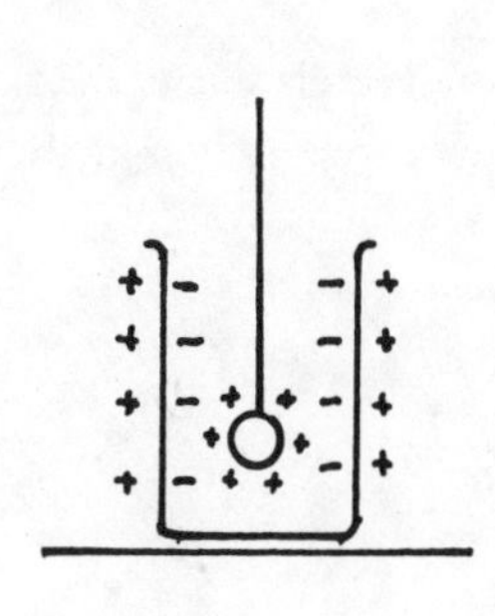

1. Total charge "inside" cup is zero.

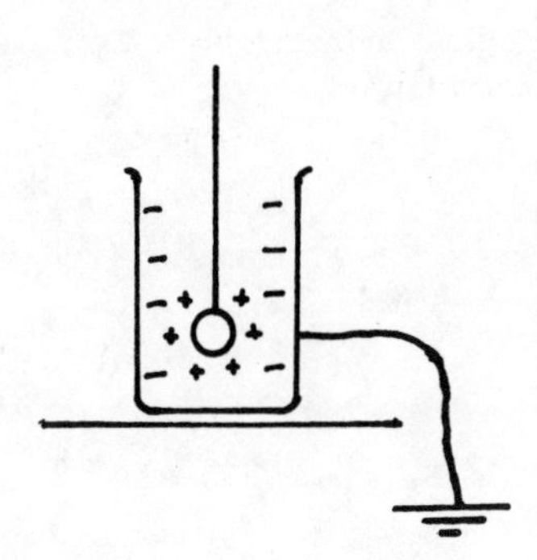

2. Total charge "inside" cup is zero.

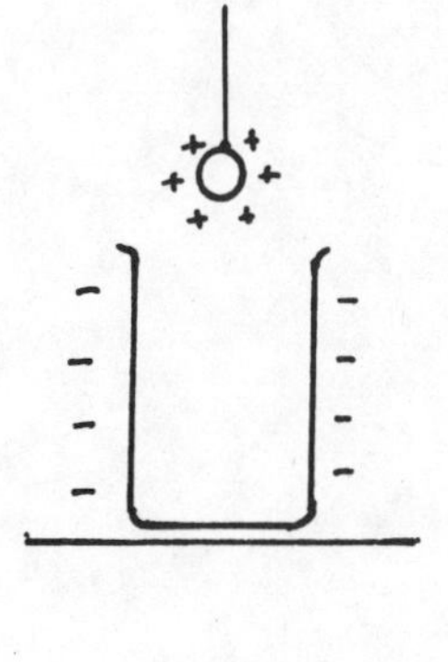

3. Total charge "inside" cup is zero.

Fig. 18-4

21. The answer is e. Since the two spheres are on insulated stands, they can neither gain nor lose charge. Since "the algebraic sum of the electric charge in any closed system remains constant," it follows that the sphere on the left must have a charge of exactly $+6\ \mu C$.

22. The answer is c. Newton's third law of motion states that if object A exerts a force on object B, then object B must exert an *equal and opposite* force on object A. This statement is *always* true. It holds for gravitational forces, electrical forces, and all other forces.

23. The equations for the two cases are:

$$F_1 = \frac{k\,q\,Q_1}{d_1^2} \quad \text{and} \quad F_2 = \frac{k\,q\,Q_2}{d_2^2}$$

Dividing the first equation by the second gives,

$$\frac{F_1}{F_2} = \frac{Q_1}{Q_2}\left(\frac{d_2}{d_1}\right)^2. \quad \text{Thus} \quad \frac{d_2}{d_1} = \sqrt{\frac{F_1 Q_2}{F_2 Q_1}}$$

$$\frac{d_2}{d_1} = \sqrt{(4)(3)} = 3.46$$

24a. $(3.6 \times 10^{18}$ elec/s$)(0.75 \times 10^{-6}$ s$) = 2.7 \times 10^{12}$ electrons

b. exactly the same number, 2.7×10^{12}

c. conservation of charge. Since electrons cannot leave the conductor between points A and B, the number crossing to the right at B must be exactly equal to the number passing point A.

d. The electron flow remains the same; it does not depend on the diameter of the wire.

e. $I = (3.6 \times 10^{18}$ elec/s$)(1.6 \times 10^{-19}$ C/elec$) = 0.58$ C/s $= 0.58$ A.

25. $Q = 16$ faradays $= (16)(96\ 500)$ C $= 1.54 \times 10^6$ C

$$\frac{(1.54 \times 10^6\ \text{C})}{(1.6 \times 10^{-19}\ \text{C/elec})}\ \frac{(1\ \text{ion})}{(2\ \text{elec})}\ \frac{1}{(6.02 \times 10^{23}\ \text{ions/mole})} = 8.0 \text{ moles of Cd}$$

mass $= (8.0$ moles$)(112$ g/mole$) = 9.0 \times 10^2$ g

CHAPTER 19

Electric Field

GOALS To explore the concepts of electric field (a vector) and electric potential (a scalar), and to study the concept of capacitance and its function in simple circuits.

OBJECTIVES After completing this chapter the student should be able to do the following:

1. Write the definition of electric field strength **E** in terms of force and charge.
2. Calculate the magnitude and direction of the electric field due to a point charge.
3. Calculate, by vector addition, the magnitude and direction of the electric field due to two or more point charges.
4. Name the three types of force that exist between material bodies. (Review Sec. 1-5.)
5. Write in mathematical form the definition of the potential difference between two points A and B.
6. Calculate any one of the quantities—V_{AB}, $W_{B\to A}$, or Q—when the other two are given.
7. Correctly identify the surface of a conductor carrying a static charge as an equipotential surface.
8. Use the energy principle to calculate the velocity of a charged particle resulting from a change in its electrical potential energy.
9. Write in words and in standard symbols the definition of potential gradient.
10. Calculate any of the quantities ΔV, E, or Δs when the other two are given.
11. Calculate the electric potential at a point due to two or more point charges. (See Sec. 18-5.)
12. Give the definition in words and in symbols of the capacitance of a pair of conductors.
13. Calculate any one of the quantities C, V, or Q when the other two are given.
14. Draw the symbol that represents a fixed capacitor in a circuit diagram.
15. Draw one of the two symbols used to represent "ground" in a circuit diagram.
16. Given the formula for the capacitance of a pair of parallel plates, $C = KA/(4\pi kd)$, identify the symbols $K, A, k,$ and d.
17. Use the formula above to calculate one of the quantities $C, A,$ or d when the others are given.
18. Write the operational definition of the dielectric constant.
19. Calculate the capacitance or charge of a parallel-plate capacitor when the dielectric constant is given.

SUMMARY

Electric field strength is defined as the force per unit charge, i.e., $\mathbf{E} = \mathbf{F}/q$. Electric field, like force, is a vector quantity. The *electric potential* difference between two points is defined as the work required to transport a charge from one point to another divided by the charge transported. In symbols, $V_{AB} = W_{B\to A}/q$. In a region of space where charges are present there will be surfaces that have the same potential at every point. Such a surface is called an *equipotential* surface. For example, an equipotential surface for an isolated point charge is any spherical surface centered on the point charge. If a conductor contains only static charge (no current is flowing), the surface of the conductor is an equipotential surface.

A pair of conductors placed near each other, usually carrying equal and opposite charges, is called a *capacitor*. If the conductors have large surface areas and are placed very close together (but not touching) large amounts of charge may be placed on the pair of conductors with only a small potential difference between them. The *capacitance* of a pair of conductors, a measure of the ability to store charge, is defined by the relationship $C = Q/V$, where Q is the charge on either conductor and V is the potential difference between the conductors. A common configuration for a capacitor is a pair of metal plates with air or some other insulator between them. The capacitance of such a parallel-plate capacitor is

$$C = \frac{KA}{4\pi kd}$$

where K is the dielectric constant, a property of the insulating medium, and k is the constant in Coulomb's law. A is the area of one of the plates and d is the distance between them.

QUESTIONS AND PROBLEMS

Secs. 19-1—19-3

1. In Sec. 8-2 of the text, gravitational field strength was defined as the gravitational force on a particle divided by the mass of the particle, symbolically,

$$\mathbf{g} = \frac{\mathbf{F}\text{(gravitational)}}{m}$$

Electric field strength is defined in an analogous way. **E** = ____________.

2. The two drawings in Fig. 19-1 show charged objects at A and B.

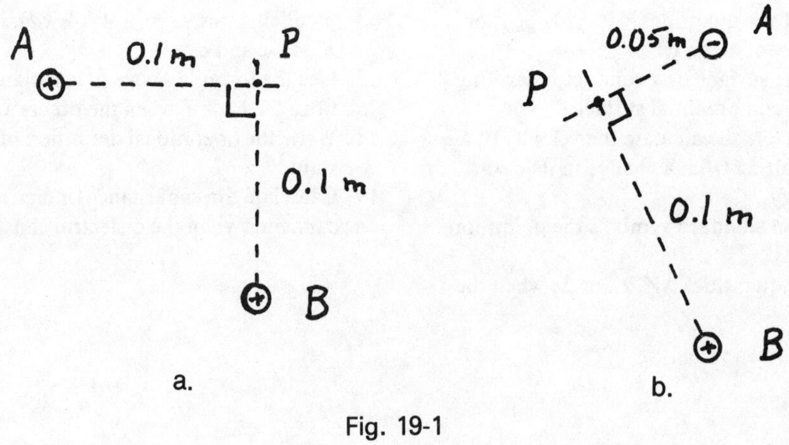

A 0.1 m P 0.1 m B

a.

P 0.05 m A 0.1 m B

b.

Fig. 19-1

At point P in Fig. 19-1a draw an arrow to represent the electric field due to the charge at A. Draw a second arrow at P to represent the electric field that is produced by the charge at B. Do the same thing in Fig. 19-1b taking note of the fact that the charge at A is negative. Now draw a third arrow at P (in both figures) to represent the resultant electric field. Label your arrows $\mathbf{E}_A$, $\mathbf{E}_B$, and $\mathbf{E}_{tot}$.

3. In Fig. 19-1a the charges at A and B are 3×10^{-6} C and 4×10^{-6} C, respectively. The point P is equidistant from A and B, the distance being 0.1 m. Calculate the magnitude of E_A, E_B, and the resultant electric field at P.

E_{tot} = ______________ .

4. In Fig. 19-1b the charges at A and B are -2×10^{-6} C and 2×10^{-6} C. The distances are 0.05 m and 0.1 m. Calculate the magnitude of the resultant electric field at P.

5. Fig. 19-2 shows the electric field in the neighborhood of a point charge. The direction of the field is outward as indicated by the arrows. What is the sign of the charge?

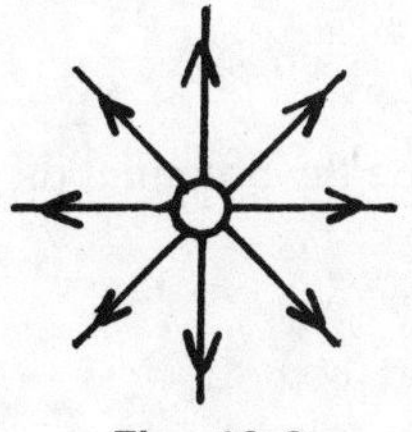

Fig. 19-2

6. Earlier in this course we studied the gravitational force. Write down the names of the other two types of force.

Secs. 19-4—19-6

7. Suppose that it requires 1.6 J of work to lift a book from the floor and place it on a table. What is the increase in potential energy of the book?

ΔPE = ______________ .

8. If the book has a mass of 2 kg, what is the increase in gravitational potential? (Gravitational potential is work divided by mass.)

9. A positive test charge at A experiences a force due to the presence of a charged body to the right in Fig. 19-3.

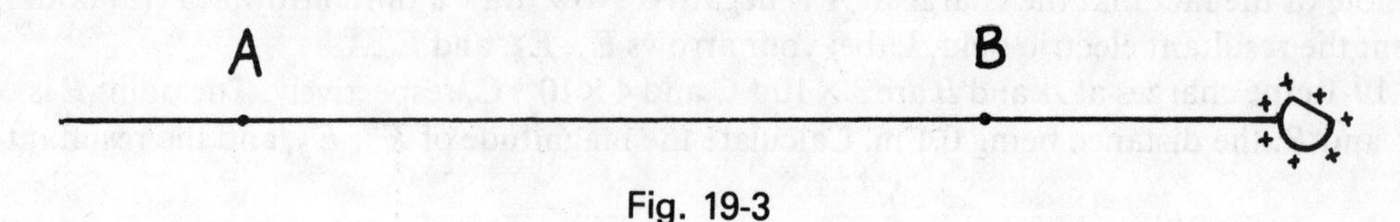

Fig. 19-3

It is found that 4×10^{-8} J of work is required to move the test charge from A to B. The test charge has a magnitude of 2×10^{-6} C.

a. Draw an arrow at A to represent the electrical force acting on the test charge when it is located there. Do the same at point B. Which arrow is longer?

b. Calculate the potential difference between points A and B.

10. In Fig. 19-4 two points P and Q are located in the neighborhood of charged objects.

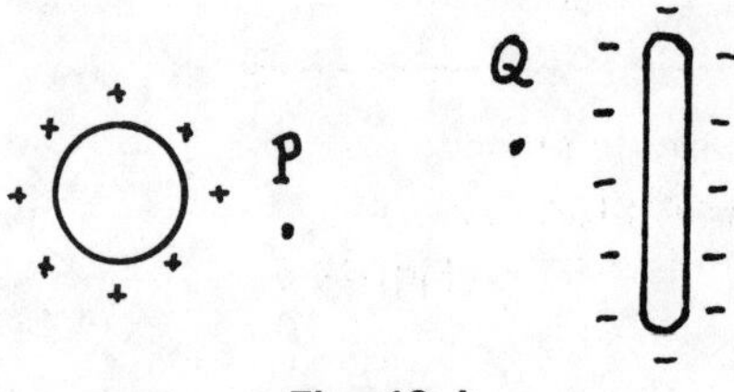

Fig. 19-4

The potential difference between P and Q is 200 volts. How much work is required to move a charge of 5 μC from Q to P? (1 μC = 10^{-6} C)

11. The "potential at a point" is a meaningful concept only if some point of zero potential has been specified. In the circuit diagram in Fig. 19-5 which point is taken to have zero potential?

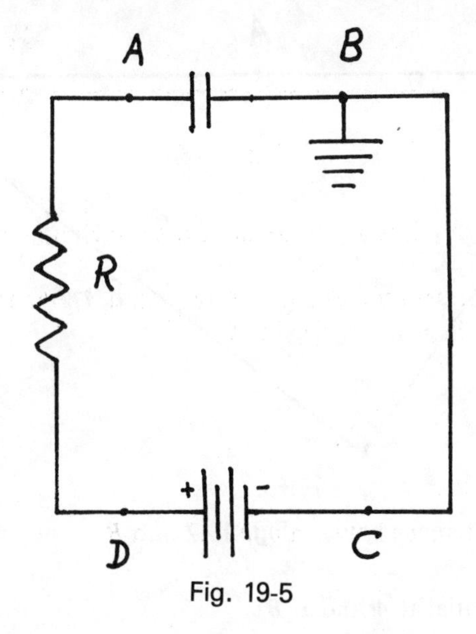

Fig. 19-5

In the case of an isolated point charge, it is convenient to assign a zero potential to a point that is a very great distance away from the charge in question. This is illustrated in Fig. 19-6.

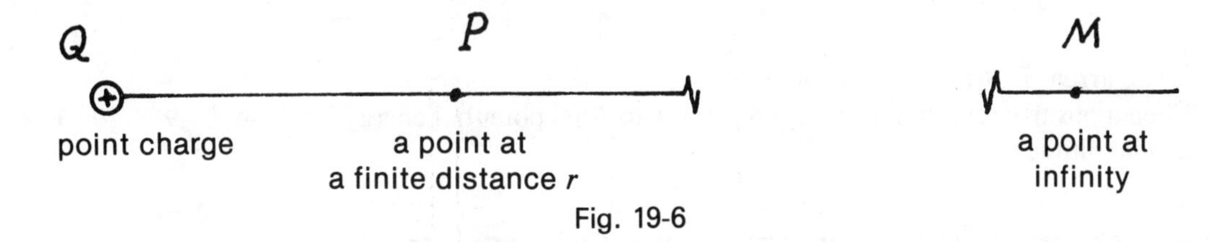

Fig. 19-6

Since the force acting on a test charge is known from Coulomb's law, it is possible to calculate (by integration) the work required to bring a positive charge q from infinity to the point P. The details of this calculation are given in Sec. 19-9. The result is

$$\frac{W_{M \to P}}{q} = \frac{kQ}{r}$$

Since the potential at M (infinity) is taken to be zero, the work per unit positive charge is the electric potential at point P. Thus

$$V_P = \frac{kQ}{r} \quad \text{(a scalar quantity)}$$

If several point charges are present, the total potential is simply the algebraic sum of the potentials due to each charge.

12. Fig. 19-7 shows two points, A and B, in the vicinity of two point charges. Point A is midway between the charges.

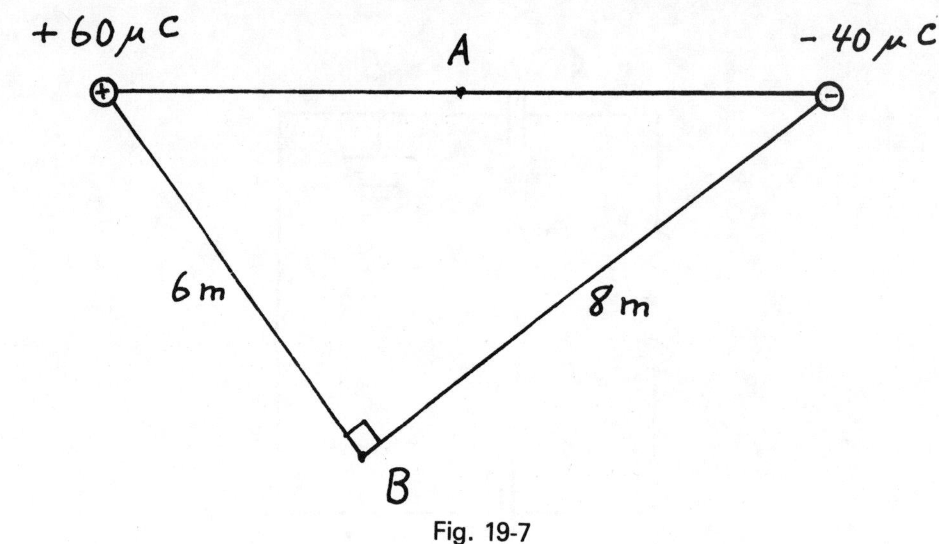

Fig. 19-7

a. Calculate the electric potential at A and at B

V_A = ____________.

V_B = ____________.

b. Calculate the magnitude of the electric field at B

E_B = ____________.

c. Put an arrow at B to show the direction of **E**.

d. When a positive test charge is moved from A to B its potential energy increases by 9×10^{-5} J. What is the magnitude of the charge?

q = ____________.

13. When the charges on a metallic conductor are at rest, all points of the conductor are at the same ____________. Thus the surface of a conductor in which no current is present is an ____________.

14. Fig. 19-8 shows a uniformly charged sphere with lines of force radiating from it. Points A and B lie on an imaginary spherical surface of radius r_1 which is concentric with the charged sphere.

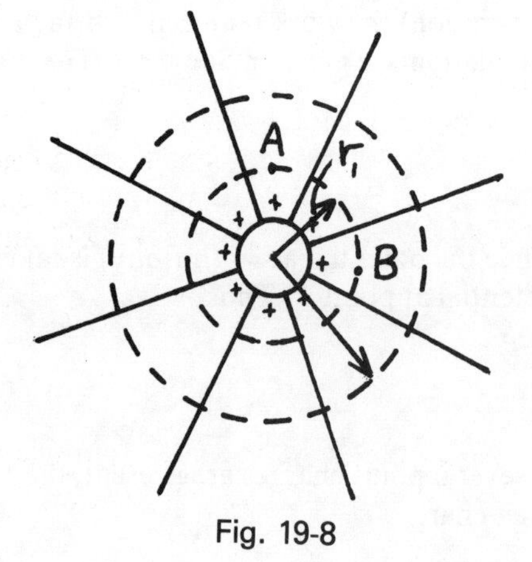

Fig. 19-8

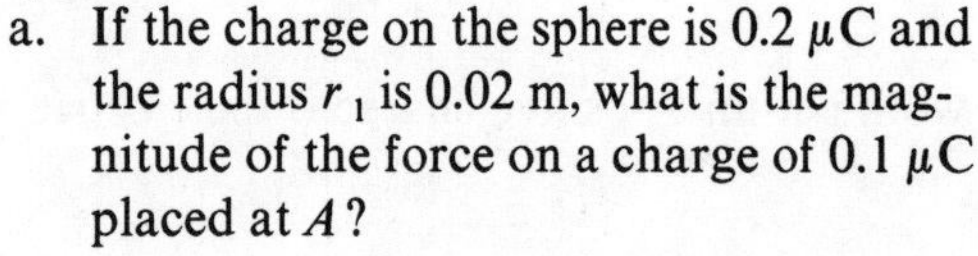

a. If the charge on the sphere is 0.2 μC and the radius r_1 is 0.02 m, what is the magnitude of the force on a charge of 0.1 μC placed at A?

F = ____________.

b. Place an arrow at A to represent the direction of this force.

c. How much work is done if the charge of 0.1 μC is moved along the spherical surface from A to B?

$W_{A \to B}$ = ____________.

d. What is the potential difference between points A and B?

V_{AB} = ____________.

e. A surface, such as this spherical surface of radius r_1, is called a(n) ____________.

15. The charged sphere shown in Prob. 14 is surrounded by a second concentric spherical surface of radius r_2. Given that $r_1 = 0.02$ m and $r_2 = 0.05$ m, calculate the potential difference between these two equipotential surfaces.

$\Delta V =$ ______________ .

16. Are lines of force necessarily perpendicular to an equipotential surface? ______________ (Yes/No). Explain why this is so.

17. A fine, iron wire 20 cm long is connected to a battery as shown in Fig. 19-9. The potential difference between the two ends of the wire is 3.0 volts.

a. What are the magnitude and direction of the electric field inside the wire?

$E =$ ______________ .

The direction of **E** is ______________ .

b. What is the potential difference between two points A and B that are separated by 4 cm?

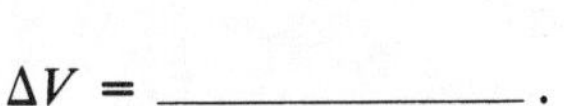

$\Delta V =$ ______________ .

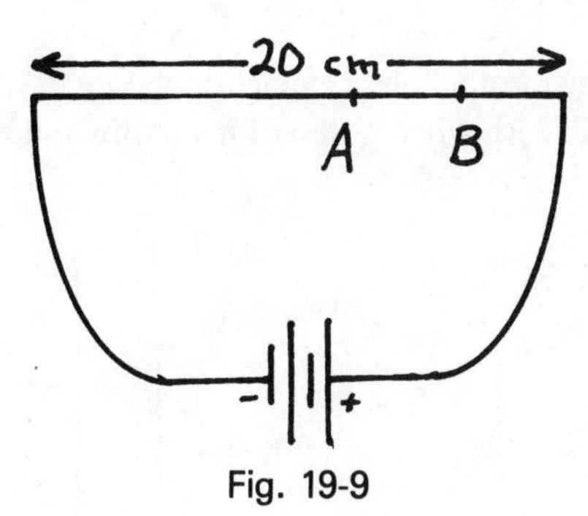

Fig. 19-9

18. Write the definition of capacitance in words and in symbols.

Capacitance = ______________ .

$C =$ ______________ .

19. Draw a circuit in which two capacitors and a battery are all connected in series.

20. Fig. 19-10 shows a pair of conductors containing equal and opposite charges. $Q = 20\ \mu C$. When a very small test charge of 4×10^{-8} C is moved from one conductor to the other, the amount of work required is 12 μJ.

a. What is the PD between the two conductors?

$\Delta V =$ ______________ .

b. Which conductor has the higher potential?

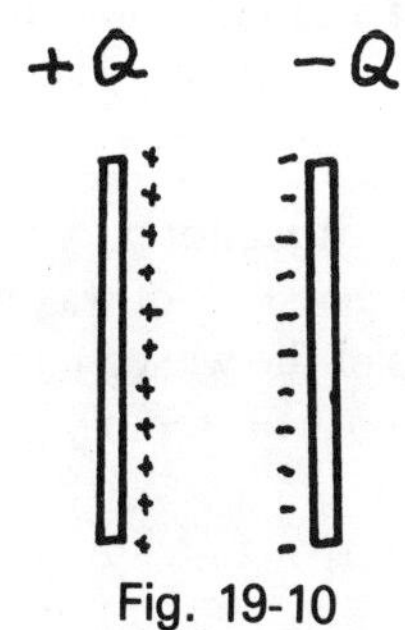

Fig. 19-10

c. What is the capacitance of this pair of conductors?

$C =$ ______________ .

d. What will the capacitance of the pair of conductors become if the charge on each is doubled?

e. Calculate the electric field strength at a point between the two conductors if their separation is 0.1 cm. (If the plates are large compared to their separation, the field will be uniform between the plates.)

$E =$ _______________.

Secs. 19-7—19-8

21. The parallel plate capacitor shown in Fig. 19-11 has a capacitance of 0.010 μF in air and a capacitance of 0.022 μF when a slab of insulating material fills up the space between the plates.

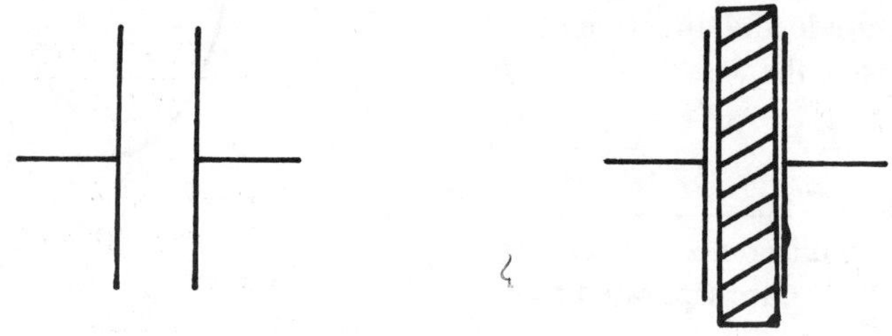

Fig. 19-11

a. What is the dielectric constant of the insulating material?

$K =$ _______________.

b. Table 19-1 in the text gives the dielectric constant of several liquids and gases. What solid dielectric has the dielectric constant found in part a?

22. Suppose that a 250-μF capacitor is connected to a 6-volt battery. How much charge is stored in the capacitor? (See Ex. 19-8 in the text.)

$Q =$ _______________.

23. A homemade capacitor may be constructed of two large pieces of aluminum foil separated by a somewhat larger sheet of waxed paper. Suppose that the aluminum sheets are 20 cm × 30 cm and that the thickness of the waxed paper is 0.03 mm. Assuming the dielectric constant to be 2.5, calculate the capacitance of the assembly.

$C =$ _______________.

Solutions

1. $\mathbf{E} = \dfrac{\mathbf{F}(\text{electrical})}{\text{charge}} = \dfrac{\mathbf{F}}{Q}$

2.

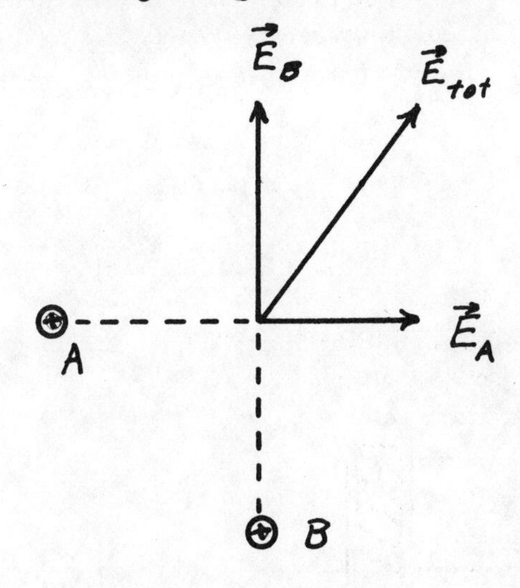

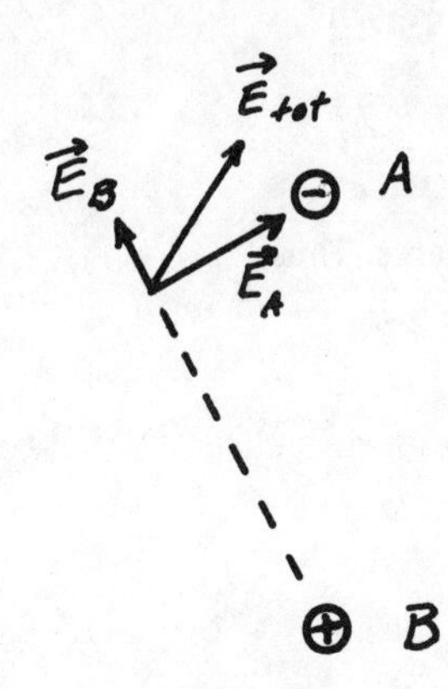

Fig. 19-12

3. $E = \dfrac{F}{Q} = \dfrac{k\,Q\,Q'/r^2}{Q} = \dfrac{k\,Q'}{r^2}$

$$E_A = \frac{(9 \times 10^9\ \text{Nm}^2/\text{C}^2)(3 \times 10^{-6}\ \text{C})}{(0.1\ \text{m})^2} = 2.7 \times 10^6\ \text{N/C}$$

$$E_B = \frac{(9 \times 10^9\ \text{Nm}^2/\text{C}^2)(4 \times 10^{-6}\ \text{C})}{(0.1\ \text{m})^2} = 3.6 \times 10^6\ \text{N/C}$$

$$E = \sqrt{E_A^2 + E_B^2} = \sqrt{(7.29 + 12.96)} \times 10^6\ \text{N/C}$$
$$= 4.5 \times 10^6\ \text{N/C}$$

4. $E_A = \dfrac{k(2 \times 10^{-6}\ \text{C})}{(0.05\ \text{m})^2} = 7.2 \times 10^6\ \text{N/C}$

$$E_B = \frac{k(2 \times 10^{-6}\ \text{C})}{(0.1\ \text{m})^2} = 1.8 \times 10^6\ \text{N/C}$$

$$E = \sqrt{E_A^2 + E_B^2} = \sqrt{(51.8 + 3.2)} \times 10^6\ \text{N/C}$$
$$= 7.42 \times 10^6\ \text{N/C}$$

5. positive
6. electric; nuclear
7. ΔPE = work done = 1.6 J
8. gravitational potential $= \dfrac{W}{m} = \dfrac{1.6\ \text{J}}{2\ \text{kg}} = 0.8\ \text{J/kg}$

9a.

Fig. 19-13

b. potential difference $= V_{AB} = \dfrac{W_{B \to A}}{Q}$

$$V_{AB} = \frac{4 \times 10^{-8}\ \text{J}}{2 \times 10^{-6}\ \text{C}} = 2 \times 10^{-2}\ \text{J/C}$$

10. $W_{Q \to P} = V_{PQ} \times Q = (200\ \text{volts})(5 \times 10^{-6}\ \text{C}) = 1 \times 10^{-3}\ \text{J}$
11. Point B

12a.

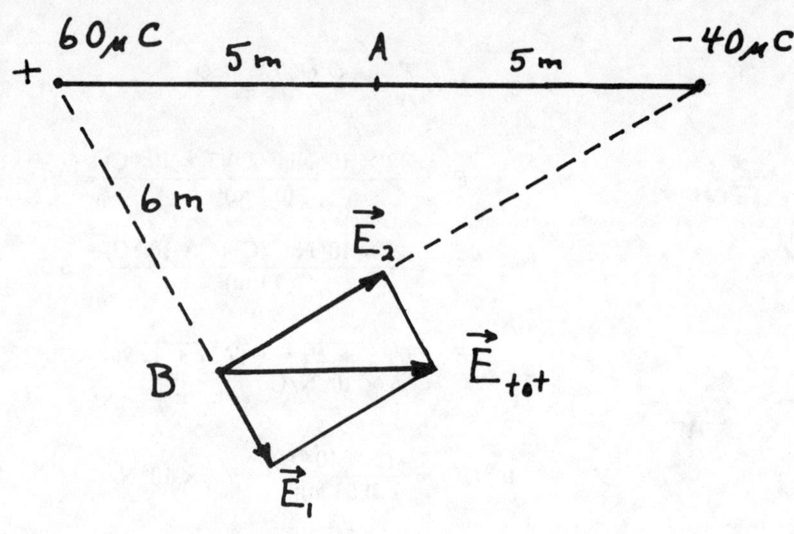

Fig. 19-14

$$V_1 = \frac{kQ}{r} = \frac{(9 \times 10^9 \text{ N}\cdot\text{m}^2/\text{C}^2)(60 \times 10^{-6} \text{ C})}{5 \text{ m}} = 1.08 \times 10^5 \text{ V}$$

$$V_2 = \frac{(9 \times 10^9 \text{ N}\cdot\text{m}^2/\text{C}^2)(-40 \times 10^{-6} \text{ C})}{5 \text{ m}} = -7.2 \times 10^4 \text{ V}$$

The total potential at A is the algebraic sum of the potentials due to each charge. Thus

$$V_A = V_1 + V_2 = (10.8 - 7.2) \times 10^4 \text{ V} = 3.6 \times 10^4 \text{ V}$$

The electric potential at B is found in the same way.

$$V_B = V_1 + V_2 = \frac{k(60 \times 10^{-6} \text{ C})}{6 \text{ m}} + \frac{k(-40 \times 10^{-6} \text{ C})}{8 \text{ m}}$$

$$= 9.0 \times 10^4 \text{ V} - 4.5 \times 10^4 \text{ V} = 4.5 \times 10^4 \text{ V}$$

b. The electric field, unlike the potential, is a vector quantity. Thus

$$\mathbf{E}_B = \mathbf{E}_1 + \mathbf{E}_2$$

$$E_1 = \frac{k\,Q_1}{r_1^2} = \frac{(9 \times 10^9 \text{ N}\cdot\text{m}^2/\text{C}^2)(60 \times 10^{-6} \text{ C})}{(6 \text{ m})^2} = 1.5 \times 10^4 \text{ N/C}$$

$$E_2 = \frac{(9 \times 10^9 \text{ N}\cdot\text{m}^2/\text{C}^2)(40 \times 10^{-6} \text{ C})}{(8 \text{ m})^2} = 5.62 \times 10^3 \text{ N/C}$$

Since $\mathbf{E}_1$ is perpendicular to $\mathbf{E}_2$,

$$E_{\text{tot}} = \sqrt{E_1^2 + E_2^2} = \sqrt{2.25 + 0.32} = 1.6 \times 10^4 \text{ N/C}$$

c. (See Fig. 19-14)

d. $V = \dfrac{\text{work}}{\text{charge}} = \dfrac{\text{PE}}{Q}$; $Q = \dfrac{\text{PE}}{V}$

$$Q = \frac{9 \times 10^{-5} \text{ J}}{(4.5 \times 10^4 - 3.6 \times 10^4)V} = 1.0 \times 10^{-8} \text{ C}$$

13. potential; equipotential surface

14a. The uniformly charged sphere produces the same field as a point charge located at the center.

$$F = \frac{(9 \times 10^9\ \mathrm{N \cdot m^2/C^2})(0.2 \times 10^{-6}\ \mathrm{C})(0.1 \times 10^{-6}\ \mathrm{C})}{(0.02\ \mathrm{m})^2} = 0.45\ \mathrm{N}$$

b. **F** is directed radially outward.

c. Since the force is always exactly perpendicular to the displacement, the work is zero.
$W_{A \to B}$ = zero

d. $V_{AB} = \dfrac{W}{Q} = 0$

e. equipotential surface

15. $\Delta V = V_1 - V_2 = \dfrac{kQ}{r_1} - \dfrac{kQ}{r_2}$

$$\Delta V = kQ\left(\frac{1}{r_1} - \frac{1}{r_2}\right) = (9 \times 10^9\ \mathrm{N \cdot m^2/C^2})(0.2 \times 10^{-6}\ \mathrm{C})$$

$$= 5.4 \times 10^4\ \mathrm{V}\left(\frac{1}{0.02\ \mathrm{m}} - \frac{1}{0.05\ \mathrm{m}}\right) = 1.62 \times 10^6\ \mathrm{V}$$

16. Yes. If the lines of force were not perpendicular to the surface, there would be a change in the potential energy of a test charge as it moved along the surface (work = $Fs \cos\theta$). Then the surface would not be an equipotential surface, which contradicts the original assumption.

17a. $E = \dfrac{\Delta V}{\Delta s} = -\dfrac{3.0\ \mathrm{V}}{0.2\ \mathrm{m}} = -15\ \mathrm{V/m}$

The direction of **E** is toward the negative terminal of the battery.

b. $\Delta V = -E\,\Delta s = -(-15\ \mathrm{V/m})(0.04\ \mathrm{m}) = 6.0\ \mathrm{V}$

18. Capacitance $= \dfrac{\text{charge on either conductor}}{PD \text{ between the conductors}}$; $C = \dfrac{Q}{V}$

19.

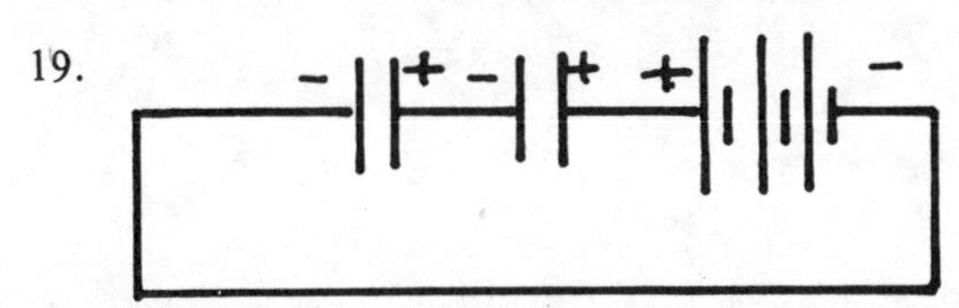

Fig. 19-15

20a. $\Delta V = \dfrac{W}{Q} = \dfrac{12 \times 10^{-6}\ \mathrm{J}}{4 \times 10^{-8}} = 300\ \mathrm{V}$

b. the positively charged conductor

c. $C = \dfrac{Q}{V} = \dfrac{20 \times 10^{-6}\ \mathrm{C}}{300\ \mathrm{V}} = 0.067\ \mu\mathrm{F}$

20d. If the charge is doubled, the potential difference will also be doubled. Thus

$$C' = \frac{2Q}{2V} = \frac{Q}{V} = C \quad \text{(same as before)}$$

The capacitance of a pair of conductors depends only on geometrical factors such as area, separation, etc.

e. $E = \dfrac{\Delta V}{\Delta s} = \dfrac{300\ \mathrm{V}}{10^{-3}\ \mathrm{m}} = 3 \times 10^5\ \mathrm{V/m}$

21a. $K = \frac{C_{med}}{C_{vac}} \cong \frac{C_{med}}{C_{air}} = \frac{0.022\ \mu F}{0.010\ \mu F} = 2.2$

b. wax (or paraffin)

22. $Q = CV = (2.5 \times 10^{-4}\ F)(6\ V) = 15 \times 10^{-4}\ C$

23. $C = \frac{KA}{4\pi kd} = \frac{2.5 \times 20\ cm \times 30\ cm \times (1\ m/100\ cm)}{4\pi(9 \times 10^9\ N \cdot m^2/C^2)(3 \times 10^{-3}\ cm)}$

$= 4.42 \times 10^{-8}\ F$

CHAPTER

20 Electric Energy

GOALS To investigate the role of energy transformation in electric circuits, and to study the concepts of resistance, resistivity, and energy storage.

OBJECTIVES After completing this chapter the student should be able to do the following:

1. Name two common devices that are sources of electromotive force.
2. Write the definition of a source of electromotive force emf ($\mathcal{E}$) and state the units in which it is expressed.
3. Describe the energy transformations that take place in a circuit containing one or more sources of electromotive force.
4. Describe the energy transformation that takes place in a *resistor* regardless of the direction of the current.
5. Write Ohm's law using the standard symbols for potential difference and current.
6. Calculate any one of the quantities R, V, or I when the other two are given.
7. Calculate the power transformed into heat in a resistor when the current is known.
8. Calculate the power dissipated in a resistor from a knowledge of the potential difference and the current.
9. Calculate the power produced in a source of emf $\mathcal{E}$ when the current and $\mathcal{E}$ are known.
10. Write the equation that defines resistivity in terms of resistance of a conductor of length L and cross section A.
11. Calculate any one of the quantities R, ρ, L, or A when the others are given.
12. Write the equation that expresses the variation of resistivity with temperature.
13. Calculate the resistance of a conductor at a given temperature when its resistance at 0°C is known.
14. Calculate the energy density of a storage battery knowing its mass, PD, and ampere-hour rating.

SUMMARY

To cause a current to flow in a circuit there must be a source of energy, some device in which nonelectrical energy is transformed reversibly into electric energy. Such a device is called a *source of electromotive force*. Examples are a battery, a solar (photovoltaic) cell, and a generator. An energy transformation also takes place in a resistor but the transformation of electrical energy to heat energy in a resistor is irreversible.

The rate at which electric energy is dissipated (i.e., transformed into heat) in a resistor depends on the resistance and on the square of the current. This relationship may be written $P = RI^2$.

The current flowing through a conductor, an iron wire for example, may be varied over a very large range. Likewise the potential difference measured between the two ends may have almost any value. However, the ratio of potential difference to current is a constant for a particular conductor provided that the temperature does not change. This constant is called the resistance, $R = V/I$.

Theoretical considerations, as well as experimental measurements, show that the resistance of a particular conductor is proportional to its length and inversely proportional to its cross-sectional area. The constant of proportionality, which depends only on the material of which the conductor is made and its temperature, is called *resistivity*. Symbolically, $R = \rho L/A$.

The variation of resistivity with temperature for metals follows a linear law over a wide range of temperatures. $\rho = \rho_0(1 + \alpha T)$ where ρ_0 represents the resistivity at 0°C and α, the temperature coefficient of resistivity, is a constant for a given material. For most metals α is positive, implying an increase of resistivity with temperature. For semiconductors, such as silicon and germanium, α is usually negative.

In our industrial, technological society, devices and systems for storing energy are becoming increasingly important. For storing small amounts of energy there exists already a variety of devices that perform quite adequately. A small mercury battery will keep a watch running for a year or more; a capacitor is an excellent short-term storage device in an electrical circuit. On the other hand, economically feasible methods for storing electrical and thermal energy in the large quantities required for transportation and home heating remain to be developed.

QUESTIONS AND PROBLEMS

Secs. 20-1—20-4

1. A cross section of a common dry cell is shown in text Fig. 19-9. The electrode in the center of the cell is a rod of carbon about 1/2 cm in diameter. The carbon rod is the ______________ (positive or negative) electrode. The entire device is a source of ______________. The outer electrode of a flashlight battery, or common dry cell, is a cylindrical can of ______________.
2. In a fresh dry cell, positive charge accumulates on the carbon electrode and negative charge accumulates on the zinc electrode. This separation of charge is brought about by what type of forces?

3. The essential action of the dry cell (or of any other type of battery) is to transform energy from ______________ *PE* to ______________.
4. A lead-acid battery (or automobile battery) and an electric generator are examples of ______________ of ______________.
5. Complete this definition: A source of electromotive force is a device within which ______________ energy can be reversibly ______________ into ______________.
6. Electromotive force $\mathcal{E}$ is measured in terms of work per unit charge (joules/coulomb). What other electrical quantity, different in concept from electromotive force, is also measured in joule/coulomb?

7. In Fig. 20-1a a battery is connected to a bulb. In Fig. 20-1b a battery is connected to a generator. The direction of electron flow is shown by arrows. Put arrows on each circuit to indicate the direction of the conventional current. Describe the energy transformations that are taking place in each case.

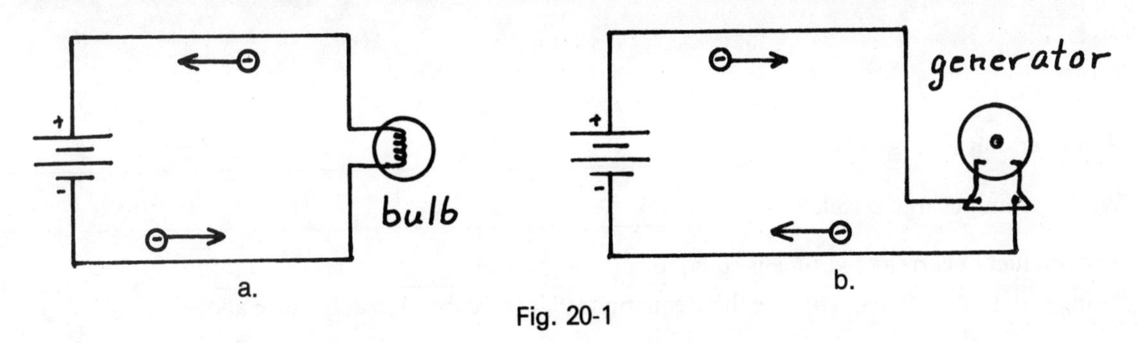

Fig. 20-1

8. Fig. 20-2 represents the filament of a light bulb. When electrons are forced to move from A to B through the filament, heat is produced.

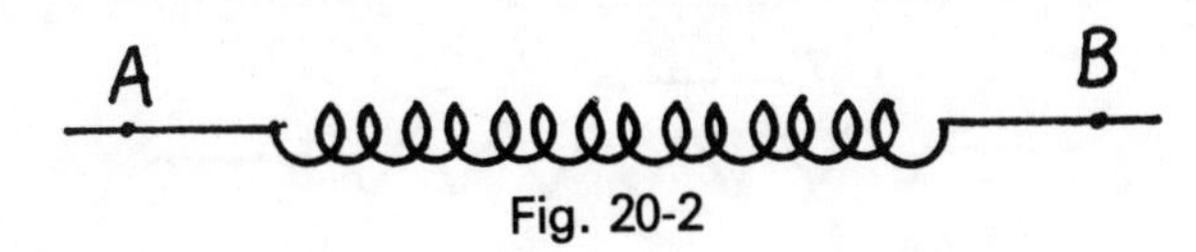

Fig. 20-2

The filament will become white hot if the electron flow is large enough. The transformation that takes place is

$$\begin{pmatrix}\text{electrical}\\ \text{energy}\end{pmatrix} \rightarrow \begin{pmatrix}\text{heat}\\ \text{energy}\end{pmatrix} \text{ (some of which is radiated as visible light)}$$

Suppose that the electron flow is reversed so that the electrons go from B to A. Will this cause the energy transformation to be reversed? ______________ (Yes/No). Explain. Is the lamp filament a *resistor*? ______________ (Yes/No). See Sec. 20-3.

9. A 60-watt light bulb operates at full brillance when the current passing through it is 0.5 A. The same bulb glows rather dimly ($P \cong 15$ watts) when a current of 0.3 A passes through it.
   a. Calculate the resistance of the filament for these two cases.

$R_1 =$ ______________ .

$R_2 =$ ______________ .

b. Is Ohm's law violated by the fact that the resistance of the filament is not the same in the two cases? ____________ (Yes/No) Explain.

10a. Work per unit time is called ____________.

b. Work per unit charge is called ____________.

c. The product of current and time is equal to ____________.

d. Using standard symbols, write the three equations that correspond to a, b, and c above.

11. The current shown in Fig. 20-3 is flowing through a resistor from M to N. Use a "+" sign and a "−" sign to show which end of the resistor is positive and which end is negative. Conventional current in a resistor flows from a ____________ potential to a ____________ potential.

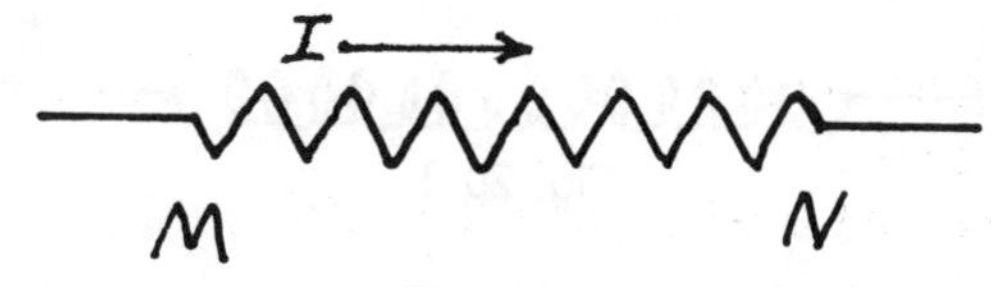

Fig. 20-3

12. Write the relationship between voltage, current, and resistance for a metallic resistor at a definite temperature. Give the name of this relationship.

13a. What is the resistance of a 100-watt light bulb operated at 120 volts?

R = ____________.

b. What current will pass through this bulb?

I = ____________.

14. The battery in Fig. 20-4 is able to maintain a steady current of 0.2 A in the circuit illustrated.

a. Put arrows below the two resistors to show the direction of the conventional current. Considering the three locations in the circuit, A, B, and C, which has the highest potential? Which has the lowest potential?

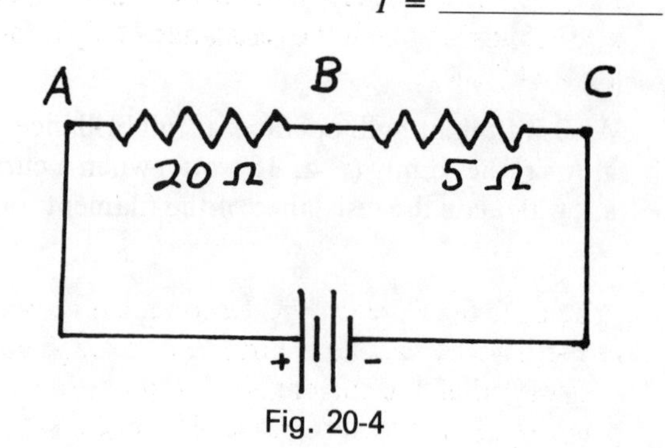

Fig. 20-4

b. Calculate the potential difference between A and B, B and C, and A and C.

V_{AB} = ____________. V_{BC} = ____________. V_{AC} = ____________.

15. Suppose that the two resistors in Prob. 14 above are replaced by a single resistor. What value of resistance will cause the same current to flow in the circuit?

R_{eff} = ____________ .

16. Find the rate at which heat is being produced (the power) in each of the resistors in Prob. 14.

P_1 = ____________ . P_2 = ____________ .

17. Suppose that the circuit in Prob. 14 contains no resistance other than that shown. At what rate is chemical energy being transformed into electrical energy in the battery?

P = ____________ .

18. A tungsten wire 2 m long has a resistance of 50 Ω. Knowing that the resistivity of tungsten is 5.0×10^{-8} Ω • m, calculate the diameter of the wire. (Hint: First show that the diameter of the wire is given by $d = 2\sqrt{(\rho L)/(\pi R^2)}$.)

d = ____________ .

19. Two pieces of aluminum wire have the same mass, but one wire is 4 times as long as the other. Let R_2 be the resistance of the longer wire and R_1 the resistance of the shorter wire. Find R_2/R_1.

R_2/R_1 = ____________ .

20. A gold wire is 50 cm long and 0.05 mm in diameter. It is desired to pass a current through the wire such that 2 watts of heat will be produced.

a. What is the resistance of the gold wire? ($\rho \cong 2.3 \times 10^{-8}$ Ω • m)

R = ____________ .

b. What should the current be?

I = ____________ .

c. What PD will exist between the ends of the wire?

V = ____________ .

21. Write the equation that relates the resistivity at a temperature T, ρ_T, to the resistivity at 0°C.

ρ_T = ____________ .

22. The temperature coefficient of resistance of platinum is approximately 4×10^{-3} (°C)$^{-1}$. Find the temperature at which the resistance of a platinum wire is 2% greater than its value at 0°C. (i.e., $R_T = (1.02)R_0$).

23. A liquid is being heated in the container shown in Fig. 20-5 by passing a current, I, through a resistor. When the current is 1.5 A the temperature of the liquid increases 3°C each minute. What value of I will cause the temperature to increase 4°C each minute?

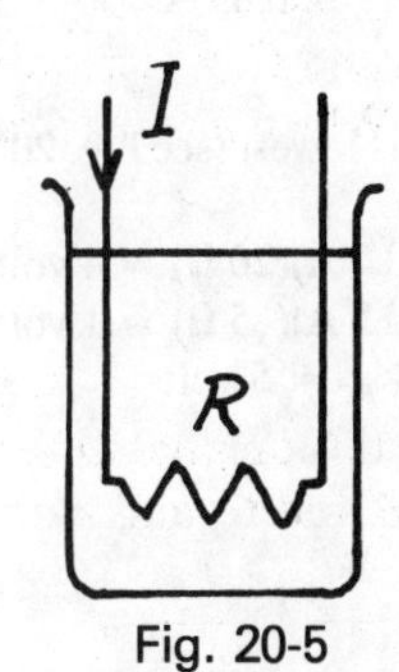

Fig. 20-5

Solutions

1. positive; electromotive force; zinc
2. chemical (It is also correct to say *electrical* forces since chemical forces are, in fact, short-range electric forces between atoms or molecules.)
3. chemical; electrical PE
4. sources; electromotive force
5. nonelectric; transformed; electric energy
6. Potential difference is also measured in joules/coulomb (or in volts).
7. In Fig. 20-1a, the battery is discharging. In the battery, chemical energy is being transformed into electrical energy. In the bulb, electric energy is being transformed into heat energy, some of which is further transformed into electromagnetic energy (light waves and heat radiation).

   In Fig. 20-1b, the battery is being charged by the generator. In the battery, electric energy is being transformed into chemical PE. In the generator, mechanical energy, used to rotate the armature of the generator, is being converted into electric energy.
8. No! The transformation of electric energy into heat in a lamp bulb is *not* reversible. A circuit element, such as this bulb, in which electric energy is dissipated no matter in which direction the charges flow, is called a resistor.

9a. $R_1 = \frac{P}{I^2} = \frac{60\text{ W}}{(0.5\text{ A})^2} = 240\ \Omega$

$R_2 = \frac{P}{I^2} = \frac{15\text{ W}}{(0.3\text{ A})^2} = 167\ \Omega$

b. No. Ohm's law states only that the resistance is constant *if* the temperature is kept constant. In this case the temperature is much less when the bulb is glowing dimly.

10a. power

b. potential difference

c. charge

d. $P = \frac{W}{t}$; $V = \frac{W}{Q}$; $Q = It$

11. high; low (see Fig. 20-6).

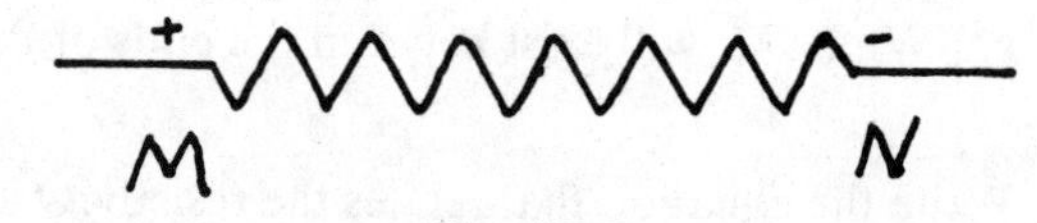

Fig. 20-6

12. $R = \frac{V}{I}$ or $I = \frac{V}{R}$; Ohm's law

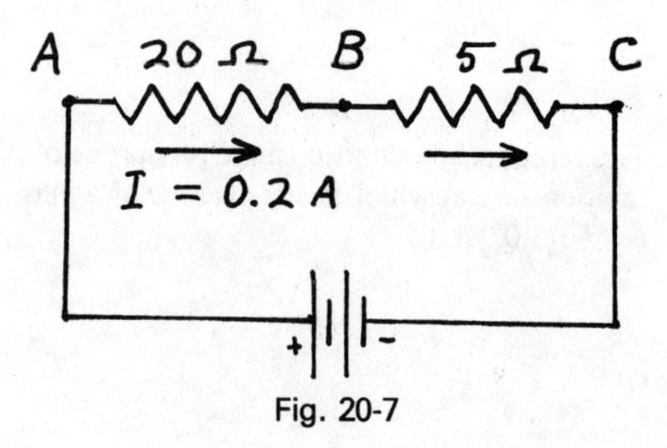

Fig. 20-7

13a. $P = I^2R = V^2/R = VI$

$R = \frac{V^2}{P} = \frac{(120\text{ V})^2}{100\text{ W}} = 144\ \Omega$

b. $I = \frac{P}{V} = \frac{100\text{ W}}{120\text{ V}} = 0.83\text{ A}$

14a. A is highest. C is lowest (see Fig. 20-7).

b. $V_{AB} = IR = (0.2\text{ A})(20\ \Omega) = 4$ volts
$V_{BC} = IR = (0.2\text{ A})(5\ \Omega) = 1$ volt
$V_{AC} = V_{AB} + V_{BC} = 5$ volts

15. $R_{\text{eff}} = \frac{V_{AC}}{I} = \frac{5\text{ V}}{0.2\text{ A}} = 25\ \Omega$

Note that $R_{\text{eff}} = R_1 + R_2$

16. $P_1 = I^2R_1 = (0.2\text{ A})^2(20\ \Omega) = 0.8\text{ W}$
$P_2 = I^2R_2 = (0.2\text{ A})^2(5\ \Omega) = 0.2\text{ W}$

17. The battery must supply all the energy that is being dissipated in the two resistors. Thus

$$P(\text{battery}) = 0.8\text{ W} + 0.2\text{ W} = 1.0\text{ W}$$

18. The resistance of a wire is given by

$$R = \frac{\rho L}{A} \quad \text{where} \quad A = \frac{\pi d^2}{4}$$

Thus $R = \frac{\rho L(4)}{\pi d^2}$ and $d = 2\sqrt{\frac{\rho L}{\pi R}}$

$$d = 2\sqrt{\frac{5 \times 10^{-8}\ \Omega \cdot \text{m}(2\text{m})}{\pi\ 50\ \Omega}} = 5.04 \times 10^{-5}\text{ m} \cong 5.0 \times 10^{-2}\text{ mm}$$

19. Since the two wires have the same mass, they must also have the same volume.

$$\pi r_1^2 L_1 = \pi r_2^2 L_2 \quad \text{or} \quad A_1 L_1 = A_2 L_2;\ A_2 = A_1 L_1 / L_2$$

$$R_1 = \frac{\rho L_1}{A_1};\ R_2 = \frac{\rho L_2}{A_2} = \frac{\rho(4L_1)}{A_1 L_1/(4L_1)} = 16\frac{\rho L_1}{A_1}$$

Thus $R_2/R_1 = 16$

20a. $R = \frac{\rho L}{A} = \frac{(2.3 \times 10^{-8}\ \Omega \cdot \text{m})(0.5\text{ m})}{(\pi/4)(5 \times 10^{-5}\text{ m})^2} = 5.86\ \Omega$

b. $P = I^2 R;\ I = \sqrt{P/R}$
$I = \sqrt{(2\text{ W})/(5.9\ \Omega)} = 0.58\text{ A}$

c. $V = IR = (0.58\text{ A})(5.9\ \Omega) = 3.4\text{ V}$

21. $\rho_T = \rho_0(1 + \alpha T)$

22. Use Eq. 20-8.

$$R_T = R_0(1 + \alpha T);\ I = \frac{R_T - R_0}{\alpha R_0}$$

$$T = \frac{(1.02)R_0 - R_0}{R_0(4 \times 10^{-3})}{}^\circ\text{C} = \frac{0.02}{4 \times 10^{-3}}{}^\circ\text{C} = 5^\circ\text{C}$$

23. The relationship between heat added to a substance and temperature change was given in Chapter 14.

Heat = $mc\Delta T$ where m is the mass, c is the specific heat, and ΔT is the temperature change. Since power is work (or energy) per unit time, we have

$$P = \frac{mc\,\Delta T}{\Delta t} \quad \text{and} \quad P = I^2 R$$

Using subscripts to distinguish the two cases,

$$I_1^2 R = \frac{mc\,\Delta T_1}{\Delta t_1};\ I_2^2 R = \frac{mc\,\Delta T_2}{\Delta t_2}$$

Thus $\left(\frac{I_2}{I_1}\right)^2 = \left(\frac{\Delta T_2}{\Delta T_1}\right)\left(\frac{\Delta t_1}{\Delta t_2}\right) = \left(\frac{4^\circ\text{C}}{3^\circ\text{C}}\right)\left(\frac{1\text{ min}}{1\text{ min}}\right)$

$$I_2 = (2/\sqrt{3})I_1 = (2/\sqrt{3})(1.5) = 1.73\text{ A}$$

REVIEW TEST FOR CHAPTERS 18–20

1. Copper and silver are both good conductors of electricity because
   a. they contain an excess of free protons each of which has a positive charge.
   b. a significant fraction of the copper ions is free to move through the crystal.
   c. their surfaces can be highly polished.
   d. their atomic numbers are both prime, 29 and 47.
   e. Some of their outer electrons are not tightly bound and can move freely through the crystal.
2. An uncharged metallic cup is placed on an insulating surface, as shown in Fig. T-39. A negatively charged sphere is let down into the cup, allowed to touch the bottom, and then removed without touching the cup again. Which of the following statements is true?
   a. The negative charge is now equally divided between the sphere and the cup.
   b. The sphere retains all of its charge, none having been transformed to the cup.
   c. The sphere has no charge; all of it was transferred to the cup.
   d. A flow of electrons to the outside of the cup during contact causes the sphere to have a positive charge.
   e. The cup is positively charged on the outside and negatively charged on the inside.

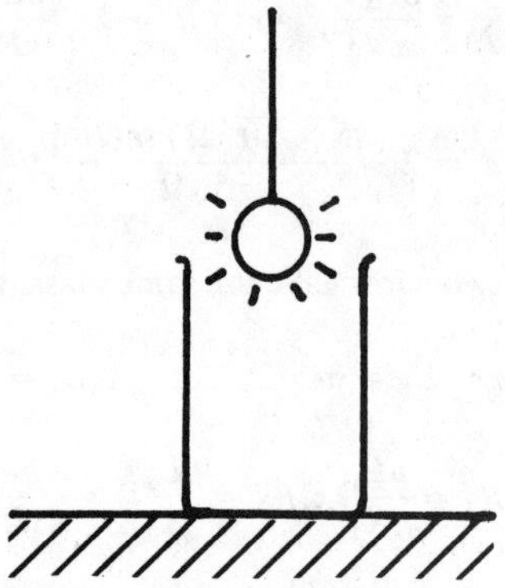

Fig. T-39

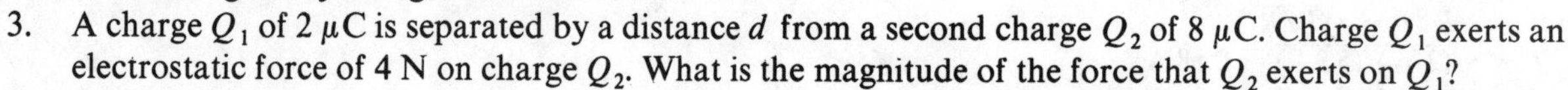

3. A charge Q_1 of 2 μC is separated by a distance d from a second charge Q_2 of 8 μC. Charge Q_1 exerts an electrostatic force of 4 N on charge Q_2. What is the magnitude of the force that Q_2 exerts on Q_1?
   a. 1 N
   b. 2 N
   c. 4 N
   d. 8 N
   e. 16 N
4. When two electric charges are separated by a distance of 0.1 m the Coulomb force is F. What is the magnitude of the force when the distance is increased to 0.3 m?
   a. $3F$
   b. $F/3$
   c. $F/2$
   d. $F/9$
   e. F
5. Two electric fields are present at a point P as shown in Fig. T-40. The magnitudes of the fields are $E_1 = 400$ V/m and $E_2 = 200$ V/m. What is the magnitude of the resultant field?

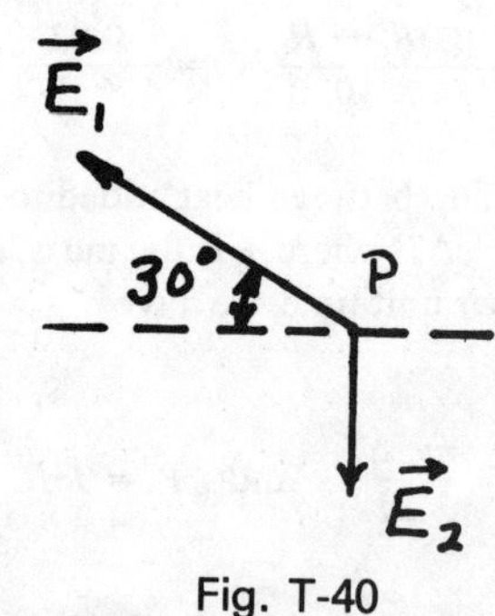

Fig. T-40

6. A small object containing a charge Q produces an electric field at a point P 3 m from the object. The magnitude of the field is 16×10^4 N/C. What is the magnitude of Q?

7. Points A and B in Fig. T-41 are located 0.1 m and 0.3 m from a small, charged sphere, Q. The electric potential at B is 27 J/C. What is the electric potential at A?
   a. 3 J/C
   b. 9 J/C
   c. 27 J/C
   d. 81 J/C
   e. 343 J/C

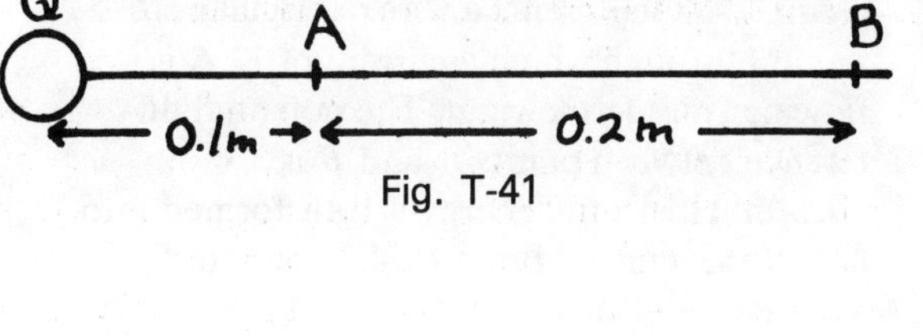

Fig. T-41

8. Points P and M are located in the vicinity of a small, charged object ($Q = 6\ \mu C$) shown in Fig. T-42. How much work is required to move a test charge of $0.2\ \mu C$ from P to M, a distance of 0.5 m?

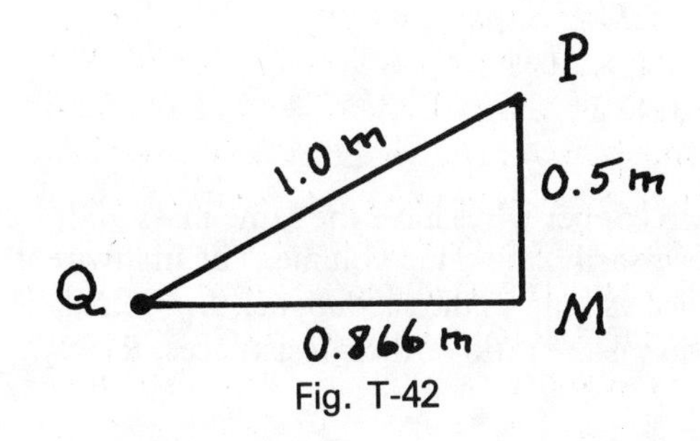

Fig. T-42

9. A current of 20 A is flowing in the wire in Fig. T-43. The potential difference between points A and B, 50 cm apart, is 0.012 V. What is the magnitude of the electric field inside the wire?

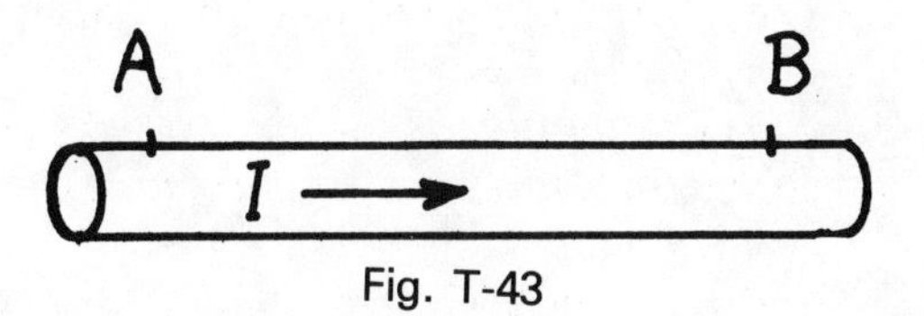

Fig. T-43

10. An electric toaster has a resistance of 8 Ω. When it is plugged into the outlet a current of 15 A passes through the toaster. How much heat energy (in joules) is produced in 5 minutes.
   a. 5.4×10^5 J
   b. 3.6×10^4 J
   c. 9×10^3 J
   d. 8.4×10^5 J
   e. 2.9×10^5 J

11. When a constant current of 1.732 A flows through a resistor immersed in water, the temperature of the water increases 3°C in 12 minutes (see Fig. T-44). What current should be used if the desired rate of heating is 2°C in 6 minutes?
   a. 5.2 A
   b. 2.0 A
   c. 2.3 A
   d. 1.5 A
   e. 1.3 A

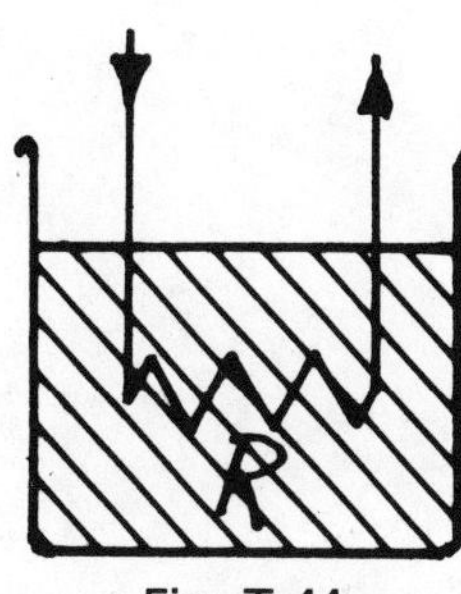

Fig. T-44

12. Figure T-45 represents a wire of circular cross section through which a current of 12 A is flowing from left to right. The potential difference between points A and B is 3 V. How much electric energy is transformed into heat if this current flows for five minutes?
a. 1.08×10^4 J
b. 1.2×10^3 J
c. 75 J
d. 1.8×10^2 J
e. 0.12 J

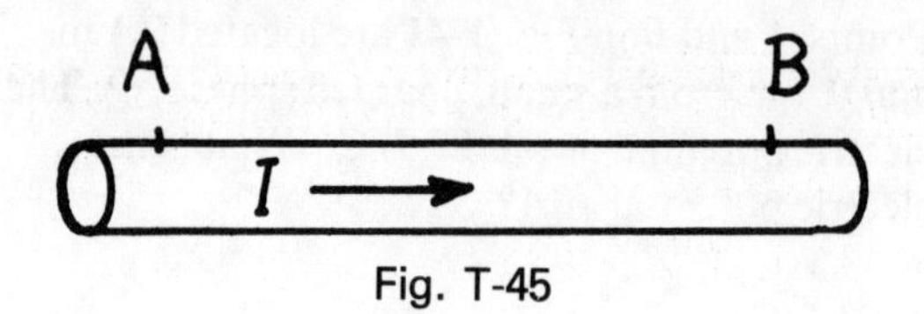

Fig. T-45

13. Two copper wires have the same mass and, necessarily, the same volume. The first wire is twice as long as the second, i.e., $L_1 = 2L_2$. What is the ratio of their resistances, R_1/R_2?
a. 2
b. 4
c. 8
d. $2\sqrt{2}$
e. $4\sqrt{2}$

CHAPTER

21 Electric Circuits

GOALS To learn to analyze some simple, direct current circuits containing capacitors, resistors, and sources of emf; to learn how to modify a galvanometer for measuring voltage or current; and to learn to design two useful measuring circuits, the Wheatstone bridge and the potentiometer.

OBJECTIVES After completing this chapter the student should be able to do the following:

1. Name the circuit element that is represented by the following electrical symbols:

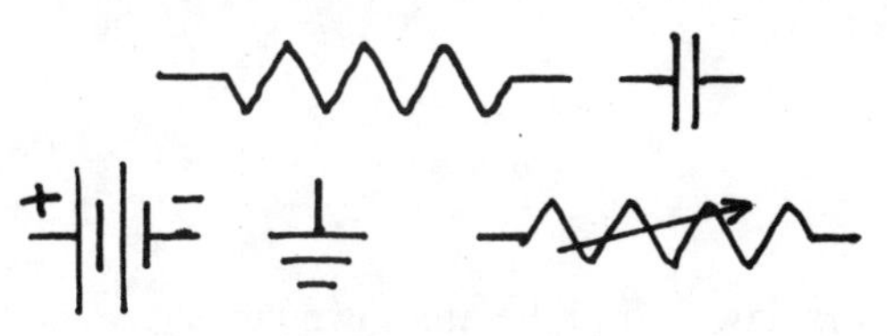

Fig. 21-1

2. State the law of conservation of charge.
3. State Kirchoff's first law.
4. State Kirchoff's second law (the loop theorem).
5. Use Kirchoff's first law to find an unknown current in a circuit.
6. Use Kirchoff's second law to find an unknown potential difference in a circuit.
7. Make a diagram of a circuit from a verbal or written description.
8. Determine the internal resistance of a battery from a knowledge of its emf and the current through an external resistance.
9. Calculate the terminal voltage of a battery if its $\mathcal{E}$ and internal resistance are known.
10. Calculate the effective resistance of a group of resistors connected in series, in parallel, or in a combination of the two.
11. Calculate the effective $\mathcal{E}$ of two or more batteries connected in series.
12. Write Ohm's law for a resistor or group of resistors in a circuit.
13. Write Ohm's law for a complete simple circuit.
14. Calculate the current in a part of a simple circuit when the resistances and $\mathcal{E}$s are known.
15. Calculate the terminal voltage of a battery or the $\mathcal{E}$ of a battery in a circuit when currents and $\mathcal{E}$s are known. known.
16. Calculate the charge on the plates of a capacitor in a simple circuit containing one or more resistors and one or more capacitors.
17. Calculate the shunt resistance required to convert a galvanometer into an ammeter.
18. Calculate the series resistance required to convert a galvanometer into a voltmeter.
19. State the essential electrical characteristic of an *ideal* ammeter and of an *ideal* voltmeter.
20. Calculate the true resistance in a circuit by the "ammeter-voltmeter" method, correcting for errors introduced by the measuring instruments.
21. Calculate the effective capacitance of capacitors connected in series, in parallel, or in a combination of the two.
22. Determine the unknown resistance in a Wheatstone bridge circuit.
23. Make a circuit diagram of a standard Wheatstone bridge.
24. Calculate an unknown $\mathcal{E}$ by means of a slide-wire potentiometer.

SUMMARY

Two conservation laws, conservation of charge and conservation of energy, lead to Kirchoff's first and second laws for electric circuits. Kirchoff's first law states that the algebraic sum of currents entering a junction is zero. Kirchoff's second law states that the algebraic sum of potential differences around any closed path is zero. These two laws of Kirchoff, together with Ohm's law, permit the determination of unknown currents, voltages, or resistances in a circuit.

Sources of emf ($\mathcal{E}$), including generators and chemical cells, have an internal resistance that is an integral part of the source of emf. Such an internal resistance cannot be measured directly but must be inferred from measurements of the current and terminal voltage. The presence of internal resistance in a cell makes it more difficult to measure the $\mathcal{E}$. If the cell is discharging, its terminal voltage V and $\mathcal{E}$ are related as follows: $V = \mathcal{E} - Ir$.

Oftentimes circuits containing several resistors can be reduced to simpler circuits by means of the formulas for the equivalent resistance of resistors in series and parallel. A simplification of the circuit to one containing a single resistance and a single $\mathcal{E}$, if such simplification is possible, facilitates the analysis of the circuit.

The galvanometer is the basic measuring instrument for direct-current circuits. By adding an appropriate resistance in series, the galvanometer may be converted to a voltmeter of any desired range. Likewise, by adding the appropriate shunt resistance across the terminals of the galvanometer, it becomes an ammeter.

When capacitors are connected in parallel the effective capacitance is simply the sum of the individual capacitances. Series-connected capacitors, on the other hand, add reciprocally. Symbolically,

$C_p = C_1 + C_2 + C_3 + \ldots$ (capacitors in parallel)
$1/C_s = 1/C_1 + 1/C_2 + 1/C_3 + \ldots$ (capacitors in series)

The Wheatstone bridge circuit is used for accurately intercomparing resistances. The potentiometer is an ideal circuit for measuring the $\mathcal{E}$ of a cell because, when properly adjusted, it draws no current from the cell being measured.

QUESTIONS AND PROBLEMS

Secs. 21-1—21-4

1. Fig. 21-2 shows junction points in a circuit. Use Kirchoff's first law to find the unknown currents. In each case put an arrow on the drawing to show the direction of the unknown current.

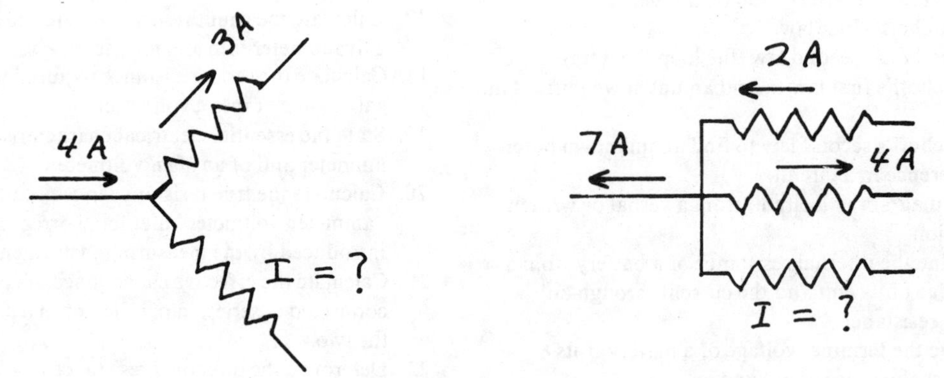

Fig. 21-2

2. In the circuit shown in Fig. 21-3, one of the ℰs is not given. Find its value by means of Kirchoff's second law.

ℰ = ______________ .

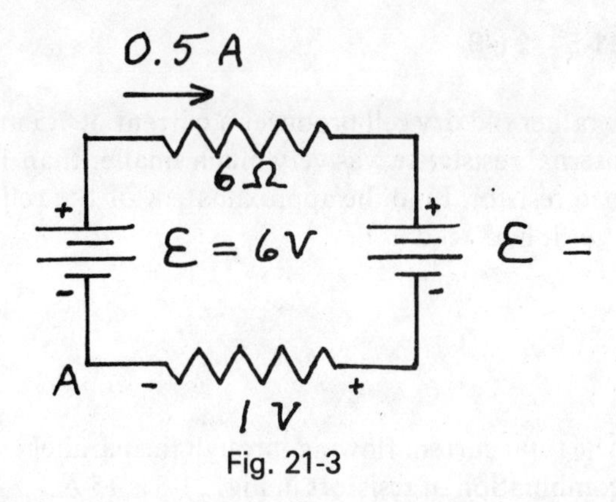

Fig. 21-3

3. Determine the current that flows through each part of the circuit shown in Fig. 21-4. It will first be necessary to find the effective resistance in the circuit.

I_1 = ______________ .

I_2 = ______________ .

I_3 = ______________ .

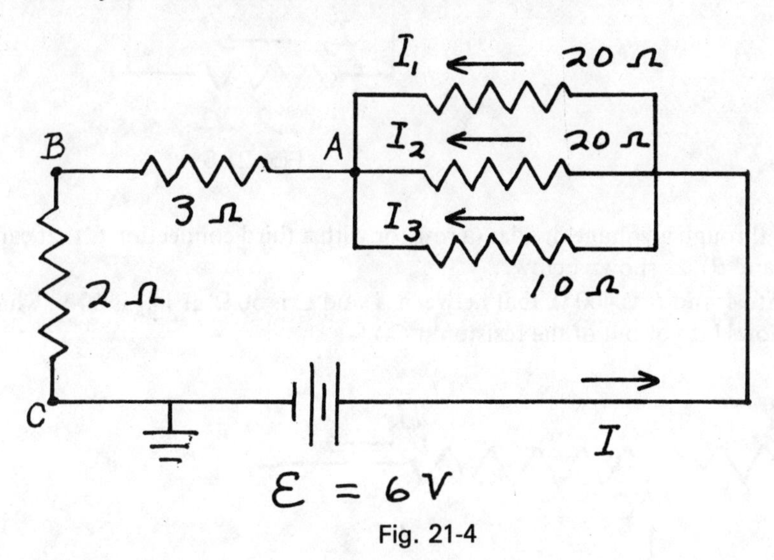

Fig. 21-4

4. Calculate the *PD* across the two resistors in series with the battery.

V_{AB} = ______________ . V_{BC} = ______________ .

5. Calculate the potential at points *A* and *C*.

V_A = ______________ . V_C = ______________ .

6. The battery in Fig. 21-5 is connected to a load whose resistance is 300 Ω. The ℰ of the battery is 6.2 V but the voltage measured across its terminals V_{AB} is only 6.0 V. What is the internal resistance of this battery?

r = ______________ .

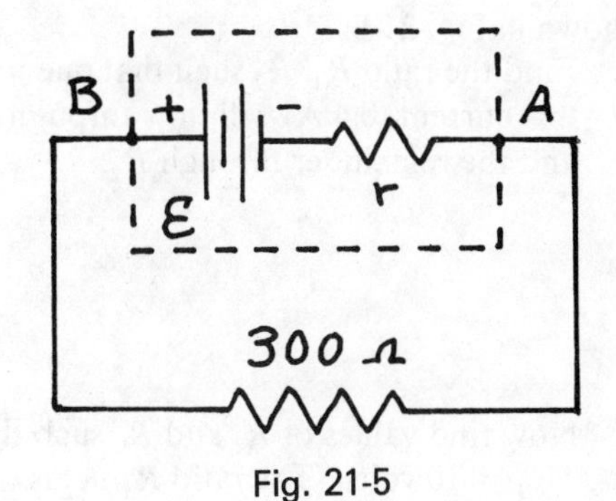

Fig. 21-5

7. A rather old dry cell produces a current of 7.5 mA through a 100 Ω resistor. When the dry cell was fresh its internal resistance was very much smaller than 100 Ω and it produced a current of 15 mA through the 100 Ω load resistor. Find the approximate $\mathcal{E}$ of the cell, assumed to remain constant, and its internal resistance after it had aged.

$\mathcal{E}$ = ______________.

r = ______________.

8. The total current flowing through the parallel combination of resistors in Fig. 21-6 is 15 A. What current flows in each resistor? (Hint: Use the fact that the *PD* across each resistor is the same.)

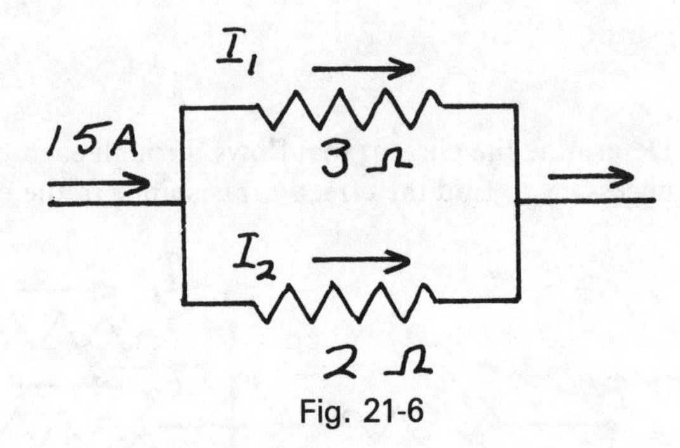

Fig. 21-6

9. An unknown current is flowing through a voltage divider (a resistor with a third connection *C* that can be moved to any point between *A* and *B*), as shown below.

The resistance between points *A* and *B* is 400 Ω, that between *A* and *C* is 50 Ω. If V_{AB} is 60 V, what is V_{AC}? (Assume that no current flows into or out of the resistor at *C*.)

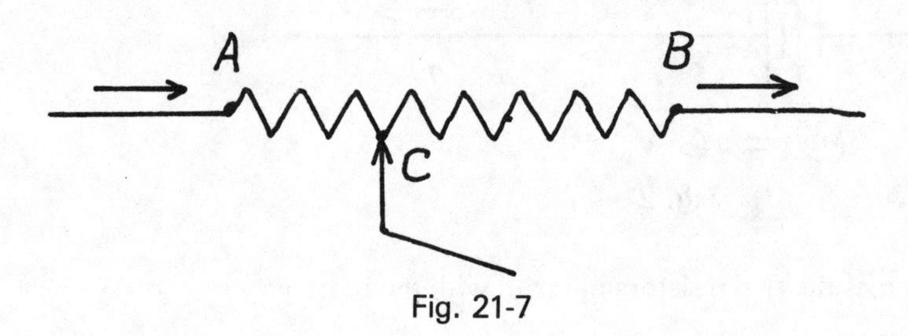

Fig. 21-7

10. A current of 4 A enters the junction point *A* shown in Fig. 21-8.

a. Find the ratio R_1/R_2 such that one tenth of this current (0.4 A) will flow through R_1 and the remainder through R_2.

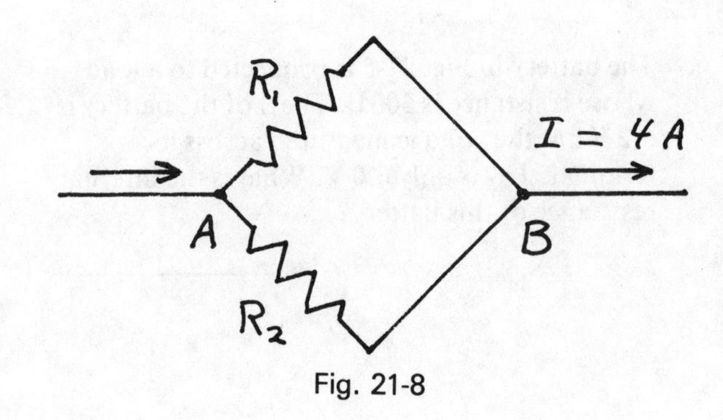

Fig. 21-8

b. Now find values of R_1 and R_2 such that V_{AB} = 10 volts. (The ratio R_1/R_2 is not changed.)

11. The $\mathcal{E}$ of an automobile battery is 12.0 V. In starting up a cold engine, the battery delivers 50 A to the starter motor for a brief period. The terminal voltage of the battery drops to 11.0 V while this current is flowing. What is the internal resistance of the battery?

$r =$ ____________.

12. A portable, gasoline-powered generator is rated at 2.2 kW. The generator is able to supply this amount of electric power while maintaining a terminal voltage of 110 volts.
   a. What current does the generator deliver at its rated power?

   $I =$ ____________.

   b. If the internal resistance of the generator is 0.2 Ω, what is its $\mathcal{E}$?

   $\mathcal{E} =$ ____________.

   c. Supposing that the internal resistance accounts for all power loss, what power must be supplied *to* the generator by the gasoline engine? Give the power in watts and in horsepower.

   $P =$ ____________.

   d. Calculate the electrical energy delivered *by* the generator at full load (2.2 kW) in a 12-hour period.

   energy = ____________.

   e. Approximately how many 60-watt bulbs could be lighted simultaneously by this generator?

   number of bulbs = ____________.

   f. Make a circuit diagram showing how the bulbs would be connected. (Remember, they are 110 volt, 60-watt bulbs.) It is not necessary to show all of them.

13. Calculate the charge on the plates of the capacitor in the circuit shown in Fig. 21-9. Remember that no steady current can flow through a capacitor. (See Ex. 21-11 in the text.)

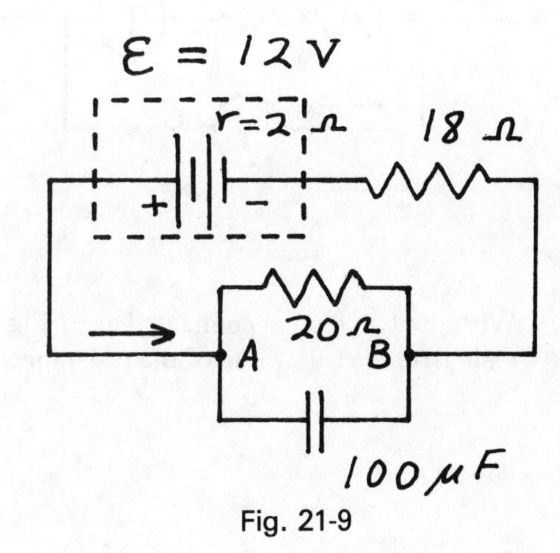

Fig. 21-9

14. A moderately sensitive galvanometer gives a full-scale deflection when the current passing through it is 50 μA. The resistance of the galvanometer coil is 500 Ω.

a. Make a drawing to show how a resistor can be connected to this galvanometer so as to make an ammeter whose full-scale current will be larger than 50 μA.

b. Calculate the value of the shunt resistor required to give a full-scale current of 20 mA.

c. Calculate the shunt required to give an ammeter reading 500 mA full scale.

15. Using the same basic galvanometer movement, show how to construct a voltmeter that will read from 0 to 0.5 volts.

16. You have available a voltmeter whose resistance is 20 kΩ and an ammeter whose resistance is 200 Ω. These two meters are to be used to determine the resistance of a resistor. If the voltage can be measured across a resistor in which a known current is flowing, the resistance can be calculated from $R = V/I$. The two drawings in Fig. 21-10 show the connections that may be made. (The battery that supplies the current is not shown.)

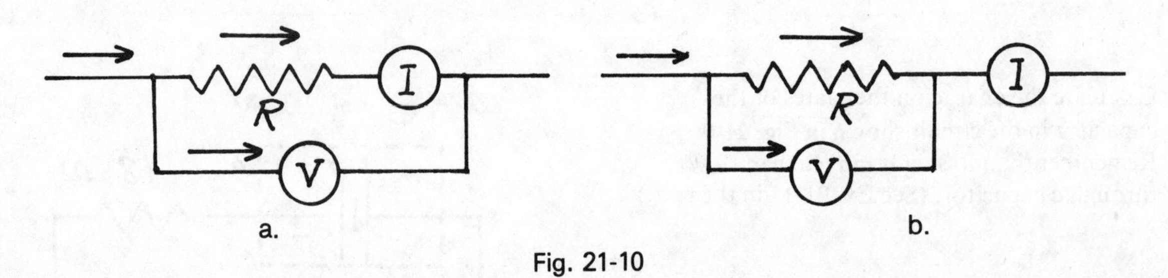

Fig. 21-10

a. With the two meters connected as in Fig. 21-10a, the meter readings are: $V = 44$ volts, $I = 0.02$ A. Calculate the true value of the resistance.

$R =$ ____________.

Calculate also the approximate value of R from V/I without corrections.

R_{appx} = ______________ .

How large is the relative error?

$\frac{\Delta R}{R}$ = ______________ .

b. With the two meters connected as in Fig. 21-10b, the meter readings are: V = 50 volts, I = 0.0275 A. Calculate the true value of the resistance for this case.

R = ______________ .

Again calculate the approximate resistance from V/I and determine the relative error.

R_{appx} = ______________ .

$\Delta R/R$ = ______________ .

17. If you were offered two voltmeters at the same price, one having a resistance of 10 kΩ and the other a resistance of 50 kΩ, both of which read 0 to 100 volts, which one would you buy? Give the reasons for your choice.

18. Two ammeters read 0 to 100 mA but one is a much higher quality instrument than the other. What statement can you make about the resistances of these two meters?

19. Capacitors of 2 μF and 6 μF are available. Show how these two capacitors can be connected in a closed circuit with a battery so that the effective capacitance is 1.5 μF.

20. Show how the same two capacitors can be connected to give an effective capacitance of 8 μF. Include the connections to the battery.

21. When the switch is closed in this circuit a charge of 40 μC flows from the battery to charge the capacitors.
   a. Calculate the effective capacitance in the circuit.

   C_{eff} = ______________.

   b. Calculate the capacitance of the unknown capacitor.

   C = ______________.

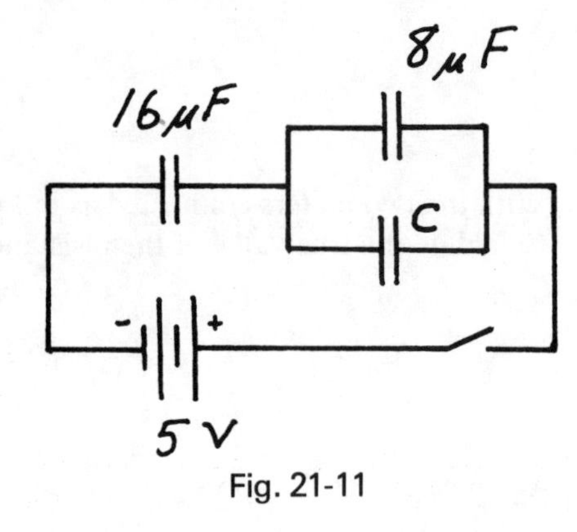

Fig. 21-11

22. In the Wheatstone bridge a galvanometer is used as a "null" indicator. This means that the galvanometer serves to show when the ______________ through it is ______________.

23. Fig. 21-12 shows a Wheatstone bridge circuit.
   a. Write the equation of balance for this bridge.

   b. The three known resistances are R_1 = 120 Ω, R_2 = 40 Ω, and R_3 = 6600 Ω. Determine R_4.

   R_4 = ______________.

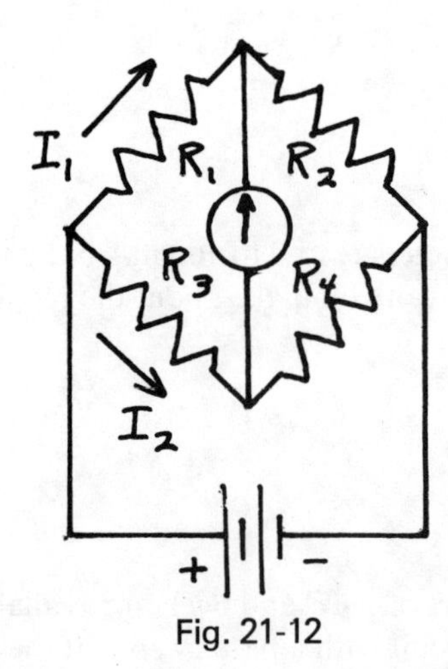

Fig. 21-12

24a. Why is it not possible to measure the emf of a cell exactly with a voltmeter?

b. What is the characteristic of a voltmeter whose indicated voltage is very nearly equal to the $\mathcal{E}$ of a cell to which it is connected?

c. Write down an equation relating V (the voltmeter reading) to $\mathcal{E}$ and r.

25. Make a circuit diagram for a potentiometer. Indicate the polarity of the working battery and the direction of the current. Assuming that the potentiometer is balanced, in what part of the circuit is there no current?

Solutions

1.

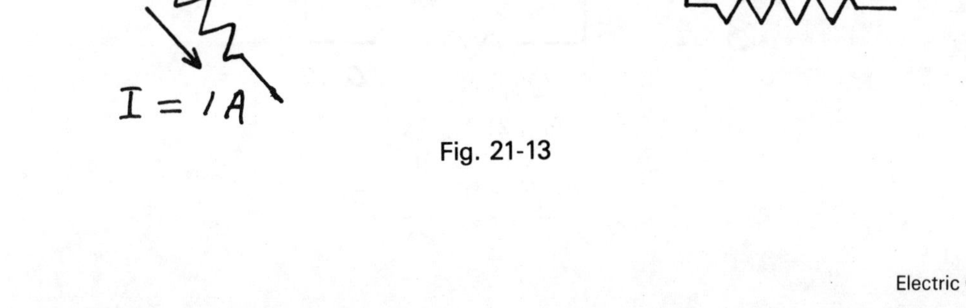

3A

4A

$I = 1A$

2 A

7 A

4A

$I = 9A$

Fig. 21-13

2. The sum of the changes in potential must be zero. Starting at point A, $+6\text{ V} - 3\text{ V} - \mathcal{E} - 1\text{ V} = 0$. Thus $\mathcal{E} = 2$ V.
3. First combine the two 20 Ω resistors that are connected in parallel.

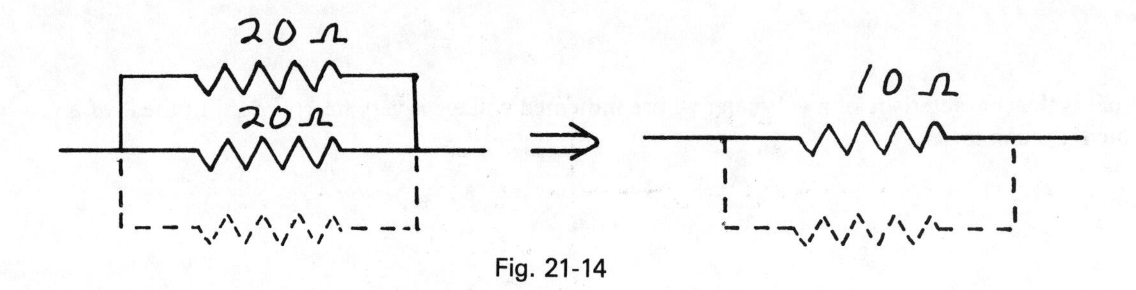

Fig. 21-14

Now combine this with the other 10 Ω resistor, also in parallel.

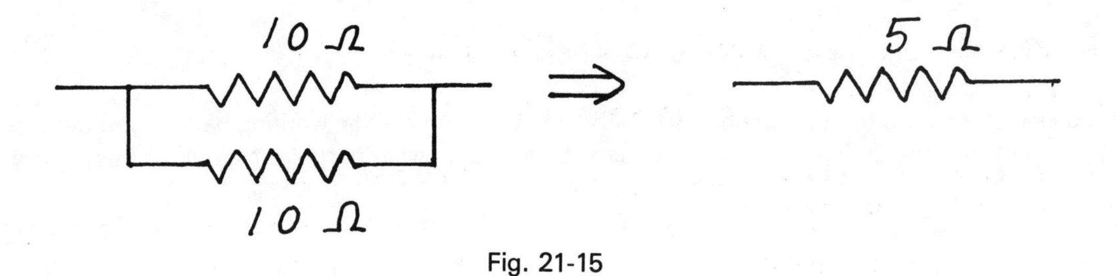

Fig. 21-15

We are left with three resistors in series.

$$R_{\text{eff}} = 2\,\Omega + 3\,\Omega + 5\,\Omega = 10\,\Omega$$

The circuit is simplified to

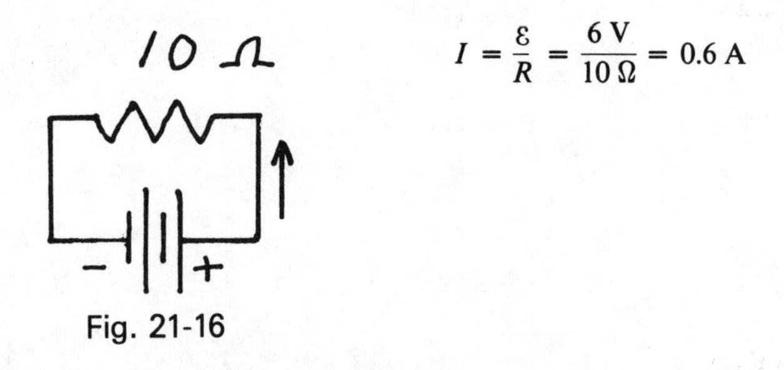

$$I = \frac{\mathcal{E}}{R} = \frac{6\text{ V}}{10\,\Omega} = 0.6\text{ A}$$

Fig. 21-16

The current that flows through the 3 Ω resistor is

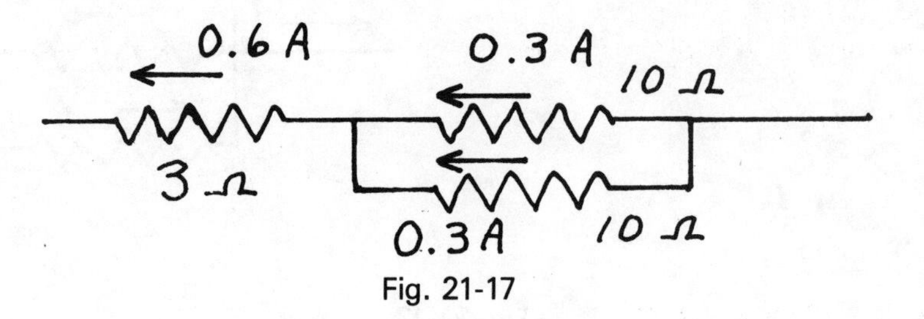

Fig. 21-17

A current of 0.3 A must flow through the 10 Ω resistor. Half of this must flow through each 20 Ω resistor. Thus the three currents in the parallel combination are 0.15 A, 0.15 A, and 0.3 A, as shown below.

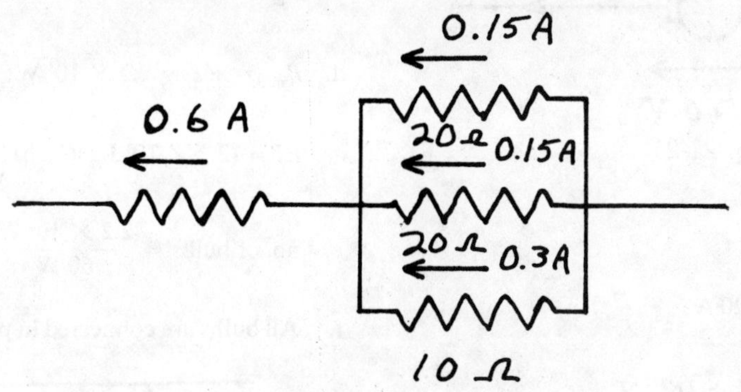

Fig. 21-18

To check this result, calculate the IR drop across each resistor. They must be the same.

$$IR = (0.15\text{ A})(20\ \Omega) = (0.3\text{ A})(10\ \Omega) = 3\text{ V}$$

4. $V_{AB} = IR = (0.6\text{ A})(3\ \Omega) = 1.8\text{ V}$
   $V_{BC} = IR = (0.6\text{ A})(2\ \Omega) = 1.2\text{ V}$
5. $V_A = 6\text{ V} - 3\text{ V} = 3\text{ V}$
   $V_C = 0\text{ V}$ (This point is grounded.)
6.

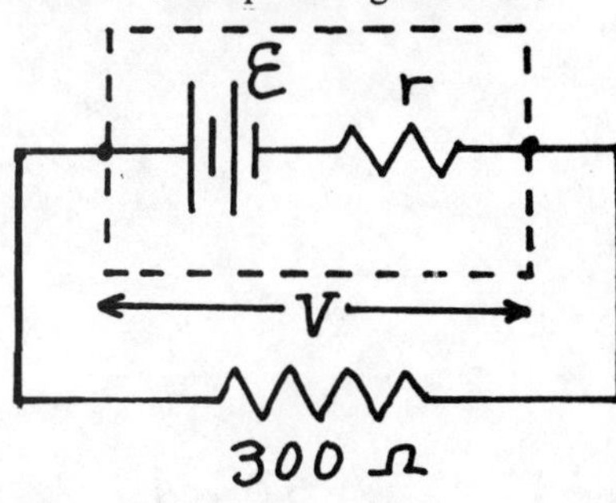

Fig. 21-19

First find the current.

$$I = \frac{V}{R} = \frac{6.0\text{ V}}{300\ \Omega} = 0.02\text{ A} = 20\text{ mA}$$

Since the PDs around the circuit must add up to zero, the PD across the internal resistance is 0.2 V.

$$r = \frac{V}{I} = \frac{0.2\text{ V}}{0.02\text{ A}} = 10\ \Omega$$

7. The potential difference across the cell terminals is

$$V = IR = (0.015\text{ A})(100\ \Omega) = 1.5\text{ V}$$
$$\mathcal{E} = 1.5\text{ V} + Ir$$

Since r is very small, $\mathcal{E} \cong 1.5$ V.
When the cell has aged,

$$V = I'R = (0.0075\text{ A})(100\ \Omega) = 0.75\text{ V}$$
$$I'r = \mathcal{E} - \text{V} = 1.5\text{ V} - 0.75\text{ V} = 0.75\text{ V}$$

$$r = \frac{0.75\text{ V}}{7.5 \times 10^{-3}\text{ A}} = 100\ \Omega$$

8. $V = I_1(3\ \Omega) = I_2(2\ \Omega)$

The effective resistance of this parallel combination is

$$R_{\text{eff}} = \frac{1}{\frac{1}{2} + \frac{1}{3}} = \frac{6}{3+2} = \frac{6}{5}\ \Omega$$

Thus $V = IR_{\text{eff}} = (15\text{ A})(6/5)\Omega = 18\text{ V}$

$$I_1 = \frac{18\text{ V}}{3\ \Omega} = 6\text{ A} \quad I_2 = \frac{18\text{ V}}{2\ \Omega} = 9\text{ A}$$

9. $V_{AB} = IR = I(400\ \Omega) = 60$ V.
   Thus $I = (60\text{ V})/(400\ \Omega) = 0.15\text{ A}$
   $V_{AC} = IR = (0.15\text{ A})(50\ \Omega) = 7.5\text{ V}$

10a. The PD across R_1 is equal to the PD across R_2. $V = IR$

$R_1(0.5\text{ A}) = R_2(3.6\text{ A})$
$R_1/R_2 = (3.6\text{ A})/(0.4\text{ A}) = 9$

b. $R_1(0.4\text{ A}) = 10\text{ V};\ R_1 = 25\ \Omega$
   $R_2(3.6\text{ A}) = 10\text{ V};\ R_2 = 2.78\ \Omega$

11.

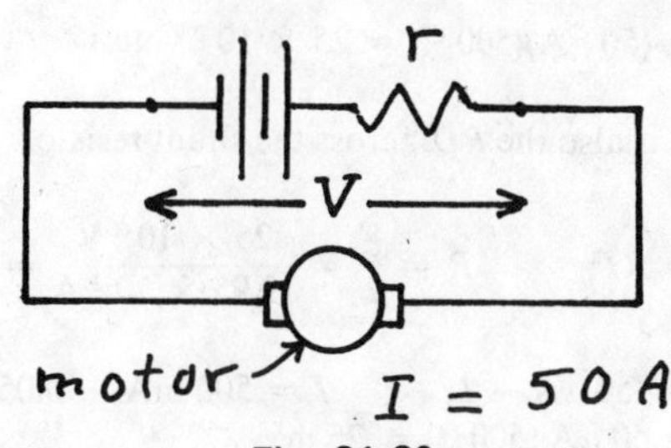

Fig. 21-20

$$V = \mathcal{E} - Ir;\ \mathcal{E} - V = Ir$$

$$r = \frac{\mathcal{E} - V}{I} = \frac{12\text{ V} - 11\text{ V}}{50\text{ A}} = 0.02\ \Omega$$

12a.

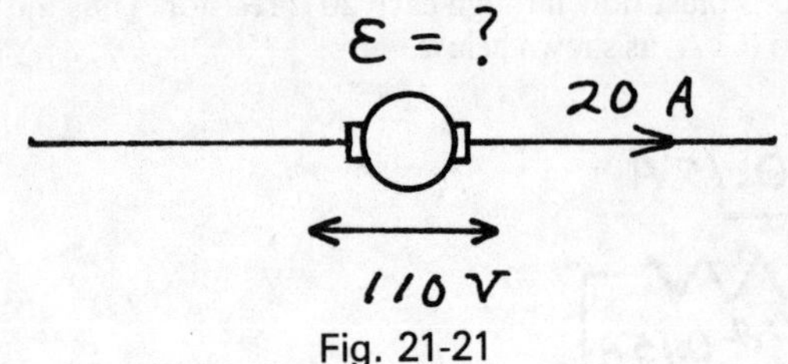

Fig. 21-21

$P = VI; I = P/V$

$$I = \frac{2.2 \times 10^3 \text{ W}}{1.1 \times 10^2 \text{ V}} = 20 \text{ A}$$

b. $V = \mathcal{E} - Ir \quad \mathcal{E} = V + Ir$
$\mathcal{E} = 110 \text{ V} + (20 \text{ A})(0.2\ \Omega) = 110 \text{ V} + 4 \text{ V} = 114 \text{ V}$

c. The *input* power depends on the ℰ of the generator. The effective *output* power of the generator depends on the *PD* between its terminals.

$P_{in} = \mathcal{E}I; \; P_{out} = VI$

$$P_{in} = (114 \text{ V})(20 \text{ A}) = (2.28 \times 10^3 \text{ W})\left(\frac{1 \text{ hp}}{746 \text{ W}}\right) = 3.0 \text{ hp}$$

d. $P_{out} = VI = 2.2 \times 10^3$ W; Work (or energy) $= Pt$

$$Pt = (2.2 \times 10^3 \text{ J/s})(12 \text{ h})\left(\frac{3600 \text{ s}}{\text{h}}\right) = 9.5 \times 10^7 \text{ J}$$

e. no. of bulbs $= \dfrac{2.2 \times 10^3 \text{ W}}{60 \text{ W}} \cong 37$

f. All bulbs are connected in parallel (see Fig. 21-22).

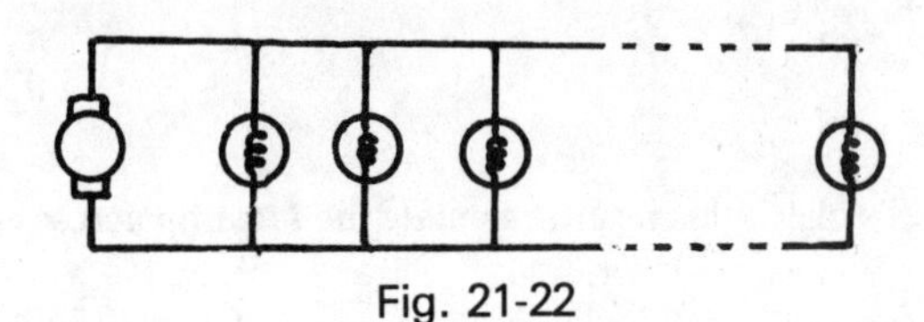
Fig. 21-22

13. Since no current flows through the capacitor, it must all flow through the 20 Ω resistor. All three resistances are, in effect, in series.

$R_{tot} = 2\ \Omega + 18\ \Omega + 20\ \Omega = 40\ \Omega$

$$I = V/R = \frac{12 \text{ V}}{40\ \Omega} = 0.3 \text{ A}$$

The *PD* across the capacitor is the same as that across the 20 Ω resistor.

$V_{AB} = (0.3 \text{ A})(20\ \Omega) = 6 \text{ V}$
$C = Q/V; Q = CV = (10^{-4} \text{ F})(6 \text{ V}) = 6 \times 10^{-4} \text{ C}$

14a. A resistor must be connected in parallel with the galvanometer (see Fig. 21-23).

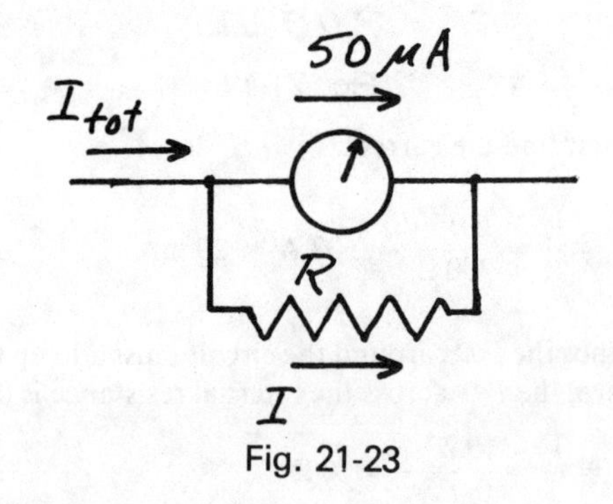

Fig. 21-23

b. $I_{tot} = 50\ \mu\text{A} + I$
$I = I_{tot} - 5 \times 10^{-5} \text{ A}$
$I = 20 \times 10^{-3} \text{ A} - 0.05 \times 10^{-3} \text{ A} = 19.95 \text{ mA}$

The *PD* across the galvanometer is

$V = (50\ \mu\text{A})(500\ \Omega) = 25 \times 10^{-3}$ volts

This is also the *PD* across the shunt resistor.

$$V = RI; \qquad R = \frac{V}{I} = \frac{25 \times 10^{-3} \text{ V}}{19.95 \times 10^{-3} \text{ A}} = 1.25\ \Omega$$

c. $I_{tot} = 50\ \mu\text{A} + I; \qquad I = 500 \text{ mA} - 0.05 \text{ mA} = 499.95 \text{ mA}$
$V = 50\ \mu\text{A}\ (500\ \Omega) = 25 \text{ mV}$

$$R = \frac{V}{I} = \frac{25 \times 10^{-3} \text{ V}}{499.95 \times 10^{-3} \text{ A}} = 0.0500\ \Omega$$

15. A resistor must be connected in series with the galvanometer, as shown in Fig. 21-24.

$IR + 25 \text{ mV} = 500 \text{ mV}$
$IR = 475 \text{ mV}$

$$R = \frac{475 \times 10^{-3} \text{ V}}{50 \times 10^{-6} \text{ A}} = 9.50 \times 10^{3} \ \Omega$$

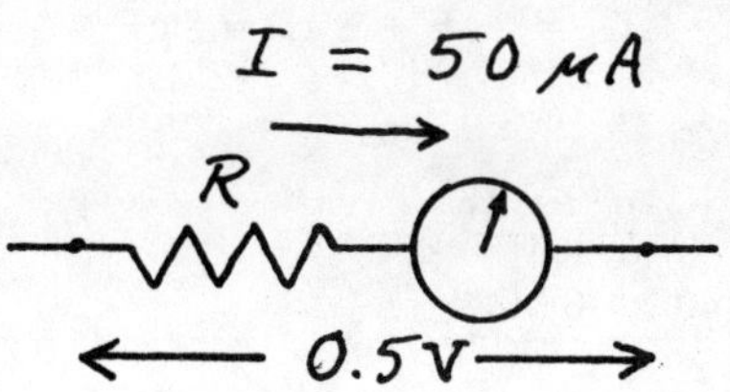

Fig. 21-24

16a. The voltmeter gives the voltage across its terminals.

$V = 44$ volts
$44 \text{ V} = IR + Ir = I(R + r)$

The ammeter gives the current that passes through it: $I = 0.02$ A.

$$R + r = \frac{44 \text{ V}}{0.02 \text{ A}} = 2200 \ \Omega$$

Thus $R = 2200 \ \Omega - 200 \ \Omega = 2000 \ \Omega$

$$R_{\text{appx}} = \frac{V}{I} = \frac{44 \text{ V}}{0.02 \text{ A}} = 2200 \ \Omega$$

The relative error is $\frac{200}{2000} = 10\%$

b. In this case, the ammeter reading must be corrected.

$I(\text{ammeter}) = I(\text{thru } R) + I(\text{voltmeter})$
$0.0275 \text{ A} = I + (50 \text{ V})/(20 \times 10^{3} \ \Omega)$
$I = 0.0275 \text{ A} - 0.0025 \text{ A} = 0.0250 \text{ A}$

$$R = V/I = \frac{50 \text{ V}}{0.0250 \text{ A}} = 2000 \ \Omega$$

$$R_{\text{appx}} = \frac{V}{I} = \frac{50 \text{ V}}{0.0275 \text{ A}} = 1818 \ \Omega$$

$$\Delta R/R = \frac{182}{2000} \cong 9\%$$

17. The meter with the higher resistance.
An ideal voltmeter is one that draws no current from the circuit to which it is connected. The higher the resistance of the meter, the closer it comes to this ideal.

18. An ideal ammeter is one that has a zero PD across its terminals. The lower the resistance of the meter, the closer it comes to this ideal. Thus the better ammeter has a lower resistance.

19. They must be connected in series.

$$\frac{1}{C} = \frac{1}{C_1} + \frac{1}{C_2}$$

$$\frac{1}{C} = \frac{1}{2 \ \mu\text{F}} + \frac{1}{6 \ \mu\text{F}} = \frac{4}{6 \ \mu\text{F}}$$

$$C = \frac{6 \ \mu\text{F}}{4} = 1.5 \ \mu\text{F}$$

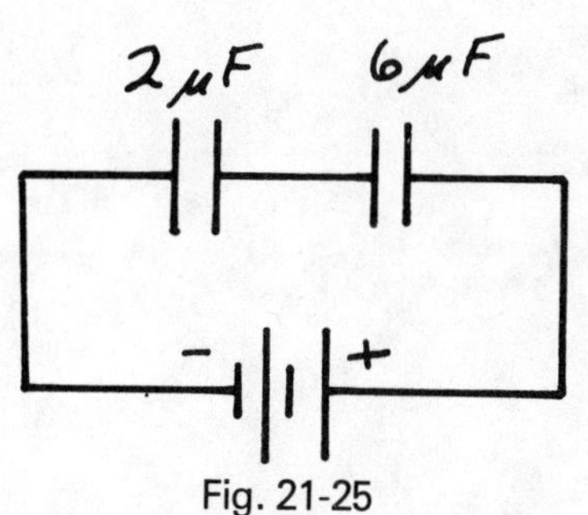

Fig. 21-25

20. $C = C_1 + C_2 = 2\ \mu F + 6\ \mu F = 8\ \mu F$

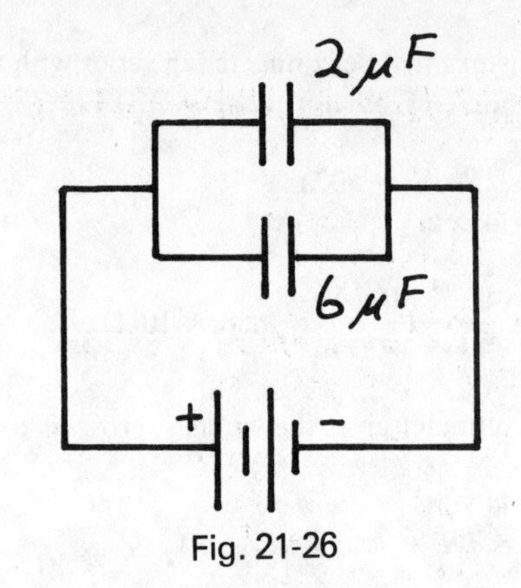

Fig. 21-26

21a. $C_{eff} = \dfrac{Q}{V}$

$$C_{eff} = \frac{40 \times 10^{-6}\ C}{5\ V} = 8\ \mu F$$

b. The two capacitors that are connected in parallel may be replaced by a single capacitor. Call it C_2. Since

$$\frac{1}{C_{eff}} = \frac{1}{C_1} + \frac{1}{C_2};\ \frac{1}{8\ \mu F} = \frac{1}{16\ \mu F} + \frac{1}{C_2};\ C_2 = 16\ \mu F$$

The unknown capacitor is connected in parallel with an 8-μF capacitor. Thus

$$C_2 = 8\ \mu F + C\ ;\ C = 8\ \mu F$$

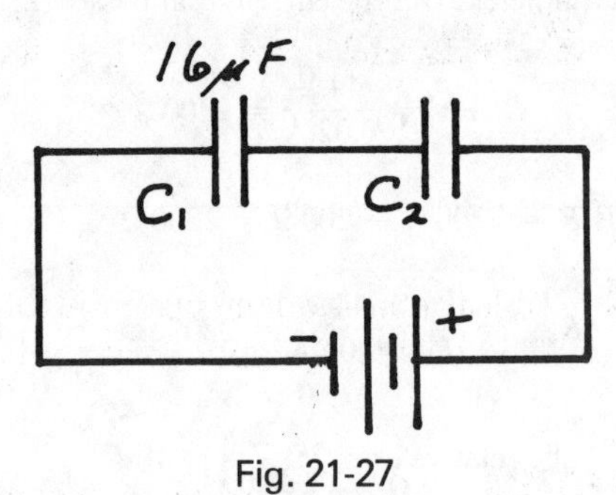

Fig. 21-27

22. current; zero

23a. $R_1/R_2 = R_3/R_4$

b. $R_4 = R_3 \dfrac{R_2}{R_1} = \dfrac{6600 \times 40}{120} = 2200\ \Omega$

24a. When the voltmeter is connected to the cell, a current flows in the cell causing its terminal voltage to be less than its $\mathcal{E}$.

b. If the voltmeter has a very high resistance (high compared to the internal resistance of the cell) then $V \cong \mathcal{E}$.

c. $V = \mathcal{E} - Ir$

25. See Fig. 21-6 in the text. The battery whose $\mathcal{E}$ is being measured should be enclosed in dotted lines. Be sure to indicate the polarities of both batteries. When balanced, no current flows through the galvanometer or through the cell whose $\mathcal{E}$ is being measured.

CHAPTER 22

Electromagnetism

GOALS To learn the fundamental principles of electromagnetism—how magnetic fields are established and how they act upon moving charges and currents.

OBJECTIVES After completing this chapter the student should be able to do the following:

1. Associate magnetic forces with the forces exerted by moving charges and electric forces with the forces between charges at rest.
2. State the definition of magnetic field in terms of the force exerted on a moving charge.
3. Given the relationship $B = F/Qv$, describe the direction of the velocity vector **v**.
4. Describe the direction of the magnetic field strength **B** in terms of the direction of a moving charge at the point in question.
5. Given information about the force experienced by moving charges, determine the magnitude and the direction of the magnetic induction **B**.
6. Given the direction of the magnetic induction, state the direction(s) in which a charge could be moving and yet experience no force.
7. Given the direction of the magnetic induction, state the direction(s) in which a charge would have to move to experience a maximum force.
8. Use the right-hand rule to determine the direction of the force **F** when the directions of **v** and **B** are given.
9. Write the vector equation that gives the magnetic force in terms of the charge, the velocity, and the magnetic induction.
10. Write the scalar equation that gives the magnitude of the magnetic force acting on a moving charge.
11. Find the magnitude and direction of the force acting on a moving charge when the magnetic induction is known.
12. Determine any one of the vectors **F**, **v**, or **B** when the two others are given.
13. Write the vector equation that gives the magnetic force in terms of the current elements and the magnetic induction.
14. Write the scalar equation that gives the magnitude of the magnetic force acting on a current element.
15. Determine any one of the quantities **F**, **I**, Δl, or **B** when the others are given.
16. Apply the relationship for force on a current element to determine the torque acting on a current loop in a magnetic field.
17. Describe the equilibrium position assumed by a current loop that is free to rotate in a magnetic field.
18. Determine the direction of a current in a loop from the loop's preferred orientation in a magnetic field.
19. Determine the direction of a magnetic field from the preferred orientation of a current loop.
20. Write the equation giving the magnitude of the magnetic induction due a current element.
21. Write the equation that gives the magnitude of the magnetic induction at the center of a circular loop carrying a current I.
22. Write the expression for the magnetic field at a distance a from a long wire carrying a current.
23. Describe the directions of the magnetic field due to a current element, a circular loop, and a long straight wire.
24. Determine the magnitude and direction of the magnetic induction due to current in a long straight conductor.

25. Calculate the magnitude of the force of attraction (or repulsion) between two parallel wires.
26. Describe the direction of the circulating current that is equivalent to a magnetized piece of iron.
27. Describe the approximate direction of the earth's magnetic field near the North Pole, the South Pole, the equator, and in the northern United States.
28. Describe the direction of the emf $\mathcal{E}$ induced in a conductor moving through a magnetic field.
29. Write the equation that relates induced $\mathcal{E}$ to the length of a conductor and its velocity through a magnetic field.
30. Calculate any one of the quantities $\mathcal{E}$, B, l, or v when the others are given.
31. Write a statement of Lenz's law.
32. Use Lenz's law to determine the direction of an induced $\mathcal{E}$ or of an induced current in a moving conductor.
33. Write the equation that defines magnetic flux Φ.
34. Write the relationship between the *weber* and the *tesla*.
35. Calculate the magnetic flux through a surface when the magnetic induction **B** makes an arbitrary angle with the surface.
36. Write either one of the two equations that relate $\mathcal{E}$ to the change in flux.
37. Solve for any one of the quantities, $\mathcal{E}$, $\Delta\Phi$, Δt, or N when the others are given.

SUMMARY

We have seen in previous chapters that charges at rest exert forces on one another and that these forces act along a line joining the two charges; the electrostatic forces are either attractive or repulsive. The new forces that come into play when the charges are in motion do not, in general, act along the line joining the two charges; the magnetic forces between moving charges are neither attractive nor repulsive. To deal satisfactorily with this new and more complicated situation, we introduce the concept of a magnetic field. A magnetic field makes its presence known by the force that it exerts on a moving charge or a current segment. The magnetic field is established by moving charges, current segments, or current loops, the latter two being special cases of moving charges. Symbolically, this relationship is

$$\begin{pmatrix}\text{moving}\\ \text{charges}\end{pmatrix} \rightarrow \begin{pmatrix}\text{magnetic}\\ \text{field}\end{pmatrix} \rightarrow \begin{pmatrix}\text{forces on}\\ \text{moving charges}\end{pmatrix}$$

These three entities are vector quantities whose mutual directions have a very strong influence on the strength of the interaction. For example, if the direction of a moving charge is parallel to a magnetic field, the magnetic force experienced by that charge is zero, whereas if the same charge is moving at the same speed *perpendicular* to the magnetic field, the magnetic force on the charge will be a maximum.

In many practical applications the moving charges that produce a magnetic field are the free electrons that flow through a conductor under the influence of an electromotive force. A loop of copper wire carrying a current, for example, produces a magnetic field in its vicinity. At the center of the loop the magnitude of the magnetic field **B** is $B = k'2\pi I/R$, where I is the current in the loop and R is its radius. The direction of **B** is parallel to the axis of the loop.

If a small current loop is placed in a uniform magnetic field, each segment of the loop will be acted upon by a magnetic force. In this case, the sum of the forces on the loop is zero. However, the sum of the torques is not zero; there will be a tendency for the loop to rotate until its axis is parallel to **B**. This is just the behavior of a compass needle, which is subject to a net torque but not to a net force. Thus a current-carrying loop acts like a small magnet. When a magnet is examined closely, on an atomic scale, it is seen that its magnetism can be ascribed entirely to a collection of small current loops in the form of spinning electrons.

One of the consequences of the magnetic force on moving charges is that a current can be made to flow in a circuit by changing the magnetic flux through the circuit. The associated $\mathcal{E}$ is equal to $\Delta\Phi/\Delta t$, the rate of change of flux with respect to time. The flux Φ is the normal component of **B** times the area of the loop. The $\mathcal{E}$ of a generator in a power station results from a rapidly changing flux in the rotating coils of the armature.

QUESTIONS AND PROBLEMS

Secs. 22-1—22-3

1. Charges at rest exert forces on each other, which are called C____________ forces or electrostatic forces. When electric charges are moving, additional forces come into play. These forces between moving charges are called ____________ forces.
2. If a small mass, placed at a point in space, is acted upon by a force, we say that a gravitational field exists at that point. If a body carrying a small electric charge experiences an electric force we say that an e____________ f____________ exists at the location of the body. In a similar fashion, a moving charge can be used to detect the presence of a(n) ____________ ____________.
3. The magnetic force on a moving charge depends on the ____________ of the charge and on its velocity, i.e., $F_{mag} \propto Qv$.
4. Since a gravitational field is defined as the force on an object divided by its mass, and an electric field is defined as the electric force divided by the charge, we might expect the magnetic field (or magnetic induction) to be defined by ____________ divided by the product ____________ times ____________, which would be written

$$B = \frac{F}{Qv}$$

5. The situation is more complicated, however, because the magnitude of **B** depends on the direction in which the charge is moving as well as its speed. In fact, if the charge Q happens to be moving parallel to **B**, the magnetic force acting on it will be ____________. The magnetic force will have its maximum value of QvB only when the charge is moving ____________ (parallel or perpendicular) to the magnetic induction **B**.
6. A magnetic field of 0.2 teslas is horizontal and directed toward the north (see Fig. 22-1). A charge of 5×10^{-19} C is moving in this field at a speed of 10^7 m/s. Calculate the magnitude of the force on the charge for the following directions of motion.
   a. The charge is moving due south.

      $F =$ ____________.

   b. The charge is moving due west.

      $F =$ ____________.

   c. The charge is moving vertically downward.

      $F =$ ____________.

   d. The charge is moving due north.

      $F =$ ____________.

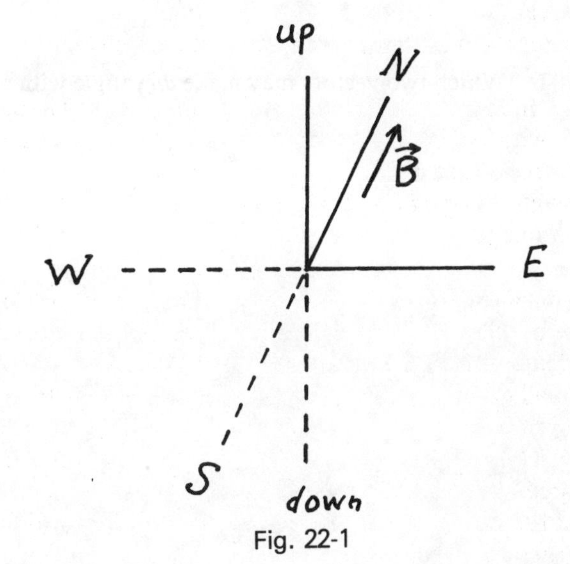

Fig. 22-1

7. In Prob. 6, the vector **B** is directed toward the north. Suppose that the vector **v** is directed toward the west. Use the right-hand rule to determine the direction of **v** × **B**.

   v × **B** is directed ____________.

8. Write the vector equation giving **F** in terms of Q, **v**, and **B**.

F = ______________ .

9. Use the right-hand rule to determine the direction of **F** in each of the four cases of Prob. 6.

a. ______________ c. ______________

b. ______________ d. ______________

10. Suppose that the positive charge of Prob. 6 is moving in a horizontal plane in the direction indicated in Fig. 22-2. Determine the magnitude and direction of the magnetic force.

F = ______________ .

Its direction is ______________ .

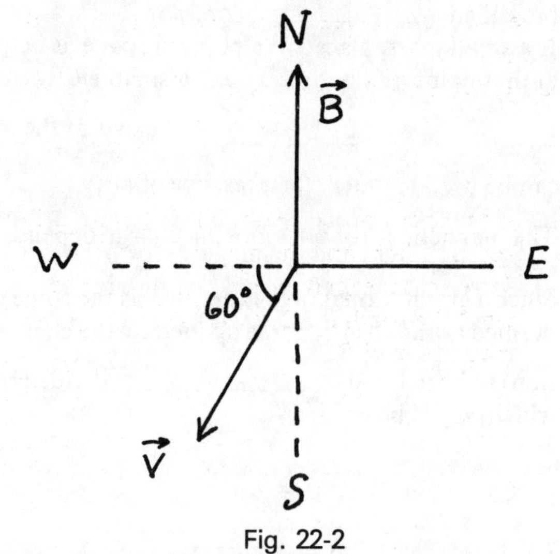

Fig. 22-2

11. The equation for the magnetic force on a moving charge is $\mathbf{F} = Q\mathbf{v} \times \mathbf{B}$.

a. Which one of these three vectors is *always* perpendicular to both of the others?

b. Which two vectors may make *any* angle with each other?

12. In the coordinate system illustrated in Fig. 22-3, the *xy*-plane is horizontal. The *z*-axis is vertical. A moving electron (negative charge), as in a TV tube, is to be used to detect the magnetic field. When the electron is directed vertically upward it experiences no force. When the electron is moving toward the south (positive *x*-direction), it experiences a force toward the east.

a. What is the direction of **B**?

b. If the speed of the electron is 0.2 times the speed of light and the magnitude of the force that it experiences when moving toward the south is 3.2×10^{-10} N, what is the magnitude of **B**?

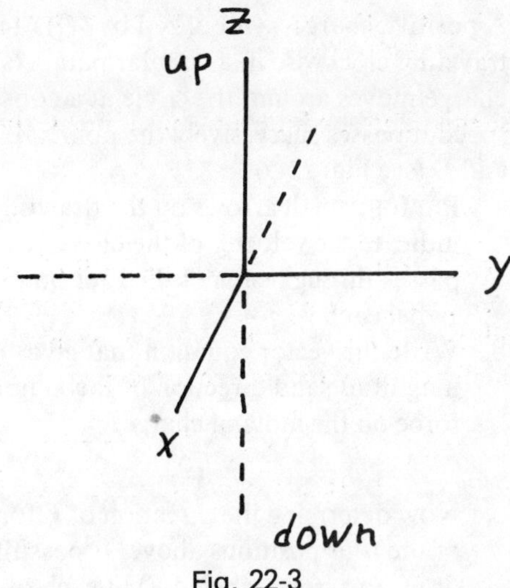

Fig. 22-3

13. The region enclosed by the irregular curve in the upper part of Fig. 22-4 contains a magnetic field **B** that is directed upward, out of the paper. What is the direction of the magnetic induction **B** in the lower part of the figure?

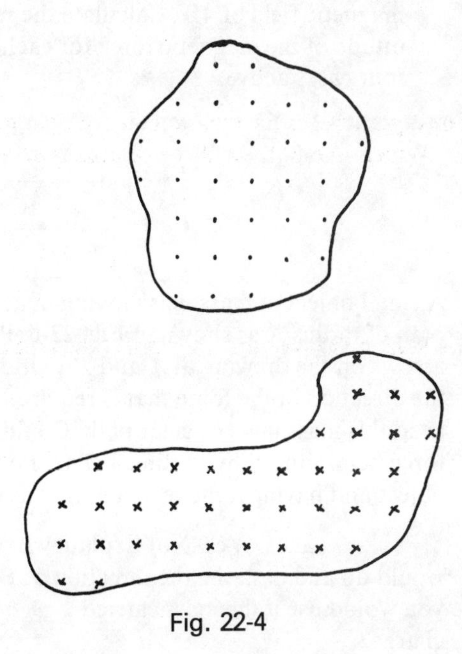
Fig. 22-4

14. A positive charge, symbolized by "⊕", is traveling clockwise in a circular path. As the charge moves around the circle at a constant speed it passes successively the points *A, B, C,* and *D* (see Fig. 22-5).

   a. Put four small arrows on the drawing to indicate the velocity of the charge as it passes through each of the four points. Label each vector **v**.

   b. Write the vector equation that gives both magnitude and direction of the magnetic force on the moving charge.

   F = ______________.

   c. Now determine the direction of **F** for each of the four positions above. If possible, draw an arrow (labeled **F**) at each of the four positions to indicate the direction of the magnetic force.

   d. Suppose that the positive charge described above has a magnitude of 1.6×10^{-19} C and that it is moving at a speed of 0.1 c in a magnetic field of 4 T. Calculate the magnitude of the magnetic force for each of the four cases above.

   F_A = ______________.

   F_B = ______________.

   F_C = ______________.

   F_D = ______________.

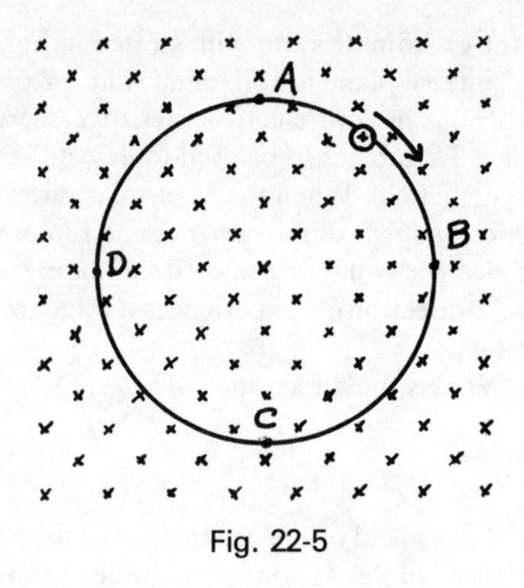

Fig. 22-5

15. A small object of mass *m* is moving in a circular path of radius *R* as shown in Fig. 22-6. Put arrows on the drawing at *A* and *B* to indicate the direction of the force that is required to keep this mass in its circular path. Could this force be produced by putting a charge on the object and having it move in a magnetic field?

   ______________ (Yes/No) Explain what you would do and describe the direction of **B** that you would use if the mass carried a *negative* charge.

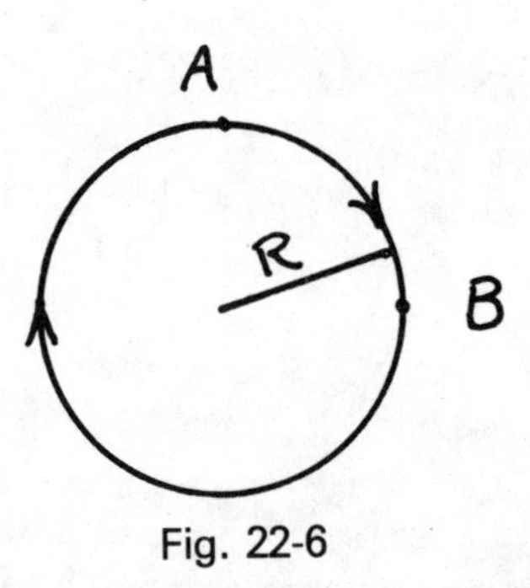

Fig. 22-6

Secs. 22-4—22-8

16. In Sec. 22-4 it is shown that a current element is acted upon by a magnetic field in the same way as a moving charge. In fact the vector quantities $Q\mathbf{v}$ and $I\Delta l$ are equal in every respect. $Q\mathbf{v} = I\Delta l$.

   a. Write the vector equation for the force on a current element in a magnetic field.

   F = ______________.

b. What is the maximum value that the magnitude of **F** may have? Write it down using the appropriate symbols.

$F_{max} =$ ________________ .

c. What is the minimum value that F may have, none of the quantities I, Δl, or B, being zero?

$F_{min} =$ ________________ .

d. What relationship between I, Δl and **B** will cause F to have its maximum value?

e. What relationship between $I\Delta l$ and **B** will cause F to have its minimum value?

f. What relationship between these two vectors will make $F = \frac{1}{2}F_{max}$? Make a drawing for this case showing the two vectors $I\Delta l$ and **B** and the angle between them.

17. A wire is carrying a current of 50 A in the direction shown in Fig. 22-7. A magnetic induction of 8 T is directed into the plane of the drawing. Consider a segment of the wire of length 2 cm.
    a. Calculate the magnitude of the magnetic force acting on this current element.

    $F =$ ________________ .

    b. Put an arrow on the drawing at P to indicate the direction of **F**.

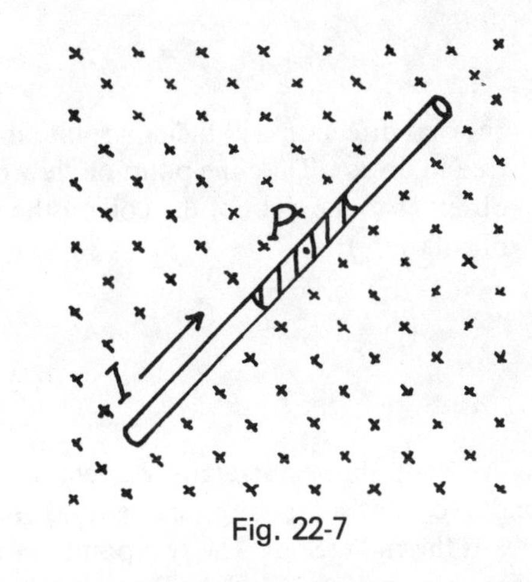

Fig. 22-7

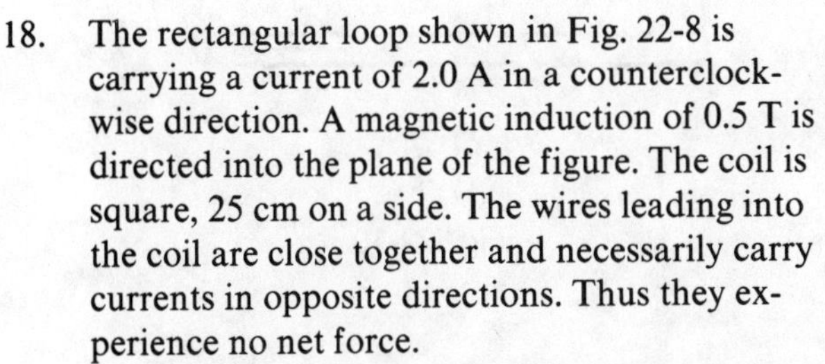

18. The rectangular loop shown in Fig. 22-8 is carrying a current of 2.0 A in a counterclockwise direction. A magnetic induction of 0.5 T is directed into the plane of the figure. The coil is square, 25 cm on a side. The wires leading into the coil are close together and necessarily carry currents in opposite directions. Thus they experience no net force.
    a. Describe the direction of the magnetic force on each of the four sides of the rectangular loop.

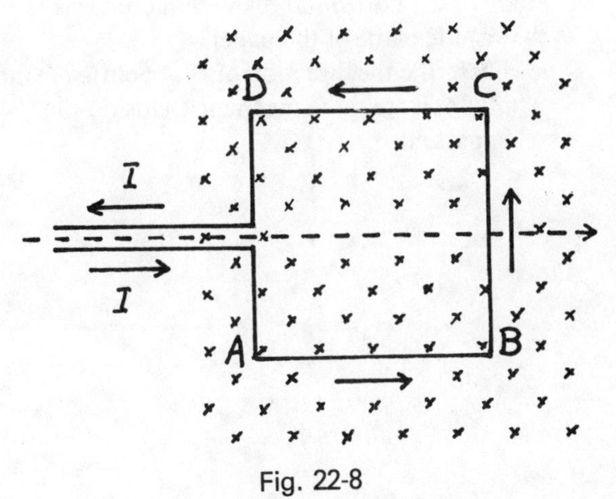

Fig. 22-8

b. If possible, put four arrows on the drawing to show the directions of these four forces. Is this a stable position for the rectangular loop?

19. The current loop in Prob. 18 is now rotated 90° about the axis (dotted line) so that the segment AB is above the plane of the paper and the segment CD is below it.
    a. What are the directions of the forces on the four sides of the current loop when it is in this position? Insofar as possible, indicate the directions of these forces by means of arrows on the drawing.
    b. Which two forces become zero in this position?

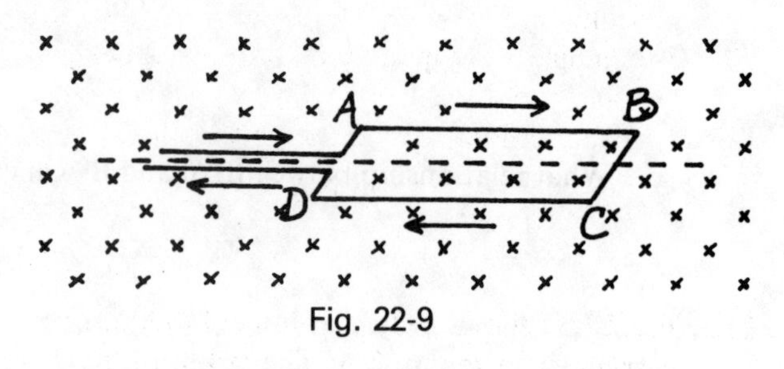

Fig. 22-9

c. Which two forces produce a net torque?

d. In what direction will the coil rotate if it is free to do so? (Take the point of view of an observer to the right of the coil on the axis of rotation.)

e. Calculate the magnitude of the resultant torque on the coil.

$\tau =$ ______________.

f. Suppose that this rectangular coil is free to rotate but that there is a little friction present so that it will not continue to oscillate and will finally come to rest. In what position will the coil then be found?

20. Figure 22-10 shows a straight segment of a conductor that is carrying a current **I** directed toward the right (east). The two points M and P are in the horizontal plane that contains **I**, that is, the plane of the paper.
    a. Describe the direction of **B** at points M and P. Indicate these magnetic fields on the diagram.

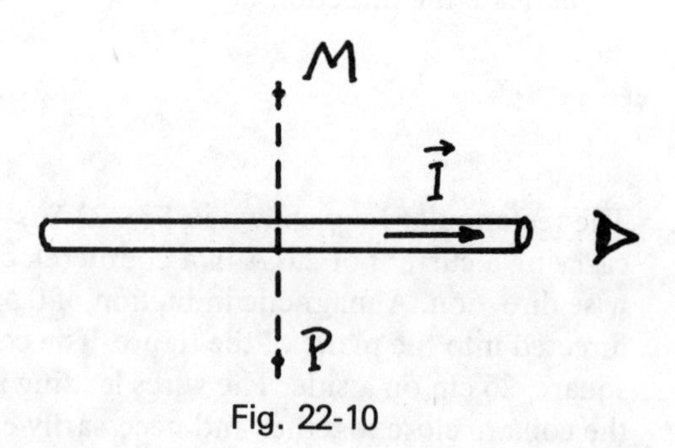

Fig. 22-10

b. In Fig. 22-11 we see the current element from the point of view of an observer looking into the current. Put arrows on the drawing at *M, N, O,* and *P* to show the directions of **B** at these four locations.

c. Now draw two or three concentric circles to show the direction of the magnetic induction in the vicinity of a current-carrying conductor.

d. Calculate the magnitude of the magnetic field at a distance of 1 cm from a straight conductor carrying a current of 50 A.

$B =$ ________________.

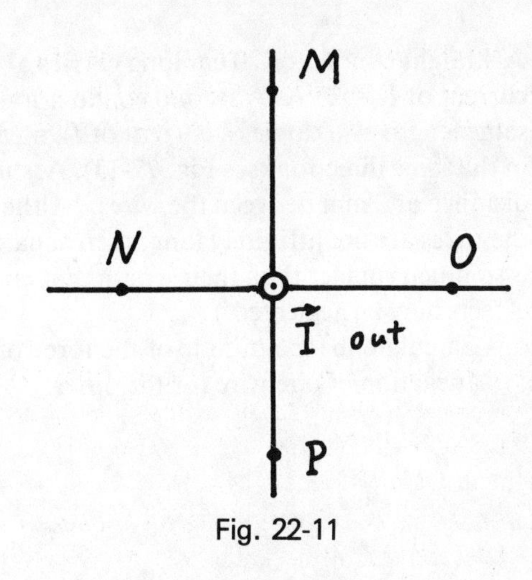

Fig. 22-11

21. Fig. 22-12 shows a circular loop of one turn carrying a current of 2 A. The radius of the loop is 5 cm.

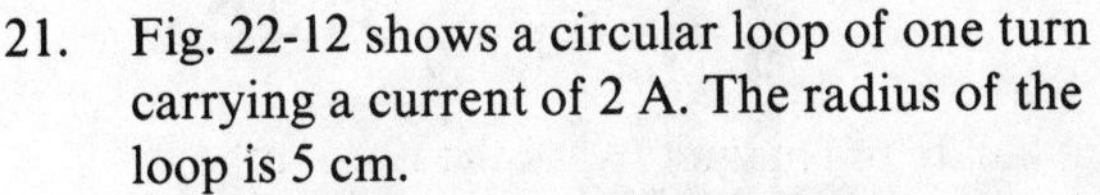

a. Calculate the magnitude of the magnetic field at the center of this loop.

$B =$ ________________.

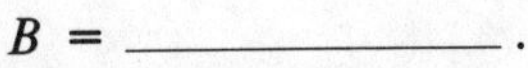

b. Indicate on the drawing the direction of **B** both inside and outside this current loop.

Fig. 22-12

22. Here are two equations that are used to calculate the magnitude of the magnetic field B:

$$B = \frac{2\pi k'I}{R} \quad \text{and} \quad B = \frac{k'2I}{a}$$

a. Describe the physical situation to which each equation applies.

b. Make a drawing to illustrate each of these situations and show I, R, and a on the drawings.

c. Give the value of k'.

$k' =$ ________________.

23. Complete the following definition of the ampere:
One ampere is the current in a long straight wire that exerts a force per unit length of exactly ________________ on a neighboring long parallel wire, ________________ distant, which carries ________________.

24. A straight conductor 20 cm long carries a current of $I_1 = 25$ A. A second conductor of the same length also carries a current of $I_2 = 25$ A in the same direction (see Fig. 22-13). Assume a distance of 2 mm between the wires. (Although the wires are not infinitely long, their separation is so much smaller than their length that end effects may be neglected.)

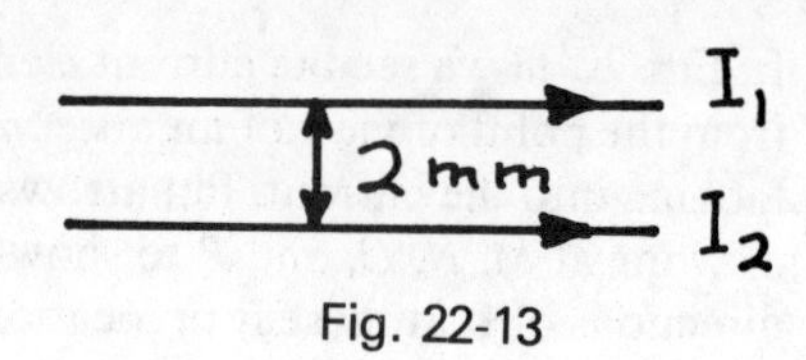

Fig. 22-13

a. Calculate the magnitude of the force of attraction of one wire for the other.

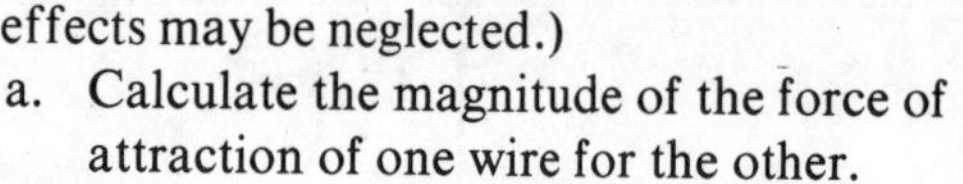

b. Suppose that the upper current is reversed and that non-conducting guides permit the upper wire to move freely in the vertical direction. Will the repulsive force be greater than the weight of the wire? If so, the upper wire should "float" at a certain distance above the lower one. Assume that each wire is copper with a diameter of 0.5 mm. The density of copper is 8.9×10^3 kg/m³. [The volume of the upper wire is Vol. $= (\pi/4) d^2 l = 3.9 \times 10^{-8}$ m³]. Calculate the weight of the upper wire and compare it with the repulsive force.

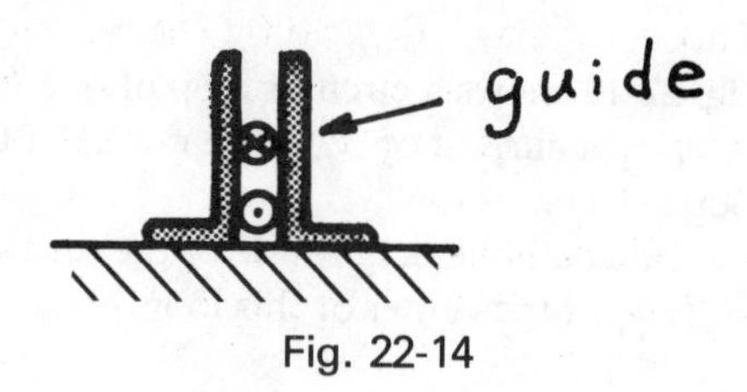

Fig. 22-14

F_{mag} = ______________.

W = ______________.

Secs. 22-9—22-12

25. Fig. 22-15 shows a bar magnet and lines of magnetic induction near one end. Which pole of the magnet is illustrated? ______________
(N or S?)

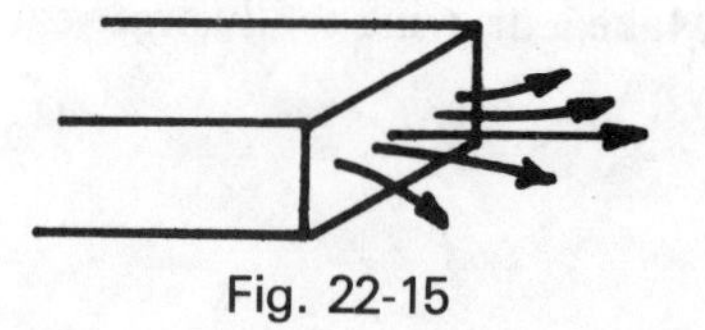
Fig. 22-15

26. Fig. 22-16 shows a long coil in which a current is flowing. The direction of the current is counterclockwise as seen by an observer to the right. Put an "N" and an "S" on the coil (side view) to show the two poles of this electromagnet.

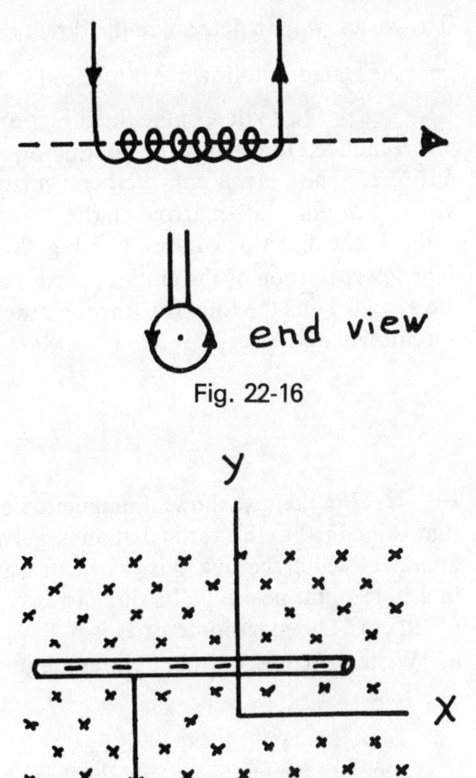

Fig. 22-16

Fig. 22-17

27. The conductor shown in Fig. 22-17 is moving through a magnetic field **B**. The free electrons that it contains are also moving through the magnetic field. The velocity of the electrons is indicated by the vector **v**.
   a. What is the direction of the magnetic force on the electrons in the conductor? (Remember that a negative charge moving in a given direction is equivalent to a positive charge moving in the opposite direction.)

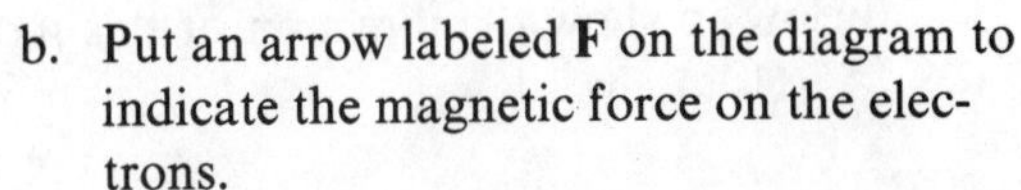

   b. Put an arrow labeled **F** on the diagram to indicate the magnetic force on the electrons.
   c. If the two ends of the conductor are connected to an external circuit, a current will flow. In the moving conductor mechanical energy is being converted into electrical energy. Thus the conductor moving through a magnetic field is a ____________ of ____________.

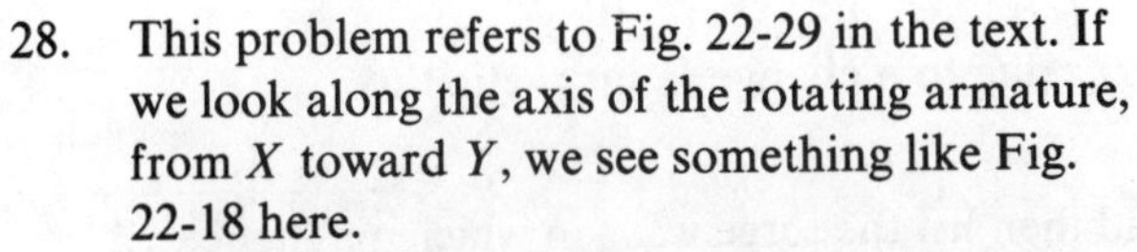

28. This problem refers to Fig. 22-29 in the text. If we look along the axis of the rotating armature, from X toward Y, we see something like Fig. 22-18 here.

   In Ex. 22-8 in the text, the $\mathcal{E}$ is calculated for the instant when the segment PQ is at A, close to the south pole.
   a. What $\mathcal{E}$ appears across the terminals of the generator when the armature has rotated 90° to position C?

   $\mathcal{E}_C =$ ____________.

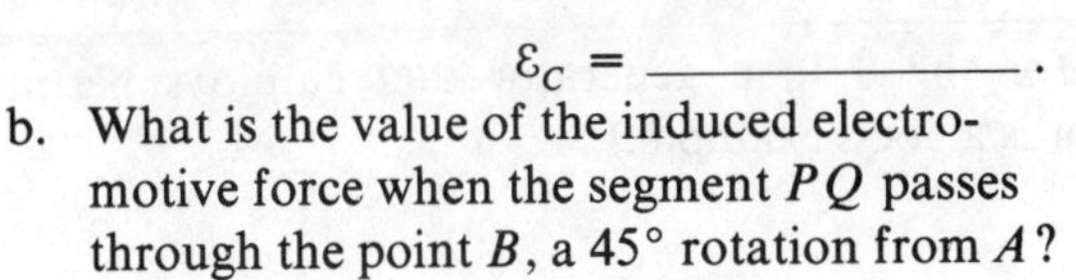

   b. What is the value of the induced electromotive force when the segment PQ passes through the point B, a 45° rotation from A?

   $\mathcal{E}_B =$ ____________.

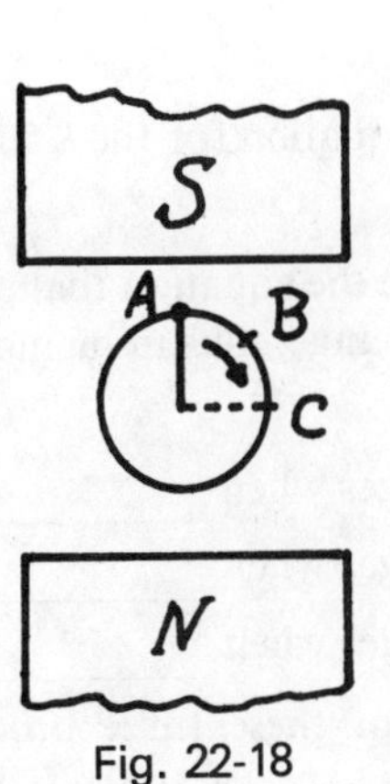

Fig. 22-18

29. The easiest way to determine the direction of an induced $\mathcal{E}$ is to use L______________ law. This law may be stated as follows: An induced $\mathcal{E}$ tends to set up a current whose action ______________.

30. Use Lenz's law to determine the direction of the $\mathcal{E}$ induced in the moving conductor of Prob. 27. The $\mathcal{E}$ is present whether a current is flowing or not. Put an arrow on the drawing to indicate the direction of the $\mathcal{E}$ (see Fig. 22-19). The lower portion of the circuit is assumed to be at rest. Thus the total flux through the circuit will decrease.

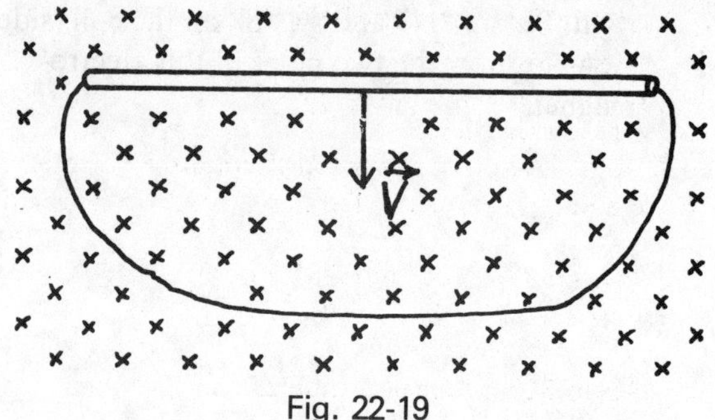

Fig. 22-19

31. Fig. 22-27 in the text shows a magnetic field that is inclined to the vertical at an angle φ. The area over which the flux Φ is to be calculated is in a horizontal position. Its dimensions are 50 cm by 20 cm. The magnitude of **B** is 5 T.

a. Write the equation that defines magnetic flux.

$\Phi =$ ______________.

b. Calculate the magnetic flux through this area for $\varphi = 30°$ and for $\varphi = 36.9°$.

$\Phi(30°) =$ ______________.

$\Phi(36.9°) =$ ______________.

c. Now suppose that the direction of **B** changes from 30° to 36.9° in a time interval of 0.2 s. The magnitudes of B and A remain the same. Calculate the change in flux $\Delta\Phi$ and the rate at which the flux is changing.

$\Delta\Phi =$ ______________.

$\Delta\Phi/\Delta t =$ ______________.

d. If the rectangular opening in the table shown in text Fig. 22-27 was bordered by a rectangular loop of copper wire, what $\mathcal{E}$ would be induced in this conducting loop?

$\mathcal{E} =$ ______________.

e. Put one or two arrows on the drawing below to show the direction of this induced $\mathcal{E}$. (Use Lenz's law.)

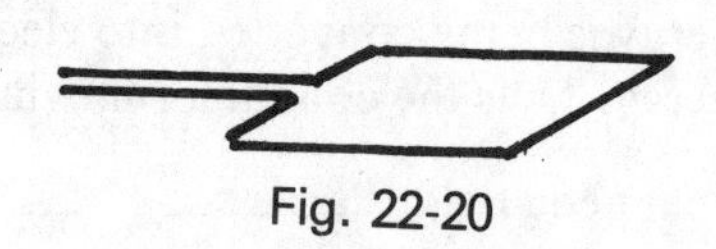

Fig. 22-20

32. Write the equation for the $\mathcal{E}$ induced in a coil of N turns due to a change in magnetic flux.

$\mathcal{E} =$ ______________.

33. Again write the equation that defines magnetic flux and then list the three ways in which magnetic flux may change, thus inducing an $\mathcal{E}$ in a circuit.

$\Phi =$ ______________.

a. Φ changes when ______________________________.

b. Φ changes when ______________________________.

c. Φ changes when ______________________________.

d. Which of these three processes is responsible for the $\mathcal{E}$ in a generator that employs permanent magnets to establish the field. (A bicycle generator is a good example.)

Solutions

1. Coulomb; magnetic
2. electric field; magnetic field
3. magnitude
4. magnetic force; charge; velocity
5. zero; perpendicular

6a. Vectors **B** and **v** are parallel, therefore F = zero.

b. **B** and **v** are perpendicular, therefore F has its maximum value.

$$F = QvB = (5 \times 10^{-19}\ \text{C})(10^7\ \text{m/s})(0.2\ \text{tesla})$$
$$= 1 \times 10^{-12}\ \text{N}$$

c. Again **B** and **v** are perpendicular.

$$F = F_{max} = 1 \times 10^{-12}\ \text{N}$$

d. **B** and **v** are parallel, F = zero.

7. downward

8. $\mathbf{F} = Q\mathbf{v} \times \mathbf{B}$

9a. F is zero; it has no direction

b. vertically downward

c. toward the east

d. no direction

10. The component of **v** perpendicular to **B** is $v \cos 60° = (10^7\ \text{m/s})(0.5)$. Thus the magnitude of the force is

$$F = (5 \times 10^{-19}\ \text{C})(0.5 \times 10^7\ \text{m/s})(0.2\ \text{T}) = 5 \times 10^{-13}\ \text{N}$$

Its direction is downward (into the plane of the drawing).

11a. **F**

b. **v** and **B**

12a. Since the electron experiences no force when it is moving vertically, the magnetic field must be vertical, either up or down.

If **B** is upward (along the positive z-axis), $\mathbf{v} \times \mathbf{B}$ is toward the west. Since Q is negative, $Q\mathbf{v} \times \mathbf{B}$ is directed toward the east. Conclusion: **B** is upward.

b. $$B = \frac{F}{Qv} = \frac{3.2 \times 10^{-10}\ \text{N}}{(1.6 \times 10^{-19}\ \text{C})(6 \times 10^7\ \text{m/s})} = 33\ \text{T}$$

13. downward; into the paper

14a. See Fig. 22-21.

b. $\mathbf{F} = Q\mathbf{v} \times \mathbf{B}$

c. See Fig. 22-21.

d. In every case, **v** is perpendicular to B so that **F** has its maximum value given by

$$F_{max} = QvB = (1.6 \times 10^{-19}\ \text{C})(3 \times 10^7\ \text{m/s})(4\ \text{T}) = 1.9 \times 10^{-11}\ \text{N}$$

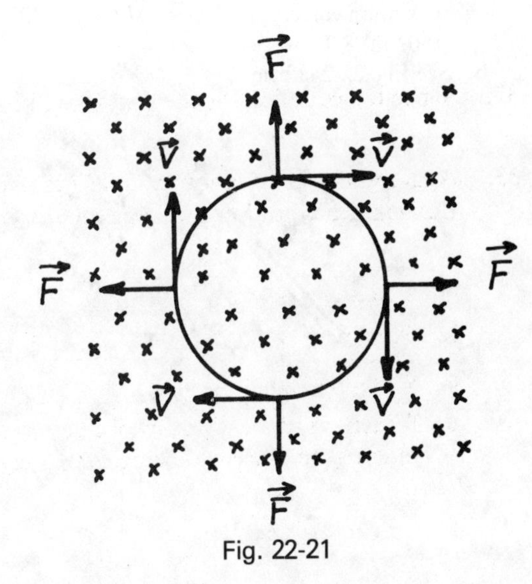

Fig. 22-21

15. Yes. If the moving object carried a *negative* charge and if the magnetic field were directed downward (perpendicular to the plane of the motion), the magnetic force would be directed radially inward.

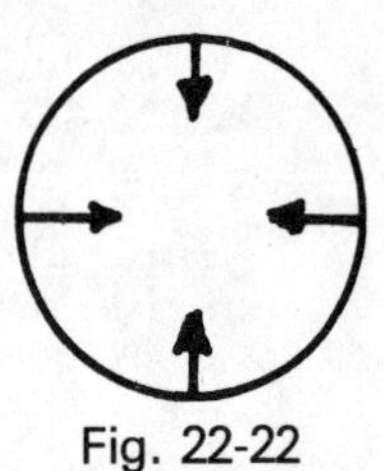

Fig. 22-22

16a. $\mathbf{F} = I\Delta l \times \mathbf{B}$

b. $F_{max} = I\Delta l\, B$

c. F_{min} = zero

d. They must be perpendicular.

e. They must be parallel or antiparallel.

f. The angle between them must be 30°. Only the component of $I\Delta l$ that is perpendicular to **B** is effective.

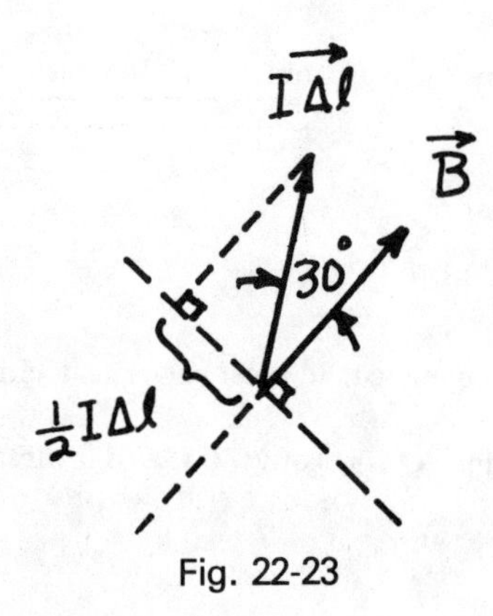

Fig. 22-23

17a. Since $I\Delta l$ and **B** are perpendicular, F has its maximum value. $F = F_{max} = I\Delta l B$ = (50 A) (0.02 m)(8 T) = 8 N

b. See Fig. 22-24a below.

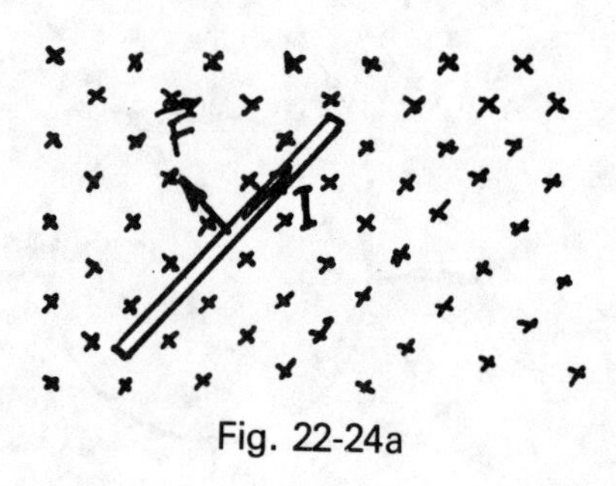

Fig. 22-24a

18a. The force on each segment is directed toward the center of the loop as shown in Fig. 22-24b below. All forces are in the plane of the drawing.

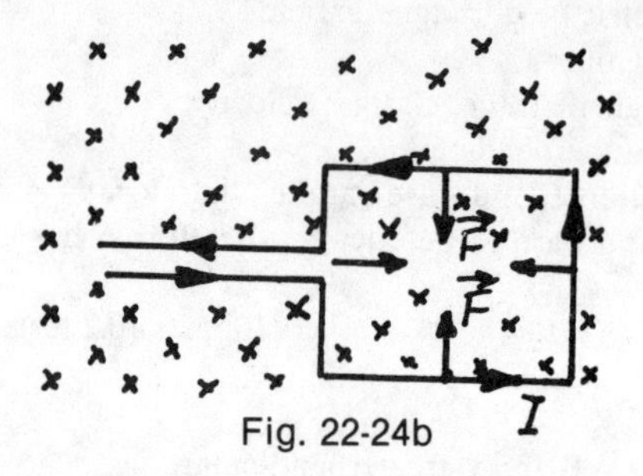

Fig. 22-24b

b. This is not a stable position. There is a tendency for the loop to rotate 180° to a stable orientation.

19a. The forces are horizontal as shown in Fig. 22-25.

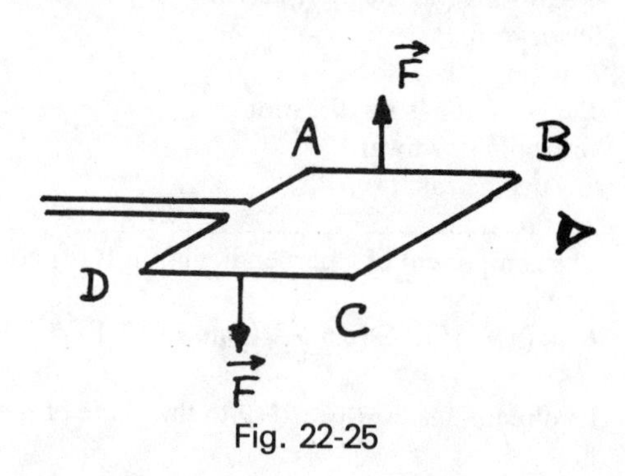

Fig. 22-25

b. The forces on the segments BC and AD become zero because these currents are parallel (or antiparallel) to **B**.

c. The forces on the segments AB and CD are oppositely directed but they do not have the same line of action. They produce a net torque about the axis.

d. clockwise. See Fig. 22-26.

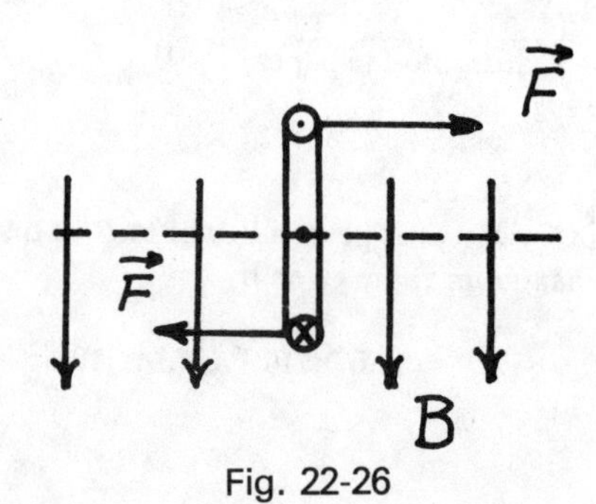

Fig. 22-26

e. $\tau = 2F(\text{lev. arm}) = 2F(0.125\text{ m})$
$= 2I\ \Delta l\, B(0.125\text{ m}) = 2(2\text{ A})(0.25\text{ m})(0.5\text{ tesla})(0.125\text{ m})$
$= 6.25 \times 10^{-2}\text{ N}\cdot\text{m}$

f. The coil will rotate into a position such that its own magnetic field will line up with the external field. This is 180° from the unstable position shown in Prob. 18.

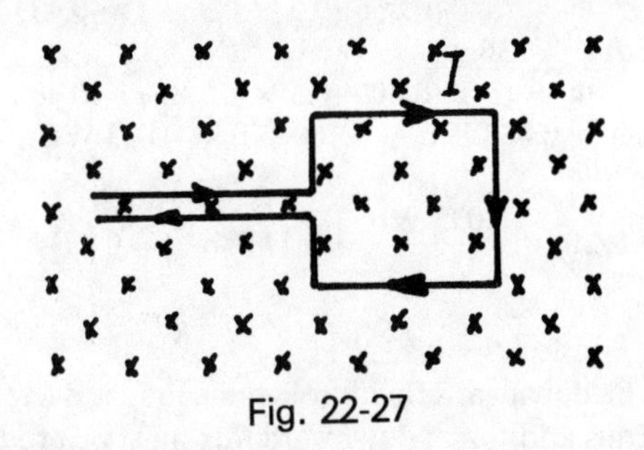

Fig. 22-27

20a. **B** is out of the drawing at M and into the drawing at P.

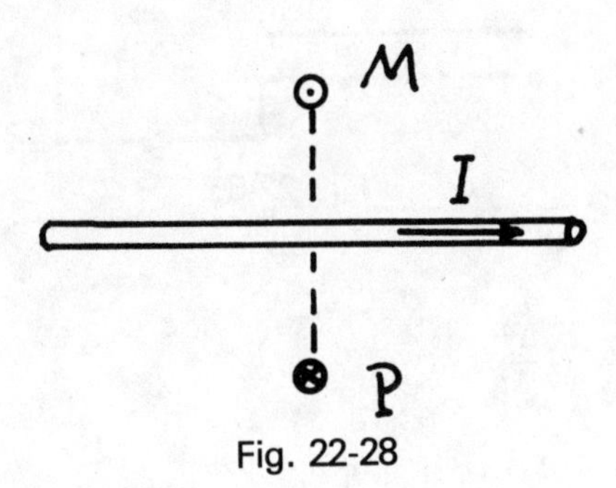

Fig. 22-28

b, c See Fig. 22-29.

Fig. 22-29

d. $B = \dfrac{k'2I}{a}$

$B = (10^{-7}\ \text{N/A}^2)(2)(50\ \text{A})/(0.01\ \text{m}) = 10^{-3}\ \text{T}$

21a. $B = \dfrac{2\pi k' I}{R} = \dfrac{2\pi(10^{-7}\ \text{N/A}^2)(2A)}{0.05\ \text{m}} = 2.5 \times 10^{-5}\ \text{T}$

b. B is into the paper within the loop and out of the paper outside the loop.

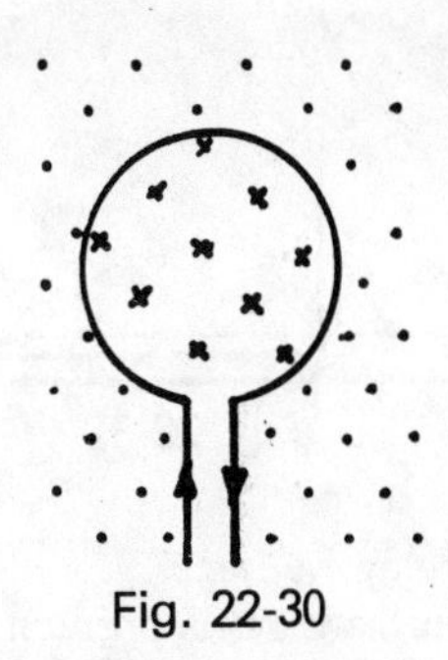
Fig. 22-30

22a. $B = \dfrac{2\pi k' I}{R}$ gives the field at the center of a circular loop.

$B = \dfrac{k'2I}{a}$ gives the field at a distance a from a long straight conductor.

b.

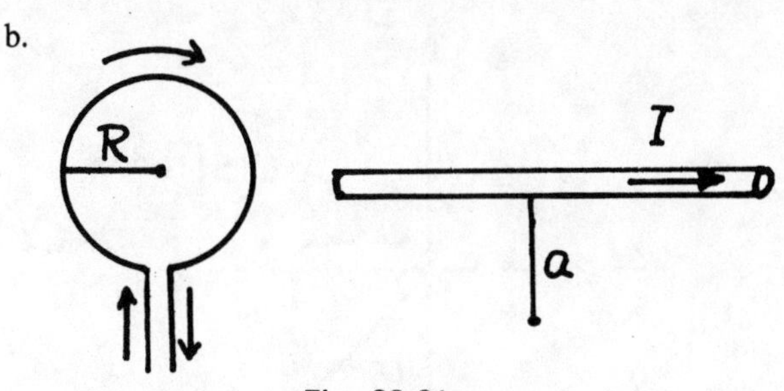

Fig. 22-31

c. $k' = 10^{-7}\ \text{N/A}^2$

23. 2×10^{-7} N/m; 1 m; an equal current

24a. The magnetic field produced by I_1 is

$$B = \frac{k'2I_1}{a}$$

The force experienced by a portion of the second wire is

$$F = I_2\,\Delta l\,B = I_2\,\Delta l\,\frac{k'2I_1}{a}$$

$$F = \frac{(25\ \text{A})(0.2\ \text{m})(10^{-7}\ \text{N/A})2(25\ \text{A})}{2 \times 10^{-3}\ \text{m}}$$
$$= 1.25 \times 10^{-2}\ \text{N}$$

b. $w = mg = dVg = (8.9 \times 10^3\ \text{kg/m}^3)(3.9 \times 10^{-8}\ \text{m}^3)(9.8\ \text{m/s}^2) = 3.4 \times 10^{-3}\ \text{N}$

The upward force is 3.7 times greater than the weight. The upper wire will "float" at a height that will make the two forces equal. This height is 7.4 mm.

25. N. Lines of magnetic field come out of a north pole.

26. The end of the coil near the observer is the north pole.
27a. The force is to the left, in the $-x$ direction.
b.

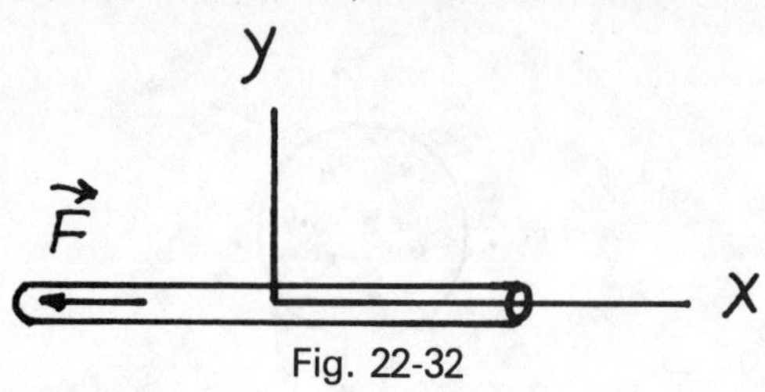

Fig. 22-32

c. seat; electromotive force
28a. When the armature is in position "C" the conductor is moving parallel to the magnetic field. **B** is directly vertically upward and **v** is directed vertically downward. The $\mathcal{E}$ is zero.
b. We must take the perpendicular component of **v**, in this case the horizontal component.

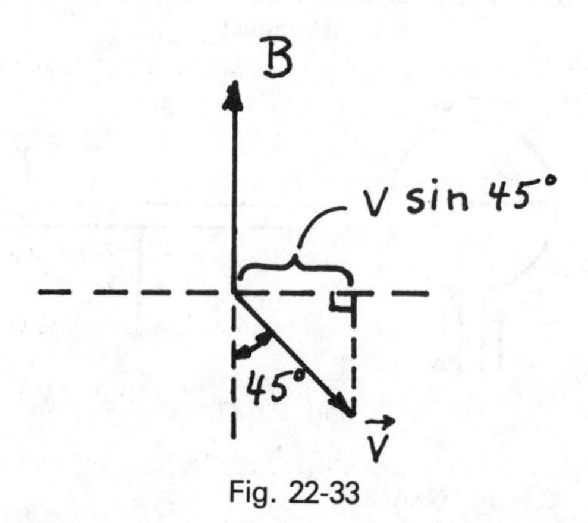

Fig. 22-33

$$\begin{aligned}\mathcal{E} &= B\,lv \sin\theta \\ &= (0.5\ \text{N/m} \cdot \text{m})(0.1\ \text{m})(5.03\ \text{m/s})(0.707) \\ &= 0.178\ \text{volts}\end{aligned}$$

29. Lenz's; opposes the change that caused it
30. In this case the downward flux will decrease as the area enclosed by the circuit decreases. This change can be counteracted only by additional downward flux. Thus a clockwise current will be set up in the circuit. The $\mathcal{E}$ will be directed toward the right in the moving conductor.

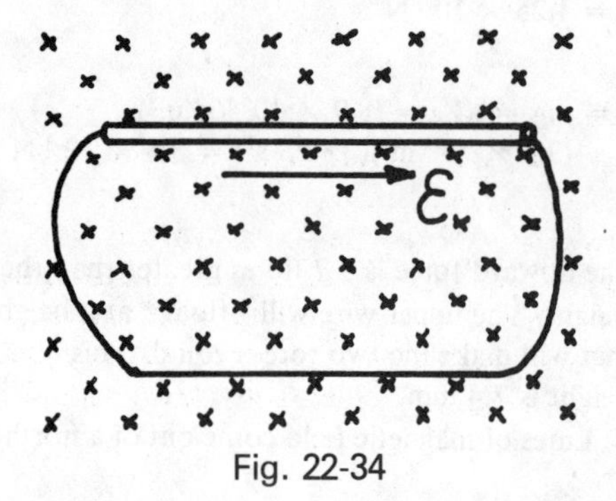

Fig. 22-34

31a. $\Phi = B \cos\varphi\, A$ where φ is the angle between **B** and the normal to the surface.
b. For $\varphi = 30°$
$\Phi = (5\ \text{T})(0.866)(0.5 \times 0.2\ \text{m}^2) = 0.433\ \text{Wb}$
For $\varphi = 36.9°$
$\Phi = (5\ \text{T})(0.800)(0.5 \times 0.2\ \text{m}^2) = 0.400\ \text{Wb}$
c. $\Delta\Phi = 0.433\ \text{Wb} - 0.400\ \text{Wb} = 0.033\ \text{Wb}$

$$\Delta\Phi/\Delta t = \frac{0.033\ \text{Wb}}{0.2\ \text{s}} = 0.17\ \text{Wb/s} = 0.17\ \text{V}$$

d. $\mathcal{E} = \Delta\Phi/\Delta t = 0.17\ \text{V}$
e. The downward flux through the loop is decreasing. Thus additional downward flux must be produced. Seen from above, the induced $\mathcal{E}$ moves positive charges clockwise:

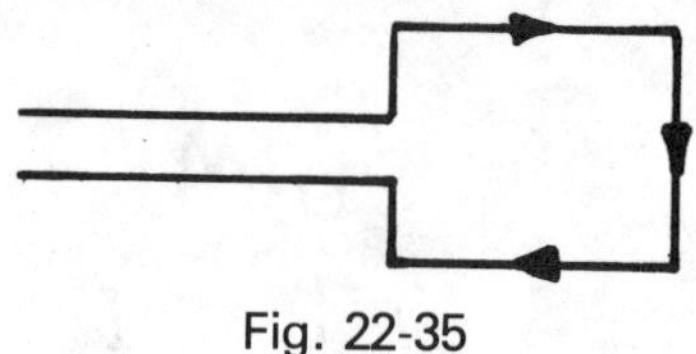
Fig. 22-35

32. $\mathcal{E} = N\dfrac{\Delta\Phi}{\Delta t}$

33. $\Phi = B \cos\varphi\, A$
33a. the magnitude of **B** changes
b. the angle between **B** and the surface changes
c. the area enclosed by the circuit changes
d. In a permanent magnet generator only the angle φ changes, B and A remain constant.

CHAPTER

23 Applied Electricity

GOALS To learn how the basic laws of electricity and magnetism are applied to useful electrical devices.

OBJECTIVES After completing this chapter the student should be able to do the following:

1. Calculate the torque acting on the coil of a galvanometer from a knowledge of the dimensions of the coil and the strength of the magnetic field in which it is located.
2. Write Hooke's law for a system having angular displacement by analogy with the case of a linear displacement ($F = k\,\Delta x$).
3. Write the equation for the angular displacement of a galvanometer coil of N turns carrying a current I and name the physical quantities represented by the symbols B, A, and k.
4. Discuss qualitatively the design parameters of a galvanometer and the choices that lead to high sensitivity.
5. Explain, in terms of magnetic forces, what causes the armature of a motor to rotate.
6. Predict the direction of rotation of a coil in a magnetic field when the direction of the current is given.
7. Describe the position of the rotating coil at the moment when the current is reversed and describe how this current reversal is accomplished.
8. Draw a graph of the current versus time for a flat coil rotating in a uniform magnetic field.
9. Calculate the starting current for a motor and the current flow at full speed.
10. Explain why a motor is more likely to blow a fuse when first turned on than when running at full speed.
11. Describe the direction of the eddy currents induced in a conductor that is moving through a magnetic field.
12. Given a graph of I versus t in one coil of a transformer, draw a graph of $\mathcal{E}$ versus t (or I versus t) in the other coil.
13. Write the relationship between $\mathcal{E}$ and number of turns for a transformer with one primary and one secondary winding.
14. Write the corresponding relationship for currents and $\mathcal{E}$s in a transformer.
15. Find one of the quantities $\mathcal{E}_s$, $\mathcal{E}_p$, N_s, or N_p when the others are given.
16. Find one of the quantities I_s, I_p, $\mathcal{E}_s$, or $\mathcal{E}_p$ when the others are given.
17. State the advantage of transmitting electric energy at high voltage.
18. Calculate the power dissipated in a transmission line (or extension cord) from a knowledge of voltage and resistance.
19. Give the definition of impedance in words and in symbols.
20. Calculate any of the quantities Z, V, or I when the others are given.
21. Write the relationship between impedance and self-inductance for a coil of negligible resistance.
22. Calculate any of the quantities Z, f, or L when the others are given.
23. Write the relationship between impedance and capacitance.
24. Calculate any one of the quantities in the above relationship when the remaining ones are known.
25. Calculate any one of the quantities L, C, or f for a resonant circuit when the other quantities are given.
26. Give the technical term for a device consisting of a pair of junctions between two dissimilar metals.
27. Draw a circuit containing two thermocouple junctions and a galvanometer. Indicate where the dissimilar metals are joined together.

28. State what must be different at the two junctions if an $\mathcal{E}$ is to be present.
29. Name the principal carriers of electric current in a silicon crystal that contains a small fraction of valence 5 atoms.
30. Name the principal carrier of electric current in a silicon crystal that contains a small fraction of valence 3 atoms.
31. Associate the terms "*n*-type" and "*p*-type" with the nature of the principal (or majority) carriers of electric current in a semiconductor.
32. Draw the conventional symbol for a diode and indicate the direction in which the diode readily conducts; i.e., the forward direction.
33. Draw the circuit symbols for an *n-p-n* and a *p-n-p* transistor labeling the three connections to each (base, emitter, collector).
34. Name the principal function of a diode and the principal function of a transistor.
35. Draw the conventional symbol for a field-effect transistor labeling the three connections (gate, drain, source).
36. Name the two methods of deflecting the electron beam in a cathode-ray tube.

SUMMARY

A current-carrying coil in a magnetic field experiences a torque. If a restoring torque is provided by a spring, the amount of rotation will be a measure of the current flowing in the coil. This is the basis for an important measuring instrument, the galvanometer. If the current is reversed at suitable intervals, a coil may be caused to rotate continuously in a magnetic field. This is the basis for the electric motor. If, on the other hand, a coil is forced to rotate in a magnetic field by an external agent, the coil will become a seat of electromotive force. In other words, the same basic configuration, a conducting coil in a magnetic field, may serve either as a motor or a generator.

When a conductor of any sort moves through a magnetic field, an $\mathcal{E}$ appears in the conductor (provided only that the motion is not exactly parallel to **B**). If the conductor happens to be a flat plate, the resulting currents will flow in circular paths through the body of the conductor. These *eddy currents* will interact with the magnetic field in such a way as to impede the motion of the plate.

If the current in a coil changes, the magnetic field produced by that current will also change. A second coil placed in the changing magnetic field of the first will become the seat of an *induced* $\mathcal{E}$. The effect becomes stronger if the two coils are placed close together and, more importantly, if they are linked by a magnetic material. This, in essence, is the principle of operation of the transformer, one of the most important components of an electric power distribution system. The magnitude of the $\mathcal{E}$ in the secondary coil depends on the $\mathcal{E}$ in the primary and on the number of turns in the two coils. The relationship is $\mathcal{E}_s = \mathcal{E}_p\,(N_s/N_p)$. The power in the secondary of a well-designed transformer will be nearly equal to the power in the primary. Thus $\mathcal{E}_s I_s \cong \mathcal{E}_p i_p$. By using a secondary with many more windings than the primary, the secondary voltage can be made high and the secondary current correspondingly low. This is advantageous in the transmission of electric power over long distances because the power lost in the transmission line, through Joule heating, is proportional to the square of the current.

Coils and capacitors are important components of alternating current circuits. The term impedance, a generalization to ac circuits of the concept of resistance, is a measure of the ac current that will flow under an applied voltage. The relationship is Z (impedance) $= V/I$. Impedance, unlike resistance, is dependent on frequency. The impedance of a coil increases with increasing frequency while the impedance of a capacitor decreases with frequency. The relationships are

$$Z = 2\pi fL \text{ (for a coil)}$$

$$Z = \frac{1}{2\pi fC} \text{ (for a capacitor)}$$

In an *ac* circuit with a capacitance and an inductance in series, resonance occurs when $f = (1/2\pi)\sqrt{1/LC}$. The impedance of an R, L, and C in series is

$$Z = \sqrt{R^2 + \left(2\pi fL - \frac{1}{2\pi fC}\right)^2}$$

If a circuit contains two junctions of dissimilar metals connected in series, and if the junctions are at different temperatures, an $\mathcal{E}$ will exist that will depend on the temperature difference. This offers an electrical means of measuring temperature usable at much higher temperatures than the conventional liquid-in-glass thermometer.

Solid state diodes and transistors are components of electronic circuits. Diodes allow currents to flow in only one direction. Transistors amplify either currents or voltages. The low power consumption of solid-state devices, their continually decreasing size, and their low cost and high reliability, have produced a technological revolution in electronics over the past thirty years.

QUESTIONS AND PROBLEMS

Secs. 23-1—23-2

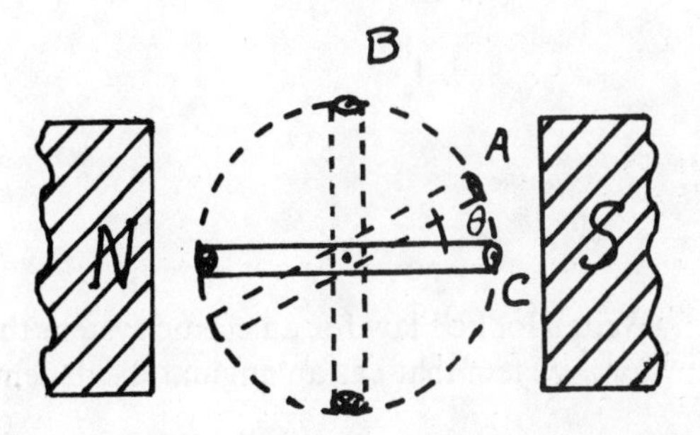

1. In Fig. 23-1, shown here, and in Fig. 23-2 of the text, a rectangular loop is placed in a uniform magnetic field between the poles of two magnets. The coil, of dimensions $a \times \ell$, is free to rotate about an axis that is perpendicular to the drawing. The lower part of the figure shows the rectangular coil in position "C" as it would be viewed from above. Three positions of the coil are to be considered. In position "C" the two longer sides of the coil (sides ℓ) are close to the poles of the magnet. The other two positions, "A" and "B," are those illustrated in Fig. 23-2 (a) and (b) of the text.

   a. What is the direction of the magnetic field between the poles of the magnet? Put one or two arrows on the drawing to show the direction of **B**.

   b. Put two small arrows on the upper drawing to show the direction of the magnetic force acting on the right and left sides of the coil.

   c. Calculate the magnitudes of the two forces from these data: $I = 0.02$ A, $B = 5$ T, $a =$ 1 cm, $\ell = 1.5$ cm.

   $F_1 =$ ____________ .

   $F_2 =$ ____________ .

   d. Does **F**, the magnetic force, change in magnitude as the coil rotates from position "C" to "A" and then to "B"?

   ____________ (Yes/No)

   e. When the coil is in position "C," what is the magnitude of the force acting on the shorter sides of the coil (sides a)?

   $F =$ ____________ .

   Suppose that the coil is in position "A" where $\theta = 60°$.

   f. Will the magnetic forces acting on the coil cause it to rotate clockwise or counterclockwise? Put a curved arrow on the drawing to indicate the direction of rotation.

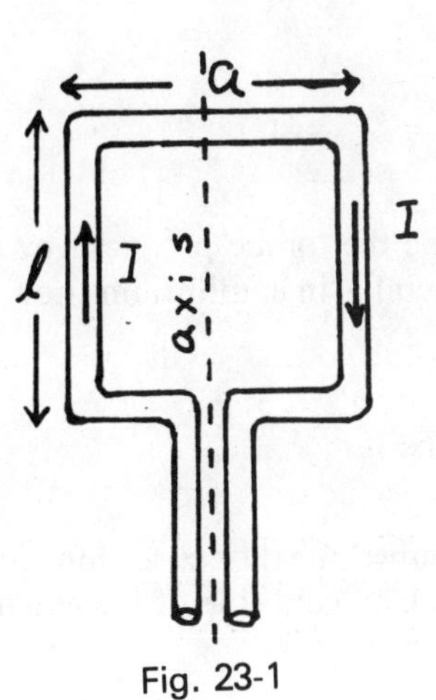

Fig. 23-1

   g. Write an equation for the torque that acts on the coil when it makes an angle θ with **B** (as in position "A"). (Be careful to use the correct moment arm for **F**.)

   $\tau =$ ____________ .

   h. Calculate the numerical value of the torque at position "A" using the data of part c. above.

   $\tau_A =$ ____________ .

   i. What is the torque on the coil when it is in position "B"?

   $\tau_B =$ ____________ .

2a. You have seen in the preceding problem that a coil placed in a magnetic field will, in general, experience a torque. If the field is uniform, i.e., everywhere constant in magnitude and direction, this torque will vary from a minimum of ______________ to a maximum of ______________. (Write it symbolically.)

b. Make a sketch in the space provided to show how the pole pieces may be shaped so that the torque on the coil will be nearly constant for any value of θ up to about 45°. It helps to have a cylindrical piece of iron in the center. Why?

3. Write Hooke's law for an elastic system that has a linear displacement Δx and the analogous relationship for a system that has an angular displacement θ.

$F =$ ______________. $\tau' =$ ______________.

4a. Returning to the rectangular coil of Prob. 1 above and supposing it to be oriented at an angle θ (position "A" in the drawing), what is the condition of rotational equilibrium?

b. Let τ be the torque produced by the magnetic forces and τ' the torque produced by the coiled hairspring. If the coil is in equilibrium, how are these torques related?

c. Now substitute the expressions for τ from 1g. and τ' from 3 and rearrange the terms so that θ is alone on the left-hand side of the equation.

$\theta =$ ______________.

d. You will notice that the angular displacement of the coil, θ, is not a linear function of the current because of the presence of the factor ______________.

e. A nonlinear current-measuring device would be inconvenient to use. How can the device be modified so that the cos θ factor will be eliminated? (See Prob. 2b above.)

5. Two rather simple modifications of the galvanometer will permit the coil to rotate continuously. First we will have to remove the ______________ that produced the restoring torque τ' and then we will have to provide a means of ______________ the current twice each revolution. Describe briefly a mechanism that will allow the current to enter the coil while the latter is rotating and cause the direction of the current to change at the appropriate time. (Make a drawing.)

6a. What energy transformation takes place in a motor?

b. What energy transformation takes place in a generator?

7. Fig. 23-2 in the text is a schematic drawing of a simple generator. We are going to modify the generator slightly by using large, flat pole pieces rather than the curved pole pieces illustrated in the text. Thus we will be dealing with a uniform magnetic field. In Fig. 23-2 the coil (or armature) is rotating clockwise. It is shown in an intermediate position such that the angle between **B** and the normal to the coil is θ.

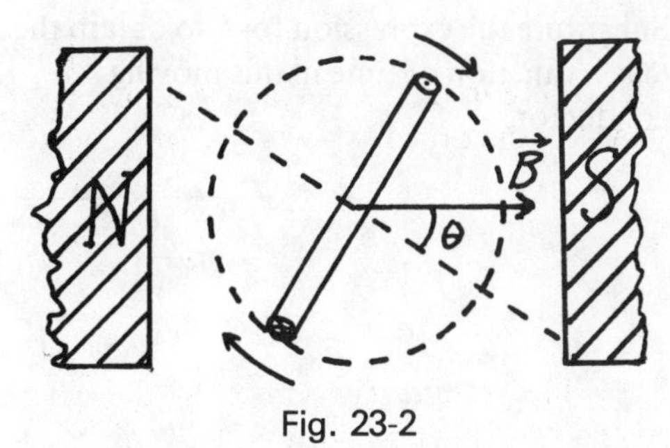

Fig. 23-2

a. Put an arrow on the upper conductor to represent the instantaneous velocity of this conductor. Label it **v**. The length of the conductor is ℓ. Write the expression for the instantaneous $\mathcal{E}$ induced in the moving conductor. Remember that the $\mathcal{E}$ depends on the relative directions of **B** and **v**; it has its maximum value only when **B** and **v** are perpendicular to each other.

$\mathcal{E}$ = ______________ .

b. For numerical calculations we will use the values of Prob. 1, namely $a = 1$ cm, $\ell = 1.5$ cm, and $B = 5$ T. We will also assume a frequency of rotation f of 60 rev/s. Write the expression for the maximum emf in terms of B, ℓ, and v.

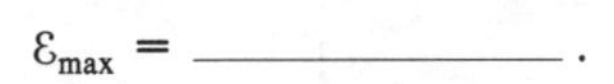

$\mathcal{E}_{max}$ = ______________ .

c. The angular speed is represented by the symbol ω. Write the relationship between angular speed and linear speed. The radius is $r = a/2$, as shown in Fig. 23-3.

v = ______________ .

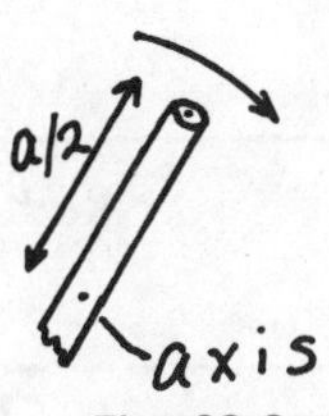

Fig. 23-3

d. Now write the relationship between angular speed and frequency of rotation f.

ω = ______________ .

e. Combine the results of (c) and (d) into a single expression for the velocity in terms of f and a.

$v =$ ______________.

f. Find the velocity of the conductor from the data given.

$v =$ ______________.

g. The armature of this generator turns at a constant speed of 60 rev/s or 120π rad/s. Thus its angular position θ is given by the linear relationship,

$\theta =$ ______________.

For convenience we take the initial displacement to be zero.

h. In (a) we found that the instantaneous $\mathcal{E}$ in the upper conductor is $\mathcal{E} = B\ell v \sin\theta$. Substitute the expression for θ to obtain the $\mathcal{E}$ as a function of time in this moving conductor.

$\mathcal{E} =$ ______________.

i. Since the first part of this expression, $B\ell v$, is a constant, the instantaneous $\mathcal{E}$ may be written

$$\mathcal{E} = \mathcal{E}_{max} \sin 2\pi f t$$

where

$$\mathcal{E}_{max} = B\ell v = B\ell(a/2)2\pi f$$

There are in fact two conductors of length ℓ both of which produce $\mathcal{E}$s in the same direction in the rotating coil. Thus the total $\mathcal{E}_{max}$ in the single rectangular loop is

$$\mathcal{E}_{max} = 2\pi f BA$$

where A stands for the product ℓa, the area enclosed by the loop. A practical generator will have a coil of many turns, N. Write the final expression for $\mathcal{E}_{max}$ and find its numerical value using the data furnished above. (Assume a coil of 200 turns.)

equation: $\mathcal{E}_{max} =$ ______________.

value: $\mathcal{E}_{max} =$ ______________.

j. In the space provided draw a graph of $\mathcal{E}$ as a function of time. Include two full periods. [Hint: $T = 1/f = (1/60)\text{s} = 0.017$ s.]

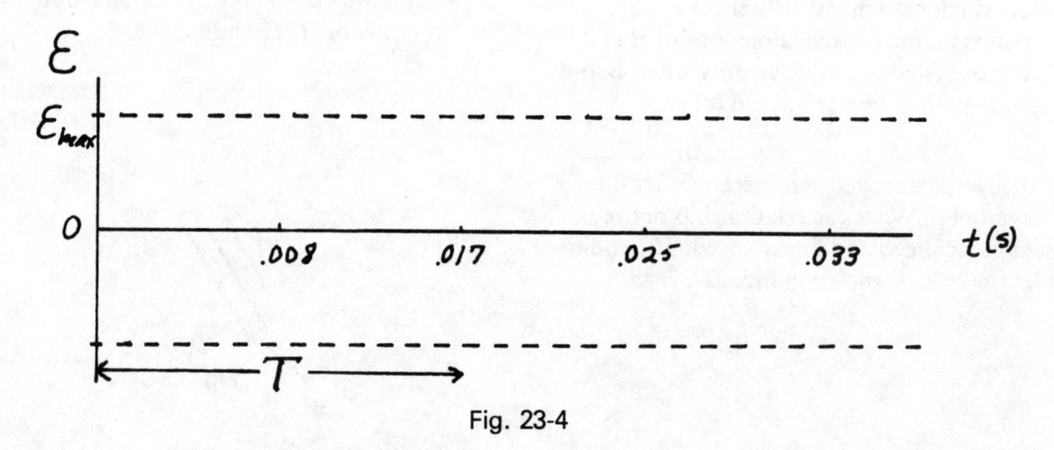

Fig. 23-4

Although practical generators are more complicated than the very simple model described, the $\mathcal{E}$s produced and the currents that flow in the power lines have exactly the sinusoidal character found above. The power-line frequency that is standard in the United States is 60 Hz. Thus the current in a lamp bulb, for example, flows in one direction for (1/120)s, half a cycle, and in the opposite direction for the next (1/120)s. It is called alternating current for this reason.

8. A certain generator produces an $\mathcal{E}$ of 60 V when it is rotating at a speed of 1800 rev/min. What $\mathcal{E}$ will be generated if the speed changes to 1200 rev/min, all other factors remaining the same? (Hint: Use the results of Prob. 7.)

Secs. 23-3—23-6

9. The dc (direct-current) resistance of a motor may be measured by passing a small current through the motor (small enough that the motor does not turn) and measuring the voltage drop across the terminals. Fig. 23-5 shows a circuit for making such a measurement. The measured current is 0.5 A.

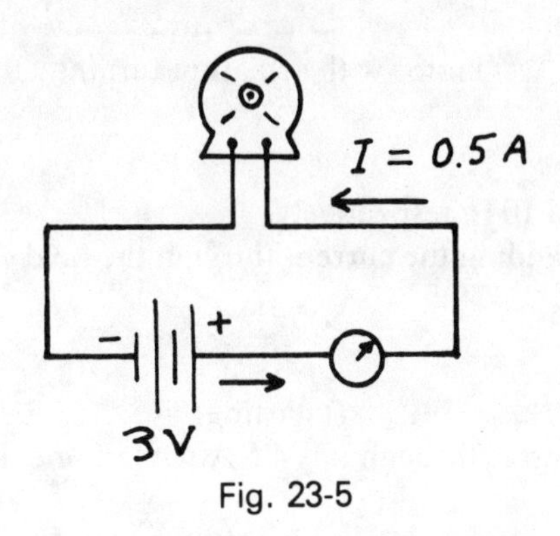

Fig. 23-5

a. What is the dc resistance of this motor?

$R =$  .

The same motor is now connected to a 120 V source bringing it up to its full speed of 1800 rev/min. The measured current flowing through the motor is 2 A.

b. What "resistance" does the motor now appear to have?

apparent resistance = ______________ .

c. How can you account for the fact that the current through the motor that is running at full speed is much less than the current that is predicted from measurements made when the armature is not rotating?

d. Determine the value of the back $\mathcal{E}$ using the simplified circuit in Fig. 23-6.

$\mathcal{E}_b =$ ______________ .

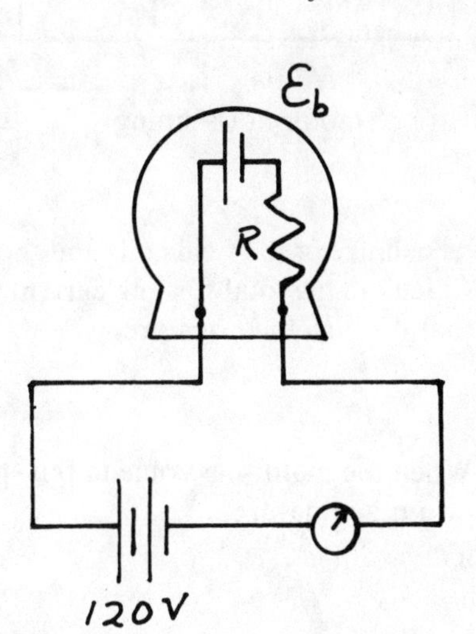

Fig. 23-6

e. What current would you expect this motor to draw from the 120 V source when it is first turned on?

starting current = ______________ .

10. What fundamental law of physics requires that the $\mathcal{E}$ produced by a motor *opposes,* rather than aids, the external source? (Hint: A similar phenomenon is discussed in Sec. 22-8.)

11. Except for some small motors operated by flashlight batteries, most motors do not employ permanent magnets. Instead, the magnetic field in which the armature rotates is provided by field coils wound on laminated iron cores. The circuit diagrams for such a motor are shown in Fig. 23-7.

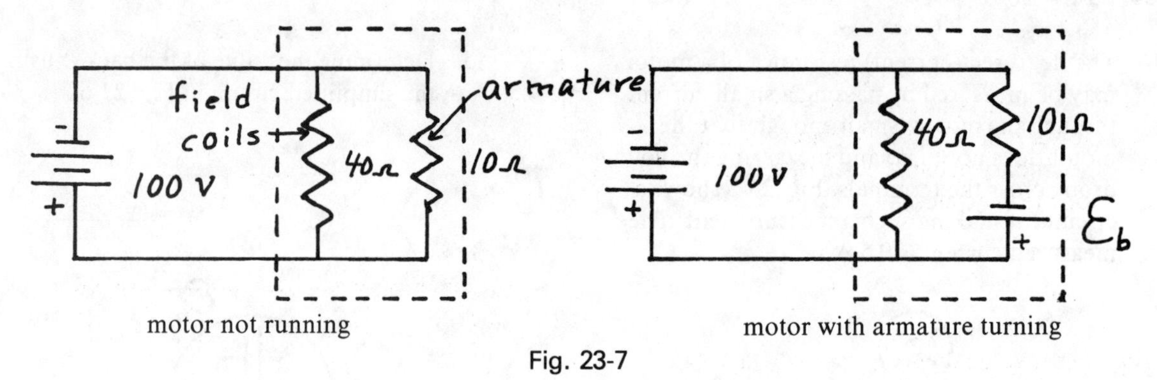

Fig. 23-7

The resistance of the field coils and the armature are 40 Ω and 10 Ω, respectively.

a. Calculate the total starting current through the motor by adding the current through the field coil to that through the armature.

I(starting) = ______________.

b. When the motor has come to full speed the current that passes through it is 4.5 A. Determine the back $\mathcal{E}$ in the armature.

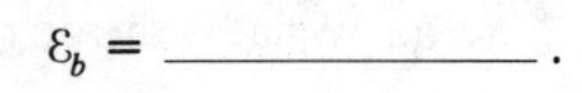

$\mathcal{E}_b$ = ______________.

12. A thin aluminum plate is moving downward between the poles of a strong permanent magnet. The magnetic field is directed into the aluminum plate as shown in Fig. 23-8.

a. Put a curved arrow on the diagram to indicate the direction of the eddy currents induced in the conductor.

b. What effect do the eddy currents have on the motion of the plate?

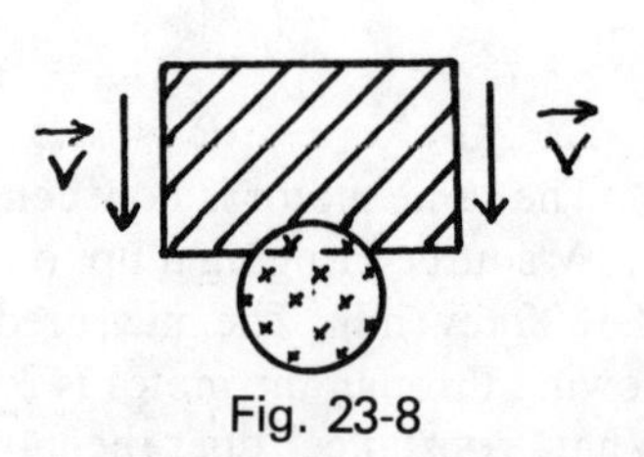

Fig. 23-8

13. Fig. 23-9 is a schematic diagram of a transformer having a primary coil and a secondary coil. The three vertical lines represent the iron core. Suppose that the primary is connected to a battery, a switch, and an ammeter, all in series. The switch is closed at t_1 and then opened again at t_2. The current in the primary is shown in the upper graph of Fig. 23-10.

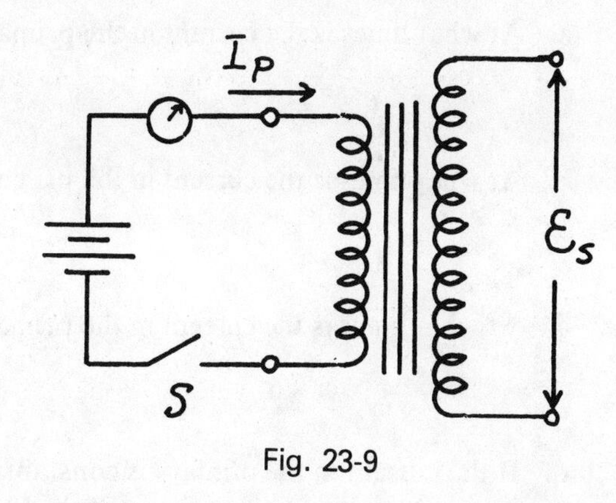

Fig. 23-9

a. Using the same time scale, show in the lower half of Fig. 23-10 how the $\mathcal{E}$ in the secondary behaves when the circuit is opened and closed.

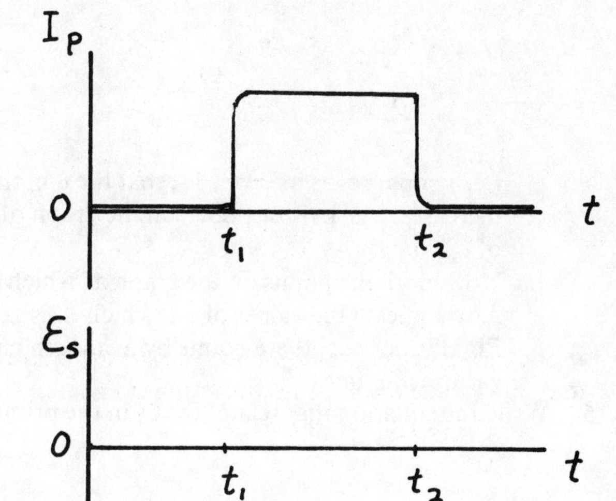

Fig. 23-10

b. What value does the $\mathcal{E}$ have when the current in the primary is a constant 20 A?

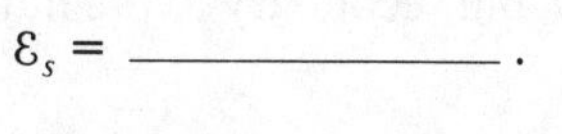
$\mathcal{E}_s$ = ____________.

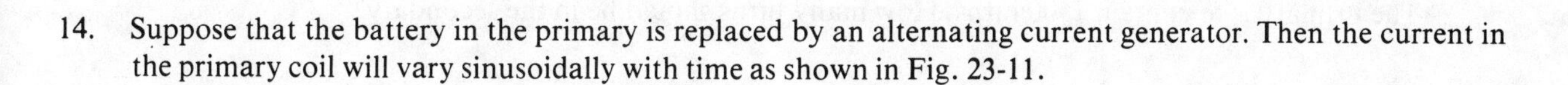
14. Suppose that the battery in the primary is replaced by an alternating current generator. Then the current in the primary coil will vary sinusoidally with time as shown in Fig. 23-11.

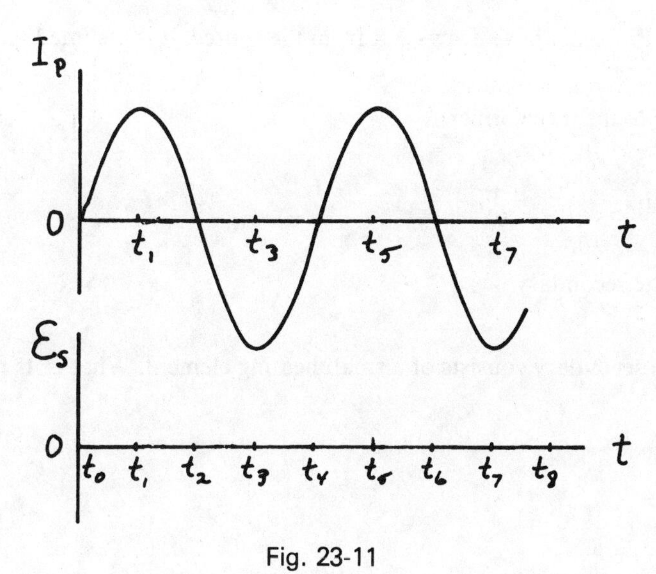

Fig. 23-11

a. At what times is the current in the primary increasing most rapidly?

b. At what times is the current in the primary decreasing most rapidly?

c. At what times is the current in the primary neither increasing nor decreasing?

d. If the current in the primary is constant momentarily, as at times t_1, t_3, and t_5, what $\mathcal{E}$ do you expect in the secondary coil? Refer to Prob. 13 if necessary.

e. Let's suppose, as in Prob. 13, that the $\mathcal{E}$ in the secondary is positive when the current in the primary increases. Mark those points on the graph of $\mathcal{E}_s$ versus t that correspond to maximum positive values of $\mathcal{E}_s$.

f. Now mark the points on the graph at which $\mathcal{E}_s$ has its maximum negative values.

g. Also indicate the values of t at which $\mathcal{E}_s$ is zero.

h. Finally, connect these points by a smooth curve to obtain a picture of how the $\mathcal{E}$ in the secondary varies with time.

15. Write the equation that relates the $\mathcal{E}$s in the primary and secondary to the number of turns in the two coils.

16. A "step-up" transformer is to be designed to produce a 220 V output when connected to a 110 V input. The primary is to contain 1500 turns. How many turns should be in the secondary?

N_s = ____________.

17. The transformer in Prob. 16 above draws 2 A from the source. It is assumed to have an efficiency of 100%. Calculate the following:

a. the power input to the transformer P_{in} = ____________.

b. the power output P_{out} = ____________.

c. the current in the secondary I_s = ____________.

d. The load in the secondary consists of a small heating element. What is its resistance?

R_L = ____________.

18. A lightweight extension cord, shown below, contains two conductors each 4 m long. Each conductor has a resistance of 0.2 Ω.

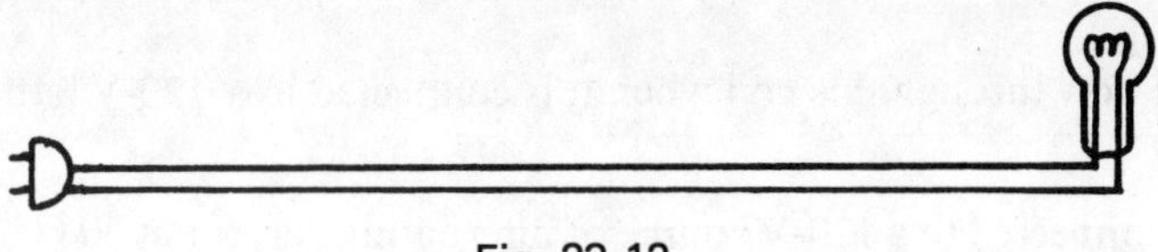

Fig. 23-12

a. If the extension cord is used for a 100-watt table lamp, what is the voltage across the lamp? (Hint: The lamp dissipates 100 watts if it has about 110 V across its terminals.)

$V =$ ______________.

b. The same extension cord is now used with an air conditioner that draws a current of 20 A. What is the voltage supplied to the air conditioner?

c. What power is dissipated in the extension cord in cases 18a and 18b?

$P_a =$ ______________.

$P_b =$ ______________.

d. What is the probable outcome if the air conditioner is used with this extension cord?

e. Suppose that an air conditioner of the same cooling capacity and the same efficiency was designed to operate on 220 V instead of 110 V. What current would it require?

$I =$ ______________.

f. If the same extension cord were used to connect a 220 V source to the air conditioner, and if the current were to remain 10 A, what power would now be dissipated in the extension cord?

$P =$ ______________.

g. What advantage do you see to the 220 V unit as compared with the 110 V unit?

h. What are the disadvantages of the higher voltage?

19. The primary coil of a transformer is wound with 600 m of No. 20 copper wire. (No. 20 copper wire has a resistance of 0.033 Ω/m.)
   a. What current will flow through this coil when it is connected to a 120-V battery?

   I(dc) = ____________.

   b. When this coil is connected to a 120-V source of alternating current at 60 Hz the current that flows is only 0.3 A. (All alternating voltages and currents are effective values unless stated otherwise.) What does the resistance of the coil now appear to be?

   Apparent resistance is ____________.

   c. If the 120-V source had a different frequency would you expect to observe the same current in the coil?

20. If the 600 m of copper wire were stretched out into a single loop 300 m long and again connected to a 120 V, 60 Hz source, the alternating current in the wire would be 6 A. When the same length of copper wire is wound on the iron core of a transformer the flow of alternating current is greatly impeded. It is reduced from 6 A to 0.3 A. The *impedance* of a length of wire depends on the frequency of the current, on the geometrical configuration of the wire, and on the nature of the core on which it is wound. 600 m of No. 20 copper wire wound on a cardboard tube would have much less impedance than the transformer coil.
   a. Write the relationship that defines impedance. Use standard symbols.

   Z = ____________.

   b. Calculate the impedance of the transformer coil described in Prob. 19.

   Z = ____________.

21. We have seen that the impedance of a coil depends on the frequency of the current. There is, however, a property of a coil, its self-inductance, which is independent of frequency, depending only on the number of turns, the nature of the core, and on how the coil is wound.
   a. Write the relationship between impedance and self-inductance.

   b. Calculate the self-inductance of the transformer coil described in Prob. 19.

   L = ____________.

22. Explain briefly, in your own words, why it is more difficult for an alternating current to flow in a coil than it is for a direct current.

23. We have seen that a tightly wound coil, which may be called an inductor, impedes the flow of an alternating current and that the impedance increases as the frequency increases. A capacitor also has an impedance but, in this case, the impedance decreases with increasing fequency.
   a. Write these two expressions.

   The inductor: Z_L = ____________. The capacitor: Z_C = ____________.

b. A capacitance and inductance are to be used in series in a circuit with an alternating current of 60 Hz. The self-inductance of the inductor is 50 mH. What is the impedance of the inductor?

Z_L = ______________.

c. What is the capacitance of a capacitor that will have the same impedance at this frequency as the inductor?

C = ______________.

24a. Which of the four graphs in Fig. 23-13 represents an ideal inductor, one without resistance?

b. Which graph represents an ideal capacitor?

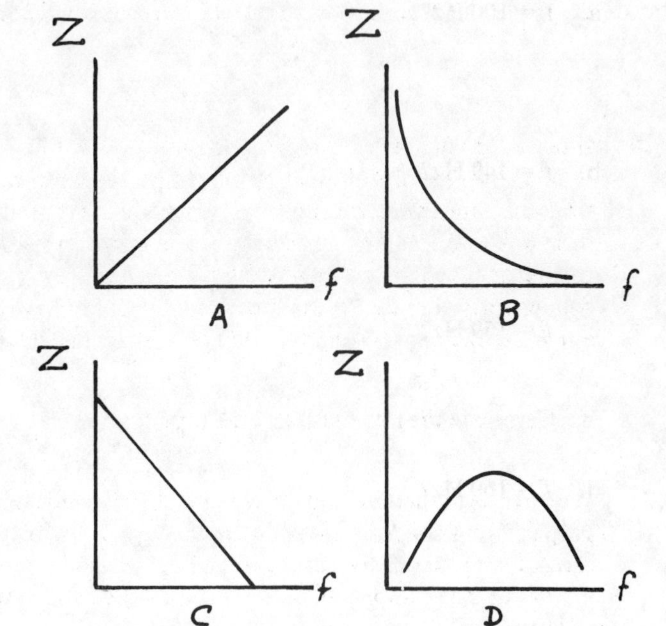

Fig. 23-13

25. A certain electromagnet has a self-inductance of 2.5 H. What current will flow through the coil of this magnet when it is connected to a 220 V, 50 Hz source?

26. A capacitor is required that will have an impedance of 800 Ω at 2000 Hz. What is the capacitance?

27. A capacitor $C = 20\ \mu F$, inductance $L = 0.05$ H, and resistance $R = 30.0\ \Omega$ are in series with an *ac* voltage source. At what frequency of the voltage source is the combined impedance of *C, L,* and *R* a minimum so that resonance takes place?

28. For the circuit described in Prob. 27, what is the impedance of the *C-L-R* series at

a. $f = 100$ Hz?

b. $f = 149$ Hz?

c. $f = 159$ Hz?

d. $f = 169$ Hz?

e. $f = 218$ Hz?

29. Draw a sketch of a thermocouple as it would be used to measure the temperature of a waterbath. Show clearly where the different metals are joined together.

30. Describe another electrical method of measuring temperature.

Sec. 23-11

31. Explain briefly why impurity atoms of valence 5 added to pure silicon will produce an n-type semiconductor.

32. What is the result of adding group III impurity atoms to pure silicon. In particular, what is the nature of the charge carrier in this case?

33. The two drawings below represent a junction diode. Put arrows above each drawing to indicate the direction in which the diode conducts easily, the forward direction.

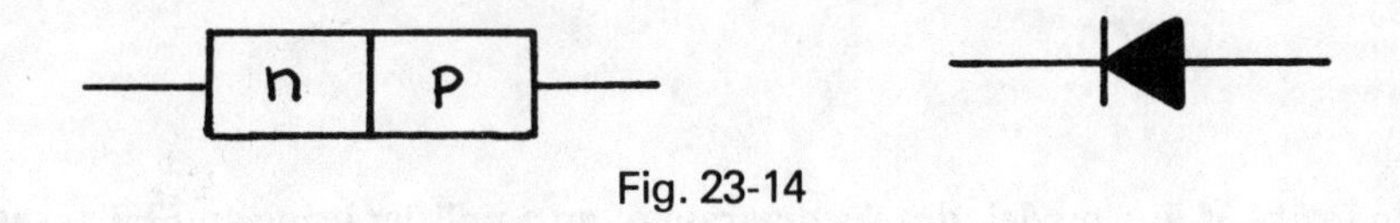

Fig. 23-14

34. A circuit contains an alternating $\mathcal{E}$, a diode, and a resistor.
   a. Make a diagram of the circuit using "⊙" to represent the $\mathcal{E}$.

   b. The left-hand graph in Fig. 23-15 represents the $\mathcal{E}$. Draw a graph to the same scale along the time axis to represent the current through the resistor.

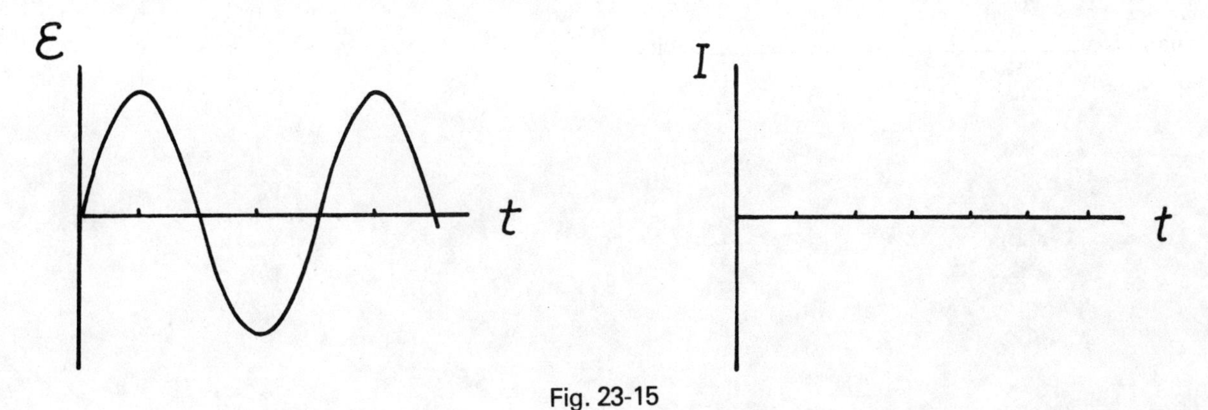

Fig. 23-15

35. Fig. 23-16 is a simplified representation of a transistor amplifier.

a. Label the three connections to the transistor using "B" for base, "C" for collector, and "E" for emitter.

b. Suppose that this amplifier has a current gain of 80. Then a change of 20 μA in the base current will produce what change in the collector current?

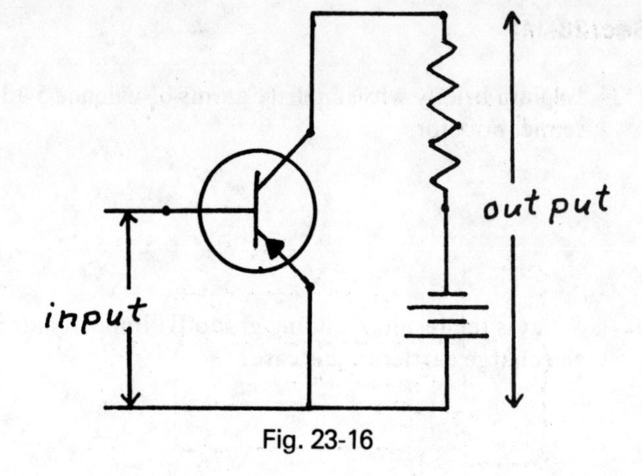

Fig. 23-16

c. What change in output voltage will result from a 20 μA change in base current if the load resistor is 2000 Ω? (Assume a current gain of 80 as in part b above.)

36. Using the circuit in Prob. 35 as a model, draw a diagram of an amplifier using an FET transistor in place of the bipolar junction transistor. Label the three connections to the transistor using "G" for gate, "D" for drain, and "S" for source. (The substrate is connected to the source.)

37. The bipolar junction transistor permits a large current in the collector circuit to be controlled by a small current in the base; it is said to be a current-controlled device. The field-effect transistor, on the other hand is a ______________________ device.

Solutions

1a. **B** is directed from the north pole to the south pole.

b. See Fig. 23-17.

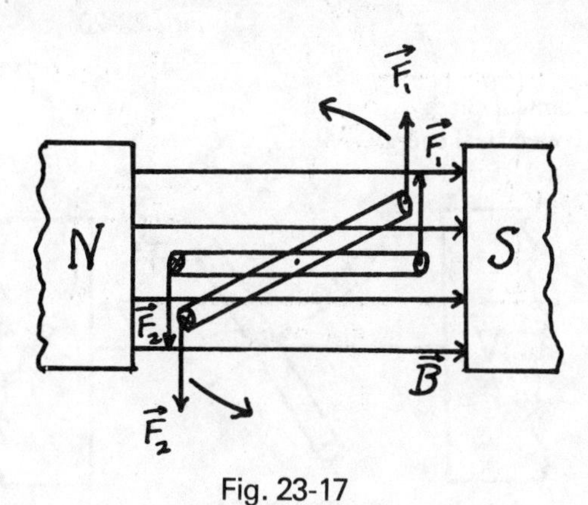

Fig. 23-17

c. Since $I\ell$ is perpendicular to **B**, F has its maximum value.

$$F_1 = I\ell B = (0.02 \text{ A})(0.015 \text{ m})(5 \text{ T}) = 1.5 \times 10^{-3} \text{ N}$$

The magnitudes of these forces are the same.

$$F_1 = F_2 = 1.5 \times 10^{-3} \text{ N}$$

d. No. The force does not change since $I\ell$ remains perpendicular to **B**.

e. The two shorter sides of the coil are parallel to **B**. Thus the magnetic force is zero.

$$F = \text{zero}$$

f. The direction of rotation is counterclockwise:

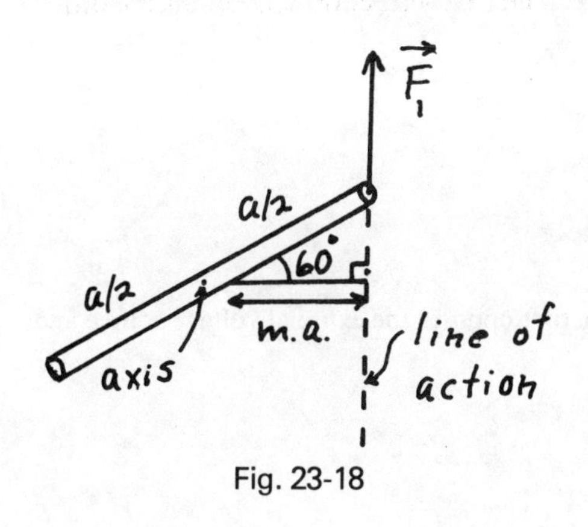

Fig. 23-18

g. $\tau_1 = I\ell B(a/2)\cos\theta$

$\tau_{tot} = I\ell B\, a\cos\theta = IBA\cos\theta$

h. $\tau_1 = F_1(a/2)\cos\theta = (1.5 \times 10^{-3} \text{ N})(5 \times 10^{-3} \text{ m})(0.5)$

$= 3.75 \times 10^{-6} \text{ N} \cdot \text{m}$

$\tau_A = 2\tau_1 = 7.5 \times 10^{-6} \text{ N} \cdot \text{m}$

i. In position "B" the moment arm becomes zero for both forces. Thus

$$\tau_B = 0$$

2a. zero, IBA

b. The iron tends to concentrate the lines of magnetic field, making **B** nearly radial between the poles.

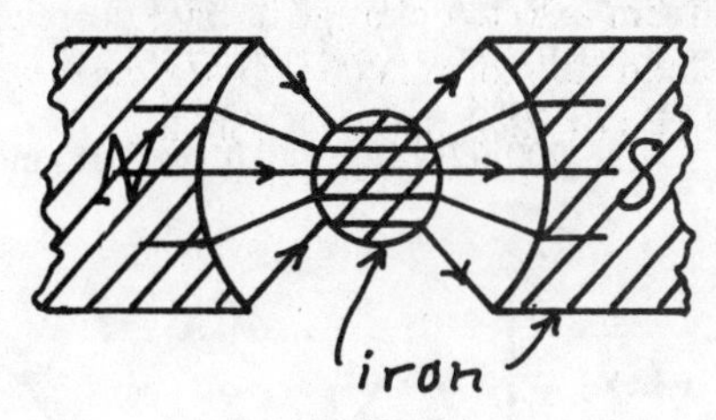

Fig. 23-19

3. $F = -k\Delta x;\ \tau' = -k\theta$

The minus sign may be inserted as a reminder that the restoring force, F, or restoring torque, τ', is opposite to the displacement.

4a. $\Sigma\tau = 0$

b. Thus, $\tau' - \tau = 0$ or $\tau' = \tau$

c. Omitting the minus sign,

$$k\theta = IBA\cos\theta$$

If there are N turns rather than one,

$$\theta = \left(\frac{NBA}{k}\right) I\cos\theta$$

d. $\cos\theta$

e. By using curved pole pieces and adding a cylindrical piece of iron in the center, the magnetic field can be made nearly radial. Then the moment arm remains $a/2$ as the coil rotates through about 45°, i.e., $\cos\theta = 1$.

5. spiral springs; reversing
Use a split-ring commutator as illustrated in text Fig. 23-4.

Fig. 23-20

6a. In a motor, electrical energy is transformed reversibly into mechanical energy.
b. In a generator, mechanical energy is transformed reversibly into electrical energy.

7a. $\mathcal{E} = B\ell v \sin\theta$
b. $\mathcal{E}_{max} = B\ell v$
c. $v = r\omega$ or $\omega = v/r$
$v = (a/2)\omega$ or $\omega = v/(a/2)$
d. $\omega = 2\pi f$
e. $v = (a/2)2\pi f = \pi af$
f. $v = \pi(0.01\text{ m})(60\text{ s}^{-1}) = 1.88\text{ m/s}$
g. $\theta = \theta_0 + \omega t = \omega t$ or $\theta = 2\pi f t$
h. $\mathcal{E} = B\ell v \sin\theta = B\ell(\pi af)\sin(2\pi f t)$
i. $\mathcal{E}_{max} = 2\pi N f BA$
$\mathcal{E}_{max} = 2\pi(200)(60\text{ s}^{-1})(5\text{ T})(0.01 \times 0.015\text{ m}^2) = 56.5\text{ volts}$

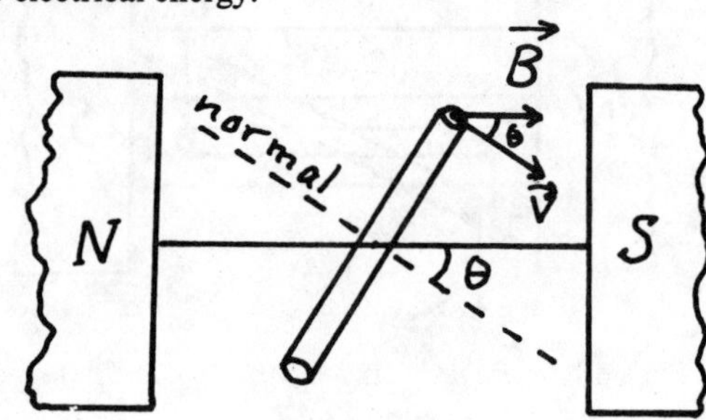

Fig. 23-21

j.

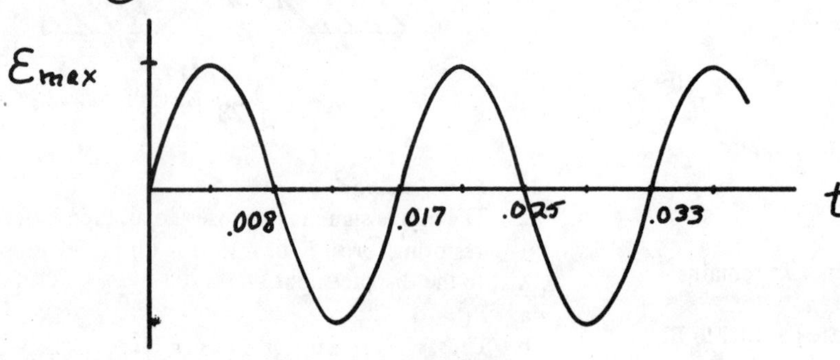

Fig. 23-22

8. The $\mathcal{E}$ of a generator is proportional to the frequency of rotation, f.

$\mathcal{E}_{max} = 2\pi NBAf$

Thus

$$\frac{\mathcal{E}_2}{\mathcal{E}_1} = \frac{f_2}{f_1} \quad \mathcal{E}_2 = 60\text{ V}\frac{1200\text{ rev/s}}{1800\text{ rev/s}} = 40\text{ V}$$

9a. $R = V/I = (3\text{ V})/(0.5\text{ A}) = 6\ \Omega$
b. $R(\text{apparent}) = V/I = (120\text{ V})/(2\text{ A}) = 60\ \Omega$
c. When the motor is running, a back $\mathcal{E}$ is produced in the armature that opposes the external voltage source and reduces the current to a much smaller value.
d. $120\text{ V} - \mathcal{E}_b = IR = (2\text{ A})(6\ \Omega) = 12\text{ V}$

Thus

$\mathcal{E}_b = 120\text{ V} - 12\text{ V} = 108\text{ V}$

e. When the motor is first turned on, the back $\mathcal{E}$ is zero. Thus the starting current is

$I = V/R = (120\text{ V})/(6\ \Omega) = 20\text{ A}$

10. The fundamental law of physics involved is the law of conservation of energy. If the induced $\mathcal{E}$ were to aid the external source, the electrical energy input to a motor that is running and doing mechanical work would be less than the energy input to a motor that is at rest and doing no work. This is a clear violation of the principle of conservation of energy. Lenz's law, which also might be invoked here, is itself a consequence of energy conservation.

11a. Let I_f and I_a be the currents through the field coils and the armature, respectively.

$$I_f = 100\text{ V}/40\ \Omega = 2.5\text{ A}$$
$$I_a = 100\text{ V}/10\ \Omega = 10\text{ A}$$

Thus $I(\text{starting}) = 12.5\text{ A}$

b. The current through the field coils is unchanged. Thus $I_a = 2$ A.

$$I_a = \frac{100\text{ V} - \mathcal{E}_b}{10} = 2\text{ A}$$

$$100\text{ V} - \mathcal{E}_b = 20\text{ V} \quad \mathcal{E}_b = 80\text{ V}$$

12a. The eddy currents in the plate are counterclockwise.

b. This causes an upward force to act on the plate, slowing its movement through the magnetic field, illustrated below.

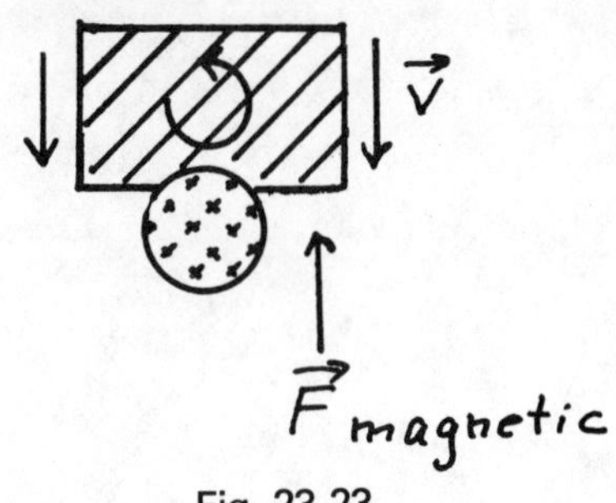

Fig. 23-23

13a.

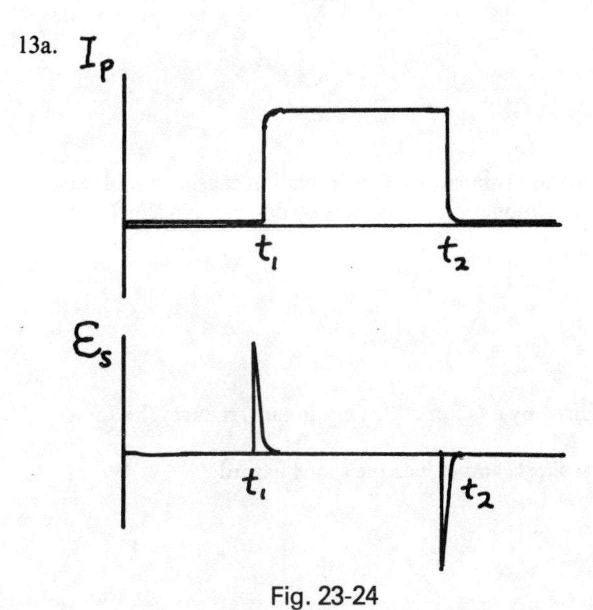

Fig. 23-24

b. $\mathcal{E}_s = 0$ when I_p is constant.

14a. t_0, t_4

b. t_2, t_6

c. t_1, t_3, t_5

d. When I_s is constant, $\mathcal{E}_s = 0$.

e. points "x" on the graph.

f. points "m" on the graph.

g. points "o" on the graph.
The smooth curve that has been drawn through the points is a sinusoidal curve displaced 1/4 of a period from the graph of I_p versus t.

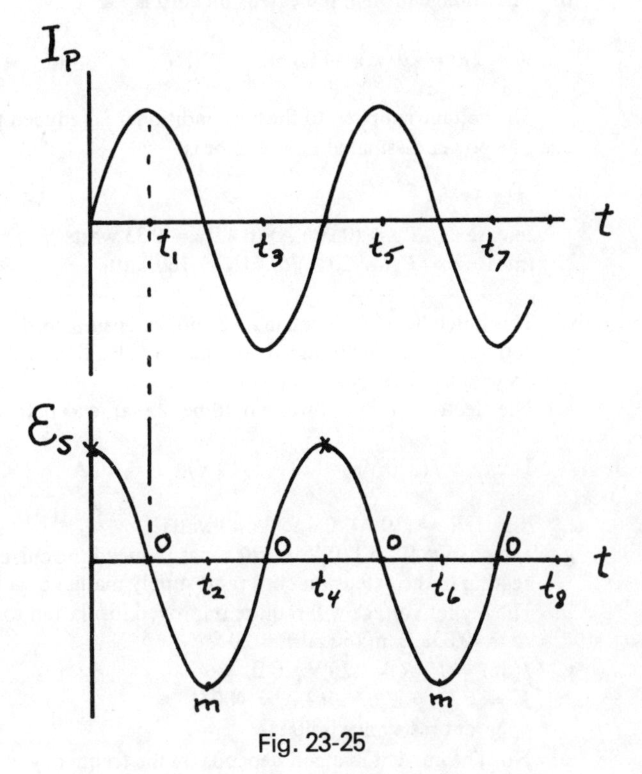

Fig. 23-25

15. $$\frac{\mathcal{E}_s}{\mathcal{E}_p} = \frac{N_s}{N_p}$$

16. $N_s = N_p(\mathcal{E}_s/\mathcal{E}_p) = 1500(220\text{ V})/(110\text{ V}) = 3000$ turns

17a. $P_{in} = \mathcal{E}_p I_p = (110 \text{ V})(2 \text{ A}) = 220$ watts
b. $P_{out} = \mathcal{E}_s I_s = 100\% \, P_{in} = 220$ watts
c. $I_s = \dfrac{P_{out}}{\mathcal{E}_s} = \dfrac{220 \text{ W}}{220 \text{ V}} = 1 \text{ A}$

d. $P_{out} = I_s^2 \text{R}$

$$R = \frac{P}{I_s^2} = \frac{220 \text{ W}}{(1 \text{ A})^2} = 220 \ \Omega$$

18a. The total resistance in the extension cord is $2 \times 0.2 \ \Omega = 0.4 \ \Omega$. The resistance of the 100-W lamp can be found as follows:

$$P = V^2/R \ ; \qquad R = V^2/P$$

$$R = \frac{(110 \text{ V})^2}{100 \text{ W}} = 121 \ \Omega$$

The resistances are in series. Thus

$$R_{tot} = 121.4 \ \Omega$$

The current in the circuit is

$$I = V/R_{tot} = (110 \text{ V})(121.4 \ \Omega) = 0.906 \text{ A}$$

and the voltage across the lamp is

$$V(\text{lamp}) = I \, R(\text{lamp}) = (0.906 \text{ A})(121 \ \Omega) = 110 \text{ V}$$

There is no significant loss of voltage due to the extension cord.
b. The voltage drop in the extension cord is

$$V = IR = (20 \text{ A})(0.4 \ \Omega) = 8 \text{ V}$$

The voltage supplied to the air conditioner is reduced to 102 V.
c. The power dissipated in a resistor is

$P = I^2 R$
In case a. $P_a = (0.906 \text{ A})^2(0.4 \ \Omega) = 0.33$ watts
In case b. $P_b = (20 \text{ A})^2(0.4 \ \Omega) = 160$ watts

d. This much heat produced in a 12-foot extension cord will almost surely burn away the insulation causing a short and perhaps a fire. The reduced voltage may also damage the air-conditioner motor by causing it to run too slowly and draw more than its normal current.
e. The electrical input power would be the same as before.

Thus $\quad I(220 \text{ V}) = (20 \text{ A})(110 \text{ V}) \quad I = 10 \text{ A}$

f. $P = I^2R = (10 \text{ A})^2(0.4 \ \Omega) = 40$ watts
g. The change from 110 V to 220 V has reduced the current required by a factor of 2. This, in turn, reduces the Joule heating in the extension cord (and supply mains) by a factor of 4.
h. The higher voltage will require improved insulation to prevent shorts and reduce the shock hazard.

19a. $R = (0.033 \ \Omega/\text{m})(600 \text{ m}) = 20 \ \Omega$
$I(\text{dc}) = V/R = 120 \text{ V}/20 \ \Omega = 6 \text{ A}$
b. $R = V/I = 120 \text{ V}/0.3 \text{ A} = 400 \ \Omega$
Apparent resistance is 400 Ω
c. No. The current in a coil depends on the frequency.

20a. $Z = \dfrac{V_{eff}}{I_{eff}}$ or $Z = \dfrac{V_{max}}{I_{max}}$ or $Z = \dfrac{V}{I}$

In alternating current circuits "I" is understood to mean "I_{eff}." Thus the subscript "eff" is usually omitted.

b. $Z = \dfrac{120\text{ V}}{0.3\text{ A}} = 400\ \Omega$

21a. $Z = 2\pi f L$

b. $L = Z/(2\pi f) = \dfrac{400\ \Omega}{2\pi\ 60\text{ Hz}} = 1.06\text{ H}$

22. Compare your answer with the last half of Sec. 23-7.

23a. $Z_L = 2\pi f L \qquad Z_C = \dfrac{1}{2\pi f C}$

b. $Z = 2\pi f L = 2\pi(60\text{ Hz})(50 \times 10^{-3}\text{ H}) = 18.8\ \Omega$

c. $C = \dfrac{1}{2\pi f Z} = \dfrac{1}{2\pi(60\text{ Hz})(18.8\ \Omega)} = 1.4 \times 10^{-4}\text{ F}$

24a. Graph A represents the ideal inductor.
b. Graph B represents the capacitor.
25. $Z = 2\pi f L = 2\ (50\text{ s}^{-1})(2.5\text{ H}) = 785\ \Omega$

$I = \dfrac{\mathcal{E}}{Z} = \dfrac{220\text{ V}}{785\ \Omega} = 0.280\text{ A}$

26. $Z = \dfrac{1}{2\pi f C} \qquad C = \dfrac{1}{2\pi f Z}$

$C = \dfrac{1}{2\pi(2 \times 10^{3}\text{ s}^{-1})(800\ \Omega)} = 9.9 \times 10^{-8}\text{ F}$

27. $f = \dfrac{1}{2\pi}\sqrt{\dfrac{1}{LC}} = \dfrac{1}{2\pi}\sqrt{\dfrac{1}{(0.05\text{ H})(20 \times 10^{-6}\text{ F})}} = 159\text{ Hz}$

28. $Z = \sqrt{R^2 + (X_L - X_C)^2} = \sqrt{R^2 + \left(2\pi f L - \dfrac{1}{2\pi f C}\right)^2}$

a. for $f = 100$ Hz, $Z = 56.8\ \Omega$
b. for $f = 149$ Hz, $Z = 30.7\ \Omega$
c. for $f = 159$ Hz, $Z = 30.0\ \Omega$
d. for $f = 169$ Hz, $Z = 30.6\ \Omega$
e. for $f = 218$ Hz, $Z = 43.8\ \Omega$

29.

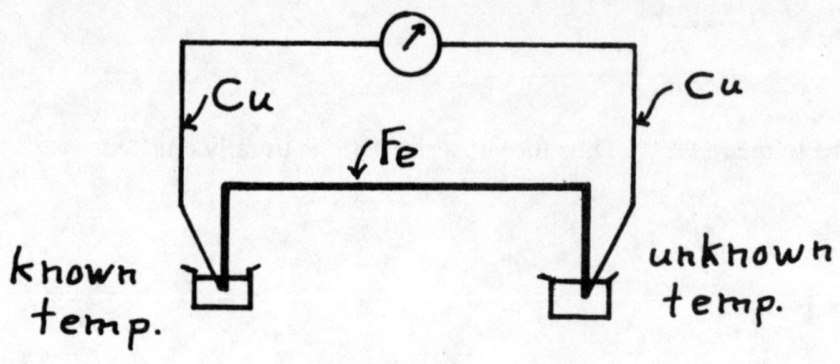

Fig. 23-26

30. Temperature may also be measured with a resistance thermometer. The resistance of a conductor varies with temperature according to

$$R_T = R_0(1 + \alpha T)$$

where α is the temperature coefficient of resistivity. The resistance measurement would be made by means of a Wheatstone bridge.

31. Only 4 of the valence electrons of the impurity atom will be tightly bound in the crystal. The extra electron will be very loosely bound and will thus be able to contribute very significantly to the conduction. When the current carriers are electrons, as in this case, the material is called *n*-type.

32. A group III impurity atom has only 3 valence electrons, not enough to fill the 4 vacancies in neighboring atoms. The result is an empty place, called a hole, which can move fairly easily through the crystal. These holes act just like positive charges moving through the crystal under the influence of an applied electric field.

33.

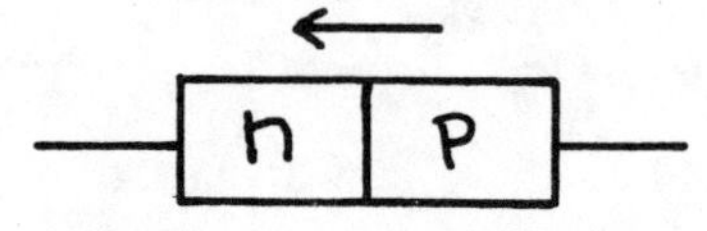

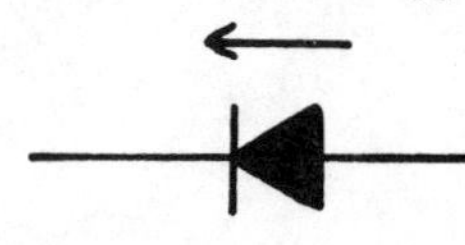

Fig. 23-27

34a.

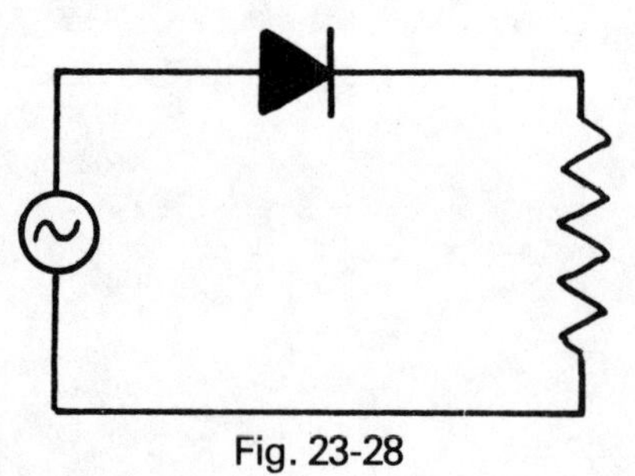

Fig. 23-28

b.

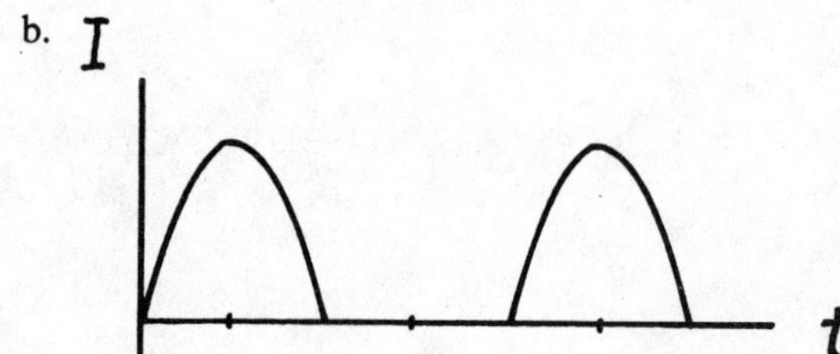

Fig. 23-29

35a.

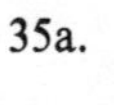

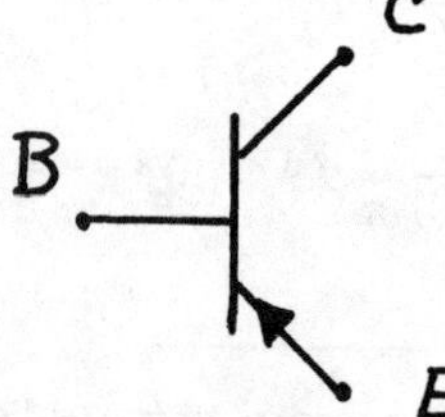

Fig. 23-30

b. $\dfrac{\text{change in } I_C}{\text{change in } I_B} = 80$

$\Delta I_C = 80\, I_B = 80(20\ \mu\text{A}) = 1.6\ \text{mA}$

c. $\Delta V = (\Delta I_C)R = (1.6\ \text{mA})(2\ \text{k}\Omega) = 3.2\ \text{V}$

36. Compare your drawing with Fig. 23-35 in the text.

37. voltage-controlled

REVIEW TEST FOR CHAPTERS 21–23

1. An electrolytic cell is known to have an $\mathcal{E}$ of 1.5 V. When this cell is connected to a lamp bulb, it is found that 5 A flows in the circuit and that the PD measured across the terminals of the cell is only 1.3 V. Find the internal resistance of the cell.

2. What current flows through the 8 Ω resistor in the circuit shown below?

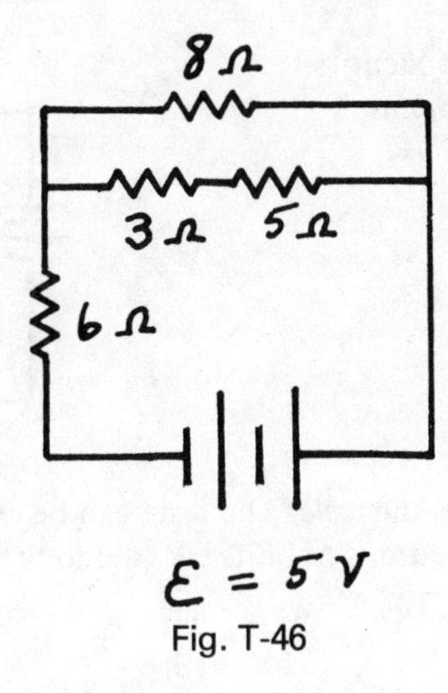

Fig. T-46

3. What is the PD across the plates of the capacitor in the circuit illustrated below?

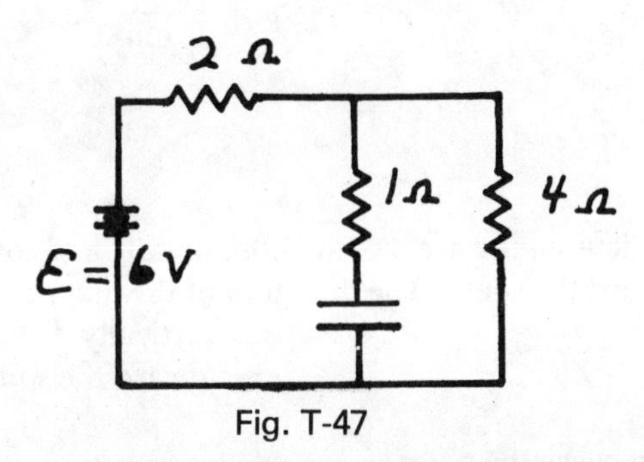

Fig. T-47

4. A galvanometer having an internal resistance of 500 Ω gives a full-scale deflection when 100 μA passes through it. A low-resistance shunt (0.5 Ω) is connected across the galvanometer. What is the full-scale deflection of the resulting ammeter?

5. When the switch is closed in the circuit in Fig. T-48, no current flows through the galvanometer. What is the value of the unknown resistor?
   a. 1.6 Ω
   b. 5.0 Ω
   c. 3.0 Ω
   d. 20 Ω
   e. 40 Ω

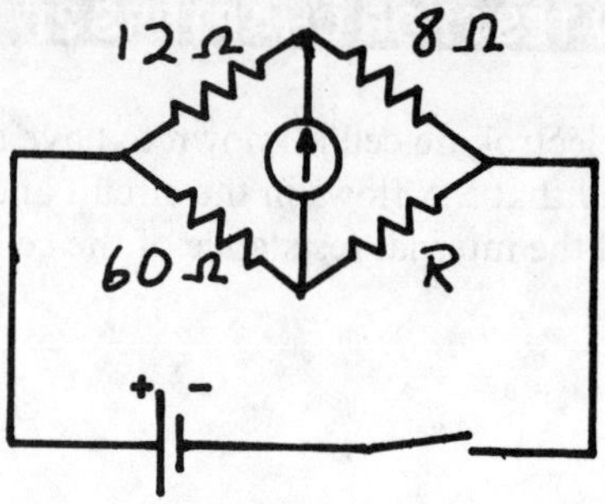

Fig. T-48

6. A 24-V battery is connected to two capacitors as shown in Fig. T-49. What is the potential difference between the plates of the 3 μF capacitor?
   a. 24 V
   b. 20 V
   c. 18 V
   d. 12 V
   e. 6 V

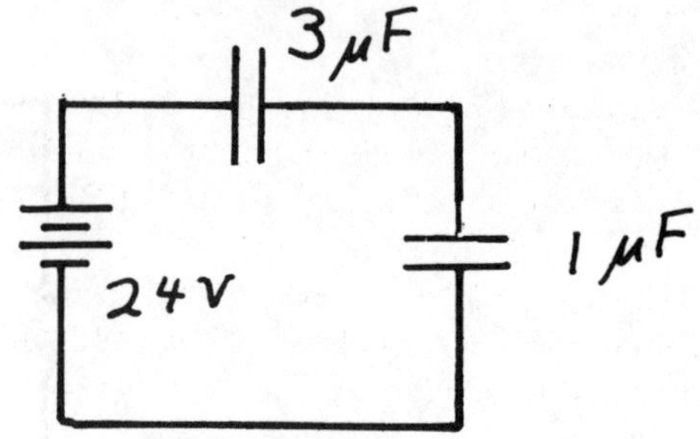

Fig. T-49

7. The SI unit for the magnetic field **B** is the *tesla*. The tesla can be expressed as a combination of other SI units. Which of the following combinations of units is equal to the tesla?
   a. $\frac{N}{C \cdot m}$
   b. $\frac{N \cdot s}{C}$
   c. $\frac{J}{C}$
   d. $\frac{kg}{C \cdot s}$
   e. $\frac{J \cdot s}{C \cdot m}$

8. The electrons in a television picture tube are moving horizontally and toward the west. A uniform magnetic field is directed toward the south. The direction of the magnetic force on the electrons is
   a. toward the north.
   b. vertically upward.
   c. toward the east.
   d. vertically downward.
   e. toward the southwest.

9. A wire is mounted vertically between the poles of two magnets, as shown in Fig. T-50. A current of 4 A is flowing upward (out of the drawing toward you). A magnetic field of 2 T is directed toward the right, as shown. Which arrow shows the direction of the magnetic force on the wire?
   a. arrow a
   b. arrow b
   c. arrow c
   d. arrow d
   e. an arrow of zero length

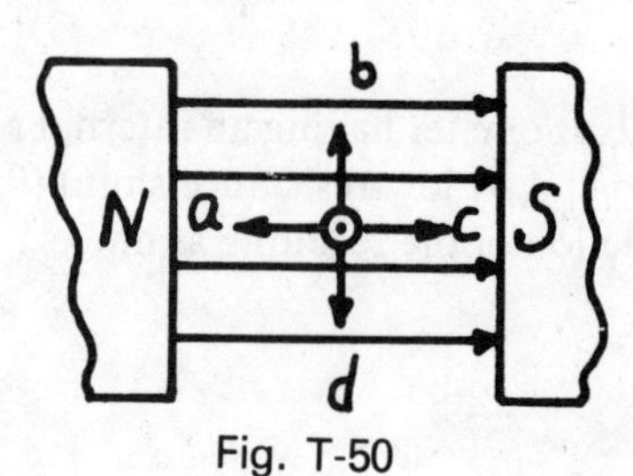

Fig. T-50

10. A circular loop 2 cm in radius produces a magnetic field of 6.28×10^{-5} T at its center (see Fig. T-51). What current is flowing in the loop?
   a. 1 A
   b. 2.2×10^{-17} A
   c. 6.28 A
   d. 2 A
   e. 3.14×10^{-3} A

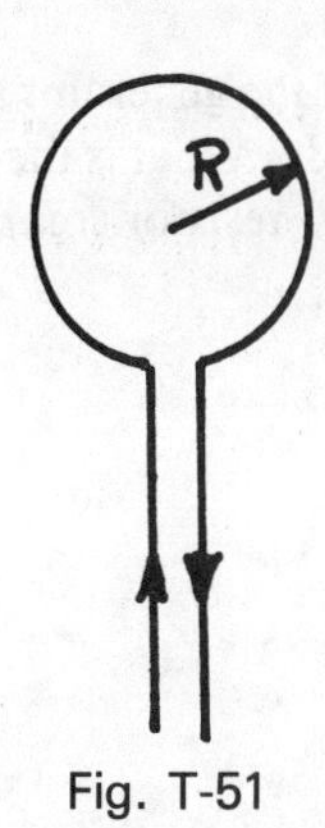

Fig. T-51

11. A conductor 25 cm long is moving through a magnetic field at a speed of 5 m/s. The velocity of the conductor makes an angle of 30° with respect to **B**. The magnitude of **B** is 4 T. What $\mathcal{E}$ is generated between the two ends of the conductor? (Fig. T-52 shows an end view of the conductor.)

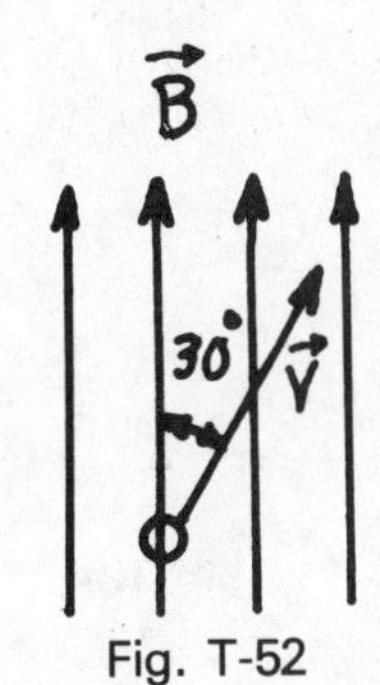

Fig. T-52

12. A small, flat, rectangular coil of 20 turns is free to rotate in the field of a permanent magnet. The longer dimension of the coil, perpendicular to the drawing, is 5 cm. The magnitude of **B** is 0.2 T. When the coil is in the position shown in Fig. T-53, and the current is 2 A, what is the magnitude of the total torque on the coil?

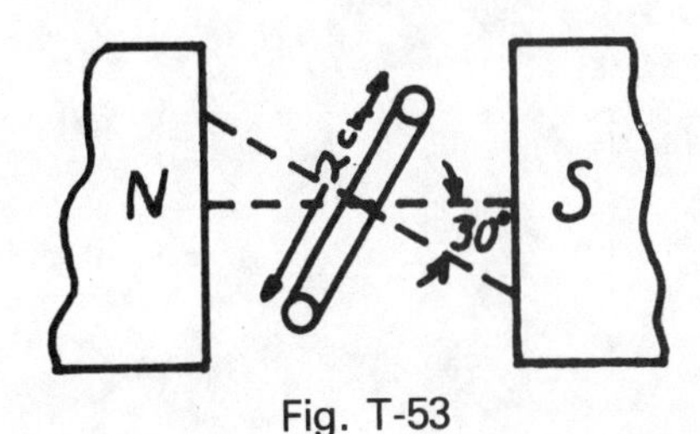

Fig. T-53

13. A transformer has 2400 windings in the primary and 80 windings in the secondary. The secondary is connected to an 8 Ω resistor. If the effective voltage across the primary is 120 V and the frequency is 60 Hz, what is the effective current in the 8 Ω resistor? (The resistance of the secondary coil is negligible.)
   a. 200 A
   b. 0.005 A
   c. 3.3 A
   d. 30 A
   e. 0.5 A
14. Which one of the following functions *cannot* be performed by a transformer?
   a. Obtain an effective $\mathcal{E}$ of 600 V when the primary is connected to an alternating current generator having an $\mathcal{E}$ of 50 V at 60 Hz.
   b. Obtain an effective current of 50 A in a secondary of 1 turn when the primary voltage is 110 V at 50 Hz.
   c. Obtain a dc voltage of 4 V across a 500-turn secondary when the 5000-turn primary is connected to a 40-V battery.
   d. Obtain an alternating voltage of 120 V in the secondary when the primary is connected to a 400 Hz generator having an $\mathcal{E}$ of 120 V.
   e. Light a 300-watt bulb with a secondary of only 200 turns.

15. The frequency of the generator shown in Fig. T-54 is 400 Hz. What is the impedance of the capacitor-plus-resistor combination?

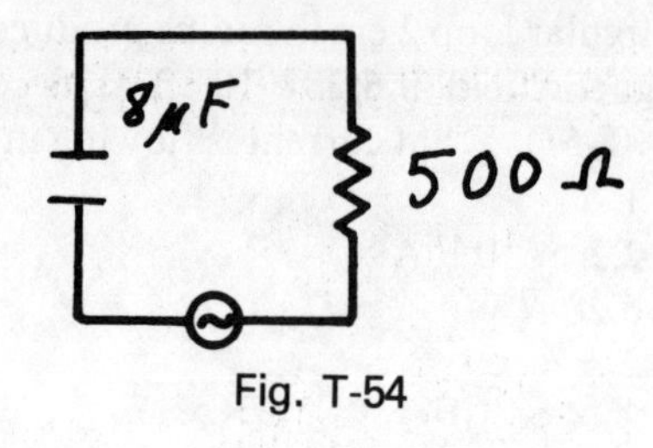

Fig. T-54

CHAPTER 24 Electromagnetic Waves

GOALS To learn how oscillations of charges in electric circuits give rise to electromagnetic radiation.

OBJECTIVES After completing this chapter the student should be able to do the following:

1. Write the expression for the potential energy stored in a charged capacitor.
2. Write the expression for the kinetic energy stored in a current-carrying coil.
3. Write the expression for the natural frequency of a circuit containing inductance and capacitance.
4. Calculate any one of the quantities, f, L, or C when the others are known.
5. State the plane of the electric field, **E**, in an electromagnetic wave when the plane of the magnetic field is given, and vice versa.
6. Write the relationship between speed, frequency, and wavelength for an electromagnetic wave.
7. Calculate any one of the quantities c, f, or λ when the others are given.
8. Give the numerical value of the speed of propagation in vacuum or in any electromagnetic wave.

SUMMARY

In earlier chapters we studied the vibration or oscillation of mechanical systems, a mass on the end of a spring, for example, or a swinging pendulum. Applying the energy principle to an electrical circuit containing inductance and capacitance, we find that charges in such a circuit can be made to oscillate and, like the mechanical system, the amplitude will be largest if the frequency of the driving force is equal to the natural resonant frequency $f = 1/(2\pi\sqrt{LC})$.

When electric charges are rapidly accelerated, energy is radiated in the form of electromagnetic waves. This happens on an atomic scale when high-speed electrons strike the screen of a television tube. Their sudden deceleration leads to x-ray radiation. Rapidly oscillating charges in a radio antenna also radiate electromagnetic energy. Regardless of the wavelengths of the electromagnetic radiation, which ranges from hundreds of meters for radio waves to less than a millionth of a meter for gamma rays, the speed of propagation through empty space is always exactly the same, 3.00×10^8 m/s. Electromagnetic waves, unlike sound waves in air, consist of transverse vibrations. Thus, electromagnetic waves may be polarized. Perhaps their most important characteristic, which again distinguishes electromagnetic waves sharply from sound waves, is that the former are readily propagated through empty space.

QUESTIONS AND PROBLEMS

Secs. 24-1—24-5

1. A capacitor of 100 μF is connected to a 6-V battery. Determine the charge on the capacitor and the energy stored.

Q = ________________.

PE = ________________.

2. A certain 1.5-V dry cell can supply 100 mA to an external load for 4 hours before it is exhausted. How many times can a 20,000 μF capacitor be charged from this battery?

N = ________________.

3. When a current of 0.2 A flows in a coil, the stored energy is 10^{-3} J. What is the inductance of the coil?

L = ________________.

4. Show how electric and magnetic measurements can be used to calculate the speed of light.

5. An electromagnetic wave is traveling along the positive x-axis shown in Fig. 24-1. The electric field **E** is in the xy plane with the **E** vector always perpendicular to the x-axis. Describe the direction of the magnetic field.

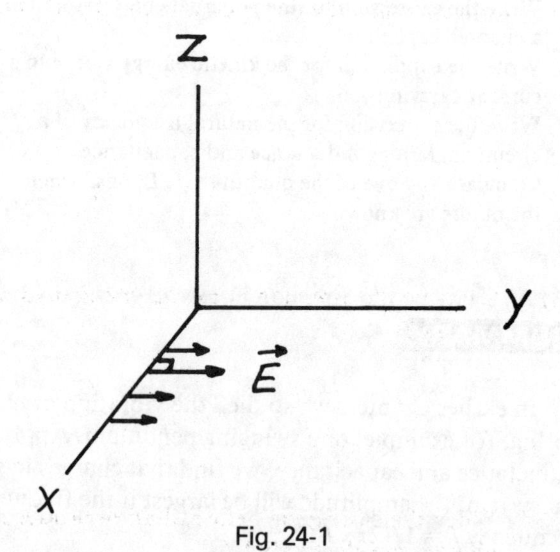

Fig. 24-1

6. A changing magnetic field causes an induced electric field. What does a changing electric field cause?

7. Are electromagnetic waves transverse or longitudinal?

8. A local FM station broadcasts at a frequency of 104 MHz. What is the wavelength of this electromagnetic wave?

9. A certain electromagnetic wave has a frequency of 6.0×10^{14} Hz. What is the common name for electromagnetic waves of this frequency? Name a possible detector of these waves. (Hint: Calculate the wavelength in Å or in nm and then consult Table 25-1 in the text.)

10. Electromagnetic waves may be produced by connecting two conductors to a high-frequency generator. Fig. 24-2 shows a generator connected to conductors lying along the positive and negative z-axes. Add several curved lines to the drawing to indicate the electric field at an instant when the upper conductor is positive.

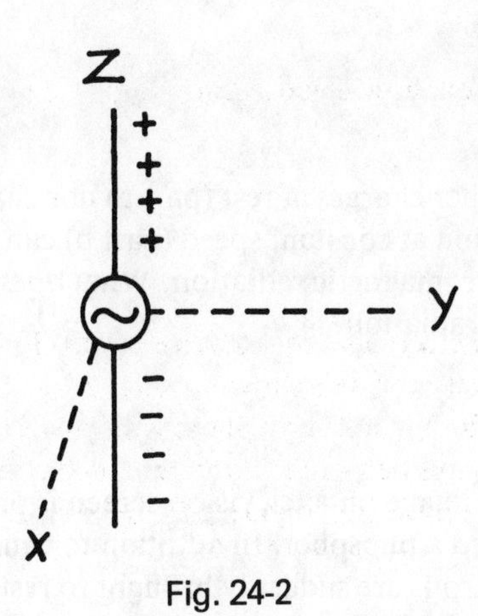

Fig. 24-2

11. Describe the direction of the magnetic field radiated by this oscillating dipole as seen by an observer on the x-axis.

12. When the electromagnetic radiation produced by this dipole antenna is viewed along the x-axis, in what plane is the electric field?

13a. The capacitor shown in Fig. 24-3 is fully charged. Is it radiating electromagnetic waves?

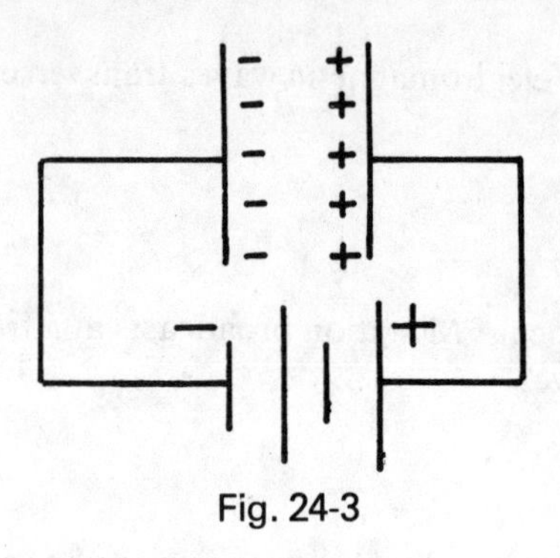

Fig. 24-3

b. A large, steady current is passing through the coil shown in Fig. 24-4. Is the coil radiating electromagnetic waves?

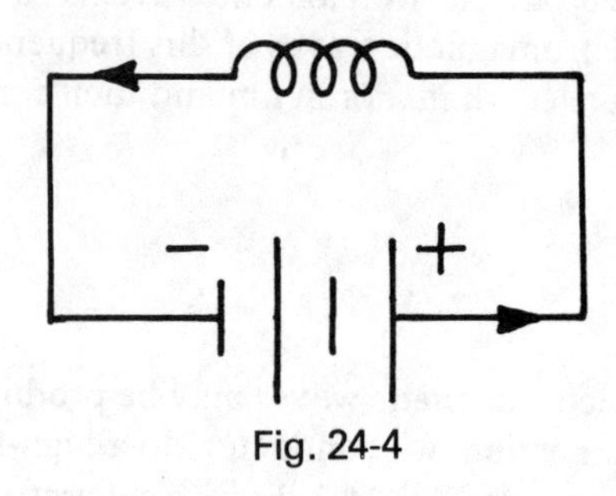

Fig. 24-4

c. Neither charges at rest (part a) nor charges in motion at constant speed (part b) can produce electromagnetic radiation. What *does* produce EM radiation?

14. The image on a television screen is produced by high-speed electrons striking a specially coated surface called a phosphor. In addition to causing the phosphor to glow, what effect can you expect when the electrons are suddenly brought to rest? Why have some people expressed concern about remaining for long periods close to the screen of a TV set?

15. The magnetron oscillator in a microwave oven produces EM waves of 20-cm wavelength. What is the effective capacitance of the oscillating circuit if the inductance is 0.5 μH?

16. We have seen that the resonant frequency of a circuit containing a capacitor and an inductor is $f = 1/(2\pi\sqrt{LC})$.

a. Rearrange this equation to show that, at resonance, the impedance of the coil is equal to the impedance of the capacitor.

16 b. Show that your equation is dimensionally correct.

c. Calculate the impedance of the coil, at the resonant frequency, in a circuit with $L = 0.1$ mH and $C = 100$ pF.

Solutions

1. $C = Q/V$; $Q = CV$

$Q = (10^{-4}\text{ F})(6\text{ V}) = 6 \times 10^{-4}\text{ C}$

$$\text{PE} = \tfrac{1}{2}(1/C)Q^2 = \frac{1}{2}\frac{(6 \times 10^{-4}\text{ C})^2}{10^{-4}\text{ F}} = 1.8 \times 10^{-3}\text{ J}$$

2. First find the total charge delivered by the battery.

$$Q = It = (0.1\text{ A})(4 \times 3600\text{ s}) = 1.44 \times 10^3\text{ C}$$

The charge stored in the capacitor is

$$Q = CV = (2 \times 10^{-2}\text{ F})(1.5\text{ V}) = 3 \times 10^{-2}\text{ C}$$

The number of times that the capacitor can be charged is

$$N = \frac{1.44 \times 10^3\text{ C}}{3 \times 10^{-2}\text{ C}} = 4.8 \times 10^4\text{ times}$$

3. $\text{KE} = \frac{1}{2}LI^2$; $L = \frac{2(\text{KE})}{I^2}$

$$L = \frac{2(10^{-3}\text{ J})}{(2 \times 10^{-1} A)^2} = 5 \times 10^{-2}\text{ H} = 50\text{ mH}$$

4. In principle, the constant k that appears in Coulomb's law can be determined from careful measurements of the force between electric charges.

$$k = \frac{Fr^2}{QQ'}$$

The constant k' that appears in equations for magnetic field, can be found by measuring the force between parallel wires carrying known currents.

$$k' = \frac{F/\Delta l}{2I_1I_2/a}$$

Then the speed of light is given by

$$c = \sqrt{k/k'}$$

5. The magnetic field direction is perpendicular to the direction of propagation and perpendicular to the electric field. Thus the vector **B** parallels the z-axis.
6. an induced magnetic field
7. transverse, i.e., perpendicular to the direction of propagation.

8. $c = f\lambda\,;\ \lambda = c/f = \dfrac{3 \times 10^8 \text{ m/s}}{104 \times 10^6 \text{ Hz}} = 2.88 \text{ m}$

9. $\lambda = c/f = \dfrac{3 \times 10^8 \text{ m/s}}{6.0 \times 10^{14} \text{ Hz}} = 5 \times 10^{-7} \text{ m} = 500 \text{ nm}$

This is the wavelength of green light. The human eye is a very sensitive detector of radiation at this frequency. Other detectors are photographic film and various photoelectric detectors.

10. 

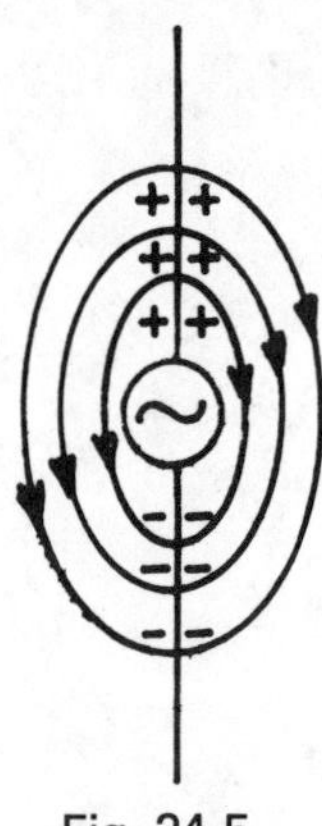

Fig. 24-5

11. The magnetic field is perpendicular to the electric field. Since the electric field is parallel to the z-axis, the magnetic field must be parallel to the y-axis.
12. the yz-plane
13a. No.
b. No.
c. accelerating charges, i.e., charges for which **v** is changing in magnitude or direction
14. The electrons are stopped within a very short distance. Thus they undergo very large decelerations. This causes them to emit EM radiation of very short wavelength, x rays. Exposure to even low-intensity x rays over a long period of time is dangerous to one's health. Contemporary TV sets are designed to absorb these x rays in the walls of the TV picture tube.
15. First find the frequency.

$$f = c/\lambda = \frac{3 \times 10^8 \text{ m/s}}{0.2 \text{ m}} = 1.5 \times 10^9 \text{ Hz}$$

$$f = (1/2\pi)\sqrt{1/LC}\,;\ C = 1/(4\pi^2 L f^2)$$

$$C = \frac{1}{4\pi^2(0.5\ \mu H)(1.5 \times 10^9 \text{ Hz})^2} = 2.25 \times 10^{-20} \text{ F}$$

16a. Begin with the equation for the resonant frequency,

$$f = \frac{1}{2\pi}\sqrt{\frac{1}{LC}} \quad \text{which gives} \quad f^2 = \frac{1}{4\pi^2 LC}$$

or $$2\pi f L = \frac{1}{2\pi f C}$$

Thus {impedance of coil} = {impedance of capacitor}

16b. To check dimensions, we begin with the definition of L.

$$[L] = \frac{[\mathcal{E}_{back}]}{[\Delta I/\Delta t]} = \frac{[volt]}{[A/s]} = \frac{[V\ s]}{[A]}$$

Thus the left-hand side has the dimensions,

$$[2\pi f L] = [s^{-1}\ V\ s/A] = [V/A] = \frac{volt}{ampere}$$

On the right-hand side,

$$[1/2\pi f C] = [s][V/Q] = \frac{[V]}{[Q/s]} = \frac{volt}{ampere}$$

c.

$$Z_{coil} = Z_{capacitor} = 2\pi f L$$
$$= 2\pi L\ (1/2\pi)\ \sqrt{1/LC} = \sqrt{L/C}$$
$$= \sqrt{10^{-4}\ H/10^{-10}\ F} = 10^3\ \Omega$$

CHAPTER

25 Geometrical Optics

GOALS To introduce geometrical methods for analyzing the properties of simple lenses, mirrors, and optical systems.

OBJECTIVES After completing this chapter the student should be able to do the following:

1. Write a definition of wave front in terms of phase of vibration.
2. Make a drawing to show how Huygens' principle may be used to find successive positions of a wave front.
3. Give the technical term describing a medium in which light travels at the same speed in all directions.
4. Give a definition of a ray of light based on the concept of energy flow.
5. State the geometrical relationship between wave fronts and rays.
6. State both parts of the law of reflection.
7. Use the law of reflection to determine the direction of a reflected ray of light.
8. State the definition of index of refraction in words and in symbols.
9. Find any one of the quantities c, n, or c_n when the others are given.
10. Write Snell's law using the commonly accepted symbols. Describe the plane of reflection and relate it to the plane of incidence.
11. Solve for any one of the quantities, n_1, n_2, θ_1, or θ_2 when the others are known.
12. Describe the conditions that cause a ray of light to be totally (100%) reflected.
13. Calculate the critical angle for total internal reflection.
14. Define the terms *principal focus, focal plane,* and *focal length* and illustrate their meaning by means of a simple drawing.
15. Distinguish between a real image and a virtual image.
16. Write the definition of magnification, and the relationship between magnification and object and image distances.
17. Calculate any one of the quantities m, h, h', q, or p when a sufficient number of the others are given.
18. Write the lens equation and state the physical significance of the symbols used.
19. Determine the position and size of an image (or of an object) for either a converging or a diverging lens by the use of the lens equation.
20. Determine the *character* (real or virtual) of an object or of an image.
21. Determine the position and size of an image (or of an object) for either a converging or a diverging lens by the use of the lens equation.
22. State the relationship between curvature and radius of a wave front.
23. Distinguish between converging and diverging wave fronts and between negative and positive curvatures.
24. Describe, qualitatively, the effect on a plane wave of converging and diverging lenses.
25. Calculate the power of a lens from the lens-maker's equation.
26. Calculate the power of a lens from its focal length and vice versa.
27. Locate the images produced by converging and diverging lenses using the method of curvatures (if Sec. 25-9 is covered).
28. Determine the size and nature of an image (either real or virtual) by the method of curvatures (if Sec. 25-9 is covered).
29. Determine the position, height, and nature (real or virtual) of an image produced by either a concave or a convex mirror (if Sec. 25-9 is covered).
30. Locate the final image produced by two or more lenses using any one of the three methods described (two analytical methods, one graphical method).

SUMMARY

A simple geometrical construction introduced by Christian Huygens allows one to predict the successive positions of a wave front. Huygens' principle states that a new position of a wave front is the envelope of small wavelets originating at the old wave front. A consistent application of this simple idea predicts both the straight-line propagation of light ordinarily observed and the bending of light (or other wave motion) as it passes through an opening that is small compared to the wavelength. Huygens' construction also predicts the correct relationship between the angles of incidence and refraction when light travels from one medium into another. This relationship, discovered experimentally by Willebrord Snell, is written $n_1 \sin\theta_1 = n_2 \sin\theta_2$ where θ_1 and θ_2 are the angles of incidence and refraction. The constants n_1 and n_2, called indices of refraction, are properties of the two transparent media. The index of refraction of a transparent material is the speed of light in vacuum divided by the speed of light in the medium. Symbolically, $n = c/c_n$. The index of refraction is a function of the frequency of the light. For this reason, the amount of bending as light passes from one medium to another will be different for light of different colors. The effect is small enough to be neglected in an approximate treatment of thin lenses.

When light traveling in a medium of index n_1 is incident upon a second medium of smaller index ($n_2 < n_1$), there is the possibility of *total* reflection. Total reflection of the incident light will take place if the angle of incidence is equal to or greater than the angle θ_c satisfying the equation, $\sin\theta_c = n_2/n_1$. Total reflection is "perfect" reflection, perfect in the sense that no energy is lost.

Lenses are sometimes described in terms of what they do to parallel light that is incident on them. A *converging* lens causes parallel rays to converge after passage through the lens. A *diverging* lens has the opposite effect, the rays spread out, or diverge, after transmission through the lens. Both types of lenses are used in optical instruments but the converging, or *positive*, lens is more common.

The principal use of a lens is to form an image of some object. The lens of your eye, for example, forms an inverted image on the retina of the printing on this page.

The relationship between object and image distances is given by the lens equation

$$\frac{1}{p} + \frac{1}{q} = \frac{1}{f}$$

where p and q are the distances of object and image from the lens. The constant f is the focal length of the lens, the image distance for an object at infinity.

Mirrors can also be used to form images and, with the proper definition of focal length ($f = R/2$), the relationship between object and image distances is the same as that for lenses.

QUESTIONS AND PROBLEMS

Secs. 25-1—25-5. You will need pencil, paper, and a ruler.

1. Fig. 25-1 represents a cross section of a wave on the surface of water at a particular instant. Points A and E are on the crests of adjacent waves.

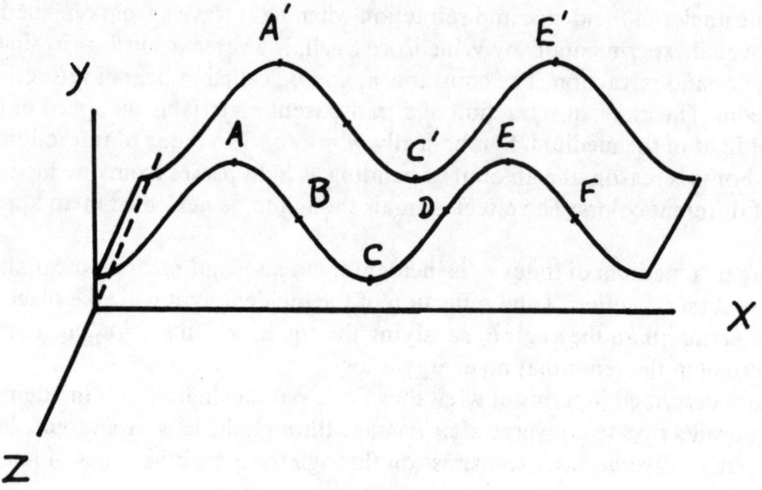

Fig. 25-1

a. What is the name and symbol for the distance from A to E?

b. The points A through F are at different positions in space; they each have a different x-coordinate. They also may be said to differ in phase. For example, the phase difference between points A and E is 2π radians (360°). What is the phase difference between points A and C? A and B?

c. Do points A and B lie on the same wave front? Can you locate a point on the same wave front as A?

d. Write the definition of *wave front*.

2. Fig. 25-2 represents an aerial view of a succession of water waves striking a breakwater containing an opening.

   a. Label the wave fronts on both sides of the obstacle. Label the rays also. What is the angle between the rays and the wave fronts?

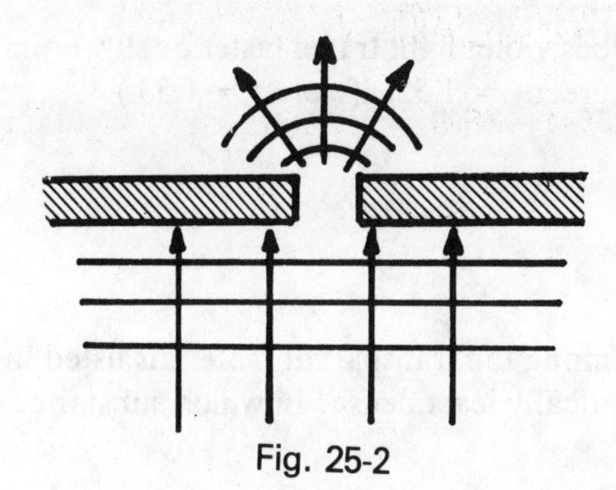

Fig. 25-2

   b. The spreading out of the water waves beyond the opening in the breakwater indicates that the wavelength of the water waves is ______________ compared with the 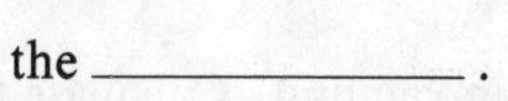 .

3. The solid curve in Fig. 25-3 indicates the position of an expanding wave front at an instant of time. Use Huygens' construction to show a later position of the wave front. Add some arrows to represent the direction of energy flow.

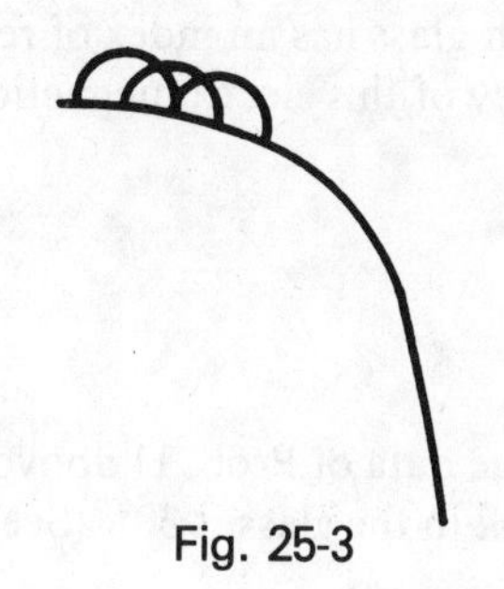

Fig. 25-3

4. Name two or three media in which light is propagated *isotropically*.

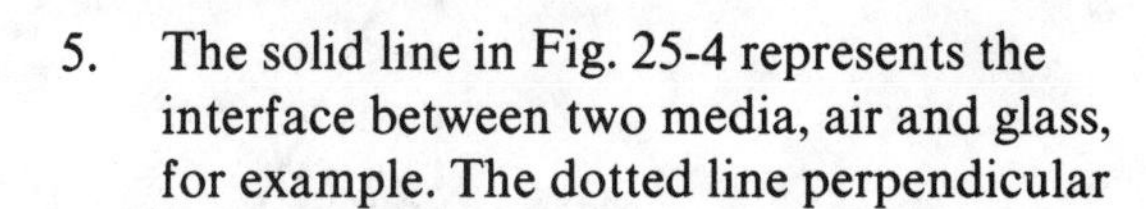

5. The solid line in Fig. 25-4 represents the interface between two media, air and glass, for example. The dotted line perpendicular to the surface is called a n______________ . Draw a line with an arrow on it to represent a ray of light incident on the surface at an angle of incidence of 30°. Have this ray strike the interface at P. Mark the 30° angle on the figure. Draw an outgoing arrow to represent the reflected ray.

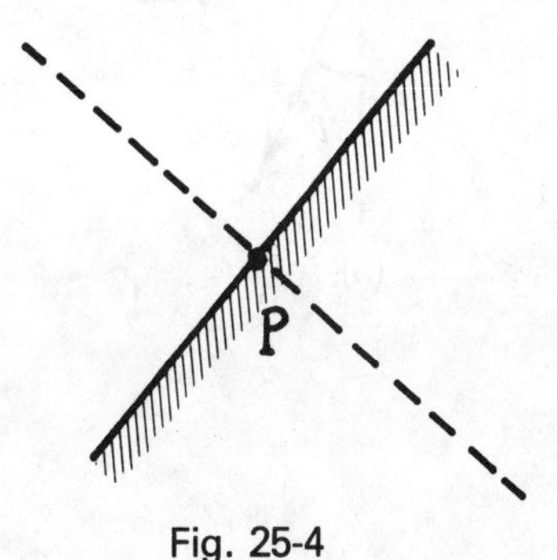

Fig. 25-4

6. State the two parts of the law of reflection.

   a. The incident ray, the reflected ray, and the normal to the surface all 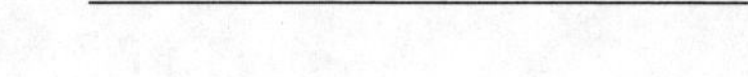 .

   b. The angle of reflection equals ______________________________ .

7. What is the speed of green light in ordinary window glass? ($n \cong 1.52$)

$c_n =$ ______________ .

8. Does violet light travel faster or slower in glass than green light? By what percentage? (Use the values n(green) = 1.52, n(violet) = 1.53.)

9. Among the transparent materials listed in Table 25-2 in the text, which is optically most dense? Which is optically least dense? In which substance does light travel at a speed of 2.25×10^8 m/s?

10. When a ray of yellow light enters a piece of glass its speed changes and its direction usually changes but its ______________ never changes.

11. A certain glass has an index of refraction of 1.60 for red light ($\lambda \cong 700$ nm in vacuum). Calculate the frequency of this electromagnetic wave, its speed of propagation in the glass, and its wavelength in glass.

12. Using the data of Prob. 11 above, calculate the angle of incidence of a ray of red light whose angle of *refraction* in the glass is 30°. (See Fig. 25-5a.)

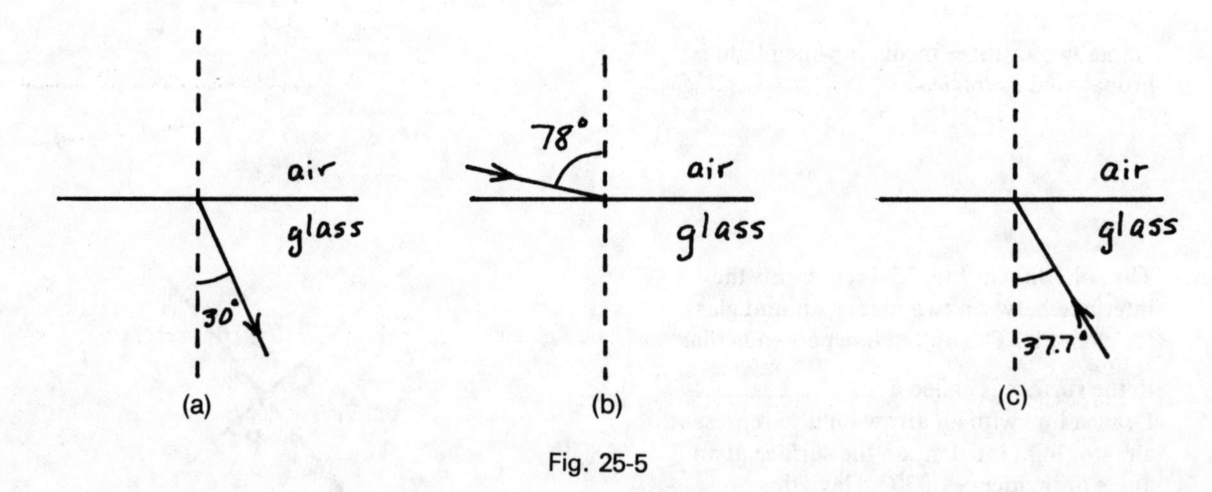

Fig. 25-5

13. Calculate the angle of refraction of a ray of red light whose angle of incidence (in air) is 78°. (Sin 78° = 0.98) (See Fig. 25-5b.)

14. Now suppose that the ray of red light traveling in glass is reversed so that it is incident on the glass-air interface from below at an angle of 37.7°. What will be the angle of refraction when this reversed ray enters the air? (See Fig. 25-5c.)

15. If a ray of red light traveling in glass ($n = 1.60$) is incident on a glass-air interface at an angle of 39°, what will happen? (Note that $1.6 \times \sin 39° = 1.007$.) What fraction of the light will be reflected?

16. What is the critical angle for a water-air interface? (Use $n = 1.33$.)

17. In the two drawings in Fig. 25-6, rays of light parallel to the axis are incident on a lens. The subsequent paths of some of the rays are shown.

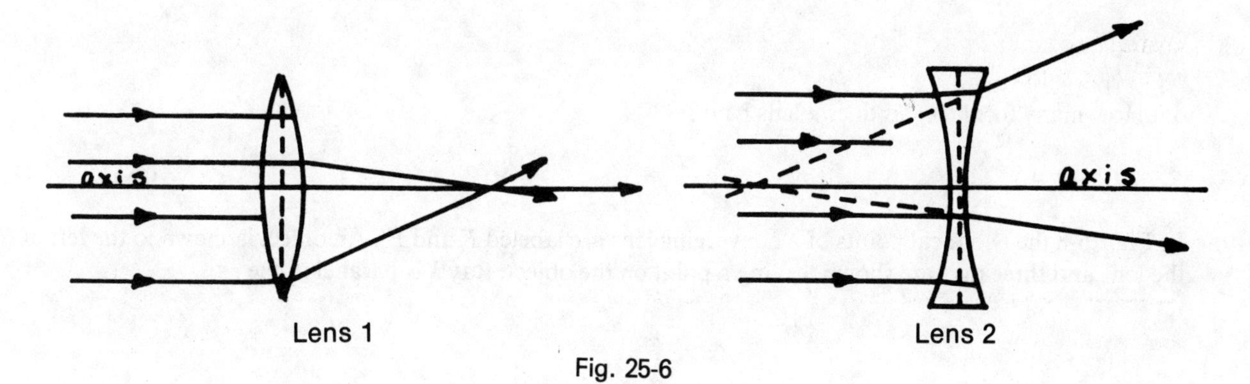

Fig. 25-6

a. Use a ruler to complete the drawings by showing the paths of the remaining rays. Represent a ray of light by a solid line and its geometrical extension by a dashed line.
b. Mark the principal focus of each lens by the letter "F."
c. Measure the focal length of each lens in centimeters.

$f_1 =$ ______________. $f_2 =$ ______________.

d. Draw a vertical line on each figure to represent the focal plane of that lens.

18. Fig. 25-7 shows a small object in front of a positive, or converging, lens. A few rays of light are shown leaving the object and passing through the lens. Complete the drawing by continuing all rays through the lens.

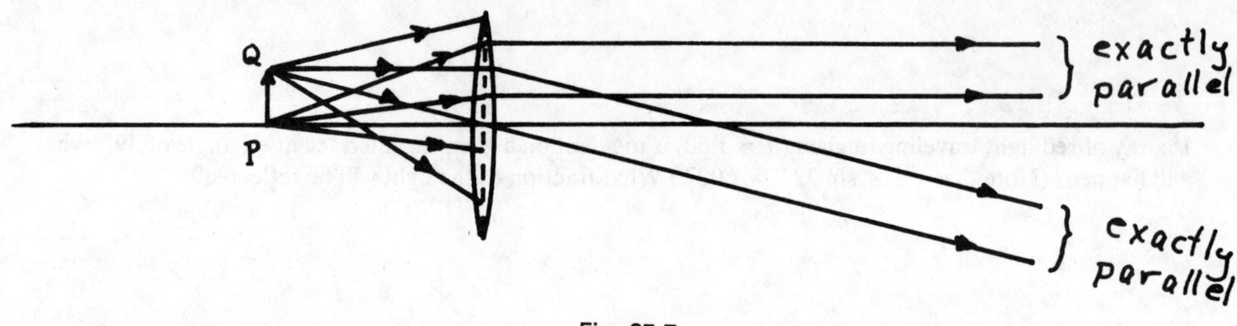

Fig. 25-7

a. Where is this object located?

b. What is the focal length of the lens? (Measure it with a ruler.)

$f =$ __________.

c. How many focal points does a lens have?

d. How many focal *lengths* does a lens have?

19. In Fig. 25-8 the two focal points of a converging lens are labeled F and F'. An object is shown to the left of the lens and three rays are shown leaving a point on the object. Ray 1 is parallel to the axis.

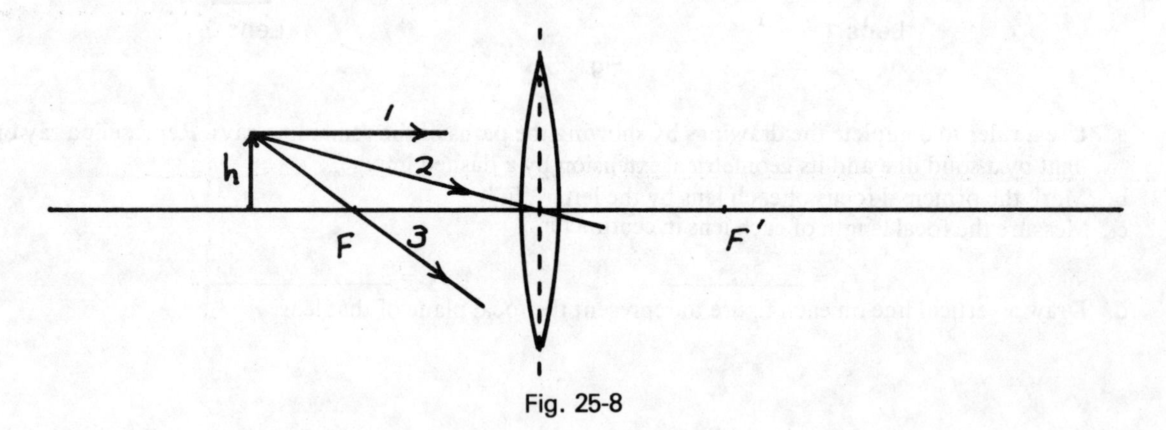

Fig. 25-8

a. Continue each ray through the lens until all three of them intersect in a common point. Draw an arrow to represent the image. Let its height be h'.
b. Carefully measure object and image distances, focal length, and heights of object and image. Record all distance in cm.

$p =$ __________. $q =$ __________. $f =$ __________.

$h =$ __________. $h' =$ __________.

c. State the nature of the image. (Real or virtual? Erect or inverted?)

d. Calculate the magnification of the system. Use the correct algebraic sign.

$m =$ ______________.

e. Taking the object distance and the focal length as known quantities, calculate the image distance.

$q =$ ______________.

f. Taking the object distance as a known quantity along with p and f, calculate the magnification.

$m =$ ______________.

g. Compare the results obtained geometrically (parts b and d) with those obtained analytically (parts e and f). Do they agree within 5%?

* * * * * * *

The next problem, illustrated in Fig. 25-9, will be more difficult because it involves a virtual image. Be careful to distinguish between real rays of light (solid lines) and mathematical extensions of rays, which represent only *apparent* paths of light (dashed lines). Two rays are enough to locate the image, a third confirms the correctness of the result. The easiest ray to draw is the one passing through the center of the lens. It is never deviated. The given quantities are: $f = 9$ cm, $p = 4$ cm, and $h = 1.2$ cm.

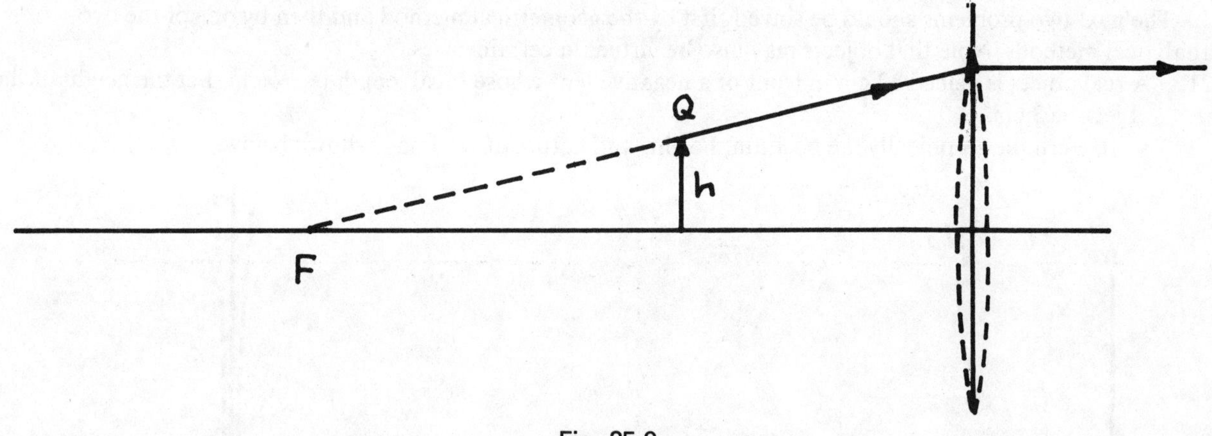

Fig. 25-9

20a. Locate the image with two rays. Confirm it with a third. Measure the relevant distances.

$q =$ ______________. $h' =$ ______________.

b. Calculate the magnification using the measured value of h'.

c. Describe completely the nature of the image.

d. Calculate the power of this lens.

power = ______________.

e. Suppose that the lens is made of glass having an index of 1.5 and that one surface is plane ($R_1 = \infty$). What is the radius of curvature of the second surface?

f. Determine the original curvature with its correct algebraic sign.

original curvature = ______________.

g. Calculate the final curvature from the relationship,

$$\{\text{final curvature}\} = \{\text{original curvature}\} + \{\text{power of lens}\}$$

What is the physical significance of the algebraic sign?

Although the light rays are actually refracted at each surface of a lens, it is sufficient to show the rays bending just once at an imaginary surface through the center of the lens. This follows from the thin-lens approximation that the thickness of the lens is negligibly small compared to the distances f, p, and q.

* * * * * * *

The next two problems should be solved first by the geometrical method and then by one of the two analytical methods. Note that objects may also be virtual in certain cases.

21. A real object is placed 12 cm in front of a negative lens whose focal length is -6 cm. Let the height of the object be 3 cm.

a. Determine graphically the position, height, and nature of the image shown below.

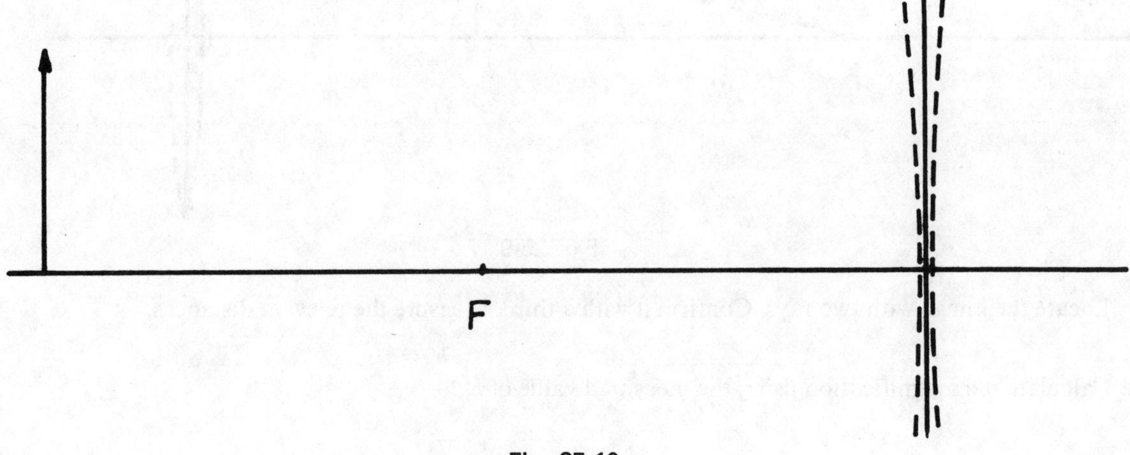

Fig. 25-10

q = ______________. h' = ______________. m = ______________.

b. Now find q and h' by one of the two analytical methods.

$q =$ ______________ . $h' =$ ______________ .

22. Using the same negative lens, the object is now placed only 5 cm to the left of the lens, as shown below. Double the scale to improve the accuracy. Let $h = 3$ cm.

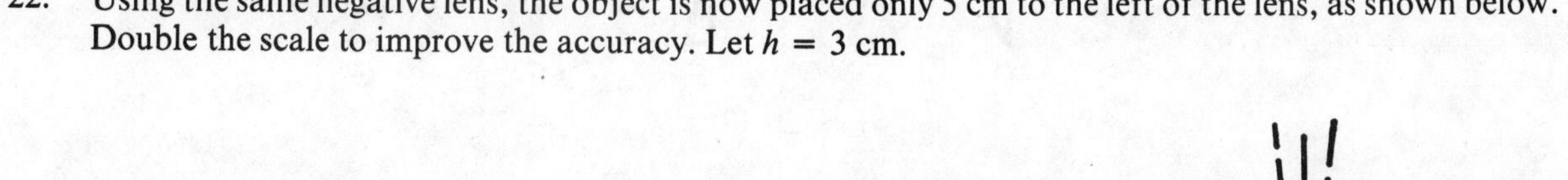

Fig. 25-11

$q =$ ______________ . $h' =$ ______________ . $m =$ ______________ .

The image is ____________ (real/virtual), ____________ (erect/inverted), ____________ (enlarged/reduced).

23. The object for the negative lens is now 4 cm to the *right* of the lens. This is necessarily a virtual object. It is produced by a second lens too far to the left to be shown in the figure below.

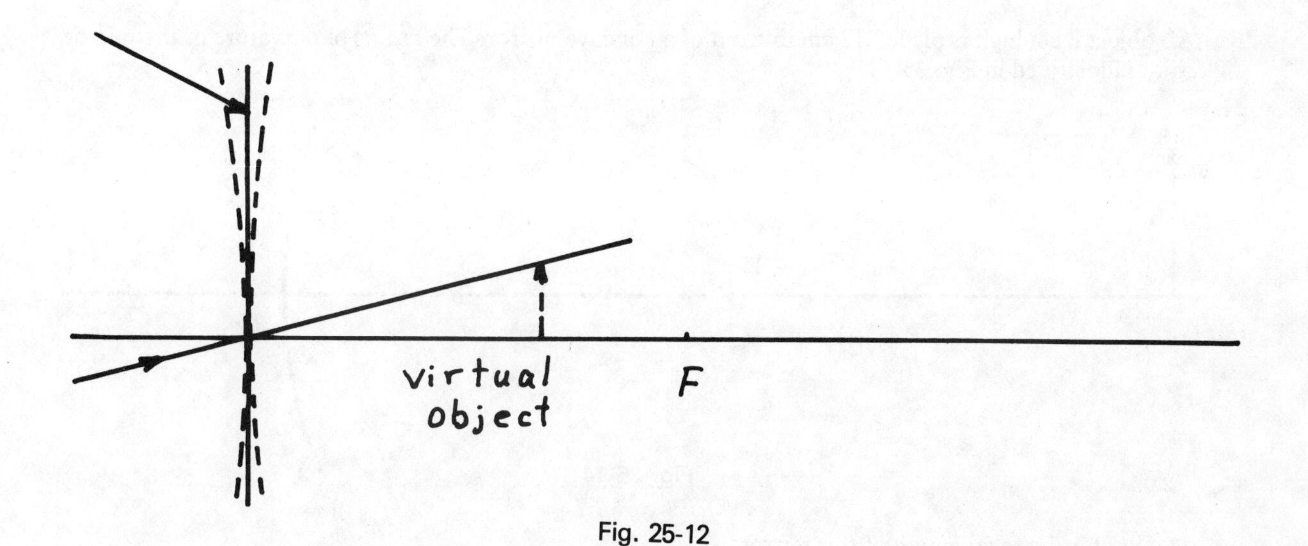

Fig. 25-12

Let the object height be 1 cm. Where is the image located? It is ______________ cm to the ______________ (right/left) of the lens. The image is ______________ (erect/inverted) and ______________ (real/virtual). Its height is ______________ .

24. A group of lenses is shown in Fig. 25-13. All are made of the same glass.

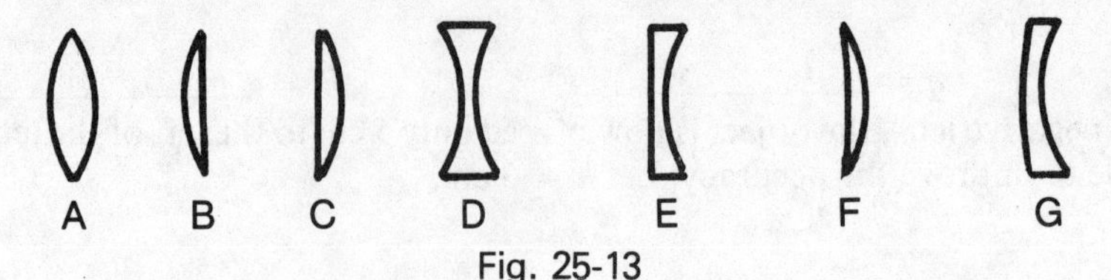

Fig. 25-13

a. Which of these lenses have a positive power?

b. Which lenses are plano-convex?

c. Which lenses are negative?

d. Which lens is plano-concave?

e. Which lens has the shortest positive focal length?

f. Which lens has the shortest negative focal length?

25. A slide projector has a projection lens of focal length 6 cm. It is to be used to project an image on a screen that is 4 m from the lens.

a. What distance is required between the slide and the lens?

b. If the height of the slide is 3 cm, what is the height of the image?

26. An object 2 cm high is placed 12 cm in front of a concave mirror. The radius of curvature of the mirror is 6 cm, as illustrated in Fig. 25-14.

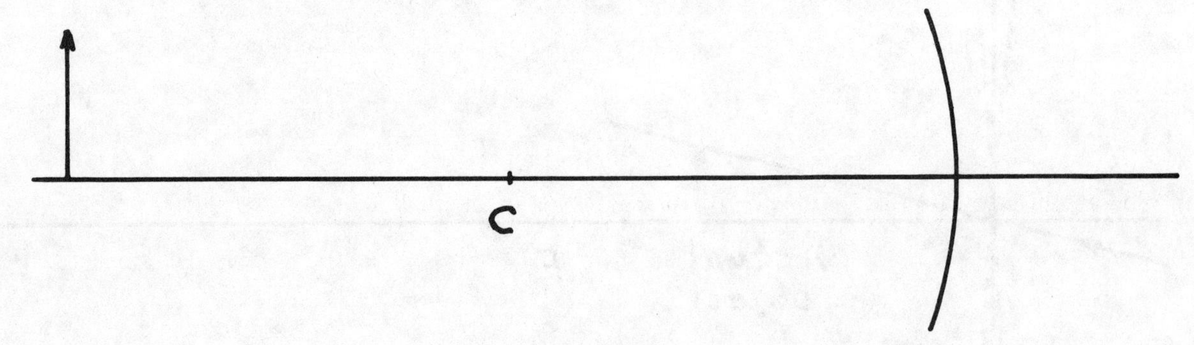

Fig. 25-14

a. What is the focal length of this mirror?

b. Complete the drawing so as to locate the image and determine its character. State the location of the image. (Hint: A ray that passes through the center of curvature will strike the mirror at exactly 90° and will reverse its direction.)

c. Is the image real or virtual?

d. Is it erect or inverted?

e. What is the height of the image?

27. If an image is formed behind a mirror, is that image necessarily virtual? Explain.

28. A convex mirror is used to form an image of a real object. The object is located 8 cm in front of the mirror, illustrated below. The mirror has a focal length whose magnitude is 6 cm. Let the height of the object be 2 cm.

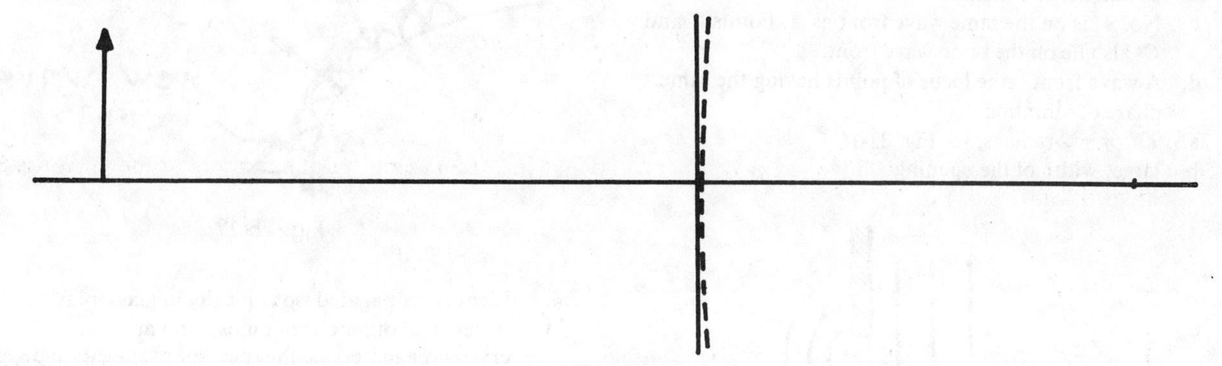

Fig. 25-15

Where is the image located? Is it real or virtual? Is it erect or inverted? What is its height?

29. Two thin, positive lenses are placed in contact. The distance between them as well as their thicknesses can be neglected. The two focal lengths are 6 cm and 4 cm. A real object, 2 cm high, is located 12 cm to the left of the lenses. Find the final image and describe its nature. [Hint: Solve the problem in two steps. First locate the image formed by the 6 cm (16.7 diopter) lens as if the other were not present. Then use this image as the virtual object for the 4 cm (25 diopter) lens.]
For the 6 cm lens alone:

$q_1 =$ ______________. $h'_1 =$ ______________.

For both lenses together:

$q =$ ______________. $h' =$ ______________.

The final image is ______________ (real/virtual) and ______________ (erect/inverted).

30. Suppose that the two lenses of Prob. 29 above are replaced by a a single lens chosen in such a way that the image is in the same position and has the same height as before.
    a. What is the required focal length?

    $f =$ ______________.

    b. How is the power of this single lens related to the power of the individual lenses in contact? Write the relationship.

Solutions

1a. wavelength; λ
b. π radians; $\pi/2$ radians
c. No. A' is on the same wave front as A. Points C and C' also lie on the same wave front.
d. A wave front is the locus of points having the same phase of vibration.

2a. 90° or $\pi/2$ radians; see Fig. 25-16
b. large; width of the opening

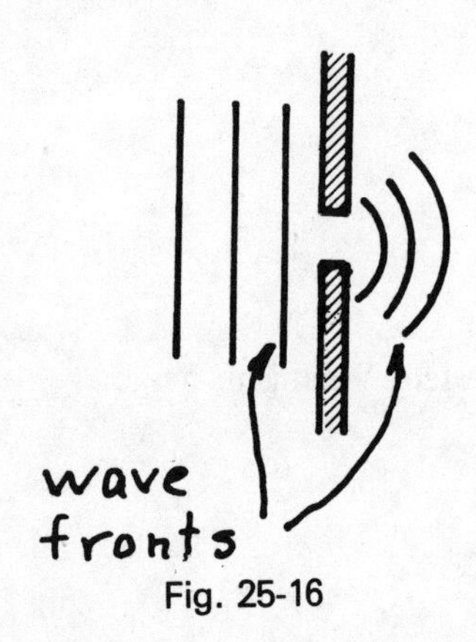

Fig. 25-16

3. The arrows in Fig. 25-17 indicate the direction of the energy flow. The new wave front is the surface that is tangent to the individual wavelets.

Fig. 25-17

4. Light is propagated isotropically in glass, plastics, water (and other clear liquids), and any gas. In crystalline materials, the speed of propagation depends on the direction.

5. normal
The angle of reflection is 30°, the same as the angle of incidence, as shown below.

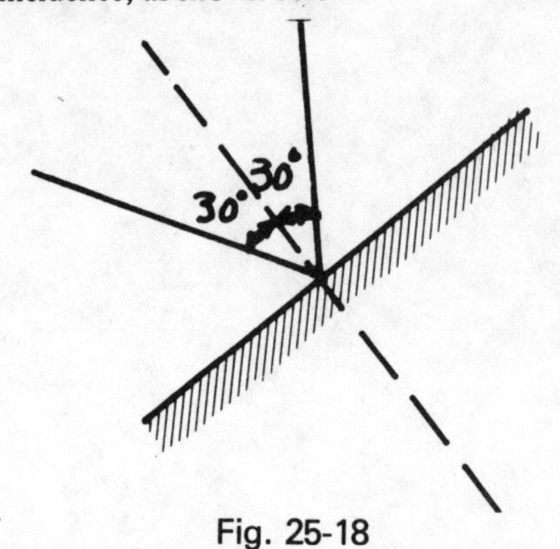

Fig. 25-18

6a. all lie in the same plane

b. equals the angle of incidence

7. $c_n = \frac{c}{n} = \frac{3.00 \times 10^8 \text{ m/s}}{1.52} = 1.97 \times 10^8 \text{ m/s}$

8. For green light, $c_n = \frac{c}{1.52} = 1.97 \times 10^8 \text{ m/s}$

For violet light, $c_n = \frac{c}{1.53} = 1.96 \times 10^8 \text{ m/s}$

Green light travels faster. The percentage difference is about 0.5%.

9. Diamond, with an index of refraction of 2.42, is optically most dense. The least dense material listed is hydrogen.

$n = \frac{c}{c_n} = (3 \times 10^8 \text{ m/s})/(2.25 \times 10^8 \text{ m/s}) = 1.33$, the index of refraction of water.

10. frequency

11. $f = \frac{c}{\lambda} = \frac{3.00 \times 10^8 \text{ m/s}}{700 \times 10^{-9} \text{ m}} = 4.29 \times 10^{14} \text{ Hz}$

$c_n = \frac{c}{n} = \frac{3.00 \times 10^8 \text{ m/s}}{1.60} = 1.88 \times 10^8 \text{ m/s}$

$\lambda_n = \frac{c_n}{f} = \frac{1.88 \times 10^8 \text{ m/s}}{4.29 \times 10^{14} \text{ Hz}} = 438 \text{ nm}$

The wavelength in glass can also be found as follows:

$\lambda_n = \frac{\lambda}{n} = \frac{7.00 \times 10^{-7} \text{ m}}{1.60} = 4.38 \times 10^{-7} \text{ m}$

12. $n_1 \sin\theta_1 = n_2 \sin\theta_2; \theta_2 = 30°, n_2 = 1.60,$
$n_1 = 1.00$
$\sin\theta_1 = (n_2/n_1)\sin\theta_2 = (1.60)(\sin 30°) = 0.80$
$\theta_1 = \sin^{-1}(0.80) = 53°$

13. $n_1 \sin\theta_1 = n_2 \sin\theta_2;\ \theta_1 = 78°, n_2 = 1.60$
$\sin\theta_2 = (n_1/n_2)\sin\theta_1 = (1.00/1.60)\sin 78°$
$\sin\theta_2 = 0.98/1.60 = 0.61$
$\theta_2 = \sin^{-1}(0.61) = 37.7°$

14. The ray will exactly reverse its path. Thus, the angle of refraction (in air) is 78° (see Fig. 25-19).

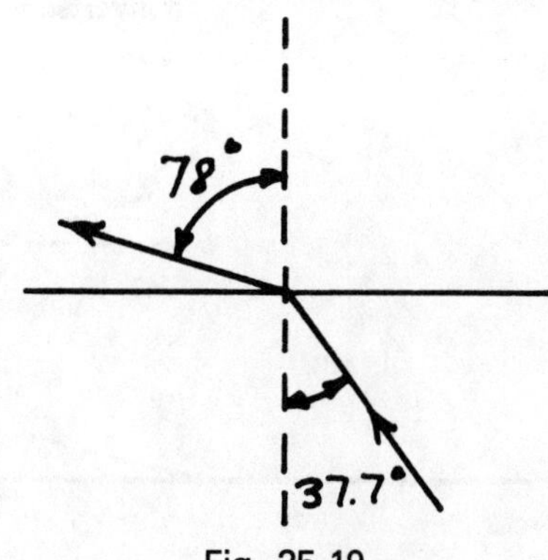

Fig. 25-19

15. There is *no* angle that satisfies $\sin \theta_1 = 1.007$. Thus there is no refracted ray; *all* of the light is reflected at the glass-air interface. The critical angle may be found as follows:

$\sin \theta_c = (n_1/n_2)\sin 90° = (1/1.60) = 0.625$
$\theta_c = \sin^{-1}(0.625) = 38.7°$

16. The critical angle is the angle of incidence for which the angle of refraction is 90°.

$1.33 \sin \theta_c = (1.00) \sin 90°$
$\sin \theta_c = 0.75; \theta_c = 48.6°$

17a., b. See Fig. 25-20. (Measurements are made on Fig. 25-6.)

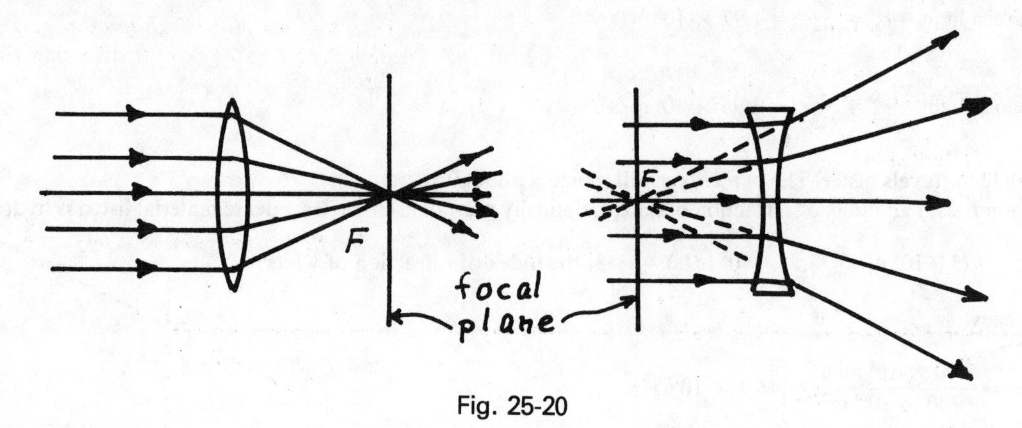

Fig. 25-20

c. $f_1 = 2.9$ cm $f_2 = -2.9$ cm
d. See Fig. 25-20.

18. All rays leaving the point P become parallel to the optic axis after passage through the lens. All rays from point Q leave the lens in the same direction as the ray that passes through the center of the lens.
   a. In the focal plane.
   b. $f = 3$ cm
   c. Two, F and F', on opposite sides of the lens.
   d. One focal length, f.

19a. Your completed figure should look like Fig. 25-17 in the text.
   b. $p = 4.0$ cm $q \cong 6.7$ cm $f \cong 2.5$ cm
   $h = 1.0$ cm $h' \cong 1.7$ cm
   c. real, inverted
   d. $m = -\frac{h'}{h} = -\frac{q}{p} = -1.7$ (Negative sign indicates an inverted image.)
   e. $\frac{1}{q} = \frac{1}{f} - \frac{1}{p} = \frac{1}{2.5 \text{ cm}} - \frac{1}{4.0 \text{ cm}} = 0.15 \text{ cm}^{-1}$
   $q = 6.67$ cm
   f. $m = -\frac{q}{p} = -\frac{6.7 \text{ cm}}{4.0 \text{ cm}} = -1.67$
   g. Carefully made drawings should give results accurate to 5 or 10%.

20.

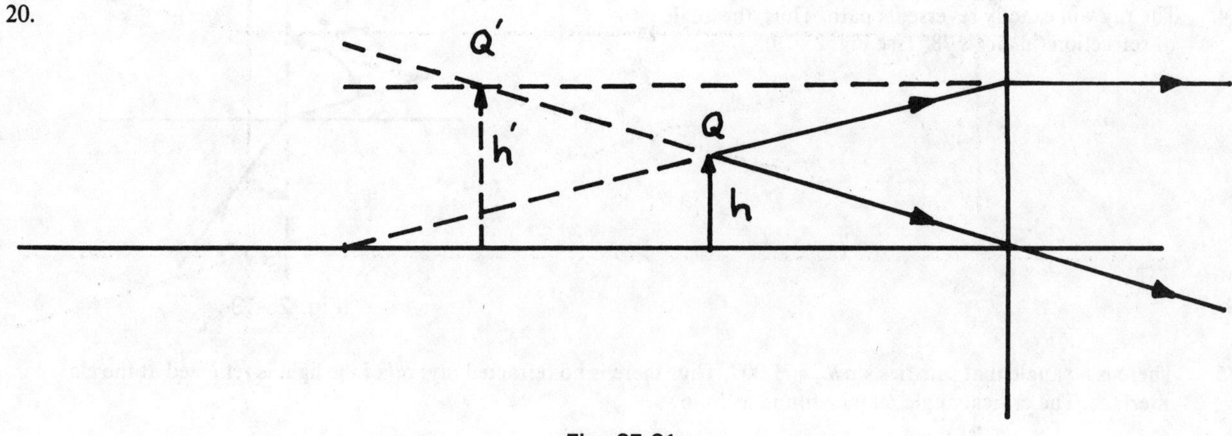

Fig. 25-21

a. Rays from Q diverge after leaving the lens, they cannot form a real image. To an observer on the right, looking into the lens, the emerging rays *appear* to have come from point Q'. Thus the virtual image is located at Q'.
$q \cong -7.2$ cm $\quad h' \cong +2.2$ cm

b. $h' \cong +2.2$ cm; $m \cong +1.8$

c. The image is erect, virtual, and magnified.

d. $\text{power} = \frac{1}{f} = \frac{1}{0.09\text{ m}} = 11.1$ diopters (or 11.1 m^{-1})

e. $$\frac{1}{f} = (n-1)\left(\frac{1}{R_1} + \frac{1}{R_2}\right) = (0.5)\left(0 + \frac{1}{R}\right) = \frac{1}{2R}$$

$$R = \frac{f}{2} = \frac{9\text{ cm}}{2} = 4.5\text{ cm}$$

f. {curvature of wave front} $= -1/p = -1/(0.04\text{ m})$
$= -25$ diopters

g. {final curvature} $= -25\text{ m}^{-1} + 11.1\text{ m}^{-1} = -13.9\text{ m}^{-1}$

The final curvature of the wave front is negative, indicating that the rays are diverging after passing through the lens.

21a.

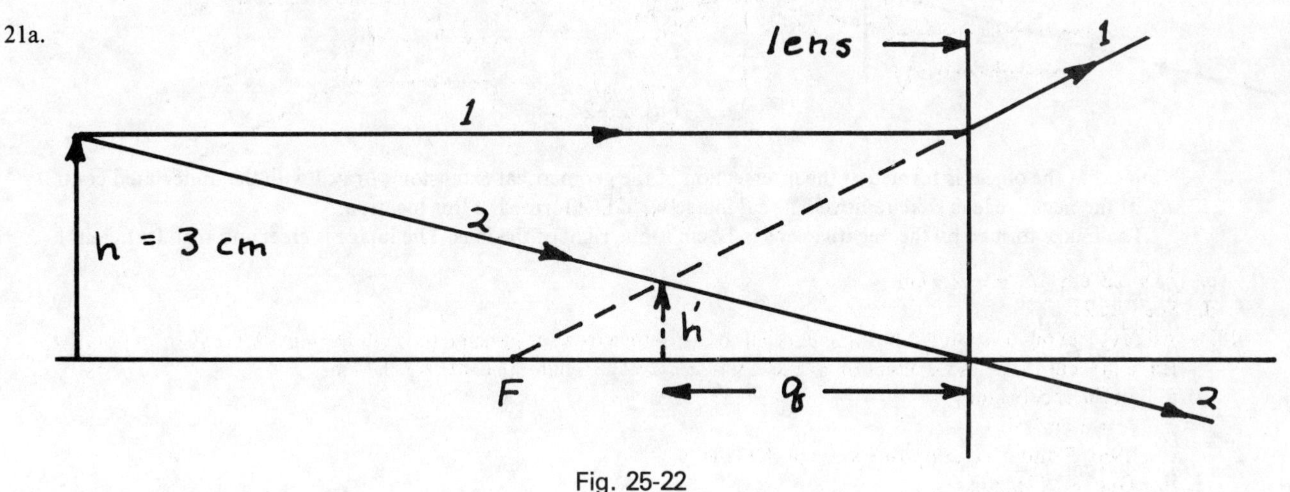

Fig. 25-22

Ray 1, originally parallel to the axis, emerges from the lens in such a direction as to *appear* to have passed through the focal point F. Ray 2 passes through the center of the lens and is undeviated. Its apparent path to the left of the lens coincides with its real path.

$q \cong -4.2$ cm $\quad h' = 1$ cm $\quad m = +1/3$

The image is virtual, erect, and reduced

b. $q = -4.0$ cm; $h' = 1.0$ cm

22.

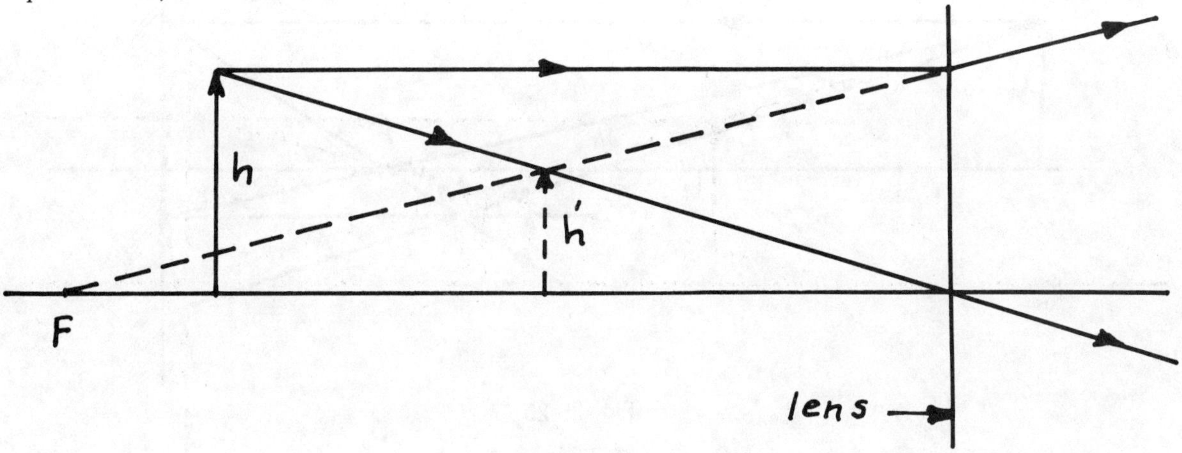

Fig. 25-23

$q \cong -(1/2)5.6\text{ cm} = -2.8$ cm; $h' \cong 1.6$ cm;
$m = 0.54$

The image is virtual, erect, and reduced.

23.

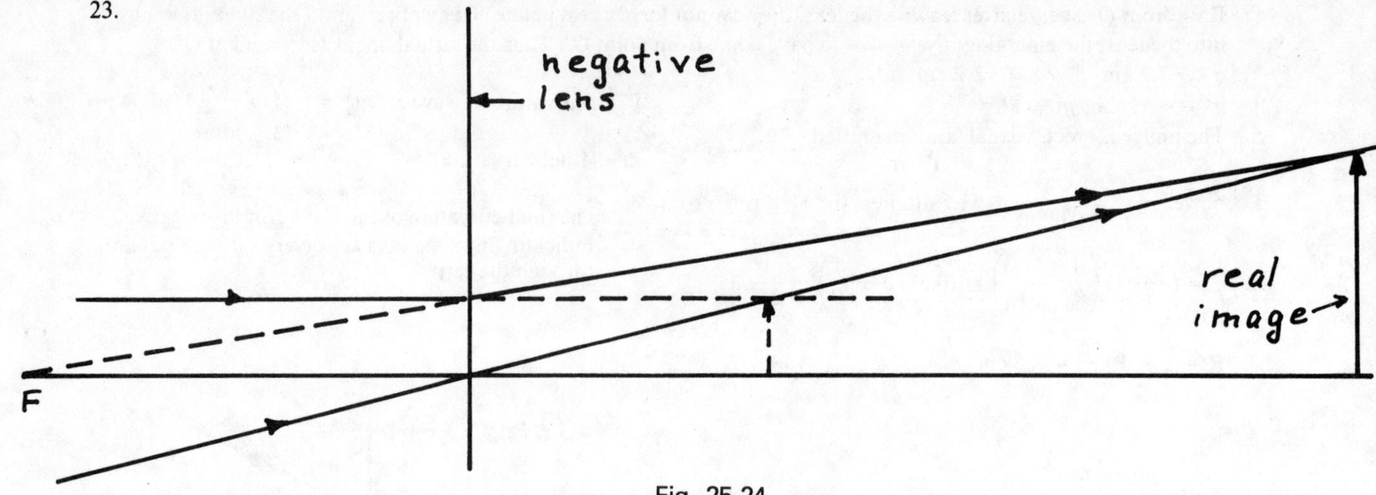

Fig. 25-24

Notice that the object is formed at the intersection of the geometrical extension of ray 1 with the undeviated central ray. If the negative lens were removed, a real image would be formed at this location.

The image formed by the negative lens is 12 cm to the right of the lens. The image is erect and real. Its height is 3 cm.

24a. *A, B, C,* and *F*
b. *B* and *C*
c. *D, E,* and *G*
d. *E*
e. *A*
f. *D*

25a. $q = 400$ cm; $f = 6$ cm

$$\frac{1}{p} = \frac{1}{f} - \frac{1}{q} = \frac{1}{6\text{ cm}} - \frac{1}{400\text{ cm}}; \; p = 6.09\text{ cm}$$

b. $m = -q/p = -400/6.09 = -65.7$
$h' = mh = (65.7)(3\text{ cm}) = 197\text{ cm}$

26.

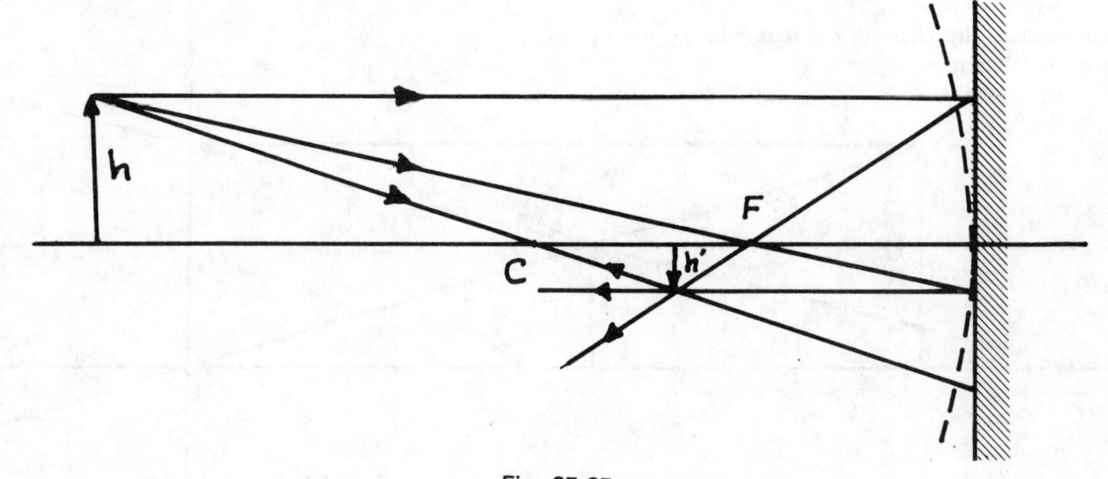

Fig. 25-25

a. $f = (1/2)R = 3$ cm
b. Three rays are shown. Only two are required to locate the image.
c. real
d. inverted
e. $h' = 0.7$ cm

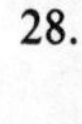

27. Yes. Since there are no light rays behind the mirror, no real image can be formed there.

28.

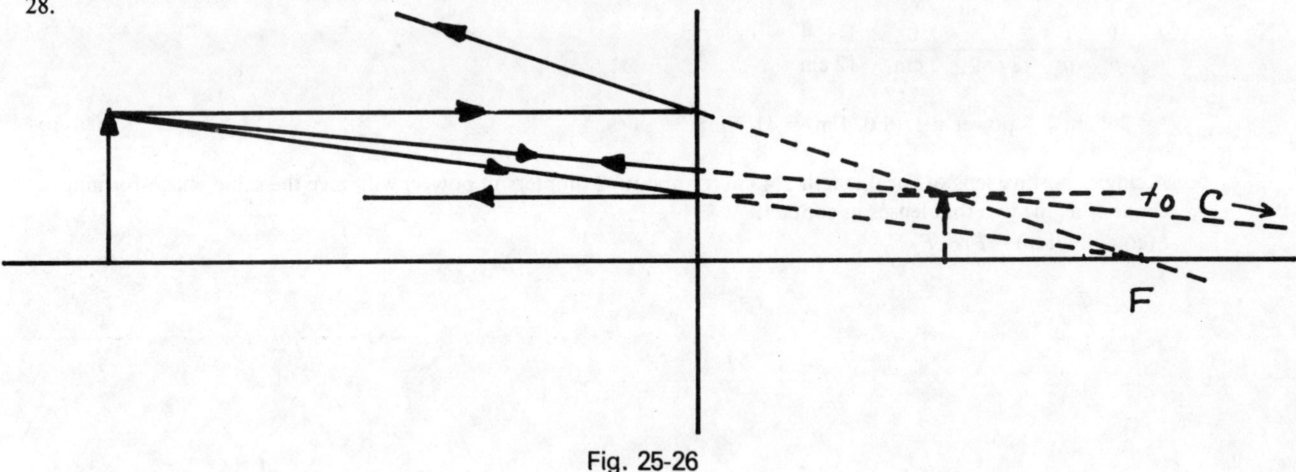

Fig. 25-26

The image is located 3.43 cm to the right of the mirror. It is virtual, erect, and 0.86 cm high.

29. We begin with the 6-cm lens. The other is assumed to be absent.

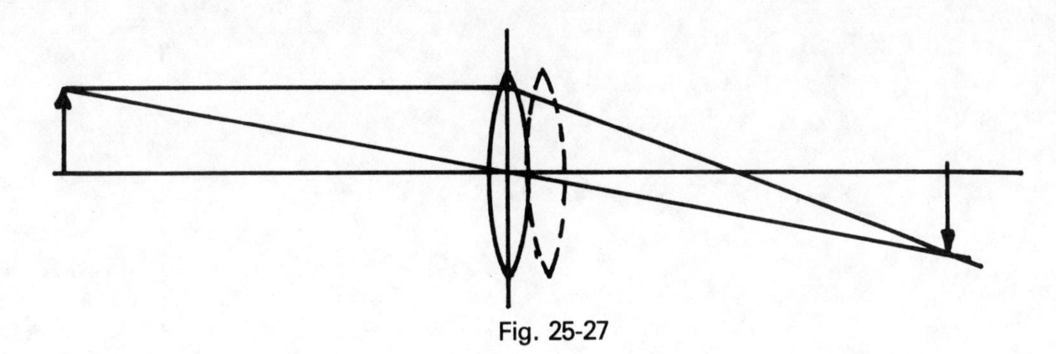

Fig. 25-27

$$\frac{1}{q} = \frac{1}{f} - \frac{1}{p} = \frac{1}{6\text{ cm}} - \frac{1}{12\text{ cm}} = \frac{1}{12\text{ cm}};\ q = 12\text{ cm} \qquad m = -\frac{q}{p} = -1 \qquad h'_1 = -2\text{ cm}$$

Now add the 4-cm lens. The object is virtual and is situated 12 cm to the right of the lens, i.e., p is negative

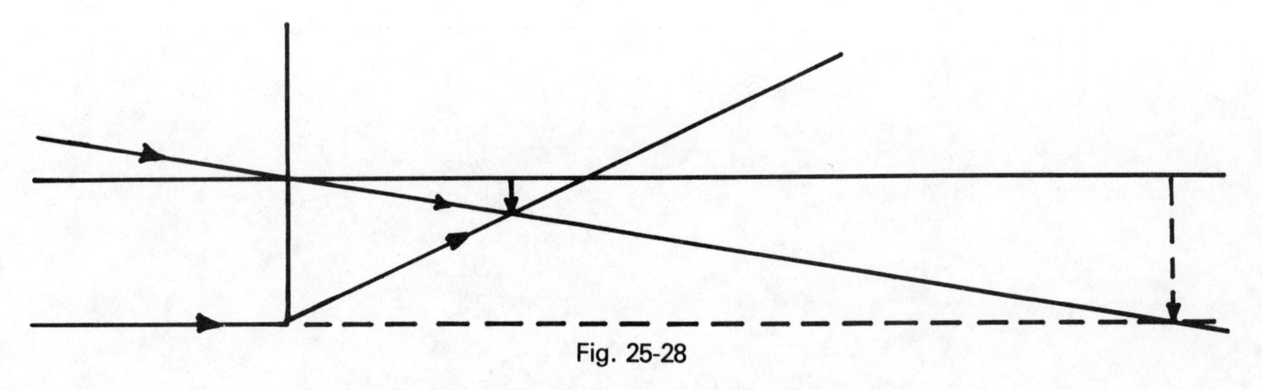

Fig. 25-28

The graphical solution gives a real, inverted image that is 3 cm to the right of the lens.

$$\frac{1}{q} = \frac{1}{f} - \frac{1}{p} = \frac{1}{4\text{ cm}} - \frac{1}{-12\text{ cm}} = \frac{3+1}{12\text{ cm}} = \frac{4}{12\text{ cm}} = \frac{1}{3\text{ cm}};\ q = 3\text{ cm}, \qquad m = -\frac{q}{p} = -\frac{3\text{ cm}}{-12\text{ cm}} = +\frac{1}{4}$$

Thus the overall magnification of the system is $m = (-1)(+1/4) = -1/4$. The height of the final image is $h' = mh = (-1/4)2\text{ cm} = -0.5\text{ cm}$.

30a. This solution is most easily found analytically, rather than graphically.

$$\frac{1}{f} = \frac{1}{p} + \frac{1}{q} = \frac{1}{12\ \text{cm}} + \frac{1}{3\ \text{cm}} = \frac{1 + 4}{12\ \text{cm}}$$

$f = 2.4$ cm power $= 1/(0.024\ \text{m}) = 41.7\ \text{m}^{-1}$

b. A single, positive lens of focal length 2.4 cm (or about 42 diopters in power) will have the same image-forming capacity as the two thin lenses in contact.
$P(\text{combination}) = P_1 + P_2$

CHAPTER 26 Wave Optics

GOALS To study the wave nature of light and to investigate those optical phenomena that can only be understood as arising from the wavelike behavior of light.

OBJECTIVES After completing this chapter the student should be able to do the following:

1. State the general condition for constructive interference between two or more light waves at a point in space.
2. Write the equation for the maxima in the interference pattern for two or more equally spaced slits.
3. Solve for any one of the quantities n, λ, a, or θ when the others are given.
4. Determine angular or linear separations of interference fringes from two or more equally spaced slits.
5. Determine the angular separation of two wavelengths produced by a diffraction grating.
6. Describe qualitatively the difference between the interference pattern for two slits and that for many slits.
7. Write the general condition for a minimum of a single-slit diffraction pattern.
8. Calculate the angular separation and the linear separation between adjacent minima in the single-slit diffraction pattern.
9. Calculate the angular half-width of the central maximum of the single-slit diffraction pattern.
10. Calculate any one of the quantities, n, λ, w, or θ when the others are given.
11. Describe qualitatively the shadow cast by a circular obstacle when the source is very small.
12. Calculate the spacing of interference fringes observed when two flat surfaces of glass form a very thin air wedge.
13. Calculate the thickness of a soap film (or other thin film) from the colors of the reflected light and vice versa.
14. Draw a simplified sketch of a Michelson interferometer showing a beam splitter, two mirrors, and the interfering beams.
15. Calculate the wavelength of light from measurements of fringe shift with a Michelson interferometer.
16. Explain why the light emitted by a single atom is plane-polarized whereas the light obtained from a large group of atoms exhibits no preferred plane.
17. Write Brewster's law and use it to determine the angle of incidence for which the reflected light will be totally polarized.
18. State why light waves can be polarized but sound waves cannot.
19. State the meaning of the term *optical activity*.
20. Give the name of a process by which coherent light (as from a laser) is used to form an image without the use of a lens.

SUMMARY

It has been known from earliest times that light travels in straight lines through homogeneous media. Newton showed in his study of mechanics that small masses, free of external forces, would necessarily move in straight lines. It was natural, then, for Newton to postulate that a beam of light consists of many small particles moving at high speed. This *corpuscular* theory could even be made to "explain" the bending of light at the interface between two media by assuming the existence of a short-range force that attracts the corpuscules toward the more dense medium, thus causing them to bend, as observed, and speed up as well.

In the sixteenth and seventeenth centuries, several experimental arrangements were discovered in which light exhibited marked deviations from straight-line propagation. Thomas Young showed that these phenomena could be readily understood on the assumption that light consists of a wave motion propagating through a medium. The wave theory of light gave such a detailed and accurate description of interference and diffraction phenomena that by the middle of the nineteenth century Newton's corpuscular theory was almost entirely discredited. In 1850 Jean Foucault measured the speed of propagation of light in air and water and showed that light slows down when passing from air to water, in direct contradiction of Newton's corpuscular theory. This gave the *coup de grace* to the old corpuscular model of Newton.

When a beam of parallel light passes through an array of two or more narrow, closely spaced slits, it deviates markedly from its original direction of propagation. A screen placed beyond the slits will show a number of equally spaced fringes with regions of strong illumination alternating with regions of little or no illumination. If only two slits are present, the fringes will be broad. They become narrower as the number of slits is increased. Regardless of the number of slits (as long as it is at least two) the positions of maximum intensity will correspond to the angles θ that satisfy the equation $n\lambda = a \sin \theta$, where a is the separation of the slits and n is an integer called the *order of interference*.

When the number of slits is very large, say 1000 or more, the array of slits is called a *diffraction grating*. The narrowness of the transmitted fringes permits the angle of diffraction θ to be measured very accurately. Since a may also be very accurately known, the diffraction grating may be used to make accurate measurements of the wavelength of light.

When the number of slits is reduced to one, the pattern received on the screen, called a *diffraction pattern*, changes significantly. The single-slit diffraction pattern consists of a broad central fringe with narrower and much weaker fringes symmetrically spaced on either side. The angles at which the intensity *minima* occur are given by $n\lambda = w \sin \theta$ where w is the width of the single slit.

Interference phenomena result when two or more beams of light, originally obtained from a single source, are recombined. Interference effects can be used to measure small distances, test very small deviations from flatness of a glass plate, or measure the index of refraction of gases, to cite only a few of the many applications.

Although all types of wave motion exhibit interference effects, only transverse waves can be polarized. Since light can be polarized in a number of ways, including reflection, we have conclusive evidence that light consists of a transverse wave motion.

The plane of polarization of a beam of light can be rotated by passage through certain materials including, in particular, sugar solutions and certain transparent plastics under stress. The rotation of the plane of polarization is called *optical activity*.

QUESTIONS AND PROBLEMS

Secs. 26-1—26-6

1. One of the simplest ways to observe interference is to use a narrow light source to illuminate a pair of narrow, closely spaced slits, as shown below.

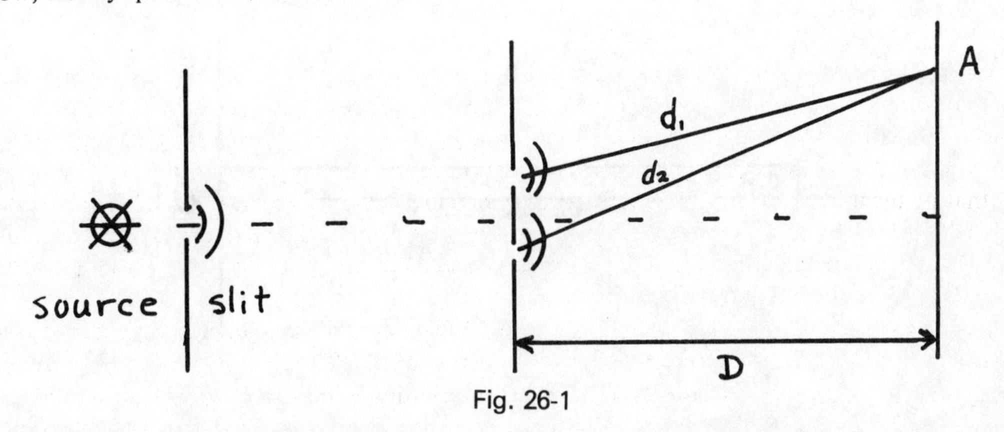

Fig. 26-1

Since the two slits are equally distant from the source slit, the waves that emerge from them are initially exactly in phase. The paths that the waves subsequently travel to reach the screen are marked d_1 and d_2.

a. How must d_1 and d_2 be related so that the waves will be in phase when they arrive at A on the screen? Give all possible cases.

b. For what values of $\Delta d = d_1 - d_2$ will the waves arrive 180° out of phase at the screen?

Δd = ______________ .

c. Suppose that the distance from the double slit to the screen is 50 cm and that the separation of the slits is 1 mm. For light of wavelength 500 nm, what will be the distance between the bright bands on the screen? (For small angles, $\sin\theta \cong \tan\theta \cong \theta$.)

Δy = ______________ .

d. What will be the angular separation in the 3rd order of wavelengths 500 nm and 600 nm? (Hint: Find the angular positions of the 3rd bright band for each wavelength and then subtract.)

2. Two slits 0.10 mm apart are illuminated by a plane wave front. Interference fringes are observed on a screen that is 80 cm from the slits. The 12th order fringe is 3.84 cm from the central image. What is the wavelength of the light used?

λ = ______________ .

3. In Fig. 26-2, a single, narrow slit is located at S, 2 m from a screen. The slit is 0.2 mm above the surface of a glass plate, which serves as a mirror. Light may reach the screen directly from S or by reflection from the mirror.

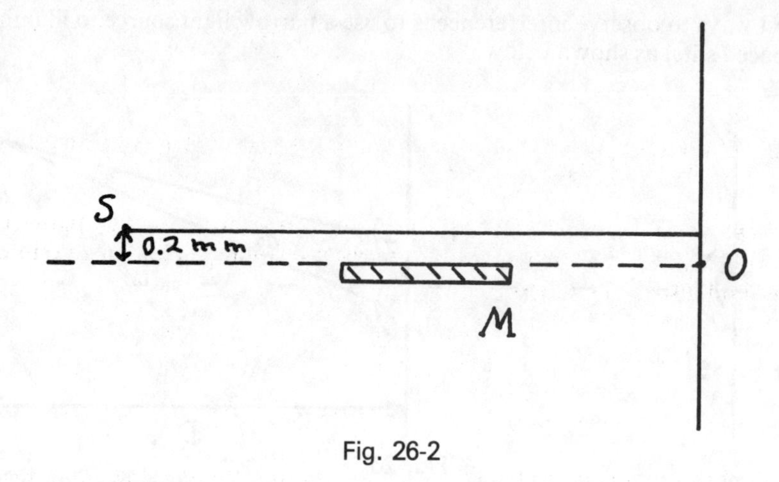

Fig. 26-2

a. The reflected light appears to come from a virtual image S'. Where is S' located?

b. Let 0 be a point on the screen that is exactly equidistant from S and S'. Taking into account the 180° change of phase that takes place when light is reflected from the glass surface, what is the resultant intensity at 0?

c. Calculate the distance from 0 to the 3rd bright band on one side. (Use $\lambda = 500$ nm.)

4. Fig. 26-16 in the text shows the interference pattern obtained from two slits and from a large number of slits. In both cases the separation of adjacent slits is the same. The vertical scales are vastly different.

a. If the bright fringes are 0.5 mm apart when two slits are used, how far apart will they be for 1000 slits? For 10 slits?

b. In what ways does the interference pattern change when the number of slits is increased above two?

5. In Fig. 26-21 in the text the intensity of light falling on a screen is plotted as a function of the path difference.

   a. This path difference is between which two rays?

   b. What is the resultant intensity in a single-slit diffraction pattern when the path difference is 3λ?

   I = ______________ .

   c. Calculate the angular width of the central fringe in the single-slit diffraction pattern when $\lambda = 600$ nm and $w = 0.02$ mm. (Note: The angular width is measured from the point of zero intensity on the left to the point of zero intensity on the right.)

   d. Since almost all of the energy is contained within the central fringe, the width of this fringe is a measure of the deviation of light from straight-line propagation. Consider a slit having a width of 0.05 mm, about the width of a human hair (see Fig. 26-3). When this slit is illuminated by plane wave fronts from a distant source, what angular deviation from rectilinear propagation will be observed? (Use the angular distance from the central maximum to the first minimum. Let $\lambda = 500$ nm.)

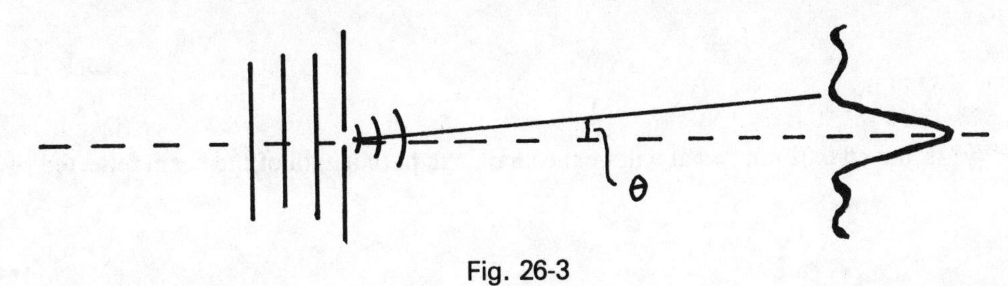

Fig. 26-3

6. The air wedge shown in Fig. 26-4 is formed between two glass plates by putting a hair under one edge of the upper plate. When the plates are illuminated from above with monochromatic (500 nm) light, parallel fringes are observed having a spacing of 0.300 mm. What is the thickness of the hair?

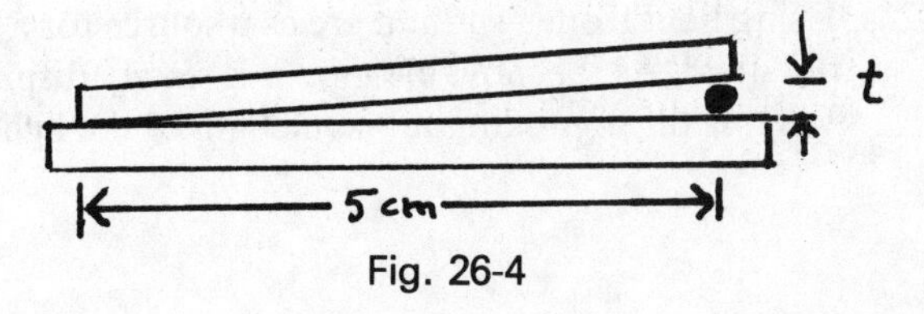

Fig. 26-4

   t = ______________ .

7. A thin oil film on the surface of a puddle of water has a bluish cast when viewed in reflected light. What is the probable thickness of the oil film? (Index of refraction of oil $\cong 1.5$. $\lambda \cong 450$ nm)

The following questions refer to the Michelson interferometer as described in the text and illustrated in text Figs. 26-30 and 26-31.

8. How many interfering beams reach the observer's eye?

9. What element in the interferometer causes a single ray from the source to be split up into two rays?

10. Draw a schematic diagram of a simplified Michelson interferometer in which the beam splitter is of negligible thickness. Since the compensating plate is not needed in this case, omit it. Which mirror is half silvered? Which mirror can be moved so as to change the path length in one arm of the interferometer?

11a. If mirror "y" is moved 0.02 mm, what will be the change in path length of the beam reflected by "y"?

b. How many fringes of green light ($\lambda = 500$ nm) will pass the field of view if mirror "y" is moved 0.02 mm?

12. Using light from a sodium arc as a source for a Michelson interferometer, clear yellow fringes may be obtained. As the movable mirror is slowly displaced 0.059 mm, 200 fringes shift across the field of view. What is the approximate wavelength of the light emitted by the sodium arc?

13. The electromagnetic radiation from a single atom consists of short pulses or wave trains lasting less than 10^{-8} s. These wave trains are plane-polarized. Why then is the radiation from a light source unpolarized?

14. Brewster's angle is given by

$$\tan \theta = n$$

If a ray of light is incident on a glass surface at Brewster's angle, both reflected and transmitted rays will be produced.

a. Which of these two (reflected or transmitted rays) will be completely polarized?

b. What will be the angle between the reflected ray and the refracted ray?

15. When light is reflected from a surface at angles other than Brewster's angle, the amount of polarization will be less than 100% but still quite appreciable. Thus the light that is reflected from the surface of a lake, for example, or from wet pavement will always be at least partially polarized.

a. What will be the predominant direction of vibration of the electric vector in light reflected from a horizontal surface?

b. How can Polaroid sunglasses reduce the "glare" associated with reflected light?

16. Why does the possibility of polarization of light prove that the electromagnetic vibrations are transverse?

Solutions

1a. The distances d_1 and d_2 must differ by a whole number of wavelengths or by zero.

b. $d_1 - d_2 = \pm\lambda/2, \pm 3\lambda/2, \pm 5\lambda/2, \ldots$

c. $n\lambda = a \sin \theta = a(y/D)$

$y = n\lambda D/a = (1)(500 \times 10^{-9}\text{ m})(0.5\text{ m})/(10^{-3}\text{ m}) = 0.25 \times 10^{-3}\text{ m}$

d. For $\lambda = 500$ nm:

$$\sin\theta = n\lambda/a = \frac{3(500 \times 10^{-9}\text{ m})}{10^{-3}\text{ m}}; \qquad \theta = 1.5 \times 10^{-3}\text{ rad}$$

For $\lambda = 600$ nm:

$$\sin\theta = n\lambda/a = \frac{3(600 \times 10^{-9}\text{ m})}{10^{-3}\text{ m}}; \qquad \theta = 1.8 \times 10^{-3}\text{ rad}$$

The angular separation is 3×10^{-4} rad.

2. $$\sin\theta = \frac{3.84\text{ cm}}{80\text{ cm}} = 4.8 \times 10^{-2}\text{ rad}$$

$$\lambda = \frac{a\sin\theta}{n} = \frac{(1 \times 10^{-4}\text{ m})(4.8 \times 10^{-2}\text{ rad})}{12}$$

$$\lambda = 4 \times 10^{-7}\text{ m} = 400\text{ nm}$$

3a. S' is 0.4 mm from S on the opposite side of the plane containing the mirror.

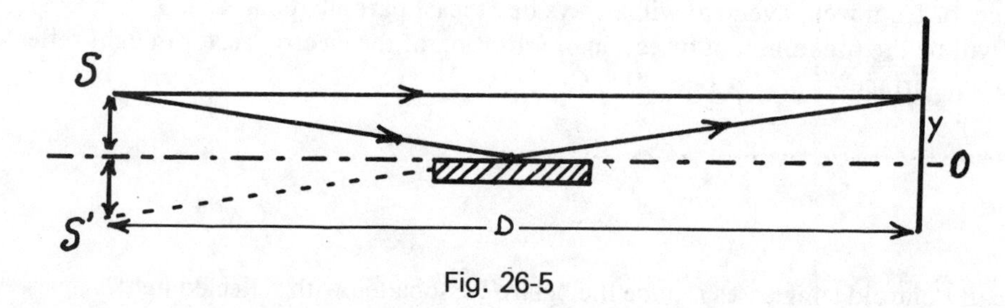

Fig. 26-5

b. Zero, since the two waves arrive exactly 180° out of phase.

c. The path difference corresponding to the 3rd bright band is (2 ½)λ. Thus

$$a\sin\theta = (2.5)\lambda; \qquad \sin\theta = \frac{y}{D}$$

$$y = D\sin\theta = D\,\frac{2.5\,\lambda}{a} = \frac{(2\text{ m})(2.5)(5 \times 10^{-7}\text{ m})}{(0.4 \times 10^{-3}\text{ m})}$$

$$y = 6.25 \times 10^{-3}\text{ m} = 6.25\text{ mm}$$

4a. 0.5 mm; 0.5 mm
The spacing of the maxima is independent of the number of slits.

b. The bright fringes or bands get narrower and their peak intensities increase. The peak amplitude is proportional to the number of slits and the peak intensity is proportional to the square of the number of slits.

5a. The two rays from opposite edges of the slit, rays 1 and 9 in text Fig. 26-18.

b. Zero, because each ray from the upper half of the single slit is canceled out by a ray from the lower half.

c. $$\text{angular width} = 2(n\lambda/w) = \frac{2(1)(6 \times 10^{-7}\text{ m})}{2 \times 10^{-5}\text{ m}}$$

$$= 6 \times 10^{-2}\text{ rad}$$

d. Since the angle is very small, $\theta \cong \sin\theta$.

$$\theta = \frac{n\lambda}{w} = \frac{(1)(5 \times 10^{-7}\text{ m})}{5 \times 10^{-5}\text{ m}} = 1 \times 10^{-2}\text{ rad} \cong 0.6°$$

The spreading out of light from a narrow slit is observable but it is very small.

6. The wedge angle is

$$\theta = \frac{\tfrac{1}{2}(\text{wavelength})}{\text{fringe spacing}} = \frac{\tfrac{1}{2}(5 \times 10^{-7}\text{ m})}{3 \times 10^{-4}\text{ m}} = 8.3 \times 10^{-4}\text{ rad}$$

From the drawing, $\theta = \dfrac{t}{5\text{ cm}}$

Thus $t = \theta(5\text{ cm}) = (8.3 \times 10^{-4})(5\text{ cm}) = 4.2 \times 10^{-3}\text{ cm}$
$= 0.042\text{ mm}$

7. If the oil film is $\lambda/4$ in thickness, the path difference will be $\lambda/2$. The 180° phase shift produced by reflection at the upper surface adds another $\lambda/2$ to the path difference. As a result, the two reflected beams—one from the upper and the other from the lower surface of the oil film—will be in phase. We conclude that the oil film has a thickness of $(450\text{ nm})/4 = 112.5\text{ nm}$ or $\sim 1.0 \times 10^{-7}\text{ m}$.
8. two
9. The mirror M at a 45° angle to the incoming ray. This mirror is partially silvered so that approximately half the light will be reflected and the other half transmitted.
10.

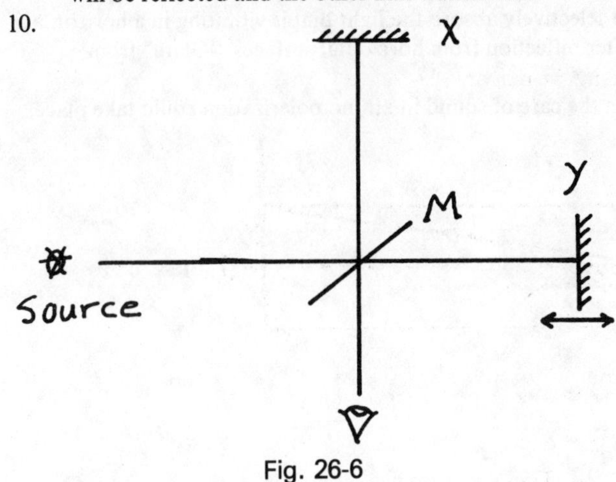

Fig. 26-6

M is half silvered. Mirror Y can be moved toward or away from M to change the path length in that arm of the interferometer.

11a. change in path length $= 2(0.02\text{ mm}) = 0.04\text{ mm}$

b. no. of fringes shifted $= \dfrac{0.04\text{ mm}}{500\text{ nm}} = \dfrac{4 \times 10^{-5}\text{ m}}{5 \times 10^{-7}\text{ m}} = 80$

12. $\lambda = \dfrac{\text{path difference}}{\text{no. of fringes}} = \dfrac{2(0.059\text{ mm})}{200}$

$$\lambda = \frac{1.18 \times 10^{-4}\text{ m}}{2 \times 10^{2}} = 0.59 \times 10^{-6}\text{ m} = 590\text{ nm}$$

13. Ordinary light sources, such as the sun, incandescent bulbs, fluorescent lamps, flames, etc., contain an enormous number of atoms with completely random orientations. The total light emitted contains radiation of every possible plane of vibration; there is no preferred plane of polarization.

14a. the reflected ray

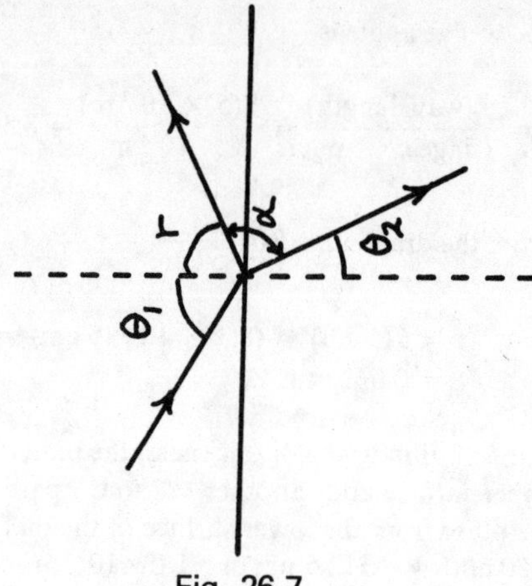

Fig. 26-7

b. Snell's law is
$n_1 \sin \theta_1 = n_2 \sin \theta_2$

Let $n_1 = 1$ (for air) and $n_2 = n$ (for glass). By the law of reflection, $\theta_1 = r$. Thus

$$(1)\sin r = n \sin \theta_2$$

But $\quad n = \tan \theta_1 = (\sin r)/(\cos r)$

$$\sin r = \frac{\sin r}{\cos r} \sin \theta_2 \, ; \; \cos r = \sin \theta_2$$

This requires that $r + \theta_2 = 90°$ and that $\alpha = 90°$

15a. The electric vector will be vibrating in a horizontal plane, i.e., parallel to the reflecting surface.

b. The lenses, made of Polaroid film, may be oriented so as to selectively absorb the light that is vibrating in a horizontal plane thus reducing the amount of light reaching the eye after reflection from horizontal surfaces. "Glare" from vertical surfaces, of course, will not be reduced.

16. If the vibrations of the wave motion were longitudinal, as in the case of sound in air, no polarization could take place.

REVIEW TEST FOR CHAPTERS 24–26

1. An inductor and a capacitor are connected in series with a switch as shown in Fig. T-55. The natural frequency of oscillation of this circuit is 1.59 kHz. At the moment when the switch is closed, the capacitor is charged to a voltage of 10 V. After what time interval will the capacitor be fully discharged?

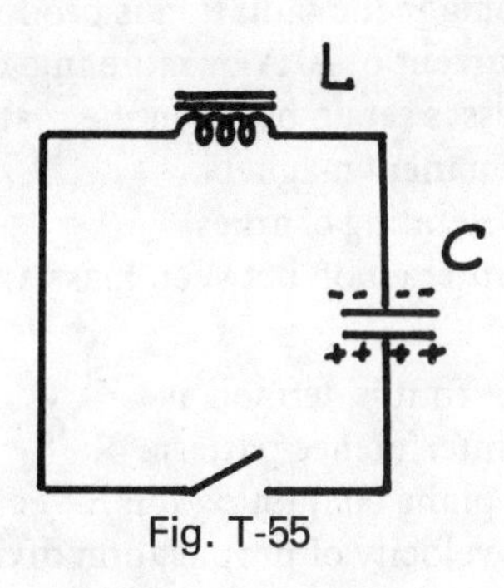

Fig. T-55

2. The capacitance in the circuit above is 5 μF. What is the inductance?

3. An alternating current is flowing in the circuit in Fig. T-56. At an instant when the current has a value of 3 A the energy in the inductor is 3.6×10^{-2} J. What is the inductance?
   a. 8 mH
   b. 4 mH
   c. 24 mH
   d. 16 mH
   e. 12 mH

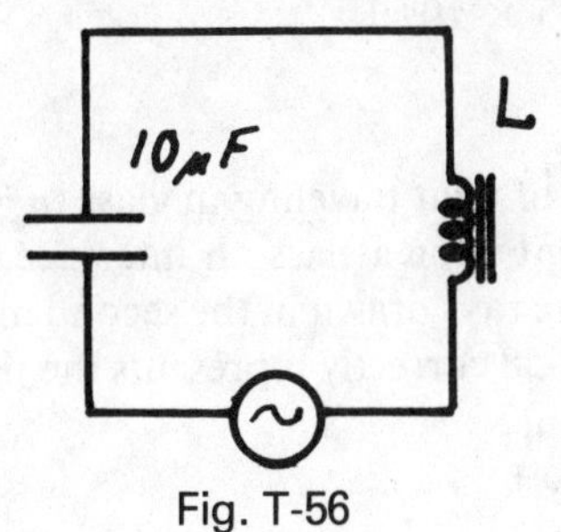

Fig. T-56

4. The resonant frequency of a series L-C circuit is 480 kHz. What will the frequency become if L is doubled and C is made four times smaller?

5. The upper circuit in Fig. T-57 has a resonant frequency f_1. When two more capacitors are added, as shown in the lower circuit, the resonant frequency becomes f_2. What is the ratio f_2/f_1?
   a. $\sqrt{3}$
   b. $1/\sqrt{3}$
   c. $\sqrt{2/3}$
   d. 1/3
   e. $\sqrt{3/2}$

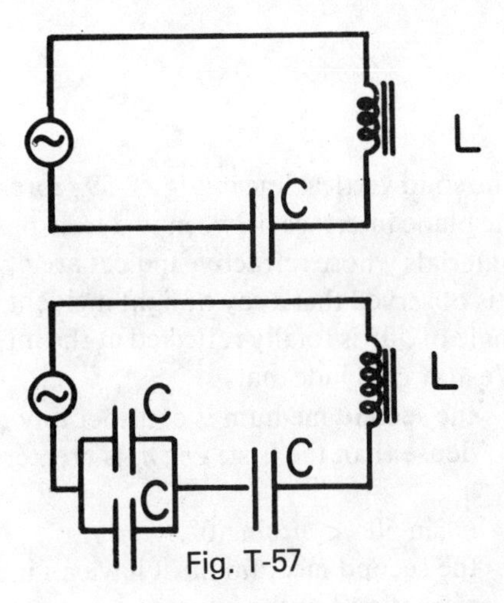

Fig. T-57

6. What is the frequency of an electromagnetic wave whose wavelength is 5×10^{-8} m?

7. Neither charges at rest nor charges in motion at a constant velocity can produce electromagnetic radiation. Electromagnetic radiation is produced by
   a. a current of 10 A or more in a coil.
   b. stresses set up between the plates of a capacitor.
   c. permanent magnets.
   d. accelerating charges.
   e. an interaction between mass and charge.

8. A wave front is defined as
   a. an interference pattern.
   b. the plane containing the E vector and the direction of propagation.
   c. the velocity of propagation divided by the frequency.
   d. the locus of points having the same phase of vibration.
   e. the region immediately in front of the wave.

9. The index of refraction of a certain glass is 1.50 for light whose wavelength in vacuum is 600 nm. What is the frequency of this light as it passes through the glass?
   a. 3.33×10^{14} Hz
   b. 5×10^{14} Hz
   c. 7.5×10^{14} Hz
   d. 180 Hz
   e. 120 Hz

10. A ray of light traveling in glass ($n = 1.60$) is incident upon a glass-air interface. Which of the five rays drawn in the second medium in Fig. T-58 correctly represents the path taken by the light?
   a. ray 1
   b. ray 2
   c. ray 3
   d. ray 4
   e. ray 5

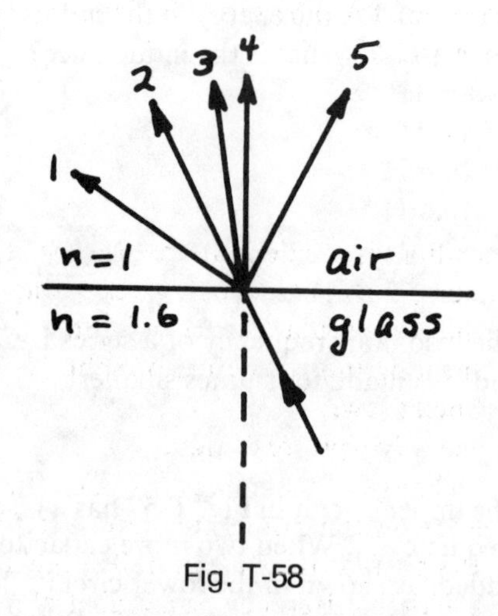

Fig. T-58

11. The solid vertical line in Fig. T-59 represents the plane interface between two transparent materials whose refractive indices are n_1 and n_2. It is observed that a ray of light incident at an angle of 50° is totally reflected at the interface. We may conclude that
   a. the second medium is considerably more dense than the first; i.e., n_2 is greater than n_1.
   b. $n_1 \sin 50° < n_2 \sin 40°$.
   c. the second medium must have an index of refraction less than one.
   d. the ratio n_2/n_1 must be less than or equal to sin 50°.
   e. the ratio n_2/n_1 must be greater than sin 50°.

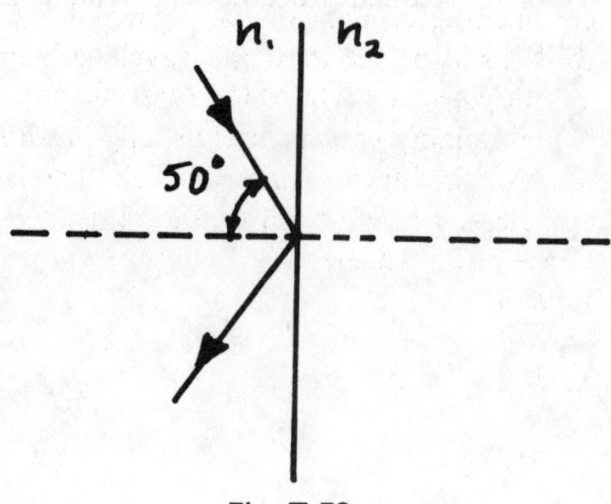

Fig. T-59

12. The image of a real object can be located by using any two of three "construction rays," rays whose direction is predictable without knowing in advance where the image is located. Which one of the rays illustrated in Fig. T-60 is not a useful construction ray?
    a. ray 1
    b. ray 2
    c. ray 3
    d. ray 4

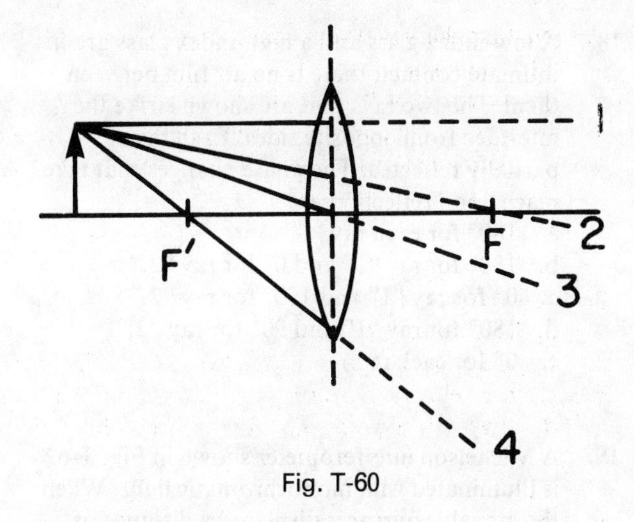

Fig. T-60

13. A real object is located 2 cm to the left of a lens whose focal length is 4 cm. The image is
    a. enlarged, inverted, virtual
    b. reduced, erect, virtual
    c. enlarged, inverted, real
    d. enlarged, erect, real
    e. enlarged, erect, virtual
14. What is the focal length of a plano-convex lens whose convex surface has a radius of curvature of 25 cm? (Use $n = 1.50$.)

15. When light is incident upon a very small pinhole, the light is seen to spread out on passing through the aperture. This observation gives strong evidence for
    a. the corpuscular theory of light.
    b. the longitudinal nature of light.
    c. Snell's law.
    d. the wave nature of light.
    e. a change in the index of refraction.
16. When two beams of light combine to form an interference pattern, the condition for a maximum is that
    a. the phase difference be 180° or less.
    b. the path difference be an integral number of wavelengths.
    c. the interfering beams be polarized.
    d. the slits be less than one wavelength apart.
    e. the wavelength to be less than 600 nm.
17. A high-quality camera lens usually has a nonreflecting coating on its outer surface and perhaps on some of the inner surfaces as well. The index of this thin film is about 1.3, between that of air (1.0) and glass (1.6). The appropriate thickness of such a nonreflecting film may be expressed in terms of λ, the mean wavelength for visible light. ($\lambda \cong 500$ nm). What thickness is required?
    a. $\lambda/8$
    b. $\lambda/4$
    c. $\lambda/2$
    d. λ
    e. $2\lambda/3$

18. A low-index glass and a high-index glass are in intimate contact; there is no air film between them. The two rays that are shown strike the interface from opposite sides. Each ray is partially reflected. The phase changes that take place upon reflection are
   a. 180° for each ray.
   b. 180° for ray "1" and 0° for ray "2."
   c. 0° for ray "1" and 180° for ray "2."
   d. 180° for ray "1" and 90° for ray "2."
   e. 0° for each ray.

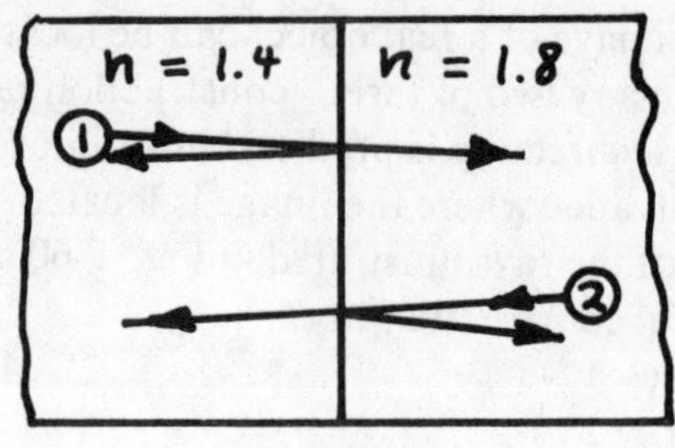

Fig. T-61

19. A Michelson interferometer shown in Fig. T-62 is illuminated with monochromatic light. When the movable mirror is displaced a distance of 0.05 mm, 200 fringes pass the field of view. Calculate the wavelength of the light.

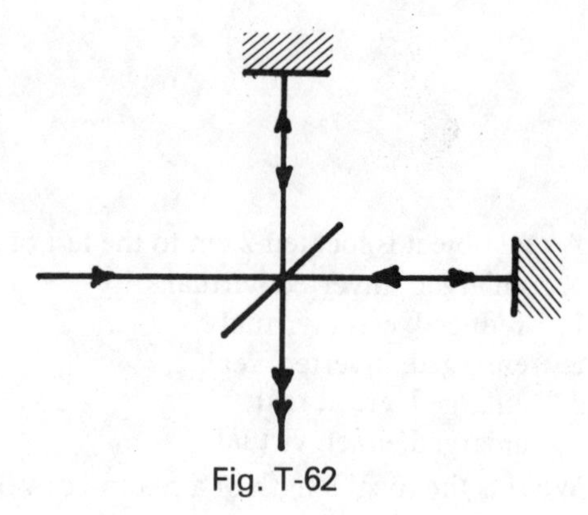
Fig. T-62

20. Light waves can be polarized, but sound waves cannot be polarized because
   a. light waves have a much smaller wavelength than sound waves.
   b. light waves travel through empty space but sound waves do not.
   c. light waves are frequency independent whereas sound waves are not.
   d. light waves are transverse whereas sound waves are longitudinal.
   e. light waves undergo a 180° phase change upon reflection but sound waves do not.

CHAPTER 27 Applied Optics

GOALS To study the construction of several useful optical instruments including the camera, eye glasses, microscopes, and telescopes.

OBJECTIVES After completing this chapter the student should be able to do the following:

1. Name two aberrations of a simple lens.
2. Discuss, qualitatively, why the focal length of a simple lens is different for different colors of light.
3. Discuss why the focal length of a front surface mirror is the same for all colors of light.
4. Define the f-number of a lens and relate it to the light energy transmitted by the lens.
5. Calculate any one of the quantities f-number, D, or f when the two others are given.
6. Name the three image-forming elements in the human eye.
7. Name the three most common vision defects of the eye.
8. Determine the power and focal length of the spectacle lens required to correct for a given degree of near-sightedness or farsightedness.
9. Calculate the angular size of a given object.
10. Define angular magnification and magnifying "power" of a simple lens.
11. Write the equation for the maximum magnification of a simple magnifier (image at near point) and the equation for the minimum magnification of a simple magnifier (image at infinity).
12. Calculate the maximum or minimum magnification of a simple lens of given focal length.
13. Calculate the overall magnification of a microscope.
14. Give the approximate expression for the angular size of a diffraction pattern formed by an optical instrument.
15. Name the property of light that causes the resolving power of a microscope to be limited.
16. Name three special types of microscopes that have been developed for viewing specimens of low contrast.
17. Name two types of microscopes that are capable of giving higher resolving power than a conventional microscope using visible light.
18. Write the expression for the magnifying power of a telescope used to observe a distant object.
19. Calculate the magnification of a telescope when the two focal lengths are known.
20. Identify drawings of Newtonian and Galilean telescopes.
21. Name the two principal types of spectroscopes.
22. Calculate the angular separation between two wavelengths in a grating spectroscope.

SUMMARY

A simple camera is similar to the human eye in that each contains a positive lens that forms a real inverted image on a light-sensitive surface. In the case of the camera, the lens-to-film distance is adjustable to permit objects at various distances to be sharply focused on the film surface. In the human eye, the lens-to-retina distance is fixed but the curvature of the lens surfaces may be changed by the ciliary muscles to bring the image into sharp focus on the retina.

Single lenses ordinarily have spherical surfaces since these are relatively easy to manufacture. However, a lens with spherical surfaces does not form a perfect image; it is said to have aberrations. The aberrations are more noticeable if light is bent through large angles in passing through the lens, as is the case for a lens of large diameter and short focal length. Some lens aberrations can be reduced by changing the radii of curvature of the lens, by using a smaller aperture, or by combining lenses of different types of glass.

Some persons do not see clearly because their eye lens forms the image of a distant object in front of the retina. They can focus on objects close by but not on objects at a distance. This vision defect, called *myopia*, can be corrected by wearing glasses with diverging lenses. The opposite vision defect, called *hypermetropia*, occurs when the image of a distant object is formed behind the retina. This problem can be corrected by glasses with converging lenses.

The amount of detail that can be perceived in an object depends on the size of the image that is formed on the retina of the observer's eye. To see detailed structure in a flower on the other side of the room, you bring it up to within a few feet of your eyes thus vastly enlarging the image of the flower that is projected on the retina. By putting a converging lens between your eye and the flower, it becomes possible to bring the flower to within a few inches of your eye while at the same time maintaining a sharp image on the retina. A positive lens used in this way is called a magnifier. The angular magnification obtained with a magnifier varies slightly because of the variable focal length of the human eye, the maximum and minimum values of magnification being $M_{max} = 1 + 25\text{ cm}/f$ and $M_{min} = 25\text{ cm}/f$ where f is the focal length of the magnifier.

The compound microscope consists of two positive lenses separated by a distance that may vary from 5 cm to 25 cm. The overall magnification of a microscope is the product of the magnification of the objective lens and that of the eyepiece. This magnification may be much greater than that of a simple magnifier but it cannot exceed several hundred no matter how carefully the lenses are designed nor how many lenses are used. The wave nature of light itself places an upper limit on the magnification of a microscope; the shorter the wavelength the higher the magnification that may be achieved.

Like the microscope, the astronomical telescope consists of two positive lenses. When the telescope is adjusted for viewing with a relaxed eye, the separation of the lenses is equal to the sum of the two focal lengths. The separation of the lenses will be greater than $F + f$ if nearby objects are being viewed. The objective lens will have a rather long focal length in the case of a telescope and a rather short focal length in the case of a microscope. The magnification of an astronomical telescope, when adjusted for relaxed viewing, is equal to the ratio of the focal lengths, $M = F/f$.

A system consisting of two positive lenses, whether a microscope or a telescope, necessarily gives an inverted image. The telescope image can be made erect by adding a third positive lens, by using an erecting prism, or by substituting a negative eyepiece. The third system is called the Galilean telescope and is employed in lightweight, inexpensive opera glasses of low magnifying power.

The spectroscope is a specialized optical instrument used for observing and measuring the wavelengths of light radiated by a source. At the heart of a spectroscope is the *dispersing element*, an optical component that causes an angular separation of the various wavelengths that are present. The most commonly used dispersing elements are the prism and the *diffraction grating*, usually a plane or concave reflecting grating. After dispersion by the prism or grating, the light, now separated into its component wavelengths, is observed by means of a small telescope. If a permanent record of the spectrum is desired, the telescope is replaced by a camera.

QUESTIONS AND PROBLEMS

Secs. 27-1—27-3

1. Rays passing through the outer regions of a simple lens do not converge to the same point as rays that pass close to the lens axis. What is the name of this aberration? Which group of rays is more strongly bent by the lens? (See text Fig. 27.1.)

2. The point at which light rays come to a focus depends on the wavelength. Which wavelengths come to a focus closer to the lens? (Longer or shorter wavelengths?) What is the name of this effect?

3. What optical element can be used to form an image that is perfectly free from chromatic aberration? (Hint: An image-forming element that does not involve refraction is discussed in Secs. 25-10 and 27-6.)

4. The intensity of light transmitted by a lens is proportional to D^2/f^2; The f-number of a lens is defined as f-number $= f/D$. Consider two camera lenses whose f-numbers are 4.5 and 6.3.
   a. Which lens transmits more light and by what ratio?

   b. What is the f-number of a lens that transmits twice as much light as the f-4.5 lens?

5. The overall power of a typical human eye is variable because of the ability of ciliary muscles to change the shape of the lens. If this power ranges from +60 diopters to +64 diopters, what is the corresponding range in focal length?

6. A normal, relaxed eye forms an image of a very distant object on the retina. The power of the normal eye, lens plus cornea and aqueous fluid, under these conditions is 60 diopters. Now suppose that the eye lens does *not* accommodate as the distant object is brought up close.
   a. In what direction does the image move? What is the visual effect that results?

   b. If a person can see distant objects clearly but cannot accommodate his eye lens for closer objects, what visual defect does he suffer from?

c. Suppose that a newspaper is to be held at a distance of 25 cm from the eye. Calculate the power and focal length of the spectacle lens required.

7. A certain person with *myopia* can focus clearly on objects not farther than 25 cm from his eye. What type of lens does he need to focus on objects at a greater distance? What power and focal length must his glasses have to permit him to see clearly at a great distance? (Assume that the eye has a minimum power of 64 diopters.)

* * * * * * *

A magnifier is a simple, positive lens that permits one to get a larger image on the retina than is possible without it. The angular magnification actually achieved depends to some extent on whether the eye lens is relaxed, as in viewing distant objects, or strained, as in viewing objects up close. The maximum and minimum angular magnifications differ by only 1.

$$M_{max} = \frac{25 \text{ cm}}{f} + 1 \qquad\qquad M_{min} = \frac{25 \text{ cm}}{f}$$

eye somewhat strained, virtual image at 25 cm — eye relaxed, virtual image at infinity

8. A magnifier has a focal length of 5 cm. What will be the maximum angular magnification obtainable with this magnifier?

M_{max} = ____________.

9. A large plano-convex lens is used for examining fingerprints. The radius of curvature of the convex surface is 5 cm and the index of refraction is 1.50. What angular magnification will be obtained if this magnifier is used with a relaxed eye?

f = ____________.

M = ____________.

Secs. 27-4—27-9

10. A compound microscope is to be constructed using an objective lens of 2.5-cm focal length and an eyepiece of 1.5-cm focal length. The object being examined is to be placed 3.0 cm from the objective and the microscope is to be adjusted for minimum eye strain.
a. Calculate the image distance for the objective lens.

b. What should be the distance between the two lenses?

c. What is the overall magnification obtained if this microscope is used as described?

d. What is the maximum magnification possible when the final image is placed at the near point (25 cm) rather than at infinity?

11. Two thin, positive lenses are available, each having a focal length of 3 cm. These lenses can be used in two ways. Placed in contact, they become a simple magnifier with greater magnifying power than either lens alone. Separated by a distance of 15 cm, they form a compound microscope, again with greater magnifying power than either lens alone.
    a. Assuming that the lenses are in contact and used with a relaxed eye (image at infinity), what is the overall magnification? (Hint: Treat the lenses one at a time as described in Sec. 25-11. Let the object for lens A be at infinity. Then the image formed by lens A, 3 cm to the right, becomes a virtual object for lens B. The final image, formed by the second lens, is located at the focal plane of the combination.)

    b. Assuming that the lenses are 15 cm apart and that the instrument is adjusted for minimum eyestrain, find the object distance for the eye lens and for the objective lens.

    c. Calculate the overall magnification of the microscope and compare it with that found for the simple magnifier formed of the same two lenses.

    d. What change in the arrangement of these two lenses would lead to a greater magnification?

12. What is the technical term that has to do with the ability of a lens, or system of lenses, to form sharp images?

13. Suppose that an objective lens of diameter 4 cm and focal length 20 cm is used to form an image of a point source at infinity, a star for example.

   a. Calculate the angular radius θ of the central part of the diffraction pattern formed by this lens. Use 500 nm as the mean wavelength of white light. (We assume, of course, that the lens is "perfect," free from all aberrations.)

   b. Calculate the diameter of the Airy disk formed in the focal plane of this lens. (The radius of the Airy disk is the radius corresponding to θ found in 13a above.)

   c. What fraction of the total light falls within the Airy disk?

14. What property of light determines the ultimate resolving power of an optical system?

15. A high-quality microscope is designed to give the sharpest possible image, limited only by diffraction effects, over a wavelength range from 400 nm to 700 nm. If the limiting angle of resolution is 5×10^{-4} rad at 600 nm, what limiting angle of resolution is to be expected at 400 nm?

16. An astronomical telescope is to be constructed with an objective of 30-cm focal length and an eyepiece of 2-cm focal length.

   a. What should be the separation of the objective and eyepiece for relaxed viewing of very distant objects?

   b. What is the magnification of this telescope when used as in 16a above?

17. The objective of an astronomical telescope, whether it be a lens or a mirror, has a long focal length because this leads to high magnification. $M = F/f$. Why does the objective also have a large diameter? Give two reasons.

18. State two advantages of a *reflecting* telescope as compared with a *refracting* telescope.

19. What are the two advantages of the Galilean telescope as compared with the astronomical telescope?

20. A Galilean telescope is to be constructed with an objective of 30-cm focal length and an eyepiece of −2-cm focal length.
   a. Draw a sketch showing the paths of typical rays through the system and the location of the intermediate image.

   b. Calculate the required separation of the lenses when the object and final image are both at infinity.

21. Name the two types of *dispersing elements* that are used in spectroscopes. (Secs. 27-8 and 27-9.)

22. What are the two essential components of the *collimator* of a spectroscope?

23. The grating of a spectroscope has 45,000 lines in a width of 7.5 cm.
   a. What will be the angle of diffraction for the 2nd order of a line at 500 nm?

b. Calculate the angular separation between wavelengths of 400 nm and 450 nm in the 3rd order.

c. What is the highest order of interference that can be obtained for a wavelength of 350 nm?

Solutions

1. Spherical aberration. The outer rays are more strongly bent and come to focus closer to the lens.
2. shorter wavelengths; chromatic aberration
3. A mirror—plane, concave, or convex—is entirely free from chromatic aberration provided that the reflection takes place at the *front* surface. This is true because the law of reflection is not wavelength dependent.

4. Light energy transmitted: $\dfrac{1}{(f\text{-no.})^2}$

$$\frac{1}{(4.5)^2} = \frac{1}{20}; \qquad \frac{1}{(6.3)^2} = \frac{1}{40}$$

a. The f-4.5 lens transmits *twice* as much light as the f-6.3 lens.

b. $\dfrac{1}{(f\text{-no.})^2} = 2\dfrac{1}{(4.5)^2} = \dfrac{1}{10}$

$f\text{-no.} = \sqrt{10} = 3.2$

5. $f_{max} = \dfrac{1}{60\ \text{m}^{-1}} = 1.67\ \text{cm}; \qquad f_{min} = \dfrac{1}{64\ \text{m}^{-1}} = 1.56\ \text{cm}$

6a. As the object is moved in toward the eye, the image moves in back of the retina. As a result, only a blurred image is seen.

b. Farsightedness (or hypermetropia). A positive lens is required.

c. We first determine the effective focal length required of the eye-plus-spectacle-lens combination. Since the image must still fall on the retina, $q = 1.67$ cm.

$$\frac{1}{f_{eff}} = \frac{1}{P} + \frac{1}{q} = \frac{1}{25\ \text{cm}} + \frac{1}{1.67\ \text{cm}} = \frac{1}{1.56\ \text{cm}} = 64\ \text{m}^{-1}$$

Thus $\quad f_{eff} = 1.56$ cm, or power (eff) $= 64\ \text{m}^{-1}$

Since the powers are additive,

$$64\ \text{m}^{-1} = 60\ \text{m}^{-1} + (\text{power of lens})$$

The required lens has a power of 4 diopters or a focal length of 25 cm.

7. The focal length is given and the object distance is known. Find the image distance.

$$\frac{1}{q} = \frac{1}{f} - \frac{1}{p} = \frac{1}{1.56 \text{ cm}} - \frac{1}{25 \text{ cm}} = 0.60 \text{ cm}^{-1} = \frac{1}{1.67 \text{ cm}}$$

$q = 1.67$ cm

When the object is moved out to infinity, the image must remain in focus on the retina. Thus

$f_{eff} = 1.67$ cm is the effective focal length of eye plus spectacle lens.
{Effective power} = {power of eye} + {power of spectacles}
$60 \text{ m}^{-1} = 64 \text{ m}^{-1}$ + {power of spectacles}

The required glasses have a negative power of 4 diopters, hence a negative focal length of 25 cm.

8. $$M = \frac{25 \text{ cm}}{f} + 1 = 5 + 1 = 6$$

9. $$\frac{1}{f} = (n-1)\left(\frac{1}{R_1} + \frac{1}{R_2}\right) = (0.5)\left(0 + \frac{1}{5 \text{ cm}}\right) = \frac{1}{10 \text{ cm}}$$

$f = 10$ cm

$$M = \frac{25 \text{ cm}}{f} = \frac{25 \text{ cm}}{10 \text{ cm}} = 2.5$$

10. focal length of objective lens = $f_1 = 2.5$ cm
focal length of eyepiece = $f_2 = 1.5$ cm

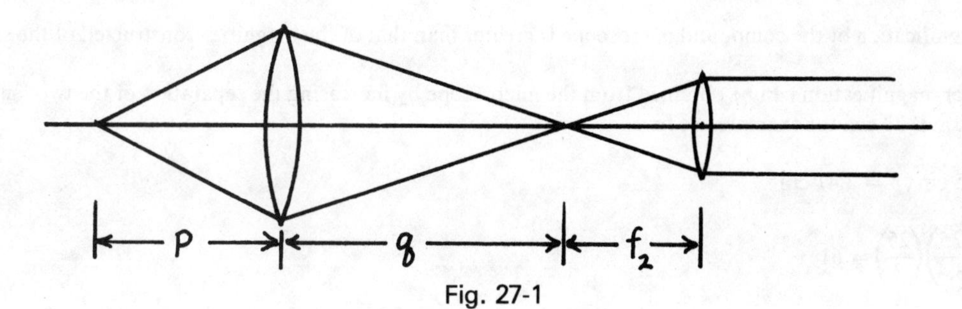

Fig. 27-1

a. $$\frac{1}{q} = \frac{1}{f} - \frac{1}{p} = \frac{1}{2.5 \text{ cm}} - \frac{1}{3.0 \text{ cm}} = \frac{3 - 2.5}{7.5 \text{ cm}}$$

$q = 15$ cm

b. separation $= q + f_2 = 16.5$ cm

c. $$m_1(\text{objective}) = \frac{q}{p} = \frac{15 \text{ cm}}{3 \text{ cm}} = 5$$

$$M_2(\text{eyepiece}) = \frac{25 \text{ cm}}{f_2} = \frac{25 \text{ cm}}{1.5 \text{ cm}} = 16.7$$

Overall magnification $= m_1 M_2 = 5 \times 16.7 = 83.3$

d. The magnification of the eyepiece increases slightly.

$$M_2 = \frac{25 \text{ cm}}{f} + 1 = 17.7$$

$$M = (5)(17.7) = 88.3$$

11a. Lenses in contact. Image formed by first lens becomes virtual object for second lens. For the second lens,

$$p_2 = -3 \text{ cm}$$

$$\frac{1}{q_2} = \frac{1}{f_2} - \frac{1}{p_2} = \frac{1}{3 \text{ cm}} - \frac{1}{-3 \text{ cm}} = \frac{2}{3 \text{ cm}} = \frac{1}{1.5 \text{ cm}}$$

$$q_2 = 1.5 \text{ cm}$$

Thus the focal length of the combination is 1.5 cm. The magnification is

$$M = \frac{25 \text{ cm}}{1.5 \text{ cm}} = 16.7$$

b. See Prob. 10 for a drawing of the microscope.

$q + f_2 = 15$ cm; $f_2 = 3$ cm (object distance for eyepiece)

Thus $q = 12$ cm

$$\frac{1}{p} = \frac{1}{f} - \frac{1}{q} = \frac{1}{3 \text{ cm}} - \frac{1}{12 \text{ cm}} = \frac{4-1}{12 \text{ cm}}$$

$p = 4$ cm (object distance for objective)

c. $M = m_1 M_2 = \left(\frac{q}{p}\right)\left(\frac{25 \text{ cm}}{f_2}\right) = \left(\frac{12 \text{ cm}}{4 \text{ cm}}\right)\left(\frac{25 \text{ cm}}{3 \text{ cm}}\right) = 25$

The magnification of the compound microscope is greater than that of the magnifier constructed of the same two lenses.

d. A greater magnification can be obtained from the microscope by increasing the separation of the two lenses. If the separation is 28 cm, for example, we find

$q = 25$ cm, $p = 3.41$ cm

$$M = \left(\frac{25}{3.4}\right)\left(\frac{25}{3}\right) = 61$$

12. resolving power (The term *resolving limit* is also used.)

13a. $\theta = \frac{\lambda}{0.82D} = \frac{500 \times 10^{-9} \text{ m}}{(0.82)(4 \times 10^{-2} \text{ m})} = 1.5 \times 10^{-5}$ radians

(or 8.6×10^{-4} deg)

b. $\theta = r/f$; $r = \theta f$
diameter $= 2\theta f = 2(1.5 \times 10^{-5})(0.2 \text{ m}) = 6 \times 10^{-6} \text{ m} = 6 \times 10^{-3}$ mm

c. 84%

14. the wavelength

15. The limiting angle of resolution is the angular radius of the Airy disk,

$$\theta = \lambda/0.82D$$

Thus the limiting angle of resolution at 400 nm becomes

$$(5 \times 10^{-4} \text{ rad}) \frac{(400 \text{ nm})}{(600 \text{ nm})} \cong 3 \times 10^{-4} \text{ rad}$$

16a. separation $= F + f = 30\text{ cm} + 2\text{ cm} = 32\text{ cm}$

b. $M = \frac{F}{f} = \frac{30\text{ cm}}{2\text{ cm}} = 15$

17. Since the size of the diffraction pattern is inversely proportional to the diameter of a lens, a large objective lens leads to sharper images.

The large diameter objective lens also transmits more energy thus giving a brighter image.

18. The reflecting telescope is free of chromatic aberration; all wavelengths come to focus in the same plane.

A second advantage is that a large-diameter mirror is easier to construct than a lens of the same diameter. The mirror can be made relatively light and more rigid.

19. The Galilean telescope gives an erect image. It is also somewhat shorter than an astronomical telescope.

20a.

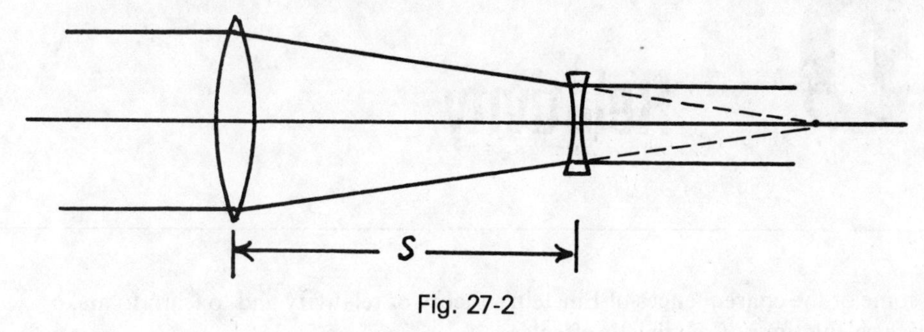

Fig. 27-2

b. The image of the objective must fall 2 cm to the right of the eyepiece. Thus the separation of the lenses is

$$S = F + f = 30\text{ cm} - 2\text{ cm} = 28\text{ cm}$$

21. prism and diffraction grating
22. slit and collimator (or lens)
23a. First calculate the grating spacing.

$$a = \frac{7.5\text{ cm}}{4.5 \times 10^4} = 1.67 \times 10^{-4}\text{ cm}$$

$$\sin\theta = \frac{n\lambda}{a} = \frac{(2)(5 \times 10^{-7}\text{ m})}{1.67 \times 10^{-6}\text{ m}} = 0.60$$

$$\theta = 36.9°$$

b. $\sin\theta_1 = \frac{(3)(4 \times 10^{-7}\text{ m})}{1.67 \times 10^{-6}\text{ m}} = 0.7200 \quad \theta_1 = 46.05°$

$$\sin\theta_2 = \frac{(3)(4.5 \times 10^{-7}\text{ m})}{1.67 \times 10^{-6}\text{ m}} = 0.8084 \quad \theta_2 = 53.94°;\ \Delta\theta = \theta_2 - \theta_1 = 53.94° - 46.05° = 7.89°$$

c. The order of interference is an integer n satisfying the equation

$$n = \frac{a \sin\theta}{\lambda}$$

The angle of diffraction corresponding to the largest value of n is unknown; however, this angle cannot exceed 90°. Thus the highest order of interference is the largest n satisfying

$$n \leq \frac{a \sin 90°}{\lambda}$$

Since $a/\lambda = 4.76$, the highest order of interference is 4.

CHAPTER 28 Relativity

GOALS To learn some of the consequences of Einstein's theory of relativity and to learn to make quantitative calculations of relativistic effects.

OBJECTIVES After completing this chapter the student should be able to do the following:

1. State whether a given frame of reference is inertial or non-inertial on the basis of its state of motion with respect to an inertial frame of reference.
2. Distinguish between inertial and non-inertial frames of reference on the basis of the validity or nonvalidity of Newton's laws of motion in that frame.
3. Determine the velocity of an object in one frame of reference from a knowledge of its velocity in a second frame in motion with respect to the first.
4. State the essential experimental result of the Michelson-Morley experiment.
5. State the two postulates upon which Einstein based his theory of relativity.
6. Calculate the time interval as measured in one frame of reference when it is given for another frame in uniform motion relative to the first.
7. Calculate the relativistic mass of a body from a knowledge of its rest mass and its speed.
8. Calculate the energy equivalent of mass from the Einstein formula.
9. State the generalized form of the law of conservation of energy.

SUMMARY

The motion of objects must necessarily be described with respect to a *frame of reference*, generally a set of three mutually perpendicular axes. Some frames of reference are more convenient than others. A description of the positions and velocities of the planets in our solar system is simplified by using a frame of reference at rest with respect to the fixed stars. In the reference system of the fixed stars, and any frame of reference moving at a constant speed with respect to that system, Newton's laws of motion are valid. Any such frame of reference is called an *inertial* frame.

A frame of reference that is accelerated with respect to an inertial frame or that is rotating with respect to an inertial frame is necessarily a non-inertial frame of reference. In a non-inertial frame of reference, an accelerating elevator, for example, fictitious forces appear; Newton's laws of motion are not valid.

Galilean relativity states that the laws of mechanics have the same form in all inertial frames of reference.

Einstein's theory of *special relativity* is based on two postulates. The first is a generalized form of the principle of Galilean relativity applying to all laws of physics. The second postulate is that measurements of the speed of light always give the same numerical value, regardless of the relative velocity of the observer and the light source.

One of the consequences that follow from the logical development of these two postulates is *time dilation.* The time interval t_0 between two events measured by an observer in a frame of reference is different from the time interval t between these same two events as measured by an observer in a frame of reference that is moving uniformly with respect to the first. The relationship between the two time intervals is $t = t_0/\sqrt{1 - v^2/c^2}$.

A second prediction of Einstein's theory of relativity is that the mass of a moving object m will be greater than the mass of an object at rest m_0. Mathematically, $m = m_0/\sqrt{1 - v^2/c^2}$.

A third consequence of Einstein's theory of relativity is the interconvertability of mass and energy. When a quantity of mass disappears, a fixed quantity of energy appears in its place, and vice versa. The mathematical relationship is $\Delta E = (\Delta m)c^2$. Mass is, in this sense, a form of energy and the law of conservation of energy can be generalized as follows:

The total equivalent amount of energy in the universe remains constant.

Length, like mass and time, depends on the relative motion between two observers. The length L measured by the moving observer is $L = L_0\sqrt{1 - v^2/c^2}$ where L_0 is the length of an object measured in the frame of reference in which the object is at rest.

QUESTIONS AND PROBLEMS

Secs. 28-1—28-2

1. Newton's second law of motion is valid in an inertial frame of reference. This law is written net **F** = m**a**.
   a. What form does Newton's second law have in another frame of reference that is moving at a constant velocity with respect to the first?

   b. What is the second frame of reference called?

2. A van is traveling along a smooth highway curve at a constant speed of 80 km/h (50 mi/h). An experiment carried out inside the van uses the van itself as a frame of reference.
   a. Does Newton's second law have the form, net **F** = m**a**, in this frame of reference? Explain your answer.

   b. What do we call a frame of reference in which Newton's law of motion, net **F** = m**a**, is not valid?

3. A large passenger ship illustrated in Fig. 28-1 is cruising due east at a speed of 16.2 knots (30 km/h). A runner crosses the deck from the starboard to the port side at a speed of 15 km/h with respect to the ship. The velocity of the runner with respect to a stationary frame of reference, the still water in this case, can be found in two steps.

a. Make a diagram showing the relationship of the three velocity vectors. Let $\mathbf{v}_{rs}$ represent the velocity of the runner with respect to the ship, $\mathbf{v}_{sw}$ the velocity of the ship with respect to the water, and $\mathbf{v}_{rw}$ the velocity of the runner with respect to the water.

b. Use this diagram and the given magnitudes to find the magnitude of the velocity of the runner with respect to the water.

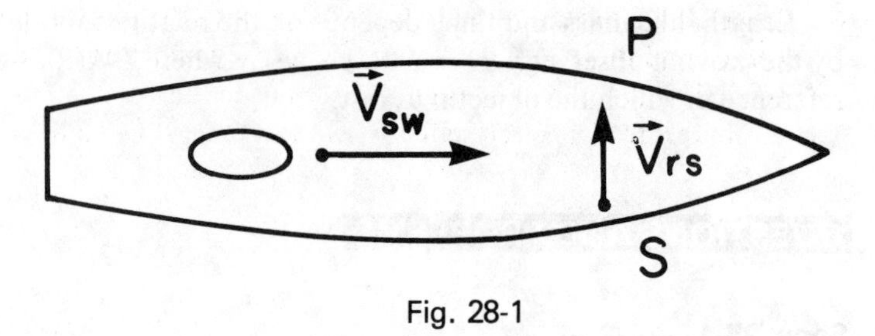

Fig. 28-1

4. A canal 100 m wide is flowing north at a speed of 1.0 km/h. A swimmer wishes to cross the canal from *A* to *B* (see Fig. 28-2) and then return to her starting point. If she is able to maintain a steady speed of 3.0 km/h, how long will it take her to make the round trip? The solution of this problem requires several steps.

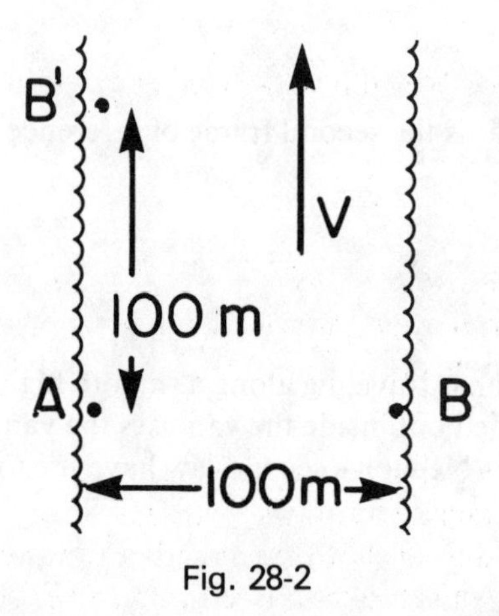

Fig. 28-2

a. Make a vector diagram showing how the velocity of the swimmer with respect to the land $\mathbf{v}_{SL}$ is related to the velocity of the swimmer with respect to the water $\mathbf{v}_{SW}$ and the velocity of the water $\mathbf{v}_{W}$.

b. Calculate the magnitude of the resultant velocity $\mathbf{v}_{SL}$.

c. Determine the time, in seconds, for the round trip.

5. The swimmer in the preceding problem starts again at point A, swims downstream a distance of 100 m to B', then turns around and swims back to the starting point. Again the total distance is 200 m and her speed with respect to the water is a constant 3.0 km/h.
   a. What is the resultant speed for the downstream part of the trip and how long does it take?
   b. What is the resultant speed for the return trip (against the flow of the canal) and how long does it take?
   c. Calculate the ratio t_x/t_y where t_x is the time to cross the canal and return and t_y is the time to travel the same distance parallel (and antiparallel) to the flow.
   d. Show that the result is given by the formula

$$t_x/t_y = \sqrt{1 - v^2/c^2}$$

6. Imagine a frame of reference that is fixed to the platform of a merry-go-round. The origin is located at the center and the x-axis lies along one of the radial struts that support the platform. The merry-go-round is revolving at a constant angular speed of 0.2 rev/s.
   a. Is this frame of reference an inertial frame? Explain your answer.

   b. Suppose that you are sitting on the floor of the merry-go-round. Then you are "at rest" in this frame of reference. Is it true that $\Sigma F_x = 0$ in this frame of reference?

   c. Are Newton's laws of motion valid in this frame of reference? Explain.

7. During what portion(s) of a jet plane's flight could the aircraft be described as an inertial frame of reference?

   List them here. ____________

   a. loading period, engines off
   b. taxiing to runway (straight line at constant speed)
   c. during slow turn at end of runway to face in takeoff direction
   d. while brakes are released and engines at maximum thrust
   e. during steep climb at a 20° angle with constant speed
   f. during power reduction at 5000 feet for noise abatement
   g. cruising at 10,000 m at a speed of 880 km/s in a straight line

8. When you hear the band from your seat in the stadium, through what medium is the sound propagated?

9. When the velocity of light is measured with very high precision in a laboratory on earth the result is $c = 2.9979248 \times 10^8$ m/s.

   Suppose an experiment of comparable precision is carried out in a future space laboratory traveling at a speed of 10^5 m/s. According to Einstein, what velocity of light would be measured in the spaceship?

   $c' =$ ____________.

10. The constant c was first discovered in experiments dealing with electricity. Somewhat later Maxwell showed that this constant was also the speed of light or of any electromagnetic wave. What was the new meaning given to this constant by Einstein?

11. State the two postulates of Einstein's theory of special relativity.

Secs. 28-3—28-8

12. The time t_0 between two events that take place at the same point in a particular frame of reference is 10 s. This time interval is called *proper time*. What is the corresponding time interval t as measured by an observer moving at the speed $0.5\ c$?

13. Suppose that the *proper time* between two events is 0.01% less than the time interval measured in another frame of reference having speed v with respect to the first. Mathematically, $t_0 = t(1 - 0.0001)$. What is the magnitude of v that will produce this difference of 0.01% in the two time intervals? (Hint: This problem and others that follow can be solved most easily by using the following approximations, valid when the quantity ϵ is small compared to 1.

$$\sqrt{1 - \epsilon} \cong 1 - \frac{\epsilon}{2} \quad \text{and} \quad \frac{1}{1 - \epsilon} \cong 1 + \epsilon \quad \text{(provided } \epsilon \ll 1.)$$

14. Imagine an interstellar space probe on a voyage to a nearby star. It passes an inhabited space station at a speed of $0.3\ c$. The length of the probe as determined by its builders while it was in their laboratory on earth is 20.00 m. What will be the length observed by the inhabitants of the space station as the space probe flies past them?

It is shown in Sec. 28-5 in the text that the two postulates of Einstein (invariance of the laws of physics and constancy of the speed of light) lead to the conclusion that the mass of an object is a function of its velocity. At normal speeds, however, the effect is extremely small.

15. The maximum speed attained by an Apollo spaceship on a lunar voyage is about 10 km/s. By what fraction does the mass of the spaceship increase in this case?

$\Delta m/m_0 =$ ______________.

16. Use the Einstein mass-energy equation to show that the energy equivalent of 1 kg of mass is 9×10^{16} J.

17. State the law of conservation of energy as applied to the universe as a whole.

Are any forms of energy to be excluded from this conservation law? ______________ . (Yes/No)

18a. Calculate the mass increase of a spaceship whose rest mass m_0 is 12×10^3 kg if the speed of the spaceship is 10 km/s.

Δm = ______________ .

b. Now calculate the energy ΔE associated with this mass increase. ΔE = ______________ .

c. Finally, calculate the KE of the spaceship and compare it to the ΔE just found. KE = ______________ .

Chapter Review

19. In a certain frame of reference it is found that Newton's laws of motion are valid. What is the technical term for such a frame of reference?

20. A person riding in an elevator observes that her weight at a given moment is about 3/4 of her true weight.
   a. What is the magnitude and direction of the acceleration of the elevator? [Hint: Make a diagram showing the forces acting on the passenger. Use an inertial frame of reference.]

   b. What conclusions, if any, can be drawn concerning the direction in which the elevator is moving?

21. What is the upper limit, if any, to the speed that can be given to a proton in a very powerful accelerator?

22. What is the upper limit, if any, to the momentum of a proton in an accelerator?

23. A blindfolded person is standing in the back of a van as it travels along the highway. When the driver applies the brakes, the van slows from 100 km/h to 10 km/h in a time of 3 s. To the passenger, whose only frame of reference is the floor of the van on which he is standing, there appears to be a strong force propelling him toward the front of the van. He resists this "force" with considerable effort and manages to remain at "at rest" in his frame of reference by pushing himself backward.
   a. Write the force equation for this situation (during the deceleration) as viewed by the passenger for whom $a = 0$ in his van reference frame.

   b. Write the equation as viewed by a stationary observer.

   c. Calculate the acceleration of the van and passenger.

   d. Assuming the passenger to weigh 80 N, calculate the magnitude of the fictitious "force" that he experiences.

   e. The observer in an inertial frame of reference doesn't need to invoke a fictitious "force" toward the front of the van. What is the magnitude and direction of the single force on the passenger as seen by the stationary observer?

Solutions

1a. Newton's law of motion has exactly the same form in the second frame of reference because the second frame has a constant velocity with respect to an inertial frame.

b. It is also an inertial frame.

2a. No. Newton's second law does not have the form net $\mathbf{F} = m\mathbf{a}$ in an accelerated frame of reference. Because the van is traveling along a circular path, its velocity is changing. The frame of reference fixed to the van is an accelerated frame.

b. A frame of reference in which Newton's laws of motion (in their usual form) are not valid is called a non-inertial frame of reference.

3a. The three velocity vectors are shown in Fig. 28-3.

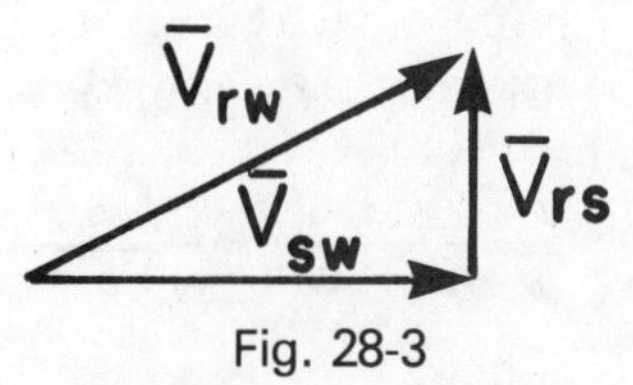

Fig. 28-3

b. The magnitude of the resultant velocity is

$$\mathbf{v}_{rw} = [(\mathbf{v}_{sw})^2 + (\mathbf{v}_{rs})^2]^{1/2} = [900 + 225]^{1/2}\ \text{km/h} = 33.5\ \text{km/h}$$

4a. It is necessary, of course, for the swimmer to head somewhat up stream so that her resultant velocity will have the required direction of due east. $\mathbf{v}_{SL} = \mathbf{v}_{SW} + \mathbf{v}_W$ with $\mathbf{v}_W$ perpendicular to $\mathbf{v}_{SL}$.

b. The magnitude of the resultant velocity is

$$\mathbf{v}_{SL} = [(v_{SW})^2 - (v_W)^2]^{1/2} = [9 - 1]^{1/2}\ \text{km/h} = 2.83\ \text{km/h}$$

c. $$t = \left(\frac{200\ \text{m}}{2.83 \times 10^3\ \text{m/h}}\right)\left(\frac{3600\ \text{s}}{1\ \text{h}}\right) = 2.55 \times 10^2\ \text{s}$$

5a. The resultant speed down stream is 3.0 km/h + 1.0 km/h = 4.0 km/h. The required time is

$$t_1 = \left(\frac{100\ \text{m}}{4.0 \times 10^3\ \text{m/h}}\right)\left(\frac{3600\ \text{s}}{1\ \text{h}}\right) = 90\ \text{s}$$

b. The resultant speed up stream is 3.0 km/h − 1.0 km/h = 2.0 km/h. The required time is

$$t_2 = \left(\frac{100\ \text{m}}{2.0 \times 10^3\ \text{m/h}}\right)\left(\frac{3600\ \text{s}}{1\ \text{h}}\right) = 180\ \text{s}$$

c. The total time, down stream and back, is

$$t_y = 90\ \text{s} + 180\ \text{s} = 270\ \text{s}$$

The ratio of the two times is

$$t_x/t_y = \frac{255\ \text{s}}{270\ \text{s}} = 0.944$$

d. This ratio can be calculated directly from the formula.

$$t_x/t_y = [1 - v^2/c^2]^{1/2} = [1 - (1/3)^2]^{1/2} = 0.944$$

6a. No. A rotating frame of reference is never an inertial frame. Fictitious forces are always present in rotating frames.

b,c. No. Newton's laws of motion are not valid in a non-inertial frame of reference. Specifically, an object that is subject to no forces will be seen to accelerate in the rotating frame of reference.

7. The jet is an inertial frame during the portions a, b, e, and g of its flight.

8. Sound waves are propagated through air; they are not propagated through a vacuum.
9. The velocity of light is exactly the same in all frames of reference.
10. It is the maximum possible velocity of any body relative to any observer.
11. The forms of the laws of physics are the same in all inertial frames of reference.
Measurements of the speed of light always give the same numerical value, regardless of the relative velocity of the observer and the source.

12. $t = \dfrac{t_0}{\sqrt{1 - v^2/c^2}} = \dfrac{10\text{ s}}{\sqrt{1 - (0.5)^2}} = \dfrac{10\text{ s}}{\sqrt{3/4}} = 11.5\text{ s}$

13. Given: $t_0 = t - (0.01\%)t = t(1 - 10^{-4})$

$$\frac{t}{t_0} = \frac{t}{t(1 - 10^{-4})} = \frac{1}{\sqrt{1 - v^2/c^2}}$$

Thus

$$1 - 10^{-4} = \sqrt{1 - v^2/c^2} \cong 1 - \tfrac{1}{2}(v^2/c^2)$$

$$10^{-4} = (1/2)(v/c)^2; \quad \frac{v}{c} = \sqrt{2} \times 10^{-2} = 1.4 \times 10^{-2}$$

$$v = (1.4 \times 10^{-2})(3 \times 10^8\text{ m/s}) = 4.2 \times 10^6\text{ m/s}$$

14. Because of the relative velocity between the space probe and the space station, the space probe appears shorter in its direction of motion. The observed length is

$$L = L_0\sqrt{1 - v^2/c^2} = (20.00\text{ m})\sqrt{1 - (0.3)^2} = 19.08\text{ m}$$

15. The mass of the moving spaceship is given by

$$m = \frac{m_0}{\sqrt{1 - v^2/c^2}}$$

Since v^2/c^2 is much smaller than one,

$$1/\sqrt{1 - v^2/c^2} = \cong [1 - (1/2)(v/c)^2]^{-1} \cong 1 + \tfrac{1}{2}(v/c)^2$$
$$m = m_0[1 + \tfrac{1}{2}(v/c)^2] = m_0 + (m_0/2)(v/c)^2$$
$$m - m_0 = \Delta m = (m_0/2)(v/c)^2 = m_0(1/18) \times 10^{-8}$$

The fractional increase in mass is

$$\frac{\Delta m}{m_0} = (1/18) \times 10^{-8} = 5.6 \times 10^{-10}$$

16. $\Delta E = (\Delta m)c^2 = 1\text{ kg}(3 \times 10^8\text{ m/s})^2 = 9 \times 10^{16}\text{ J}$

17. The total equivalent amount of energy in the universe remains constant. No forms of energy are excluded.
18a. $\Delta m = m_0(5.6 \times 10^{-10}) = (12 \times 10^3\text{ kg})(5.6 \times 10^{-10}) = 6.7 \times 10^{-6}\text{ g}$
b. $\Delta E = (\Delta m)c^2 = (6.7 \times 10^{-6}\text{ kg})(9 \times 10^{16}\text{ m}^2/\text{s}^2)$
$\Delta E \cong 6.0 \times 10^{11}\text{ J}$
c. $KE = \tfrac{1}{2}mv^2 = \tfrac{1}{2}(12 \times 10^3\text{ kg})(1 \times 10^4\text{ m/s})^2 = 6.0 \times 10^{11}\text{ J}$

The two results agree, as they should.

19. A frame of reference in which Newton's laws of motion are valid is called an *inertial* frame.

20. Let P represent the upward force that the elevator exerts on the passenger. Take the upward direction to be positive. See Fig. 28-4.

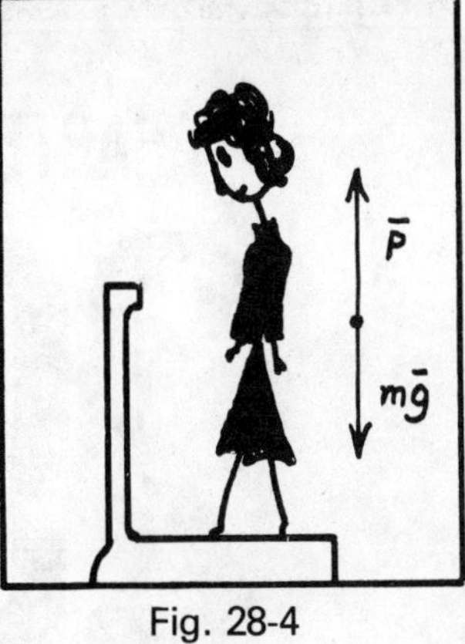

Fig. 28-4

a. $P - mg = ma$; $P = (3/4)\ mg$

$(3/4)\ mg - mg = ma$; $\quad a = -(1/4)\ g = -2.45\ m/s^2$

The negative sign indicates that the acceleration of 2.45 m/s² is downward.

b. No conclusions whatsoever can be drawn concerning the direction of the velocity.

21. The upper limit to the speed of any material particle is c, the velocity of light.

22. Since the mass of a particle is not limited, the momentum mv is not limited.

23a. Let F_{im} represent the imaginary or fictitious force experienced by the passenger. In the accelerated frame of reference the force equation could be written

$$F - F_{im} = 0$$

b. In the inertial frame of reference

$$F = ma$$

c. $a = \dfrac{\Delta v}{\Delta t} = \dfrac{10\ km/h - 100\ km/h}{3\ s}\left(\dfrac{1\ h}{3600\ s}\right) = -8.3\ m/s^2$

d. $ma = \dfrac{W}{g}a = \dfrac{80\ N}{9.8\ m/s^2}(8.3\ m/s^2) = 68\ N$

e. The observer in the inertial frame sees only a single force of 68 N acting on the passenger *toward the rear* of the van.

CHAPTER

29 Electrons and Photons

GOALS To introduce the concept of wave-particle duality of matter and radiation and to explore some of its consequences.

OBJECTIVES After completing this chapter the student should be able to do the following:

1. State the name and magnitude of the quantum of electric charge.
2. Describe the essential elements of an experiment in which the quantum of charge is measured.
3. Describe the essential features of an experiment to measure the charge-to-mass ratio of an electron.
4. Calculate the velocity of a charged object that has been accelerated by an electric field.
5. Calculate the acceleration and the magnetic force for a charged particle moving in a magnetic field.
6. Determine the charge-to-mass ratio for a charged particle moving through a uniform magnetic field.
7. Predict whether light of a given frequency will cause electrons to be ejected when the threshold frequency is known for that surface.
8. State the relationship that exists between intensity of light and number of photons emitted from a surface.
9. Name the property of light that determines whether electrons will be ejected from a surface.
10. Write the equation that expresses symbolically Planck's quantum hypothesis.
11. State the name of the quantum of radiation and write the equation that gives its energy.
12. Calculate one of the quantities—frequency, wavelength, energy, or Planck's constant—when the others are given.
13. Write Einstein's photoelectric equation and identify each term that appears in it.
14. Give the definition of the electron volt and calculate its equivalent in joules.
15. Calculate the KE of electrons ejected from a given metal when the wavelength of the incident light is known.
16. Calculate the work function of a surface from observations of the photoelectric effect.
17. Determine the threshold wavelength for a surface when its work function is known.
18. Name the two fundamental attributes of the electron and give the magnitude associated with each.
19. Write the equations relating a photon's mass to its energy and a photon's momentum to its wavelength.
20. Name the conservation laws that must be invoked in analyzing the collision of a photon with an electron.
21. Calculate the energy of a scattered photon from a knowledge of the KE of the recoil electron.
22. Describe the Compton effect in one or two sentences and state what characteristic of the photon is involved.
23. Name the two aspects of light included in the expression *dual nature of light*.
24. Describe experiments that illustrate the corpuscular nature of light and experiments in which the wave nature of light is in evidence.
25. Name the properties of a photon that lead to its being described as a particle.
26. Calculate the de Broglie wavelength of a moving particle.
27. Write down, symbolically, Heisenberg's uncertainty principle and identify each term appearing in it.
28. Calculate the uncertainty in a particle's position when the uncertainty in its momentum is known and vice versa.

SUMMARY

The electron is one of the building blocks of an atom. The average charge of an electron can be determined experimentally in electrolysis. Millikan devised the oil-drop experiment to measure the charge of single electrons. The result is 1.60×10^{-19} C.

The magnetic deflection method is one of the several means for determining the mass of the electron from a knowledge of its charge and the radius of its path in a magnetic field.

When light falls on a metallic surface, electrons are emitted from the metal. Careful measurements of the ejected electrons reveal that their energies depend on the frequency of the incident light and that their number per unit time depends only on the intensity of the light. These results constitute very strong evidence for the corpuscular nature of light. The light corpuscle, called a *photon*, is shown by this experiment and others to have an energy proportional to its frequency. Symbolically, $E = hf$ where h is Planck's constant and f is the frequency. The KE of the photoelectrons is given by $KE = hf - w$, where w is the work function of the surface.

The corpuscular nature of the photon is further evidenced by the Compton effect in which the momentum of the photon plays a major role as the photon collides with an electron. The photon's momentum is equal to Planck's constant divided by the wavelength, $p = h/\lambda$.

The wave-particle duality of the photon, a particle of zero rest mass, also characterizes particles of nonzero rest mass, the electron and the proton, in particular. The wavelength associated with a moving mass is given by $\lambda = h/p$. The wavelength of a macroscopic object, a rifle bullet for example, is so small as to be unobservable. Although the wavelength of an electron or a neutron is very small, it can be measured by means of diffraction from crystals. The small wavelength of an electron is taken advantage of in the electron microscope, permitting a much higher magnification than is possible with an optical microscope.

One of the consequences of the dual nature of light is that a lower limit is placed on the precision with which a photon's momentum and position may be simultaneously measured. Letting Δp and Δx represent the uncertainties in the measurements of momentum and position, their product cannot be significantly less than Planck's constant, h. Symbolically, $\Delta p \Delta x \approx h$, a relationship that is called Heisenberg's uncertainty principle.

QUESTIONS AND PROBLEMS

The following questions and problems are presented approximately in the order in which the topics are presented in the text. The recommended procedure is to read one or two sections in the text, solve the problems related to that material, and then read another portion of the text, etc.

1. The text states that "Even in the most delicate weighing experiments, we are dealing with some 10^{16} ions." Taking copper as an example, what is the sensitivity (in micrograms) of a balance that is capable of responding to 10^{16} ions?

2. How is it possible for a drop of oil to be in static equilibrium in the Millikan oil-drop apparatus? Draw a picture of the drop showing the forces that act on it.

3. A small oil drop containing an excess of two electrons is held stationary when the potential difference between the plates is 400 V. The separation of the plates is 1 cm.
   a. Which plate is positive?

   b. What is the electric field acting on the charge?

   c. How much does the oil drop weigh?

   d. What is the diameter of the oil drop? (Assume a density of 0.9 g/cm³.)

   e. Can you justify the statement that "diffraction effects prevent direct measurement of its radius"?

4. State in one sentence the essential result of the Millikan oil-drop experiment.

5. Why is it not possible to measure the mass of an electron with a sensitive balance? (Refer to Prob. 1.)

6. Describe briefly two mechanisms by which electrons may be caused to leave the surface of a conductor. (See the first two paragraphs of Sec. 29-2.)

7a. In Fig. 29-3 in the text, the voltage V, applied between the cathode and anode, is 2000 V. What is the magnitude of the velocity of the electrons as they pass through the small hole in the anode? (Use $m_0 = 9.11 \times 10^{-31}$ kg.)

$v =$ _______________.

b. The electrons, after being accelerated by the PD of 2000 V, move in a circular path of radius 10 cm. What is the magnitude of their acceleration?

$a =$ _______________.

c. In what direction are the electrons being accelerated?

d. What force is required to produce this acceleration?

$F =$ _______________.

e. How is it possible for a magnetic field to produce exactly the force required? What must be the magnitude of **B**?

$B =$ _______________.

f. What is the direction of **B**?

8. When the speed of the electron is very small compared to the speed of light, e/m has the value of 1.759×10^{11} C/kg. What value of e/m would be observed if the speed of the electron were 0.9 c?

9. In the magnetic deflection method for measuring e/m, the radius of the electron's path was 50 cm and the accelerating voltage was 250 V.

   a. What was the magnitude of the magnetic field?

   b. If the electrons moved counterclockwise in a horizontal plane, what was the direction of the magnetic field?

10. Calculate the energy of a photon of violet light (λ = 4000 Å) and of red light (λ = 6000 Å) in joules and in electron volts.

E(violet) = _______________ = _______________

E(red) = _______________ = _______________

11a. The work function for sodium metal is 2.46 eV. A thin sodium surface is illuminated by three different colors of light, red, green, and violet (λ = 6000 Å, 5000 Å, and 4000 Å), one after the other. In which case(s) will photoelectrons be ejected?

b. What is the maximum energy that the ejected electrons will have?

12. What is the threshold wavelength for photoemission from the surface of tungsten? (The work function of tungsten is about 4.5 eV.)

13. Give the name of (a) the quantum of charge and (b) the quantum of radiation. Which of these two has a rest mass of zero? Which one cannot travel at the speed c? Which one can *only* travel at the speed c?

14a. Write the equation for the momentum of a photon.

b. Calculate the momentum and the mass of a photon whose energy is 3.10 eV.

p = ____________.

m = ____________.

c. The rest mass of the electron is 9.11×10^{-31} kg. Calculate the energy in eV of a photon whose relativistic mass is 9.11×10^{-31} kg.

E = ____________.

d. What is the frequency of a photon whose relativistic mass is 9.11×10^{-31} kg?

f = ____________.

15a. Calculate the momentum of an x-ray photon whose wavelength is 0.154 nm.

p = ____________.

b. Calculate the relativistic mass of this photon.

m = ____________.

16. Calculate the momentum of a photon of yellow light whose wavelength is 600 nm.

17a. The calculation in Example 29-7 in the text shows that the de Broglie wavelength associated with a Ping-Pong ball is much too small to be observable. What about the rapidly moving pellet from an air gun? Calculate the de Broglie wavelength using m = 0.03 g and v = 300 m/s.

λ = ______________.

b. If the air-gun pellet above is shot through a slit whose width is slightly larger than the diameter ($w \cong 2$ mm), will diffraction effects be observed?

18. Calculate the KE and the wavelength of a neutron whose speed is 10^4 m/s.

KE = ______________.

λ = ______________.

19. The reovirus particles shown in the photograph of Fig. 29-11 in the text are about 50 nm in diameter.
 a. Why is it not possible to see these particles with a high-quality, high-magnification optical microscope?

 b. Supposing that the electrons used to obtain text Fig. 29-11 had a wavelength of 2 nm, what was their energy in eV?

E = ______________.

 c. Through what PD were they accelerated?

V = ______________.

20. Write a mathematical statement of Heisenberg's uncertainty principle.

21. The motions of astronomical objects can be predicted accurately. For example, astronomers can state with high precision exactly where the moon will be at 12:31 P.M. on July 25, 1991. On the other hand, the motion of a particular electron through an accelerator cannot be predicted with precision even over short periods of time. Explain why these situations are so different.

22. In the Compton effect a photon is incident upon an electron. After the collision has taken place, a scattered photon is observed as well as the recoiling electron. In this process which of the following quantities are conserved (or remain constant): momentum, frequency of photon, or energy?

23. In an experiment with fast-moving protons it has been possible to determine the momentum of the proton with an uncertainty of about $\pm 0.5 \times 10^{-32}$ kg·m/s. If the position of the proton were simultaneously measured along the direction of motion, what uncertainty would be expected?

Solutions

1. $(10^{16}\text{ ions})\dfrac{63.5\text{ g}}{6 \times 10^{23}\text{ ions}} \cong 1.0\ \mu\text{g}$

   The sensitivity of the balance is 1 μg.

2. 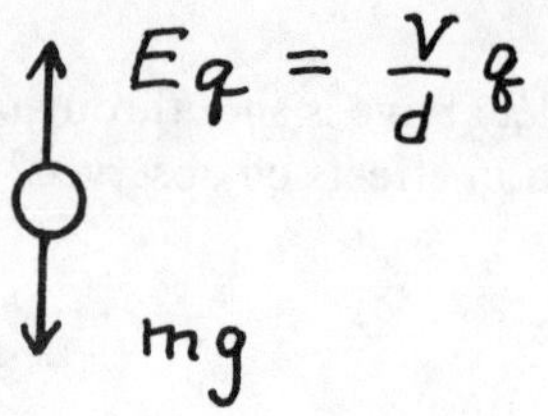 $Eq = \frac{V}{d}q$

   mg

   Fig. 29-1

3a. The upper plate is positive.

b. $E = V/d = 400\text{ V}/0.01\text{ m} = 4 \times 10^4\text{ V/m}$

c. $W = Eq = (4 \times 10^4\text{ V/m})(2 \times 1.6 \times 10^{-19}\text{ C})$
   $= 1.28 \times 10^{-14}\text{ N}$

d. $\text{Vol} = (4/3)\pi r^3 = \text{mass/density} = \dfrac{W/g}{d}$

$$r^3 = \frac{3(W/g)}{4\pi d} = \frac{3(1.28 \times 10^{-14}\text{ N})}{(9.8\text{ m/s}^2)(4)(0.9 \times 10^3\text{ kg/m}^3)\pi}$$

$r^3 = 3.46 \times 10^{-19}\text{ m}^3$; $r = 7.02 \times 10^{-7}\text{ m}$
$d = 14.0 \times 10^{-7}\text{ m}$

e. This oil drop is only about three wavelengths (of green light) in diameter. This is too small to be resolved with a microscope.

4. The Millikan oil-drop experiment showed that electric charged is quantized, and that the quantum of charge, called the electron, has a magnitude of 1.60×10^{-19} C.

5. The smallest mass that can be measured with a very sensitive balance is about 1 microgram. The mass of the electron is about 10^{-21} micrograms.

6. If the conductor is brought to a high temperature, a "white heat," a few electrons will be "boiled off." This is thermionic emission.

   If the surface is bombarded by energetic ions, some electrons will be ejected.

   A third method is to irradiate the surface with light of a sufficiently short wavelength.

7a. $v = \sqrt{\dfrac{2Ve}{m}} = \sqrt{\dfrac{2(2 \times 10^3\text{ V})(1.6 \times 10^{19}\text{ C})}{9.11 \times 10^{-31}\text{ kg}}}$

   $= 2.65 \times 10^7\text{ m/s}$

b. $a = v^2/R = (7.03 \times 10^{14})/(0.1\text{ m}) = 7.03 \times 10^{15}\text{ m/s}^2$

c. toward the center of the circular path

d. $F = ma = (9.11 \times 10^{-31}\text{ kg})(7.03 \times 10^{15}\text{ m/s}^2)$
   $= 6.40 \times 10^{-15}\text{ N}$

e. The magnetic force on a moving charge is perpendicular to the field **B** and perpendicular to the velocity. This force has just the direction required for uniform circular motion.

$$B = \frac{ma}{ev} = \frac{(9.11 \times 10^{-31}\text{ kg})(7.03 \times 10^{15}\text{ m/s}^2)}{(1.6 \times 10^{-19}\text{ C})(2.65 \times 10^7\text{ m/s})} = 1.51 \times 10^{-3}\text{ T}$$

f. **B** must be perpendicular to the plane of the circular path. If the electron is seen as moving clockwise, the direction of **B** is the direction of observation. (See text Fig. 29-3.)

8. $m = m_0/\sqrt{1 - (v/c)^2}$; $m_0/m = \sqrt{1 - (v/c)^2}$
$m_0/m = \sqrt{1 - (0.9)^2} = \sqrt{0.19} = 0.436$
$e/m = (e/m_0)(m_0/m) = (1.759 \times 10^{11}\ \text{C/kg})(0.436)$
$= 0.767 \times 10^{11}\ \text{C/kg}$

9. Given: $R = 0.5$ m, $V = 250$ V

$$\frac{e}{m} = \frac{2V}{B^2R^2};\ B^2 = \frac{2V}{(e/m)R^2}$$

a. $$B^2 = \frac{2(250\ \text{V})}{(1.76 \times 10^{11}\ \text{C/kg})(0.5\ \text{m})^2} = 1.14 \times 10^{-8}\ \text{T}^2$$

$B = 1.07 \times 10^{-4}$ T

b. **B** was upward, toward the observer.

10. For violet light:

$E = hc/\lambda = (6.63 \times 10^{-34}\ \text{J} \cdot \text{s})(3 \times 10^8\ \text{m/s})/(4 \times 10^{-7}\ \text{m}) = 4.97 \times 10^{-19}$ J
$E = (4.97 \times 10^{-19}\ \text{J})/(1.6 \times 10^{-19}\ \text{J/eV}) = 3.11$ eV

For red light:

$E = (6.63 \times 10^{-34}\ \text{J} \cdot \text{s})(3 \times 10^8\ \text{m/s})/(6 \times 10^{-7}\ \text{m}) = 3.32 \times 10^{-19}$ J
$E = (3.32 \times 10^{-19}\ \text{J})/(1.6 \times 10^{-19}\ \text{J/eV}) = 2.07$ eV

11a. The energies of the photons are given by: $E = hc/\lambda$. Thus $E = 2.07$ eV, 2.48 eV, and 3.11 eV for red, green, and violet. Photoelectrons will be ejected by green light and by violet light.

b. The maximum energy of the photoelectrons will be $hf - 2.46$ eV $= 0.65$ eV and 0.02 eV for violet light and green light, respectively.

12. At threshold the energy of the photon is just equal to the work function.

$$\lambda = \frac{1240\ \text{eV} \cdot \text{nm}}{4.50\ \text{eV}} = 276\ \text{nm}$$

13. The quantum of charge is the electron; it cannot travel at the speed of light. The quantum of radiation is the photon; it travels only at the speed of light. The photon has a rest mass of zero.

14a. $p = \frac{hf}{c}$ or $p = \frac{h}{\lambda}$

b. $$m = E/c^2 = \frac{(3.10\ \text{eV})(1.60 \times 10^{-19}\ \text{J/eV})}{(3 \times 10^8\ \text{m/s})^2} = 5.51 \times 10^{-36}\ \text{kg}$$

$p = mc = (5.51 \times 10^{-36}\ \text{kg})\ (3 \times 10^8\ \text{m/s}) = 1.65 \times 10^{-27}\ \text{kg} \cdot \text{m/s}$

or $$p = \frac{h}{\lambda} = \frac{6.63 \times 10^{-34}\ \text{J} \cdot \text{s}}{4 \times 10^{-7}\ \text{m}} = 1.65 \times 10^{-27}\ \text{kg} \cdot \text{m/s}$$

c. $E = mc^2 = (9.11 \times 10^{-31}\ \text{kg})(3 \times 10^8\ \text{m/s})^2/(1.6 \times 10^{-19}\ \text{J/eV})$
$= 5.12 \times 10^5$ eV

d. $$f = E/h = \frac{(9.11 \times 10^{-31}\ \text{kg})(3 \times 10^8\ \text{m/s})^2}{6.63 \times 10^{-34}\ \text{J} \cdot \text{s}} = 1.24 \times 10^{20}\ \text{Hz}$$

15a. $p = h/\lambda = (6.63 \times 10^{-34}\ \text{J} \cdot \text{s})/(1.54 \times 10^{-10}\ \text{m}) = 4.31 \times 10^{-24}\ \text{kg} \cdot \text{m/s}$
b. $m = p/c = (4.31 \times 10^{-24}\ \text{kg} \cdot \text{m/s})/(3 \times 10^{8}\ \text{m/s}) = 1.44 \times 10^{-32}\ \text{kg}$

16. $$p = h/\lambda = \frac{6.63 \times 10^{-34}\ \text{J} \cdot \text{s}}{6 \times 10^{-7}\ \text{m}} = 1.11 \times 10^{-27}\ \text{kg} \cdot \text{m/s}$$

17a. $$\lambda = \frac{h}{mf} = \frac{6.63 \times 10^{-34}\ \text{J} \cdot \text{s}}{(3 \times 10^{-5}\ \text{kg})(300\ \text{m/s})} = 7.37 \times 10^{-32}\ \text{m}$$

b. No. The slit is enormously wide compared to the de Broglie wavelength.

18. (See text Ex. 29-8 for details of calculation.)

$\text{KE} = 5.22 \times 10^{-1}\ \text{eV}$
$\lambda = 0.0397\ \text{nm}$

19a. Even an optically perfect microscope, one totally free of aberrations, cannot resolve objects that are smaller than the wavelength of the light employed. These virus particles are ten times smaller than the wavelength of visible light.

b. $$E = \frac{1240\ \text{eV} \cdot \text{nm}}{20\ \text{Å}} = 620\ \text{eV}$$

c. $V = 620$ volts

20. $\Delta p \Delta x \cong h$; or actually, to be quite correct, $\Delta p \Delta x \geq h/4\pi$

The second statement applies to a more precise, statistical definition of "uncertainty."

21. In the case of large-scale objects, whether planets, baseballs, or steel pellets, the de Broglie wavelength is very small compared with the dimensions of the object. Since the uncertainty in position is of the order of the de Broglie wavelength, large objects can be located very precisely. For a high-speed electron, on the other hand, the de Broglie wavelength is large compared with the size of the electron thus making the electron's position uncertain.

22. For the system as a whole (photon plus electron) both momentum and energy are conserved. The incident photon gives up some of its energy to the electron. As a result the scattered photon has less energy and, therefore, a lower frequency.

23. We will use the approximate statement of the uncertainty principle given in Eq. 29-12 of the text.

$$\Delta p \Delta x \approx h \quad \text{or } \Delta x \approx \frac{h}{\Delta p} = \frac{6.6 \times 10^{-34}\ \text{J} \cdot \text{s}}{5 \times 10^{-33}\ \text{kg} \cdot \text{m/s}} = 13\ \text{cm}$$

The best we can hope for is an uncertainty of 10 cm to 20 cm in the position of the proton.

REVIEW TEST FOR CHAPTERS 27–29

1. A simple lens, whose surfaces are spherical, does not form a perfect image. Three of the aberrations of a simple lens are spherical aberrations, off-axis astigmatism, and chromatic aberration. When a plano-convex lens is used to form an image of an object illuminated by the red light ($\lambda = 633$ nm) of a helium-neon laser, which of these aberrations will *not* be evident?

2. Two lenses are available for use with a certain camera. The first lens has an effective diameter of 4 cm and a focal length of 20 cm. The second lens has an effective diameter of 6 cm and a focal length of 60 cm. Find the ratio of the intensities of the images produced by these two lenses I_1/I_2.

3. A simple magnifier consists of double convex lens of which each surface has a radius of curvature of 5 cm. What is the maximum angular magnification that can be obtained with this magnifier? (Hint: First find the focal length of the lens from the lens-maker's equation.)

4. A crude microscope may be constructed of two identical lenses. Suppose that the lenses are separated by a distance of 22 cm and that each lens has a focal length of 2.0 cm. Calculate the overall magnifying power of this microscope. (The final image is at infinity.)

5. An astronomical telescope has an objective lens of 60-cm focal length and an eyepiece of 3-cm focal length. What is the magnifying power of this telescope?

6. A pilot wishes to reach a destination that is 200 km due south of her home field within the shortest possible time. She can maintain a constant air speed of 150 km/h, but she will have to take into account a 50 km/h wind from the west. What will her effective ground speed be and how long will it take to make the trip? (Hint: A diagram showing the relationship of the three velocity vectors will be helpful.)

7. A physicist measures the speed of light in a laboratory on earth obtaining the value $c = 3.0 \times 10^8$ m/s. Using the same apparatus, she carries out the measurement again aboard a spaceship moving at a speed $v = c/2 = 1.5 \times 10^8$ m/s. The second experiment, if carried out carefully, will yield the value
   a. $c/4$
   b. $c + c/2$
   c. $c/2$
   d. c
   e. $2c/\sqrt{3}$
8. An electron in an accelerator is traveling at one half the speed of light. The ratio of the mass of the electron to its rest mass, written m/m_0, is
   a. 0.90
   b. 1.01
   c. 1.15
   d. 1.33
   e. 1.50
9. A small grain of pure substance has a mass of 20 μg. If all this material could be converted into energy, how much energy would be produced?

10. In the Millikan oil-drop experiment, one measures the net electric charge on a small drop of oil suspended between a pair of horizontal plates. Assuming that measurements are made with reasonable care, which one of the following values of the net charge is *not* a possible outcome?
    a. -8.0×10^{-19} C
    b. -2.4×10^{-19} C
    c. -3.2×10^{-19} C
    d. $+8.0 \times 10^{-19}$ C
    e. $+1.6 \times 10^{-19}$ C
11. In an experiment to measure the ratio e/m, negatively charged particles (electrons) are caused to move in a horizontal circular path of radius R as illustrated in Fig. T-63. This is accomplished by
    a. applying a vertical magnetic field with **B** out of the drawing (upward).
    b. applying a vertical electric field with **E** into the drawing.
    c. applying crossed electric and magnetic fields.
    d. applying a vertical magnetic field with **B** into the drawing (downward).
    e. applying a horizontal electric field.

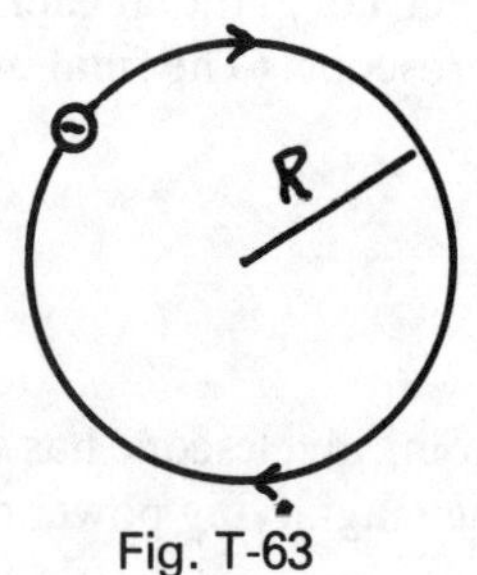

Fig. T-63

12. A beam of light is falling on a clean metallic surface. Which of the following properties determines whether electrons will be ejected from the metal?
    a. The angle of incidence and the intensity.
    b. The plane of polarization.
    c. The angle of incidence, the plane of polarization, and the frequency.
    d. Only the intensity.
    e. Only the frequency.

13. The "duality of matter" refers to
    a. the fact that matter has both mass and volume.
    b. the ability of matter on a microscopic scale to be in two places at once.
    c. the fact that electrons have both KE and PE.
    d. the dual model of matter that ascribes to ordinary matter both corpuscular and wavelike properties.
    e. the dual nature of the atomic nucleus, containing both protons and neutrons.
14. In an experiment with fast-moving protons, it has been possible to determine the momentum of the proton with an uncertainty of about $\pm 0.5 \times 10^{-32}$ kg • m/s. If the position of the proton were simultaneously measured along the direction of motion,
    a. the uncertainty would be 1.3×10^{-2} mm.
    b. the proton would be knocked out of its circular path.
    c. the Heisenberg principle would not apply.
    d. the uncertainty would be at least of the order of a few centimeters and perhaps much more.
    e. the uncertainty would be of the order of a few centimeters and perhaps much less.

CHAPTER

30 The Outer Atom

GOALS To introduce the principal features of the electronic structure of atoms, including the Bohr model for hydrogen.

OBJECTIVES After completing this chapter the student should be able to do the following:

1. Describe the main features of the Rutherford nuclear model of the atom giving orders of magnitude of the distances involved.
2. Name at least two ways in which atoms can be given enough energy to cause them to emit light.
3. Describe the essential character of the spectrum of a gas as compared to that of a liquid or solid.
4. Name the state of a substance (solid, liquid, or gas) in which the spectrum of the substance can be used for identification.
5. Write a mathematical statement of Bohr's postulate concerning the energy of a photon emitted by an atom.
6. Calculate the frequency and the wavelength of a quantum of radiation emitted by an atom in changing from one energy state to another.
7. Use an energy-level diagram of a simple atom to predict the energies of photons emitted and absorbed by that atom.
8. Write Bohr's postulate concerning the allowed values of the angular momentum of an electron in an atom.
9. Calculate the wavelength of a spectral line from a knowledge of the energy levels involved.
10. Combine expressions for the Coulomb force on an electron and the centripetal force to obtain an expression for the radius of an allowed orbit.
11. Use the expression for the energy of an electron in a Bohr orbit to calculate energy levels and radiated wavelengths for the hydrogen atom and for singly ionized helium.
12. Describe the terms *emission spectrum* and *absorption spectrum* and state the relationship between the two.
13. Use an energy-level diagram to predict the wavelengths of emitted photons after absorption of a photon of known energy.
14. State the order of magnitude of the lifetime of a ground state, an ordinary excited state, and a metastable state.
15. State the meaning of the term *ionization potential* as applied to an atom.
16. Use an energy-level diagram to predict whether photons of a given energy will be absorbed by an atom.
17. Describe briefly and qualitatively the difference between the Bohr model and quantum mechanical model as regards the location of the electron in a hydrogen atom.
18. Explain the terms *normal population* and *inverted population* as used to describe the number of atoms in different energy states.
19. State two of the three conditions that are necessary for laser action to take place.

SUMMARY

Most of our present knowledge of *atomic structure*, the arrangement of the electrons surrounding the nucleus, has been obtained from the light that is emitted or absorbed by the atom. However, the first convincing evidence that the positive charge of an atom is concentrated in an extremely small volume at the center was obtained by Rutherford in a quite different way. Rutherford arranged for a narrow stream of *alpha* particles (helium nuclei) to impinge upon a thin metallic foil. The rather large deflections of a few of the alpha particles led Rutherford to conclude that the positive charge occupied a volume whose diameter was about 10^{-14} m, some 10,000 times less than the diameter of the atom.

When a substance is in the vapor phase, its atoms are far enough apart from one another that they can act independently. When a vapor is heated to incandescence, the individual atoms radiate electromagnetic waves of well-defined frequencies, characteristic of the atom involved. The set of frequencies, or wavelengths, emitted by an atom furnishes detailed information about the structure of the electron cloud surrounding the nucleus of that species of atom.

Atoms in the liquid or solid phase are packed so closely together that they interact strongly with one another. As a result, the radiation emitted by an incandescent solid or liquid consists of a continuous spectrum that is a function only of the temperature of the material.

Hydrogen, the simplest of all atoms, consists of one electron bound to a single proton, its nucleus. The spectrum of light emitted by a hydrogen atom contains only a few frequencies related to one another by a simple mathematical law. Bohr developed a model for the hydrogen atom that accounted remarkably well for the observed spectrum. Bohr's model was based on three assumptions, or postulates:

(1) The energy of the atom may have any of a set of discrete values, E_1, E_2, E_3, ...

(2) The energy of the atom may suddenly change from one of these *energy states* to another, emitting or absorbing a photon in the process. The energy of the emitted photon is exactly equal to the difference in energy of the two states involved. Thus, $hf = E_2 - E_1$.

(3) The allowed orbits, corresponding to the energy states, are those for which the angular momentum of the electron is an integral multiple of Planck's constant divided by 2π. Symbolically, ang. mom. $= n(h/2\pi)$.

With these three postulates Bohr was able to calculate the frequencies of all the wavelengths emitted by the hydrogen atom. The formula is

$$f = \frac{(13.6 \text{ eV})}{h}\left(\frac{1}{n^2} - \frac{1}{m^2}\right)$$

where h is Planck's constant and n and m are integers ($m > n$).

When an atom goes from a higher energy state to a lower one, a photon is *emitted*. The opposite transition, from a lower energy state to a higher one, may be caused by the *absorption* of a photon. The frequencies absorbed by an atom are fewer than the frequencies emitted because most absorbing atoms are in the lowest, or *ground*, state and can undergo transitions only to higher energy states. The relationships between energy states, emitted radiations and absorbed radiations, is most clearly shown by an *energy-level diagram*.

The Bohr model has been superseded by quantum-mechanical descriptions of atoms. In quantum mechanics the location of the electron is described not in terms of orbits, but rather as a probability distribution; the electrons in an atom are seen as resembling clouds more nearly than point particles.

Under very special conditions a photon impinging on an atom may *stimulate* the emission of a second photon of the same frequency and phase. If a large enough number of atoms are present in the upper level of the transition, the result will be a copious output of *coherent* light at that frequency. The device that produces coherent radiation in this way is called a *laser*. Invented in physics laboratories relatively recently, the laser is the basis for a multibillion dollar industry.

QUESTIONS AND PROBLEMS

The questions and problems that follow are designed to help you learn some of the important ideas of atomic structure. An adequate treatment of the subject is difficult to present, even if the entire semester were devoted to it. The interested student is directed to the references that appear in the text at the end of the chapter.

1. What is an α particle?

2. When Rutherford allowed fast-moving α particles to impinge on very thin metallic foil he found that most of the α particles passed through the foil as if it were not there but that a few underwent much larger deflections than expected. What model of atomic structure would account for these results?

3. What is the approximate diameter of the "center of charge" that Rutherford found? (Give only the order of magnitude.)

4. The head of a pin is about 1 mm in diameter and its length is about 25 mm. Using the head of a pin to represent the nucleus of an atom, at what distance would an electron be found? In other words, what is the radius of an atomic model that uses a pinhead to represent the nucleus?

5. By what factor does the volume of an atom exceed the volume occupied by the nucleus? Is it correct to say that an atom is mostly empty space?

6. Fig. 30-2 in the text contains emission spectra of several elements. The instrument used was similar to the spectroscope illustrated in text Fig. 27-31 with a photographic film placed at the focal plane of L_1.
   a. In order to obtain these spectra, the elements had to be in what state or phase (solid, liquid, or gas)?

   b. Which of these elements would you expect to emit a yellowish light when placed in a flame or in an electric arc?

7. The opening photograph for this chapter contains emission spectra, absorption spectra, and a wavelength scale. Compare the unknown emission spectra in the photograph with the six spectra in text Fig. 30-2 to determine if the unknown element could be Hg, Na, He, Ne, or H. Use the wavelength scale that accompanies the spectra.

8. What property of a solid determines the nature of the light that it gives off (mass, volume, shape, or temperature)?

9. Write Bohr's three postulates as succinctly as possible. (Two of the postulates can be written as equations.)

   (1) ______________________________.

   (2) ______________________________.

   (3) ______________________________.

10. The following questions can be answered by referring to the energy-level diagram for hydrogren, text Fig. 30-3.

    a. Which of the following are allowed energy levels for a hydrogren atom? Cross out the energies that are not allowed.

       9.5 eV, 12.75 eV, 11.8 eV, 0.0 eV, 13.06 eV, 12.5 eV, 13.6 eV.

    b. What is the energy in eV of the emission line H_γ?

       $\Delta E =$ ______________.

    c. What is the frequency of the red line in the Balmer series (H_α)?

       $f =$ ______________.

    d. What is the wavelength of the first line of the Lyman series (the lowest energy transition)?

       $\lambda =$ ______________.

11. Calculate the two lowest values of angular momentum for a hydrogen atom according to Bohr's postulate.

12. Calculate the radius of the first Bohr orbit for doubly ionized lithium. (The atomic number of lithium is 3.)

13. In the case of hydrogen, the radius of the first Bohr orbit, calculated in Sec. 30-3, is 0.529 Å (0.0529 nm).

    a. Will there be a Bohr orbit whose radius is 2×0.529 Å? How about 3×0.529 Å?

    b. What is the principal quantum number (the integer n) corresponding to the orbit whose radius is 4.761×10^{-10} m?

       $n =$ ______________.

14. Use the Bohr energy formula to calculate the energy of a photon emitted by an atom of He in a transition from the state $n = 4$ to the state $n = 1$.

15. Assume, for the purposes of this problem, that the energy-level diagram for mercury in Fig. 30-7 in the text is complete, i.e., that all allowed energy levels below 8.82 eV are shown in that figure.
 a. An evacuated quartz bulb contains mercury vapor at room temperature. A beam of light traversing the mercury vapor contains photons of the following energies: 4.50 eV, 4.67 eV, 4.75 eV, 4.80 eV, and 4.89 eV.
 The mercury vapor can absorb certain of these photons but not others. Why not?

 List the energies corresponding to photons that *cannot* be absorbed.

 b. Calculate the wavelength of an emission line between the state E_6 and E_3.

 $\lambda =$ ____________.

 c. Give estimates of the lifetimes of the energy levels E_1, E_2, and E_3. (Note that E_2 is metastable.)

16. When an atom is in a metastable state, how can it lose energy and fall back to its ground state?

17. How much energy (in eV) is required to completely remove one electron from a mercury atom?

18. In Chapter 17 we showed that the average KE of a molecule of a gas is given by

$$\text{KE} = \frac{3}{2}\frac{nR}{N}\text{T}$$

where n is the number of moles, R is the gas constant, and N is the number of molecules. Since nR/N is a universal constant, the KE per molecule may be simply written

$$\text{KE} = \frac{3}{2}kT$$

k having the value of 1.38×10^{-23} J/K.

a. What is the average energy in eV of the molecules of a gas at room temperature? (Use $T_C = 23°C$, i.e., $T = 300$ K.)

b. In a bulb containing mercury vapor at room temperature, what fraction of the atoms would you expect to find in the first excited state, E_2?

19. Suppose that an electrical discharge is sustained in a glass tube containing mercury vapor so that some mercury atoms are in the excited energy states E_2, E_3, E_4, etc.
    a. If a photon whose energy is $(7.70 - 5.43)\text{eV} = 2.27$ eV passes through the mercury vapor, what transitions are possible? (Use the labels for energy levels given in Fig. 30-7 in the text.)

    b. What are the technical terms for these two processes?

    c. Is the principle of energy conservation violated when one photon strikes an atom and two photons leave? Explain.

20. Suppose the energy state E_7 of mercury contains 100 times more atoms than the energy state E_3. We say that state E_7 is 100 times more ______________ than state E_3.
    a. What is the technical term to describe this situation?

    b. If photons of energy 3.04 eV pass through a large number of these mercury atoms, which of the two possible transitions will be more likely to occur?

c. What does it mean, in this case, to say that there will be amplification?

d. When the 3.04 eV photon strikes a mercury atom in the E_7 state and causes a second photon to be produced by stimulated emission, what is the phase difference between the two photons?

21. Write the three conditions required for laser action.

Solutions

1. An α particle is the nucleus of a helium atom. It consists of two protons and two neutrons tightly bound together.
2. All of the positive charge of the atom must be confined to a very small volume at the center of the atom.
3. 10^{-14} m
4. The ratio of the diameter of an atom to the diameter of the nucleus is $(10^{-10}\text{ m})/(10^{-14}\text{ m}) = 10^4$. Thus the distance from the nucleus (pinhead) to an electron would be about $(0.5 \times 10^{-3}\text{ m})(10^4) = 5$ meters, more than 15 feet.
5. Vol. $\propto r^3$, thus the volume of the atom is $(10^4)^3 = 10^{12}$ times the volume of the nucleus. The space occupied by the nucleus is insignificant compared with that of the atom, but that space is occupied by the electron "cloud."

6a. the gaseous phase
b. Sodium. The prominent emission lines are in the yellow.

7. The spectrum recorded in the photograph is in the wavelength range 390 nm to 420 nm. None of the five elements whose spectra are shown in text Fig. 30-2 have a large number of strong lines in this region. The photograph shows the spectrum of the sun, a bright background with dark absorption lines, and, above and below it, the spectrum of iron.
8. The emission spectrum of an incandescent solid depends only on temperature.
9. (1) The atom can exist only in certain energy states, $E_1, E_2, E_3, \ldots$
(2) The atom radiates only in changing from one energy state to another, the energy of the emitted photon being

$$hf = E_2 - E_1$$

(3) The only orbits allowed for the electron are those for which its angular momentum $= n\,h/2\pi$ where $n = 1, 2, 3, \ldots$

10a. Allowed levels are 12.75 eV, 13.06 eV, and 13.60 eV.
b. 13.06 eV − 10.20 eV = 2.86 eV
c. $hf = 12.09\text{ eV} - 10.20\text{ eV} = 1.89\text{ eV}$

$$f = \left(\frac{1.80\text{ eV}}{6.63 \times 10^{-34}\text{ J}\cdot\text{s}}\right)\left(\frac{1.6 \times 10^{-19}\text{ J}}{1\text{ eV}}\right) = 4.56 \times 10^{14}\text{ Hz}$$

d. $hf = h\dfrac{c}{\lambda} = \Delta E$

$$\lambda = \frac{hc}{\Delta E} = \frac{(6.63 \times 10^{-34}\text{ J}\cdot\text{s})(3 \times 10^{8}\text{ m/s})}{(10.2\text{ eV})(1.6 \times 10^{-19}\text{ J/eV})} = 1.22 \times 10^{-7}\text{ m}$$

11. angular momentum $= nh/2\pi$
$h/2\pi = 1.055 \times 10^{-34}\text{ J}\cdot\text{s}$; $2(h/2\pi) = 2.110 \times 10^{-34}\text{ J}\cdot\text{s}$

12. $r = \dfrac{n^2h^2}{4\pi^2 mk\, Z\, e^2}$ $\quad n = 1$, $Z = 3$

$$r = \frac{(1)^2(6.63 \times 10^{-34}\ \mathrm{J \cdot s})^2}{(4\pi^2)(9.11 \times 10^{-31}\ \mathrm{kg})(9.00 \times 10^9\ \mathrm{N \cdot m^2/C^2})(3)(1.60 \times 10^{-19}\ \mathrm{C})^2}$$

$$= 0.177 \times 10^{-10}\ \mathrm{m} = 0.177\ \text{Å} = 0.0177\ \mathrm{nm}$$

13a. The radii of the Bohr orbits are proportional to n^2. Thus the allowed radii are:

4(0.529)Å = 2.116 Å = 0.212 nm
9(0.529)Å = 4.761 Å = 0.476 nm, etc.

b. $\dfrac{0.476\ \mathrm{nm}}{0.0529\ \mathrm{nm}} = 9$ which corresponds to $n = 3$

14. $hf = 13.60\ \mathrm{eV}(2)^2(1/n^2 - 1/m^2)$
$hf = 13.60\ \mathrm{eV}(2)^2(1/1 - 1/4^2) = 51\ \mathrm{eV}$

15a. A photon can be absorbed only if its energy corresponds exactly with the difference between two energy levels of the atom.* At room temperature almost all atoms are in the ground state. Thus the allowed energies of absorption are: 4.67 eV, 4.89 eV, 5.46 eV, etc.
The photons with energies of 4.50 eV, 4.75 eV, and 4.80 eV cannot be absorbed.

b. $E_6 - E_3 = 7.73\ \mathrm{eV} - 4.89\ \mathrm{eV} = 2.84\ \mathrm{eV}$

$$\lambda = \frac{hc}{\Delta E} = \frac{(6.63 \times 10^{-34}\ \mathrm{J \cdot s})(3 \times 10^8\ \mathrm{m/s})}{(2.84\ \mathrm{eV})(1.6 \times 10^{-19}\ \mathrm{J/eV})}$$

$$= 3.48 \times 10^{-7}\ \mathrm{m}$$

c. Order of magnitude estimates of these lifetimes are: E_1 (ground state), hours or days; E_2 (metastable state), 10^{-3} s to 1 s; E_3 (ordinary excited state), 10^{-8} to 10^{-9} s.

16. It can lose its energy in a collision with another atom or with the walls of the container. It cannot lose its energy by radiating a photon.

17. $E = 10.38$ eV (This is the ionization energy.)

18a. $\mathrm{KE} = \dfrac{3}{2}\ \dfrac{(1.38 \times 10^{-23}\ \mathrm{J/K})(300\ \mathrm{K})}{1.6 \times 10^{-19}\ \mathrm{J/eV}} = 3.88 \times 10^{-2}\ \mathrm{eV}$

b. The fraction of atoms in the excited state will be essentially zero. Very few atoms will have energies as high as 10 times the average. Since the energy of the first excited state is more than 100 times the average KE, it will be unpopulated at room temperature.

19a. A 2.27 eV photon may induce the transition E_4 to E_6, an absorption, or the transition E_6 to E_4, a stimulated emission.

b. $E_4 \rightarrow E_6$ absorption; $E_6 \rightarrow E_4$ stimulated emission

c. No. The energy of the second photon comes from potential energy stored in the excited atom.

20. populated

a. This is an inverted population since the higher energy state would normally have a smaller population.

b. $E_7 - E_3 = 7.90\ \mathrm{eV} - 4.86\ \mathrm{eV} = 3.04\ \mathrm{eV}$
The transition E_7 to E_3 (stimulated emission) is 100 times more probable because there are 100 times more atoms in state E_7 than in state E_3.

c. The number of photons of energy 3.04 eV will increase.

d. Zero. The emitted photon is exactly in phase with the photon that stimulated it.

21. The upper energy state must be metastable. There must be an inverted population. The atoms undergoing laser action must, ordinarily, be in a resonant cavity so that the radiation is reflected back and forth many times.

---

* This condition is necessary but not sufficient. Certain transitions between energy may be forbidden for other reasons.

CHAPTER 31

Atomic Structure

GOALS To study the principal features of a more detailed model of atomic structure based on quantum mechanics.

OBJECTIVES After completing this chapter the student should be able to do the following:

1. Name the four quantum numbers required for a complete description of the state of an atom and give the symbol for each.
2. List the range of values that are possible for each of the four quantum numbers.
3. Name the quantum number that becomes especially important when the atom is placed in an external magnetic field.
4. State the Pauli exclusion principle.
5. Use the Pauli exclusion principle to show that two electrons fill the state $n = 1$ and that eight electrons fill the state $n = 2$.
6. Use the rules regarding the allowed values of quantum numbers l, m_l, and m_s to determine the number of electrons in a given shell.
7. Write the spectroscopic notation for the ground state of an atom using the table of electronic configurations for the ground state.
8. Identify an element from its ground state configuration with the aid of a table of electronic configurations.
9. State which of the electrons surrounding the nucleus are more easily removed.
10. State how x-ray photons differ from photons of visible light.
11. Describe, qualitatively, how x-ray photons are produced.
12. Calculate the minimum x-ray wavelength from a knowledge of the voltage used in the x-ray tube.
13. Calculate the approximate energy and wavelength of an x-ray photon from a knowledge of the atomic number and the principal quantum numbers.

SUMMARY

Although the Bohr model gives an adequate description of the hydrogen atom, it fails to predict the energy levels of even the simplest atom having more than one electron. The new *quantum mechanics* was phenomenally successful in explaining the detailed structure of even complex atoms. In addition, the new theory led to an understanding of such diverse phenomena as chemical bonding, superconductivity, and transistor action in semiconductors.

In the quantum mechanical description, an atom is characterized by four *quantum numbers*, three associated with the distribution of electron charge in the atom and a fourth associated with the electron's *spin*. The names and symbols of the four quantum numbers are: (1) *principal quantum number n*, (2) *orbital quantum number l*,

(3) *magnetic quantum number* m_l, and (4) *spin quantum number* m_s. The energy of an atom depends on n, l, and m_s. Since these quantum numbers may take on different values, a large number of different energy states results. When the atom is in the presence of an external magnetic field, the allowed orientations of the atom, given by m_l, give rise to slightly different energies, a phenomenon called the *Zeeman effect*.

The Pauli exclusion principle states that no two electrons in the atom can have all their quantum numbers the same. As a result the electrons are grouped into *shells*, labeled according to the value of the principal quantum number n. The K shell ($n = 1$) cannot have more than 2 electrons, the L shell ($n = 2$) not more than 8, and the M shell ($n = 3$) not more than 18 electrons.

The arrangement of the electrons into shells and subshells accounts for the periodic arrangement of the elements according to their chemical properties. When the outer shell is completely filled, as is the case for the rare gases He, Ne, Ar, and Kr, all electrons are tightly bound and are not available for forming bonds with other atoms. This accounts for the chemical inertness of the rare gases.

Electrons in the inner shells of an atom are very tightly bound. An atom, from which an inner shell electron has been removed, has a large potential energy. When such an atom makes a transition to a lower energy state, the photon emitted has a large energy and a short wavelength. These *x-ray* photons penetrate solid matter much more easily than do photons of visible light. Because their wavelength is of the order of magnitude of the inter-atomic distance in a crystal (≈ 0.1 nm $= 1$ Å), x-rays can be diffracted by a crystal, in the same way that visible light is diffracted by a grating.

QUESTIONS AND PROBLEMS

1. In the quantum mechanical model of an atom, three quantum numbers are needed to describe the probability cloud for the electron's position. Give the name of each of these quantum numbers and the symbol for each.

2. If the principal quantum number for a particular electron is 7, what are the values that l may have?

3. Ionized helium ${}_4He^+$ is hydrogenlike because its nucleus is surrounded by a single electron. Calculate the energy of the photon emitted when the electron of this atom makes a transition from the state $n = 4$ to the state $n = 3$. What is the wavelength of this photon?

4. List all the values allowed for l and m_l when $n = 4$.

5. We have seen in Chapter 22 that a current loop acts like a small magnet; it has north and south poles and experiences a torque when placed in an external magnetic field. Fig. 31-1 shows a current loop in four positions with respect to an external field **B**. (The plane of the loop is perpendicular to the drawing.) In which position does the system have the most energy? In which position does it have the least energy?

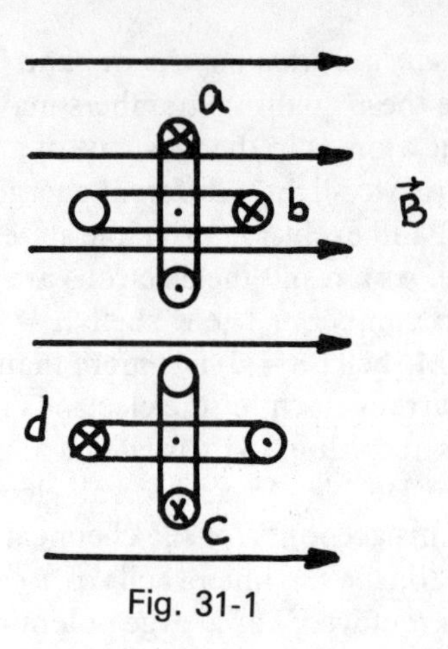

Fig. 31-1

6. Give the symbol for the quantum number associated with orientation of electron spin. What values are allowed for this quantum number?

7. The *M* shell has a principal quantum number *n* equal to 3. What are the allowed values of *l* for the *M* shell? What are the allowed values of m_l when $l = 2$? Tabulate the values of l, m_l, and m_s for the *M* shell and show that the total number of electrons required to completely fill this shell is 18.

8. The spectroscopic notation for $l = 0$ is *s*. For $l = 1$ it is *p*. What letters are used to represent *l* values of 2, 3, and 4?

9. The ground state of beryllium has the configuration $1s^22s^2$, which indicates that there are 2 electrons in the 1*s* state ($n = 1$, $l = 0$) and 2 electrons in the 2*s* state ($n = 2$, $l = 0$). What is the configuration of the ground state of boron, the element that follows beryllium?

10. Use Table 31-2 in the next to identify the element whose ground state configuration is $1s^22s^22p^4$.

11. Identify the element whose ground state configuration is $1s^22s^22p^63s^23p^1$.

12. The accelerating voltage in an x-ray tube has been set to 18,500 eV. What is the smallest wavelength of the x-rays produced?

13. Calculate the approximate energy and wavelength of the K_β line of $_{74}W$ (tungsten).

Solutions

1.

Name	*Symbol*
principal quantum number	n
orbital quantum number	l
magnetic quantum number	m_l

2. If $n = 7$, then $l = 0, 1, 2, 3, 4, 5, 6$.
3. The energy of a hydrogenlike atom is given by the Bohr formula,

$$E_n = \frac{-13.60(Z)^2}{n^2}\text{eV}$$

For $_4He^+$ $Z = 2$.

$$E_4 = -(13.6)(2)^2/(4)^2 \text{ eV} = -3.40 \text{ eV}$$
$$E_3 = -(13.6)(2)^2/(3)^2 \text{ eV} = -6.04 \text{ eV}$$
$$\Delta E = E_4 - E_3 = 2.64 \text{ eV}$$

$$\Delta E = h\left(\frac{c}{\lambda}\right)$$

$$\lambda = \frac{hc}{\Delta E} = \frac{(6.63 \times 10^{-34} \text{ J} \cdot \text{s})(3.00 \times 10^8 \text{ m/s})}{(2.64 \text{ eV})(1.60 \times 10^{-19} \text{ J/eV})}$$

$$= 4.71 \times 10^{-7} \text{ m}$$

4. If $n = 4$, $l = 0, 1, 2, 3$, and $m_l = -3, -2, -1, 0, 1, 2, 3$.
5. The energy of the system is greatest when the field produced by the loop is opposite to (antiparallel) the external field, case *a*. The energy is least when the field produced by the loop is aligned with (parallel to) the external field, case *c*.
6. m_s. The spin magnetic quantum number, m_s, may have one of two values, either $+\frac{1}{2}$ or $-\frac{1}{2}$.

7. The allowed values of the four quantum numbers may be tabulated as follows:

n	l	m_l	m_s
3	0	0	$+\frac{1}{2}, -\frac{1}{2}$
3	1	−1	$+\frac{1}{2}, -\frac{1}{2}$
3	1	0	$+\frac{1}{2}, -\frac{1}{2}$
3	1	1	$+\frac{1}{2}, -\frac{1}{2}$
3	2	−2	$+\frac{1}{2}, -\frac{1}{2}$
3	2	−1	$+\frac{1}{2}, -\frac{1}{2}$
3	2	0	$+\frac{1}{2}, -\frac{1}{2}$
3	2	1	$+\frac{1}{2}, -\frac{1}{2}$
3	2	2	$+\frac{1}{2}, -\frac{1}{2}$

There are 18 possible combinations of quantum numbers with $n = 3$. Thus 18 electrons are required to fill the M shell.

8. $l \rightarrow$ 0 1 2 3 4
notation → *s p d f g*

9. The electron that is added in going from beryllium to boron must have $n = 2$ and $l = 1$. Thus the ground state configuration for boron is $1s^2 2s^2 2p$.

10. The element is oxygen.

11. The element is aluminum.

12. The shortest wavelength corresponds to the largest energy of the photon. The largest energy that the x-ray photon can acquire is the total energy of the bombarding electron, 18,500 eV in this case. Thus

$$\frac{hc}{\lambda} = 18{,}500 \text{ eV}; \quad \lambda = \frac{hc}{18{,}500 \text{ eV}} = \frac{(6.63 \times 10^{-34} \text{ J·s})(3.00 \times 10^8 \text{ m/s})}{(1.85 \times 10^4 \text{ eV})(1.60 \times 10^{-19} \text{ J/eV})}$$

$$\lambda = 6.7 \times 10^{-11} \text{ m} = 0.067 \text{ nm}$$

13. The energy of the x-ray photon is given by the approximate formula,

$$E_{K\beta} = (13.6 \text{ eV})(Z - 1)^2 \left(\frac{1}{1^2} - \frac{1}{3^2}\right)$$

For tungsten $Z = 74$. Thus

$$E_{K\beta} = (13.6 \text{ eV})(73)^2(1 - 1/9) = 6.4 \times 10^4 \text{ eV}$$

The corresponding wavelength is

$$\lambda = \frac{1240 \text{ eV} \cdot \text{nm}}{64{,}000 \text{ eV}} = 0.019 \text{ nm}$$

CHAPTER 32 The Nucleus

GOALS To study the fundamental nuclear properties, including natural and artificial radioactivity.

OBJECTIVES After completing this chapter the student should be able to do the following:

1. Name and identify the three kinds of radiation emitted by such radioactive substances as uranium, polonium, and radium.
2. State which of these radiations are deflected by magnetic and electric fields.
3. Name the most penetrating of these types of radiation and the least penetrating.
4. Determine the atomic number, the mass number, and the number of neutrons from the symbol of a nuclide.
5. State how one isotope of an element differs from another, and state also the properties they have in common.
6. State the definition of unified atomic mass unit and give its standard symbol.
7. Calculate the mass (in kg) of 1 u and its energy equivalent in MeV.
8. Give the technical term for the work required to "take apart" an assemblage of particles.
9. Calculate the binding energy (in MeV) of a nucleon in a particular nucleus with the aid of a table of nuclear masses.
10. Calculate the binding energy per nucleon with the aid of a table of nuclear masses.
11. State the qualitative relationship between binding energy and stability of a nucleon.
12. Determine the number of atoms of a species that remain after a given lapse of time, when the half-life is known.
13. Write the mathematical expression for the number ΔN of nuclei that decay during a short time interval Δt.
14. Write the equation that represents the law of radioactive decay, the formula for N as a function of time.
15. Write the equations that relate half-life $T_{1/2}$ and activity to the disintegration constant λ.
16. Calculate the activity of a sample from its activity at a previous time or from the number of nuclei present.
17. Calculate one of the quantities A, λ, $T_{1/2}$, or N when a sufficient number of the others are known.
18. Given the equation representing a nuclear process, identify the nature of the process and name the emitted radiation (or particle).
19. Name the two states of charge of the electron and give the symbol for each.
20. Name the three ways in which a light nucleus decays.
21. Name three additional modes of decay that may occur with very heavy nuclei.
22. Name the four conservation laws that hold for all nuclear processes.
23. Give the name of the particle of zero rest mass that accounts for the missing energy in β decay.
24. Identify all terms in the equation that represents β decay.
25. State the magnitude of the quantum of *orbital* angular momentum and the magnitude of the quantum of *spin* angular momentum.
26. Give the chemical symbol for the common, light particles, alpha particle, proton, deuterium, neutron, and gamma-ray photon.
27. Relate the terms "stable" and "unstable" to the half-lives of nuclei.

28. Write a nuclear reaction either in the abbreviated form or in the full form from a verbal description and vice versa.
29. Identify an unknown nucleus (or nucleon or other particle) in the equation for a nuclear reaction.
30. Name two of the four devices that permit the paths of charged particles to be observed.

SUMMARY

A few dozen heavy elements found in small quantities on earth, particularly radium and polonium, emit radiation spontaneously. Three types of emission occur, *alpha rays, beta rays*, and *gamma rays*. Alpha rays are helium nuclei; beta rays consist of fast-moving electrons; and gamma rays are high-energy photons. These three types of "rays" may be distinguished by their penetrating power and by their path in a magnetic field.

The particles of which a nucleus is composed, neutrons and protons, are called *nucleons*. Since the proton and the electron have equal and opposite charges, a neutral atom must contain equal numbers of protons and electrons. This is the *atomic number* Z. Elements of different atomic numbers have different chemical properties. The total number of nucleons, neutrons plus protons, is called the *mass number* A. A *nuclide* is a given combination of protons and neutrons. Two nuclides having the same Z but different A are called *isotopes*.

As a general rule, for the lighter elements, the stable nuclides contain about equal numbers of protons and neutrons. Boron, atomic number 5, has two stable isotopes, ${}^{11}_{5}B$ with 6 neutrons and ${}^{10}_{5}B$ with 5 neutrons. The nuclide ${}^{8}_{5}B$ with only 3 neutrons is very unstable, any given amount being reduced by half in less than a second.

The masses of individual isotopes can be measured with a *mass spectrometer*. The standard for these measurements is ${}^{12}_{6}C$, the neutral carbon atom of mass number 12, whose mass is taken to be 12.00000 *unified atomic mass units* (u). The energy equivalent of 1 u is 931 MeV.

The *binding energy* (BE) of a system of particles is the amount of energy required to disassemble the system into its constituent particles. The BE of a hydrogen atom is 13.60 eV; it takes that much energy to remove the single electron from its ground state close to the nucleus. The BE of a nucleon in a stable nucleus is many times greater. The energy required to remove a neutron from the ${}^{12}_{6}C$ nucleus, for example, is more than a million times that required to break apart the hydrogen atom.

Unstable nuclides decay in various ways. Some emit β rays (negative electrons); others emit α particles or capture an electron from an inner shell. These processes are often accompanied by the emission of one or more γ rays. Radioactive decay is a statistical process. The decay time of an individual nucleus is unpredictable. However, the decay of a large number of nuclei follows an exact mathematical law. Radioactive decay is governed by five conservation laws: those dealing with charge, mass-energy, linear momentum, angular momentum, and parity. The law of conservation of parity, unlike the other four, is not universally applicable. In certain "weak interactions" parity is not conserved.

As a result of natural radioactive decay, one element is changed into another. These *transmutations* can also be produced artificially by bombarding a target nucleus with suitable *projectiles*. Among the projectiles used are α particles, protons, deuterons, and neutrons.

QUESTIONS AND PROBLEMS

The following problems, conscientiously performed, should help you achieve the Objectives. For the most part, the questions and problems follow the order of the text.

1. Naturally radioactive substances emit three distinct types of radiation, called alpha rays, beta rays, and gamma rays. Complete the table below, filling in all missing entries.

Name of Radiation	Symbol	Charge	Mass	Stopped by
alpha				sheet of paper
	β			
		0		

2. Which of these three "rays" has the largest mass? Which travels at the highest speed?

3. The charge of a moving particle may be determined by observing how it is deflected in a magnetic field. Fig. 32-1 shows a radioactive sample at the bottom of a small hole that has been drilled in a lead block. The various rays are emitted in all directions but are absorbed in the lead except for those few rays that happen to be moving straight upward.

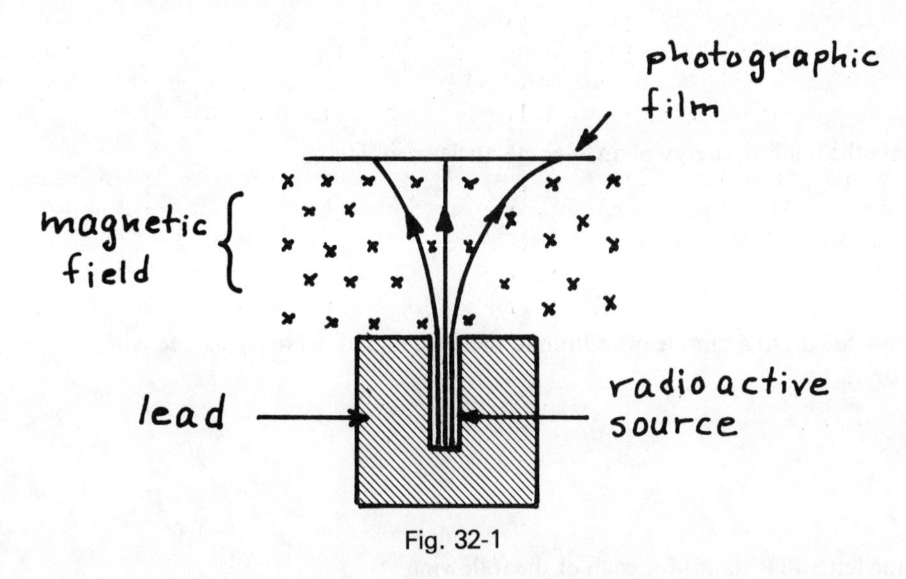

Fig. 32-1

a. Which of the three emissions is not deflected by the magnetic field? Why?

b. Label the three rays using the greek letters introduced in the table.
c. Which "ray" consists of helium nuclei?

4. Iron has several naturally occurring isotopes. All of the isotopes of iron have the same ______________.
   They differ only in ______________.
5. According to the general rule for stability of a nuclide, would you expect $^{8}_{2}He$ to be stable?

6. Complete the following definition of the unified atomic mass unit.

   1 u = (1/12) the mass of ______________.

7a. Carbon-12 contains 12 nucleons. The mass of $^{12}_{6}C$ is exactly 12.0000 u. Does this imply that the mass of a nucleon is exactly equal to 1.0000 u? Explain.

b. Neutral He contains exactly 1/3 as many neutrons and protons as $^{12}_{6}C$. Do you expect $^{4}_{2}He$ to have a mass of exactly 4.0000 u?

c. Calculate the binding energy of an average nucleon in $^{4}_{2}He$.

8. What can you do to a sample of radium to increase its radioactivity, i.e., to cause it to emit more α particles per second?

9. Write the letter that stands for each of the following:

   disintegration constant ______________

   half-life ______________

   activity ______________

10. Suppose that we have 0.1 mg of $^{234}_{90}Th$ on January 1, 1987.
   a. How many thorium atoms are present at the beginning?

   b. How many atoms will remain on January 20, 1988 (384 days later)? (Hint: Study Ex. 32-3 in the text.)

11. How long does it take for a sample of ^{32}P ($T_{1/2}$ = 14 days) to lose 1/3 its activity?

12. A 90-g sample of pure carbon contains 1 part in 10^{12} of radioactive ^{14}C ($T_{1/2}$ = 5730 years). How many disintegrations occur per second?

13. When a nucleus of $^{55}_{24}Cr$ decays it emits a negatron, i.e., an "ordinary" electron with a negative charge. What change must take place in the nucleus to compensate for this loss of negative charge?

* * * * * * *

The equation that represents the radioactive decay of an isotope of uranium is written as follows:

$$^{238}_{92}U \rightarrow ^{234}_{90}Th + \_\_\_\_\_\_\_\_ \quad (4.5 \times 10^9 \text{ y})$$

The missing portion of this equation can be found by applying two rules:
(1) The mass number A must be the same on both sides of the equation. (conservation of mass)
(2) The charge number Z must be the same on both sides of the equation. (conservation of charge)
Uranium 238 has 238 nucleons, thorium 234 has 234 nucleons. Thus 4 more nucleons must be accounted for. So we can put ^4X in the blank. Since the charge is also conserved, we must have a subscript 2 on the right-hand side. The blank now becomes ^4_2X. This nuclide is, of course, the common isotope of helium, ^4_2He.

14. Use this approach to complete the nuclear reactions below.
   a. $^{234}_{90}Th \rightarrow ^{234}_{91}Pa + \_\_\_\_\_\_\_\_$ (24 d)
   b. $^{234}_{91}Pa \rightarrow ^{0}_{-1}e + \_\_\_\_\_\_\_\_$ (1.2 min)

15. One of the important nuclear reactions is called β decay. What are the three processes included in this term?

(1) ______________________________.

(2) ______________________________.

(3) ______________________________.

16. The electron may be considered to be a particle with two charge states. Give the names and symbols for these two states.

______________ ______________ ______________ ______________

name name

17. Here are four nuclear reactions:

(1) $^{25}_{11}Na \rightarrow ^{25}_{A}X + ^{0}_{-1}e$
(2) $^{230}_{90}Th \rightarrow ^{4}_{2}He + ^{226}_{88}Ra$
(3) $^{14}_{7}N + ^{4}_{2}He \rightarrow ^{1}_{1}H + ^{17}_{1}O$
(4) $^{51}_{24}Cr + ^{0}_{-1}e \rightarrow ^{51}_{23}V$

a. In which reaction is an α particle emitted?

b. Which reaction leads to emission of a negatron?

c. Which reaction involves bombardment by an α particle?

d. What is the atomic number and chemical symbol of the daughter nucleus in reaction (1)?

e. Which reaction begins with an electron capture? From where does the electron come?

f. In which reaction is a proton emitted?

g. Which reactions may be described as β decays?

18. What are the two ways in which high-energy photons may be produced in radioactive decay processes? (Hint: One of these methods was discussed in the preceding chapter.)

19. The following table summarizes the modes of decay or disintegration of nuclei. Fill in the missing entries.

Decay Mode	Symbol of Particle Emitted or Captured
α emission	
	β^- or ${}_{-1}^{0}e$
positron emission	
electron capture	
	two large fragments
transition of nucleus from excited state to stable state	

20. Use the conservation laws to identify the missing term in the reaction

$${}_{24}^{55}Cr \rightarrow {}_{25}^{55}Mn + \_\_\_\_\_\_\_\_\_\_$$

21. What change of state of a nucleon is involved in the preceding disintegration?

22. Name two nuclear processes in which neutrinos are emitted.

23. Neutrinos have neither charge nor rest mass. What physical properties do they possess?

24. Stable isotopes of the lighter elements have about 1 neutron for every proton. The heavy elements tend to have somewhat more neutrons than protons; in other words, for them the mass number A tends to be somewhat greater than twice the atomic number Z. How is this ratio changed when an α particle is emitted? Take ${}^{218}Po$ as an example.

$\frac{A}{Z}$ = __________ for ${}^{218}Po$

$\frac{A}{Z}$ = __________ for __________

25. What steps are involved when the nucleus ${}^{A}_{Z}X$ decays to ${}^{A-4}_{Z}X$? (Hint: See Fig. 32-5 in the text for some examples.)

26. The chart in text Fig. 32-5 shows that ^{218}Po decays to ^{214}Po by an α emission followed by two negatron emissions.
   a. What other sequence could (in theory) accomplish the same result?

   b. What would the intermediate nuclei be for this alternative sequence?

27. Write the full equations for the nuclear reactions symbolized by $^{50}Cr(n,\gamma)^{51}Cr$ and $^{9}Be(\alpha,n)^{12}C$.

28. Compute the energy of the negatron emitted when a free neutron decays to a proton plus a negatron.

29. Compute the energy of the negatron emitted during the decay of ^{14}C.

30. What emissions from a radioactive source are readily detected by a Geiger counter?

31. What is the name of a device that will detect γ rays with high efficiency?

32. Name four devices that make it possible to observe and record the paths of charged particles.

(1) ____________ (2) ____________

(3) ____________ (4) ____________

33. Fill in the missing symbols in these reactions:

(1) $^{10}B + ^{2}_{1}H \rightarrow ^{11}C +$ ____________

(2) $^{35}Cl + ^{1}_{0}n \rightarrow ^{1}_{1}H +$ ____________

(3) $^{x}Mn(p,n)^{55}X$

(4) $^{x}X + ^{2}_{1}H \rightarrow 2^{1}_{0}n + ^{65}Zn$

(5) $^{59}Co(n,?)^{60}Co$

Solutions

1.

Name of Radiation	Symbol	Charge	Mass	Stopped by
alpha rays	α or $^{4}_{2}He^{2+}$	+2e	4.003 u	sheet of paper
beta rays	β or $^{0}_{-1}e$	-1e	0.00055 u	1 mm of lead
gamma rays	γ	0	0	sheet of lead

2. the α particle; gamma rays at 3×10^8 m/s

3a. γ rays They carry no charge.

b. Left to right they are α, γ, β.

c. α rays

4. atomic number; mass number
(Some other nuclear properties, spin in particular, are different for different isotopes.)

5. No. It has too many neutrons in proportion to its two protons.

6. The neutral $^{12}_{6}C$ atom

7a. No. Small differences in mass of bound and free nucleons are accounted for by the very large binding energies of nucleons in a nucleus.

b. No, because average binding energies are not the same for all nuclei.

c.

Before	Mass
$^{4}_{2}He$	4.002603 u
After	
$2^{1}_{1}H = 2(1.007825)$	2.015650 u
$2^{1}_{0}n = 2(1.008665)$	2.017330 u
	4.032980 u
	−4.002603 u
Mass difference =	0.030377 u
(0.030377 u)(931 MeV/u) =	28.3 MeV
BE per nucleon = 28.3/4 =	7.07 MeV

8. absolutely nothing

9. $\lambda, T_{1/2}, A$

10a. $\dfrac{0.1 \times 10^{-3}\text{ g}}{x} = \dfrac{234\text{ g}}{6 \times 10^{23}\text{ atoms}}; \quad x = 2.56 \times 10^{17}$ atoms

b.

Days	Fraction Remaining
0	1
24	1/2
48 = 2(24)	$1/2^2$
72 = 3(24)	$1/2^3$
96 = 4(24)	$1/2^4$
...	...
384 = 16(24)	$1/2^{16} = 1/65536$

$$\frac{2.56 \times 10^{17} \text{ atoms}}{65536} = 3.9 \times 10^{12} \text{ atoms}$$

11. The basic equation for radioactive decay can be written

$$N = N_0 e^{-\lambda t} \quad \text{or} \quad A = A_0 e^{-\lambda t}$$

where A is the activity and λ is the disintegration constant.
Let t be the time at which $A = (2/3)A_0$.

$$(2/3)A_0 = A_0 e^{-\lambda t}\text{ ; } 2/3 = e^{-\lambda t}$$
$$3/2 = e^{\lambda t} \quad \log_e 3/2 = \lambda t$$
$$t = (1/\lambda) \log_e 1.5 = 0.405/\lambda$$

But $T_{1/2} = 0.693/\lambda$, thus

$$t = T_{1/2} \frac{\log_e(3/2)}{0.693} = 14 \text{ days} \frac{0.405}{0.693} = 8.19 \text{ days}$$

12. First find the number of atoms.

$$\frac{x}{90 \text{ g}} = \frac{6.0 \times 10^{23} \text{ atoms}}{12.0 \text{ g}}; \; x = 4.5 \times 10^{24} \text{ atoms}$$

The number of ^{14}C atoms is 4.5×10^{12}. The number of disintegrations per second is

$$A = \lambda N$$

where

$$\lambda = 0.693/T_{1/2} = \left(\frac{0.693}{5730 \text{ y}}\right)\left(\frac{1 \text{ y}}{365 \text{ day}}\right)\left(\frac{1 \text{ day}}{86400 \text{ s}}\right) = 3.84 \times 10^{-12} \text{ s}^{-1}$$

Thus

$$A = (3.84 \times 10^{-12} \text{ s}^{-1})(4.5 \times 10^{12}) = 17.3 \text{ s}^{-1}$$

13. One of the neutrons must change into a proton.

14a. $^{0}_{-1}e$

b. $^{234}_{92}U$

15. (1) emission of a negatron
(2) emission of a positron
(3) electron capture

16. negatron $^{0}_{-1}e$; positron $^{0}_{+1}e$

17a. 2

b. 1

c. 3

d. $A = 12$, thus $^{25}_{12}Mg$

e. 4; the electron probably comes from the K-shell

f. 3

g. 1 or 4

18a. X rays may be produced by outer electrons falling into a vacancy in an inner shell.

b. Gamma rays may be the result of a nucleus undergoing a transition from an excited state to a state of lower energy.

19.

Decay	Symbol of Particle Emitted or Captured
α emission	α or ${}^{4}_{2}He^{2+}$
negatron emission	β^- or ${}^{0}_{-1}e$
positron emission	β^+ or ${}^{0}_{+1}e$
electron capture	${}^{0}_{-1}e$
fission	two large fragments
nuclear transition	γ ray

20. The missing term is ${}^{0}_{-1}e$, the negative electron. Note, however, that energy conservation indicates that a neutrino $\bar{\nu}_e$ is also given off.

21. A neutron is changed to a proton. That is, the neutral charge state of a nucleon is changed to the positive charge state.

22. beta decay and electron capture

23. They have angular momentum (spin) and energy.

24. It is changed very little since the α particle contains 2 of each.

$$\frac{A}{Z} = \frac{218}{84} = 2.60$$

$$\frac{A}{Z} = \frac{214}{82} = 2.61 \text{ for } {}^{214}Pb$$

25. Emission of one α particle and two negatrons.

26a. Two negatron emissions followed by the emission of an α particle.

b. ${}^{218}At$ then ${}^{218}Rn$ (The sequence β, α, β is also conceivable.)

27. ${}^{50}_{24}Cr + {}^{1}_{0}n \rightarrow {}^{51}_{24}Cr + {}^{0}_{0}\gamma$
${}^{9}_{4}Be + {}^{4}_{2}He \rightarrow {}^{12}_{6}C + {}^{1}_{0}n$

28. ${}^{1}_{0}n \rightarrow {}^{1}_{1}H + {}^{0}_{-1}e$

Mass Before — Mass After
1.008665 u — 1.007825 u + 0.000549 u = 1.008374 u
Mass difference is = (0.000291 u)(931 MeV/u) = 0.271 MeV

The KE of the bombarding neutron is negligible, as is the energy of the proton. Thus the negatron has 0.271 MeV of KE.

29. ${}^{14}_{6}C \rightarrow {}^{14}_{7}N + {}^{0}_{-1}e$

Mass Before — Mass After
14.003242 u — 14.003074 u + 0.000549 u = 14.003623 u
The mass difference is 0.000381 u = 0.355 MeV

30. negatrons; positrons and α particles; x rays and gamma rays are detected with much lower efficiency.
31. scintillation counter
32. (1) cloud chamber (2) bubble chamber
(3) spark chamber (4) photographic emulsion
33. (1) $^{10}_{5}B + ^{2}_{1}H \rightarrow ^{11}_{6}C + ^{1}_{0}n$
(2) $^{35}_{17}Cl + ^{1}_{0}n \rightarrow ^{1}_{1}H + ^{35}_{16}S$
(3) $^{55}_{25}Mn + ^{1}_{1}H \rightarrow ^{1}_{0}n + ^{55}_{26}Fe$
(4) $^{65}_{29}Cu + ^{2}_{1}H \rightarrow 2^{1}_{0}n + ^{65}_{30}Zn$
(5) $^{59}_{27}Co + ^{1}_{0}n \rightarrow ^{0}_{0}\gamma + ^{60}_{27}Co$

REVIEW TEST FOR CHAPTERS 30–32

1. Fig. T-64 shows the supposed paths of α particles that pass in the vicinity of the gold nuclei. Which of these paths does not represent a probable occurrence during the experiment?

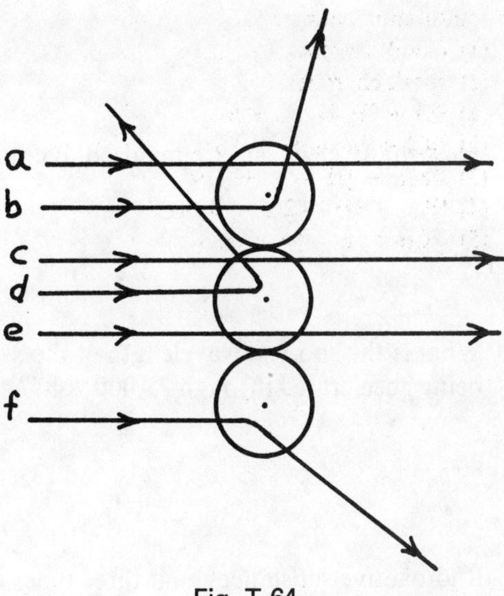

Fig. T-64

2. The spectra emitted by several molten metals is examined with a spectroscope. What may we expect to learn from these spectra?
   a. The nature of the molten metal—silver, iron, platinum, tin, etc.
   b. The interatomic spacing.
   c. The atomic weight of the emitting metal.
   d. The temperature of the emitting metal.
3. The drawing to the right shows the energy levels of hydrogen. The excitation energy is given to the right. The arrows represent three of the transitions of the Balmer series. Calculate the longest wavelength of the Balmer series.

   $\lambda =$ _______________.

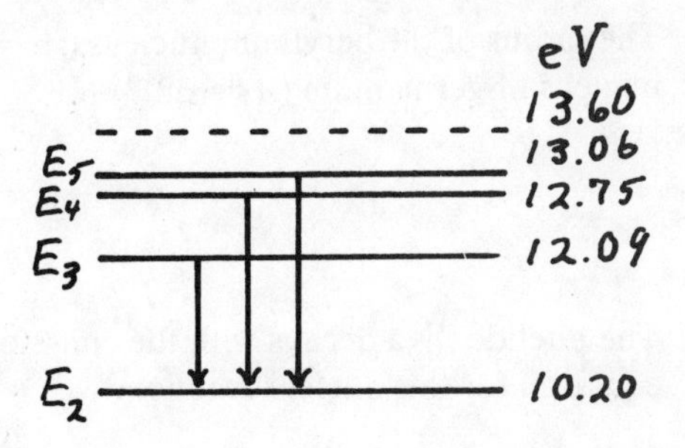

4. Use the Bohr formula for hydrogen to calculate the total energy (PE + KE) of the level E_6 and then the corresponding excitation energy.

5. An electron in a hydrogen atom is completely described by four quantum numbers, n, ℓ, m_ℓ, and m_s. If the principal quantum number has the value 4 and the orbital quantum number has the value 3, what are the values of m_ℓ?

E_1 —— 0.00

Fig. T-65

6. Write a statement of the Pauli exclusion principle.

7. The ground state of a certain atom is $1s^2 2s^2 2p^6 3s^1$. How many electrons does this atom have?

8. How many electrons are required to completely fill the *L* shell? (The *L* shell corresponds to $n = 2$.)

9. What is the shortest wavelength of the x rays produced when electrons strike a molybdenum target after being accelerated through 25,000 volts?

10. Radioactive substances emit three types of "radiation" designated by the Greek letters α, β, and γ. Identify each of these "rays" in modern terminology.

11. The radius of the beryllium nucleus ($A = 4$) is about 1.9×10^{-15} m. What is the approximate radius of the nucleus of germanium ($A = 32$)?

12. The nuclide $^{25}_{11}Na$ decays with the emission of a negatron, the half-life being 60 s. Write the complete equation for this nuclear reaction.

13. The reaction above, called beta decay, involves the emission of an additional particle of zero rest mass. What is the name of this particle and why is its existence postulated in connection with beta decay?

14. Supposing 8×10^{12} atoms of $^{25}_{11}Na$ to be present at 2:00 P.M., how many remain at 2.07 P.M.?

15. Natural chromium, as found in the earth's crust, contains four stable isotopes, ^{50}Cr, ^{52}Cr, ^{53}Cr, and ^{54}Cr. Suppose that chromium oxide is made in the same way from each of these four isotopes. How would these four compounds compare in color? Solubility? Melting point? Chemical activity?

CHAPTER 33 Applied Nuclear Physics

GOALS To study some of the applications of nuclear processes including fission and radioactivity.

OBJECTIVES After completing this chapter the student should be able to do the following:

1. Name two radioactive substances that are used in radiation therapy.
2. Name the radioactive nuclide that is used extensively in dating archeological finds.
3. Give the technical term for the splitting of a nucleus into relatively large fragments.
4. Write the equation for a fission reaction with the aid of a table of nuclides.
5. Name the particle that "triggers" the fission reaction in uranium-235.
6. Describe briefly how a chain reaction may take place in a nuclear reactor.
7. Calculate the "geometrical cross sections" of nuclei and of atoms in m^2 and in barns.
8. State why heavy water is used as a moderator in a nuclear reactor rather than ordinary water.
9. Describe, briefly, the concept of a breeder reactor.
10. Distinguish between the terms *fission* and *fusion* giving at least one example of each.
11. Describe the main features of the nuclear reactions that are the sun's source of energy. (What fuel is "burned"? What is the end product?)
12. Calculate the average kinetic energy of a particle from the temperature of the surrounding medium.
13. Name the "fuel" used in a fusion reactor and the final end product.
14. State two advantages of a fusion reactor and two disadvantages.
15. Define the *curie*.
16. Calculate the counting rate for a radioactive sample of known activity.
17. Define the *radiation absorbed dose* (rad).
18. Name three sources of natural radiation.
19. Name the two categories of biological radiation damage.
20. Explain the statement, "Genetic damage seems to have no threshold."

SUMMARY

Radioactive isotopes, both those occurring naturally and those artificially produced, are used as *tracers* for the study of many chemical, biological, and even physical processes. Other fields of application of radioactive isotopes are radiation therapy in medicine and radioactive dating in archeology.

Two nuclear processes that are important in energy production are *fission*, the splitting up of heavy nuclei, and *fusion*, the combining of light nuclei. The only three long-lived heavy isotopes that undergo fission readily by slow neutrons are ^{235}U, ^{239}Pu, and ^{233}U. The first nuclide, ^{235}U, which constitutes 0.7% of the uranium found in nature, is the energy source in the current generation of nuclear reactors.

The fission process produces energy because the mass of the products is less than the mass of the original nucleus. When a nucleus fissions, as a result of an interaction with a slow neutron, several additional neutrons are produced. After being slowed down in a suitable moderator, these neutrons can induce fission reactions in other nuclei. This chain reaction can be controlled by increasing or decreasing the amount of moderator present. The large amount of heat given off can produce steam to power electric generators.

The fusion process leads to a net energy gain because the mass of the nucleus formed, He, is less than the mass of the four protons that are combined to produce it. The fusion process has the advantage of using a fuel that is so abundant as to be virtually inexhaustible. A second, and very important, advantage is that the end products of the fusion reaction are nonradioactive. There is no problem of disposal of radioactive wastes.

The technical difficulties of realizing a practical fusion reactor are enormous, however, because the reaction will take place only at extremely high temperatures and pressures.

All types of radiation, but particularly the emissions from radioactive materials, have biological effects. At very low rates of radiation, such as that from cosmic rays, air, and water, the parts of the body (other than the reproductive cells) are able to repair the damage that occurs. The reproductive cells, on the other hand, may undergo irreversible damage at even the lowest radiation levels.

Because high-energy particles have correspondingly short wavelengths ($\lambda \approx hc/E$), they are able to "see" very small-scale structures, including the internal structure of a nucleus or even of a single proton.

QUESTIONS AND PROBLEMS

Secs. 33-1—33-3

1. Compare the effect of γ rays on cancerous cells and on normal cells.

2. Name three radioactive isotopes that are useful in cancer treatment.

3. Ex. 33-1 in the text describes an experiment in which red blood cells are labeled with the radioactive isotope ^{51}Cr, which has a half-life of 27.8 d. The activity of 5-cm³ samples of blood was measured four times with the results stated. Suppose that on the eighth day (at t = 8.0 d) a 10-cm³ sample was found to have an activity A of 266 counts/min.
   a. Following the method given in Ex. 33-1, calculate both the corrected activity A_0 and $\log A_0$ for a 5-cm³ sample.

   b. Use the data for t = 0.0 d and t = 8.0 d to determine the slope of the graph of $\log A_0$ versus t. From this slope calculate the half-life of a red cell in the bloodstream.

   c. Write the equation that describes the radioactive decay of ^{51}Cr. (Consult Table 32-1 in the text for the mode of decay.)

4. What is the age of the earth as determined from measurements of radioactive isotopes in rocks?

5. Explain briefly how $^{14}_{6}C$ can be used to date the bones found in an archeological site.

6. One milligram of carbon is obtained from bone fragments found in an ancient burial site. The measured activity of this 1.0-mg sample indicates that it contains about 7.8×10^5 atoms of ^{14}C. What is the approximate age of the bone from which this carbon was obtained?

7. What is the name of a nuclear process in which a large nucleus, uranium or plutonium especially, is split into two large fragments?

8. Supply the missing data in the following equation for a fission reaction.

$$^{235}_{92}U + ^{1}_{0}n \rightarrow ^{116}_{48}Cd + \underline{\hspace{4cm}} + 16\,^{1}_{0}n$$

9a. Calculate the energy released in the fission reaction of Prob. 8 above. The nuclear masses are:

$$^{235}_{92}U - 235.0439\text{ u} \,;\; ^{116}_{48}Cd - 115.9050\text{ u} \,;\; ^{104}_{44}Ru - 103.9055\text{ u}$$

b. What is the total energy (in joules) released when 10 g of ^{235}U undergo this fission reaction?

c. How many tonnes of coal must be burned to produce the same amount of energy? (Depending on its quality, one tonne of coal may release about 2.9×10^{10} J when burned.)

10. Show that the reaction $^{238}U\,(n,\gamma)^{239}U$ is energetically possible. (The masses are ^{238}U — 238.0508 u; ^{239}U — 239.0543 u.)

11. The following questions can be answered from the data on nuclear cross sections in Table 33-1 in the text.
    a. What is the cross section of the proton for capture of slow neutrons? Express it in barns and in m^2.

    b. Which of the nuclides listed would be most useful for absorbing thermal neutrons in a reactor? Give the reason for your choice.

    c. Why are control rods in a nuclear reactor sometimes made of ^{10}B?

    d. Ordinary uranium from natural ores contains more than 99% ^{238}U and less than 1% ^{235}U. Why is reactor fuel enriched in the rare isotope ^{235}U rather than using natural uranium?

12. Suppose that a mass *m* of pure ^{235}U is not quite enough to sustain a chain reaction. What will be the result of putting together two pieces of ^{235}U metal, each of mass *m*?

13a. Calculate the "geometrical cross section" (i.e., the physical size of a nucleus) of ^{113}Cd and compare it with the cross section for capture of slow neutrons. (Use 10^{-14} m for the diameter of the Cd nucleus.)

b. A neutron need not come in contact with a ^{113}Cd nucleus in order to be captured. Estimate the maximum distance at which neutron capture may take place.

14. Neutrons that are released during fission are moving too fast to be captured effectively. They must be slowed down by collisions with other particles, preferably particles of nearly the same mass. Why then is heavy water used as a moderator in some reactors rather than ordinary water? (Consult Table 33-1 in the text.)

15. For a proton whose total energy is 28.00 GeV, calculate
   a. the rest-mass energy.

   b. the kinetic energy.

16. What is the source of the sun's energy? What is the end product of the various reactions?

17. The reaction $^{1}_{1}H + ^{1}_{0}n \rightarrow ^{2}_{1}H$ is an example of a reaction in which light nuclei are joined to form heavier nuclei.
   a. What technical term is used to describe this type of nuclear reaction?

   b. Calculate the amount by which the mass of the reactants exceeds the mass of the product nucleus.

c. How much energy (in joules) is released when 1.0 g of hydrogen undergoes this reaction?

18. Calculate the average energy, in MeV, of a particle at the center of a star where the temperature is 6 million K.

Secs. 33-4—33-5

19. Why is it so much easier to produce a controlled fission reaction than to produce a controlled fusion reaction?

20. Define the *curie* and the *rad.*

21. Complete the following table of relative biological effectiveness of various kinds of radiation.

Type of Radiation	Relative Biological Effectiveness
x rays, gamma rays	1
β rays, protons	
thermal neutrons	
fast neutrons	
α particles	

22. The two categories of biological radiation damage are *somatic* damage and *genetic* damage. In what important way do these two categories differ, for example, with regard to very low levels of radiation?

Solutions

1. Gamma rays do damage to all living cells, but cancer cells are more susceptible than normal cells.
2. radium; ^{60}Co; ^{131}I
3. The observed activity for a 5 cm³ sample is

$$A = \tfrac{1}{2}(266)\ \text{min}^{-1} = 133\ \text{counts/min}$$

a. This must be corrected to what it would have been at the beginning of the experiment.

$$A_0 = A e^{\lambda t} = (133)\, e^{(0.0249)(8.0)} = 162\ \text{min}^{-1}$$
$$\log A_0 = \log(162) = 2.21$$

b. $$\text{slope} = \frac{\Delta(\log A_0)}{\Delta t} = -\frac{2.32 - 2.21}{8.0} = -0.0138\ \text{d}^{-1}$$

The half-life of a red blood cell is

$$T_{1/2} = \frac{-\log 2}{\text{slope}} = \frac{0.30}{0.0138} \approx 22\ \text{d}$$

c. $^{51}_{24}Cr + ^{0}_{-1}e = ^{51}_{23}V$

4. 5×10^9 years.
5. During the life of the organism, it contained ^{14}C in the ratio of one ^{14}C atom for every 10^{12} atoms of ^{12}C. After death the proportion of ^{14}C decreases in a predictable way, one half of the original ^{14}C disintegrating every 5730 years. If, for example, a sample of bone contains 1 atom of ^{14}C for every 8×10^{12} atoms of ^{12}C, the bone must be three ^{14}C half-lives old, or $3(5730) = 1.7 \times 10^4$ years old since $8 = 2^3$.
6. First calculate the number of carbon atoms in 1 mg of carbon.

$$(1.0 \times 10^{-3}\ \text{g})\frac{(6.02 \times 10^{23}\ \text{atoms})}{12\ \text{g}} = 5.02 \times 10^{19}\ \text{atoms}$$

When this material was a part of a living organism it contained

$$(5.02 \times 10^{19}\ \text{atoms of C})/(1 \times 10^{12}\ \text{atoms of C/atom of }^{14}\text{C}) = 5.02 \times 10^{7}\ \text{atoms of }^{14}\text{C}$$

The amount of ^{14}C has decreased by a factor of

$$\frac{5.02 \times 10^7\ \text{atoms}}{7.8 \times 10^5\ \text{atoms}} \cong 64 = 2^6$$

Thus about six half-lives have passed since this material ceased to live. The approximate age is $6(5730\ \text{years}) = 3.4 \times 10^4$ years.

7. fission
8. $^{220}_{44}Ru$

9a.

Before	After
235.0439 u	115.9050 u
1.0087 u	103.9055 u
	16.1392 u
236.0526 u	235.9497 u

$$E = (0.1029\ \text{u})(931\ \text{MeV/u}) = 95.80\ \text{MeV}$$

b. First find the number of atoms

$$\frac{10\ \text{g}}{x} = \frac{235\ \text{g/mole}}{6 \times 10^{23}\ \text{atoms/mole}}; \quad x = 2.56 \times 10^{22}$$

$$(2.56 \times 10^{22}\ \text{atoms})(95.8\ \text{MeV/atom}) = 2.45 \times 10^{24}\ \text{MeV}$$
$$(2.45 \times 10^{24}\ \text{MeV})(1.60 \times 10^{-13}\ \text{J/MeV}) = 3.91 \times 10^{11}\ \text{J}$$

c. $(3.91 \times 10^{11}\ \text{J})/(2.9 \times 10^{10}\ \text{J/tonne}) = 13.5$ tonnes

10. $^{238}\text{U} + {}^{1}\text{n} \rightarrow {}^{239}\text{U} + \gamma$

Before:	After:
238.0508 u	239.0543 u
1.0087 u	
239.0595 u	

The reaction is certainly possible. The energy available for the gamma ray is 0.0052 u = 4.8 MeV.

11a. 0.33 barns $= 0.33 \times 10^{-28}\ \text{m}^2$

b. Since thermal neutrons are slow neutrons, the best absorber is ^{113}Cd, which has a cross section for neutron capture of 27×10^3 barns.

c. ^{10}B is also a good absorber of thermal neutrons.

d. The abundant isotope ^{238}U does not undergo fission in a reactor because it does not capture slow neutrons.

12. A violent nuclear explosion. This is the basic principle of the atomic bomb.

13a. cross section $= \dfrac{\pi d^2}{4} \cong 10^{-28}\ \text{m}^2 = 1$ barn

The capture cross section is about 27,000 times greater.

b. $\pi r^2 = 2.7 \times 10^{-24}\ \text{m}^2$; $r \cong 10^{-12}$ m

This distance is more than 100 times the nuclear radius.

14. The function of a moderator is to slow the neutrons, not capture them. When ordinary water is used, many of the neutrons are lost in the reaction.

$$^{1}_{1}\text{H} + {}^{1}_{0}\text{n} \rightarrow {}^{2}_{1}\text{H} + \gamma$$

Heavy water, on the other hand, has deuterium, $^{2}_{1}$H, replacing hydrogen. The neutron capture cross section for $^{2}_{1}$H is only 0.0006 barns.

15a. Let m_0 = rest mass of proton.
$m_0 = (1.007277\ \text{u})(1.66 \times 10^{-27}\ \text{kg/u}) = 1.672 \times 10^{-27}$ kg
$E = m_0c^2 = (1.672 \times 10^{-27}\ \text{kg})(3 \times 10^8\ \text{m/s})^2 = 1.50 \times 10^{-10}$ J
$E = (1.50 \times 10^{-10}\ \text{J})/(1.6 \times 10^{-19}\ \text{J/eV}) = 9.39 \times 10^8\ \text{eV} = 939$ MeV

b. $\text{KE} = E_{\text{tot}} - m_0c^2$
$= 28 \times 10^9\ \text{eV} - 0.94 \times 10^9\ \text{eV} \cong 27$ GeV

16. The sun obtains its energy through fusion reactions in which protons are fused into α particles.

17a. This is a fusion reaction.

b.

Before:	After:
1.007825 u	2.014102 u
1.008665 u	
2.016490 u	

The "loss" of mass is 0.002388 u

c. $(2.388\ \text{u} \times 10^{-3})\ (931\ \text{MeV/u})\ (1.60 \times 10^{13}\ \text{J/MeV}) = 3.56 \times 10^{-13}$ J

Each proton that undergoes this reaction leads to an energy of 3.56×10^{-13} J.

$(1.0\ \text{g})(6.02 \times 10^{23}\ \text{atoms/g}) = 6.02 \times 10^{23}$ atoms (or protons)

The energy released per gram of hydrogen is

$(3.56 \times 10^{-13}\ \text{J/proton})(6.02 \times 10^{23}\ \text{protons}) = 2.1 \times 10^{11}$ J

18. In Chapter 16 it was shown that the KE per atom is

$$KE = \frac{3}{2}RT/N$$

Thus

$$KE = \frac{(3/2)(8.31\ J/mole \cdot K)(6 \times 10^6\ K)}{6.02 \times 10^{23}\ atoms/mole}$$

$$= (1.24 \times 10^{-16}\ J/atom)(1\ eV/1.6 \times 10^{-19}\ J)$$

$$= 7.76 \times 10^2\ eV \text{ per particle} = 0.766\ keV \text{ per particle}$$

19. The fission reaction is triggered rather easily by slow-moving (thermal) neutrons. The neutrons, having no charge, can approach close to the nucleus. They are not repelled by the electrostatic force. In the fusion reaction, on the other hand, it is necessary to bring two charged particles quite close together in spite of a strong electrostatic repulsion. This is possible only if the particles have very large energies corresponding to temperatures in the range of millions of kelvins.

20. One curie is equal to 3.7×10^{10} disintegrations per second. One radiation absorbed dose (rad) is the absorption of 100 ergs per gram of tissue. 1 rad = 10^{-2} J/kg.

21.

Type of Radiation	Relative Biological Effectiveness
x rays, gamma rays	1
β rays, protons	1
thermal neutrons	3
fast neutrons	10
α particles	10–20

22. There may be a threshold for somatic damage. If this threshold exists, it means that below this radiation level no permanent damage occurs. In the case of genetic damage, there is no such threshold, even an extremely small dose to a gonadal chromosome produces irreversible genetic damage.

CHAPTER

34 Particle Physics

GOALS To study accelerators and the subatomic particles that can be created in them.

OBJECTIVES After completing this chapter the student should be able to do the following:

1. Use the equations for magnetic force and centripetal force to obtain an expression for the speed of a proton in a cyclotron.
2. Show that the time required for a proton to make half a revolution in a cyclotron is independent of the radius of the proton's orbit if relativistic effects are neglected.
3. Calculate the energy required of an electron that is to probe the internal structure of a nucleus.
4. Name two applications of high-energy particles.
5. Name the four types of fields and the field particles associated with each.
6. Rank the four interactions in order of decreasing strength.
7. Give the general term for the light, generally stable particles of spin 1/2.
8. Name the group of heavy particles, subject to the strong interaction, to which the nucleons belong.
9. Explain briefly why 10^{-23} s is a "natural" time for a nuclear collision.
10. Describe briefly the differences between the photon and the neutrino, both particles of zero rest mass.

SUMMARY

The production of unstable isotopes and the investigation of nuclear properties require that nuclei be bombarded with fast-moving particles, usually electrons or protons. Several types of machines, called *accelerators*, have been developed for the purpose of bringing charged particles up to very high energies. Linear accelerators, including the Van de Graaff generator, accelerate particles by the use of strong electric fields. Cyclotrons and synchrotrons cause the accelerated particles to move in circular paths by means of magnetic fields. The amount of energy required increases very rapidly as the speed of the particle approaches the speed of light. An electron moving at one-half the speed of light has a KE of a little less than 10^5 eV. One hundred times as much energy is required to bring the electron up to a speed of 0.999 c.

The amount of energy available in a collision between two nucleons is much greater if the particles are approaching each other at high speeds. Colliding-beam experiments have permitted the study of interactions that take place only at extremely high energies.

Under certain conditions, the interaction of a high-energy particle with a nucleon may lead to the creation of new particles. This is an instance of energy being converted into mass as opposed to the more familiar conversion of mass into energy in a nuclear reactor. In all such processes, the four conservation laws including that of mass-energy are strictly obeyed. The subatomic particles, of which more than 200 have been discovered, may be classified according to the way they interact. The photon interacts via the *electromagnetic* field; the graviton (not yet observed) interacts via the *gravitational* field. A hundred times stronger than the electromagnetic interaction is the *strong* or nuclear interaction for which the field particle is the *meson*. The fourth type of interaction is called the *weak* interaction. Its relative strength, 10^{-15}, is intermediate between those of the electromagnetic interaction, 10^{-2}, and the gravitational interaction, 10^{-40}.

Those particles that are not subject to the strong interaction are called *leptons*, of which the best known is the electron. Particles that are subject to the strong interaction are collectively known as *hadrons*, of which the two subgroups are the *mesons* and the *baryons*. The proton and the neutron are baryons.

The earth's outer atmosphere is continually bombarded by high-energy particles, principally protons, from outer space. Some of these *primary cosmic ray particles* have extremely high energies whose origins are still unknown. The less-energetic cosmic ray particles that penetrate to sea level are products of the interactions of the primary particles with nuclei in the upper atmosphere.

QUESTIONS AND PROBLEMS

Secs. 34-1—34-6

1. When a charged particle is injected into a magnetic field in such a way that **v** is perpendicular to **B**, the path of the particle is necessarily a circle.
   a. Write the expression for the magnetic force F_m on the moving charge and the expression for the centripetal force F_c.

   $F_m =$ ________________ . $F_c =$ ________________ .

   b. Use the fact that $F_m = F_c$ to obtain an expression for the speed v of the charged particle.

   $v =$ ________________ .

   c. It is desired to double the speed of a particle without changing the radius of its path. How is this possible?

   d. Obtain an expression for the time for half a revolution.

   $T/2 =$ ________________ .

2. How long does it take a proton to make half a revolution in a magnetic field of 2 T, assuming that the particle is moving slowly enough that its mass does not differ significantly from the rest mass?

   $T/2 =$ ________________ .

3. The Fermilab accelerator at Batavia, Illinois, is capable of producing a proton beam in which the number of protons per second is 3.3×10^{12} and the energy of each proton is 1000 GeV.
   a. Calculate the current carried by this proton beam.

   b. Suppose that the proton beam is allowed to impinge on a solid target for one second. How much energy (in joules and in calories) would the target absorb? (Neglect radiation losses.)

c. Compare the result above with the energy required to raise the temperature of 100 g of copper from 25°C to 2595°C, its boiling point. What is the likely result of placing a solid target in the path of the 1000 GeV proton beam? (The specific heat of copper is 0.093 cal/g • °C.)

4. In the scattering experiment performed by Rutherford, the 9-MeV α particles that were used had a velocity of 2.1×10^7 m/s and a wavelength of 4.8×10^{-15} m. Calculate the velocity and wavelength that would have resulted if the energy of the α particles had been 12 MeV.

5. Calculate the energy in MeV associated with a photon whose wavelength is 2.4×10^{-12} m.

6. The photon considered in Prob. 5 above has slightly more energy than an electron at rest (m_0 is equivalent to 0.51 MeV). However, the conversion of a photon to an electron is not possible. Explain why.

7. Of the four types of interactions to which matter is subject—strong, electromagnetic, weak, gravitational—which are long-range interactions inversely proportional to the square of the distance?

8. For what type of interaction is the meson the field particle? The photon?

Consult text Tables 34-2 and 34-3 for the answers to the next four questions.

9. Name the stable lepton having a spin of 1/2 and a rest mass of 0.511 MeV.

10. To what group of subatomic particles do the neutrinos belong?

11. Which of the baryons has the greatest rest mass?

12. Aside from the proton (which does not decay), which baryon has the longest half-life?

13. The delay of the free neutron is symbolized by

$$n \rightarrow p + e^- + \bar{\nu}_e$$

Name the three particles that appear when the neutron decays.

14. What is the principal component of the primary cosmic ray particles?

15. How do cosmic rays at sea level differ from those observed at very high altitudes?

16. The photon and the neutrino are both stable particles of zero rest mass. How do they differ?

Solutions

1a. $F_m = qvB$; $F_c = mv^2/r$

b. $v = \dfrac{Bqr}{m}$

c. If the magnitude of the magnetic induction is doubled, a particle of speed $2v$ will move in a circle of the same radius.

d. $T/2 = \dfrac{\text{distance}}{\text{speed}} = \dfrac{\frac{1}{2}(2\pi r)}{(Bqr)/m} = \dfrac{\pi m}{Bq}$

2. $T/2 = \dfrac{\pi m}{Bq} = \dfrac{\pi(1.67 \times 10^{-27}\text{ kg})}{(2T)(1.60 \times 10^{19}\text{ C})} = 1.64 \times 10^{-8}\text{ s}$

3a. $I = (3.3 \times 10^{12}\text{ protons}/s)(1.60 \times 10^{-19}\text{ C/proton})$
$= 0.53 \times 10^{-6}\text{ A} \cong 0.5\ \mu\text{A}$

b. The energy absorbed per second is

$$(3.3 \times 10^{12} \text{ protons})(1000 \times 10^{9} \text{ eV/proton})(1.6 \times 10^{-19} \text{ J/eV}) = 5.3 \times 10^{5} \text{ J}$$
$$= (5.3 \times 10^{5} \text{ J})/(4.184 \text{ J/cal}) = 1.3 \times 10^{5} \text{ cal.}$$

c. $Q = mc\Delta t = (10^{2} \text{ g})(0.093 \text{ cal/g} \cdot {}^{\circ}\text{C})(2570^{\circ}\text{C})$
$Q = 2.4 \times 10^{4} \text{ cal}$

The energy that would be absorbed per second from the proton beam is ten times that required to bring this block of copper up to its boiling point. The result of such an experiment would be the sudden, violent vaporization of the copper.

4. $\text{KE} = \frac{1}{2} m_0 v^2$
$\text{KE} = (12 \times 10^{6} \text{ eV})(1.60 \times 10^{-19} \text{ J/eV}) = 1.9 \times 10^{-12} \text{ J}$

$$v = \sqrt{\frac{2(\text{KE})}{m_0}} = \sqrt{\frac{2(1.9 \times 10^{-12} \text{ J})}{(4.003 \text{ u})(1.66 \times 10^{-27} \text{ kg/u})}}$$

$$v = 2.4 \times 10^{7} \text{ m/s}$$

$$\lambda = \frac{h}{m_0 v} = \frac{6.63 \times 10^{-34} \text{ J} \cdot \text{s}}{(6.64 \times 10^{-27} \text{ kg})(2.4 \times 10^{7} \text{ m/s})} = 4.2 \times 10^{-15} \text{ m}$$

5. $$E = \frac{hc}{\lambda} = \frac{(6.63 \times 10^{-34} \text{ J} \cdot \text{s})(3.0 \times 10^{8} \text{ m/s})}{2.4 \times 10^{-12} \text{ m}}$$

$$E = (8.3 \times 10^{-14} \text{ J})/(1.60 \times 10^{-19} \text{ J/eV})$$
$$= 5.2 \times 10^{5} \text{ eV} = 0.52 \text{ MeV}$$

6. The conversion of a photon to an electron is not possible because it would violate the law of conservation of charge.
7. electromagnetic and gravitational
8. strong; electromagnetic
9. electron
10. leptons
11. the omega particle with a rest mass of 1672 MeV
12. the neutron
13. the proton, the electron, and the electron's antineutrino
14. very high-energy photons
15. The primary cosmic ray particles do not, as a rule, reach sea level. They encounter nuclei in the atmosphere and produce many secondary particles including negatrons, positrons, photons, neutrons, and heavier particles.
16. The photon is the field particle associated with the electromagnetic interaction. The neutrino is a lepton like the electron; it is not a field particle. The photon has a spin of 1, whereas the neutrino has a spin of ½. The photon interacts strongly with matter; the neutrino, on the other hand, hardly interacts at all. Most neutrinos striking the earth simply pass straight through and emerge on the other side.

REVIEW TEST FOR CHAPTERS 18–34

1. A small conducting sphere, which has been positively charged, is lowered into a tin can, allowed to touch the bottom, and then withdrawn without touching the can again. The tin can rests on an insulating surface, shown in Fig. T-66. After this experiment is performed, the conducting sphere
   a. still has some positive charge.
   b. has a small negative charge.
   c. has retained all of its original positive charge.
   d. has a negative charge of the same magnitude as its original positive charge.
   e. has a charge of exactly zero.

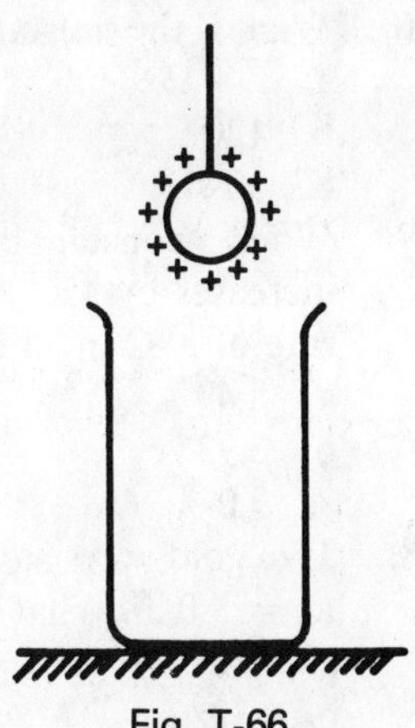

Fig. T-66

2. When two electric charges are separated by a distance of 0.1 m, the Coulomb force is F. What is the magnitude of the force when the distance is increased to 0.3 m?
   a. $3F$
   b. $F/3$
   c. $F/2$
   d. $F/9$
   e. F

3. The electric field at P due to Q_1 alone has a magnitude of 5×10^3 N/C (see Fig. T-67). When a second charge Q_2 is brought up and placed at M, the magnitude of the total electric field at P becomes 5.83×10^3 N/C. What would be the magnitude of the electric field at P if Q_1 were removed and Q_2 left in position at M?
   a. 0.83×10^3 N/C
   b. 0.415×10^3 N/C
   c. 0.91×10^3 N/C
   d. 9.0×10^3 N/C
   e. 3.0×10^3 N/C

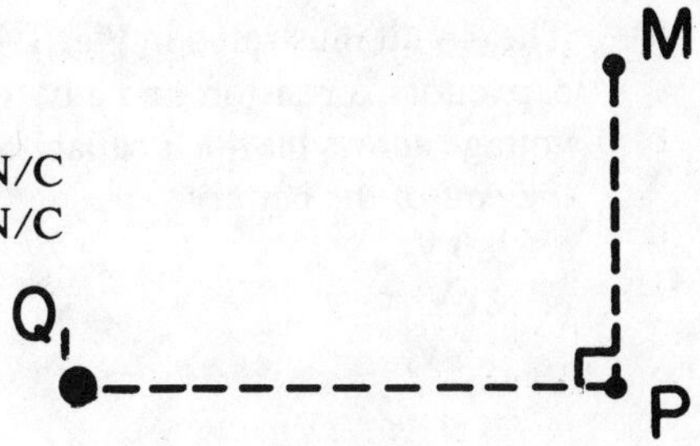

Fig. T-67

4. A point P is located in the neighborhood of two charges as shown in Fig. T-68. The total potential at P is known to be 100 V, whereas the potential due to Q_1 alone is only 60 V. What is the potential at P due to Q_2?
   a. 40 V
   b. 160 V
   c. 80 V
   d. −40 V
   e. 60 V

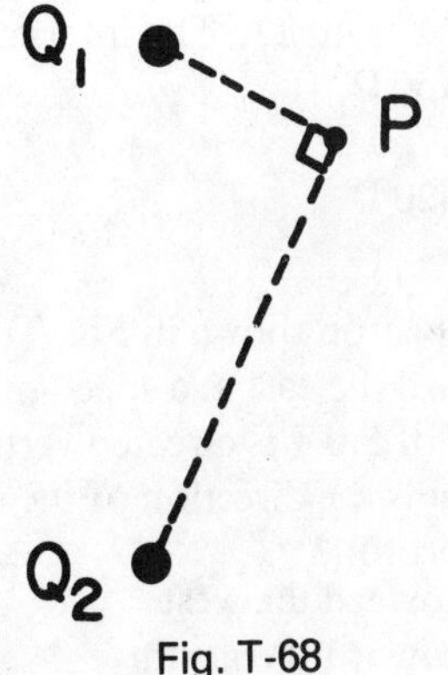

Fig. T-68

5. A 2 μF capacitor and a 6 μF capacitor are connected in series with a 24 V battery. What is the potential difference across the 2 μF capacitor?
   a. 24 V
   b. 6 V
   c. 18 V
   d. 8 V
   e. 16 V

6. What is the resistance of a 60-W light bulb which is lit at full brilliance by a current of 0.5 A?
   a. 240 Ω
   b. 120 Ω
   c. 30 Ω
   d. 15 Ω
   e. 180 Ω

7. When a constant current of 6 A flows through a resistor immersed in water, the temperature of the water increases by 4°C in 20 min. What current should be used if the same quantity of water is to be heated at the rate of 8°C in 30 min? The water is in an insulated container, thus preventing heat loss to the surroundings.
   a. 8.49 A
   b. 6.93 A
   c. 9.0 A
   d. 5.20 A
   e. 12.0 A

8. Two gold wires have the same mass and necessarily, the same volume. Their lengths are $L_1 = 2.0$ m and $L_2 = 3.0$ m. What is the ratio of their resistances, R_2/R_1?
   a. 3/2
   b. $\sqrt{3/2}$
   c. 9/4
   d. $(3/2)(\sqrt{3/2})$
   e. 1/1

9. The emf of the battery in Fig. T-69 is known to be 6 V. When it is connected to an 11 Ω resistor a current of 0.5 A flows in the circuit. What is the internal resistance of the battery?
   a. 12 Ω
   b. 3 Ω
   c. 11 Ω
   d. 1 Ω
   e. zero

I = 0.5A

11Ω

Fig. T-69

10. The circuit illustrated in Fig. T-70 contains 3 capacitors, a resistor, and a battery. If the voltage across the 4 μF capacitor is 3 V, what is the emf of the battery?
   a. 9.0 V
   b. 21 V
   c. 4.5 V
   d. 5.25 V
   e. 3.0 V

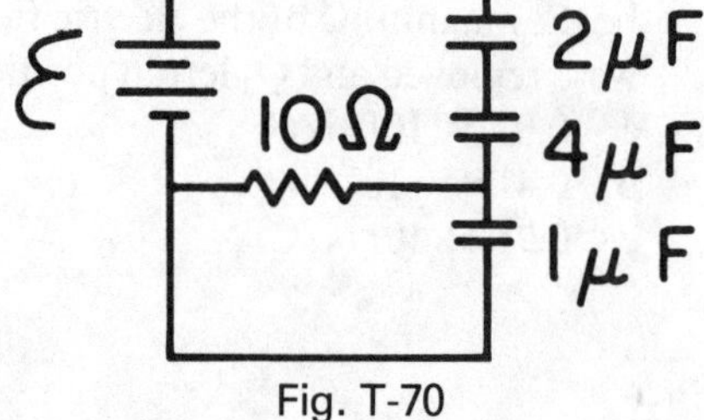

Fig. T-70

11. A certain galvanometer gives a full-scale deflection at a current of 25 μA. When a 700 Ω resistor is connected across the terminals of the galvanometer, the *total* current corresponding to full-scale deflection becomes 40 μA. The internal resistance of the galvanometer is
   a. 750 Ω
   b. 1750 Ω
   c. 420 Ω
   d. 280 Ω
   e. 357 Ω

12. The proton shown in Fig. T-71 is moving toward the east at a speed of 0.9 c. A magnetic field of 5.0 T is directed vertically downward. What is the direction of the magnetic force on the proton?
   a. toward the west
   b. toward the north
   c. toward the south
   d. toward the east
   e. vertically upward

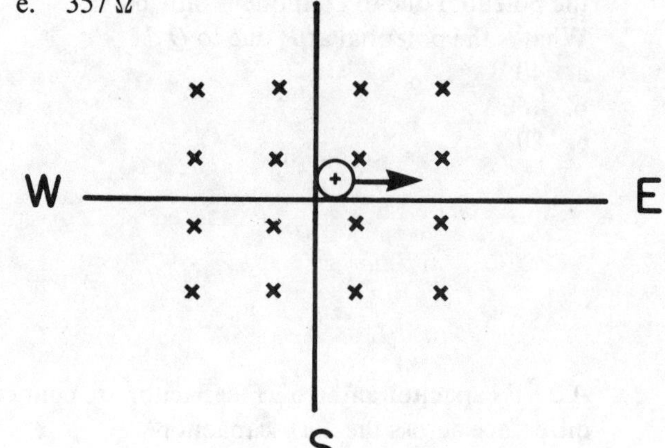

Fig. T-71

13. A magnetic field of 0.4 T is directed upward, out of Fig. T-72. A 25-cm copper rod is moving to the right at a speed of 50 m/s. The resistance of the rod is $2 \times 10^{-3}\ \Omega$. What emf is induced between the two ends of the rod?
   a. 2.5×10^{3} V
   b. 5 V
   c. 1×10^{-2} V
   d. 2×10^{-4} A
   e. zero

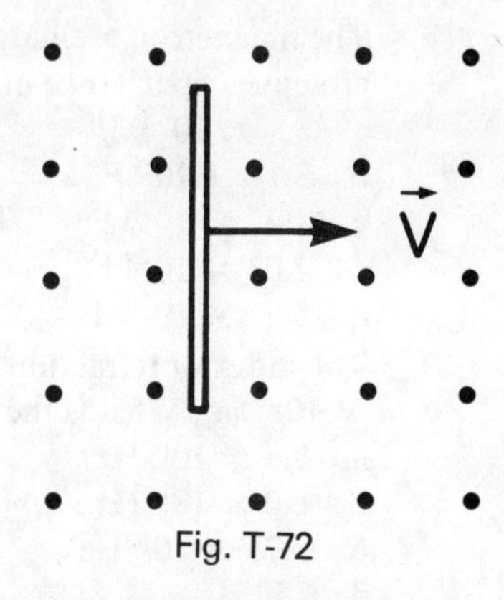

Fig. T-72

14. A galvanometer consists of a coil that may rotate through an angle in a magnetic field. Which of the following combinations of characteristics would lead to the greatest sensitivity?
   a. large number of turns, large magnetic field, small area of coil
   b. large area of coil, large spring constant, small magnetic field
   c. small spring constant, large value of the product NA, large magnetic field
   d. large magnetic field, small coil area, small spring constant
   e. large spring constant, small value of NA, small magnetic field

15. A transformer has 2400 windings in the primary and 160 windings in the secondary. The secondary is connected to an 8 Ω resistor. If the voltage across the primary is 120 V and the frequency is 60 Hz, what is the current in the 8 Ω resistor? (The resistance of the secondary coil is negligible.)
   a. 1.0 A
   b. 0.5 A
   c. 60 A
   d. 6.6 A
   e. 0.01 A

16. When a battery is used in the circuit in Fig. T-73, heat is developed in the resistor at the rate of 40 J/s. What is the effective emf of an ac generator (50 Hz), which will produce heat in the resistor at the same rate?
   a. 28 V
   b. 40 V
   c. 14 V
   d. 10 V
   e. 20 V

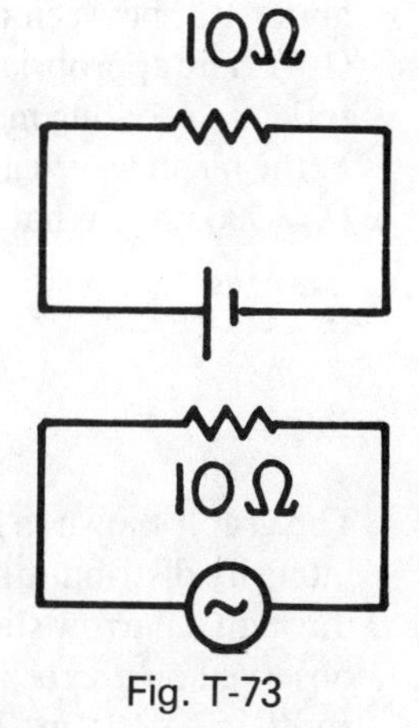

Fig. T-73

17. The resonant frequency of the circuit in Fig. T-74 is 24 MHz. What will the resultant frequency become if L and C are both doubled?
   a. 48 MHz
   b. 12 MHz
   c. 6 MHz
   d. 96 MHz
   e. 24 MHz

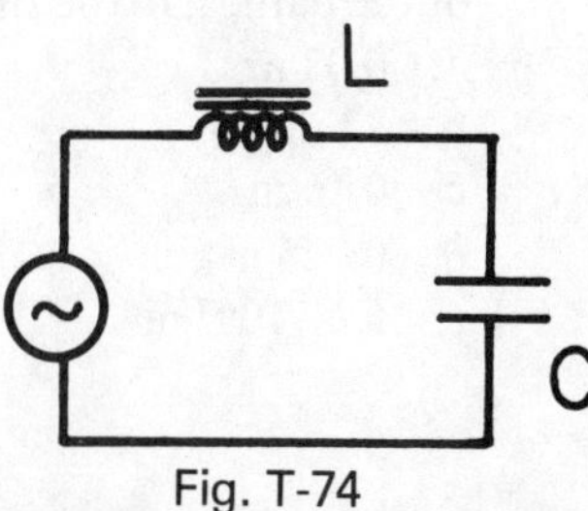

Fig. T-74

18. The magnetron oscillator in a microwave oven produces EM waves of 12-cm wavelength. What is the effective capacitance of the oscillating circuit if the inductance is 0.5 μH?
   a. 8.1×10^{-15} F
   b. 3.17×10^{23} F
   c. 1.27×10^{-4} F
   d. 2.03×10^{-5} F
   e. 5.07×10^{-18} F
19. The index of refraction of a flint glass is 1.60 for light whose wavelength in vacuum is 500 nm (5.0×10^{-7} m). What is the frequency of this light as it travels in the glass?
   a. 9.6×10^{14} Hz
   b. 6.0×10^{14} Hz
   c. 3.75×10^{14} Hz
   d. 150 Hz
   e. 3×10^{8} Hz
20. An erect object is placed at a distance of 10 cm to the left of a diverging lens whose focal length is -5 cm. The image produced by the lens is
   a. real, erect, and reduced.
   b. virtual, erect, and enlarged.
   c. real, inverted, and enlarged.
   d. virtual, erect, and reduced.
   e. virtual, inverted, and reduced.
21. If a real image is formed by a plane mirror, then
   a. the light striking the mirror surface is converging.
   b. the light striking the mirror surface is diverging.
   c. the light striking the mirror surface is parallel.
   d. the light leaving the mirror surface is diverging.
   e. the object had to have been at infinity.
22. A high-quality camera lens, shown in Fig. T-75, usually has a nonreflecting coating on its outer surface. The index of refraction of this film is about 1.3, between that of air (1.0) and glass (1.6). The appropriate thickness of such a nonreflecting coating may be expressed in terms of λ, the mean wavelength for visible light ($\lambda \cong 500$ nm). What thickness is required?
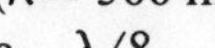
   a. $\lambda/8$
   b. $\lambda/4$
   c. $\lambda/2$
   d. λ
   e. $2\lambda/3$

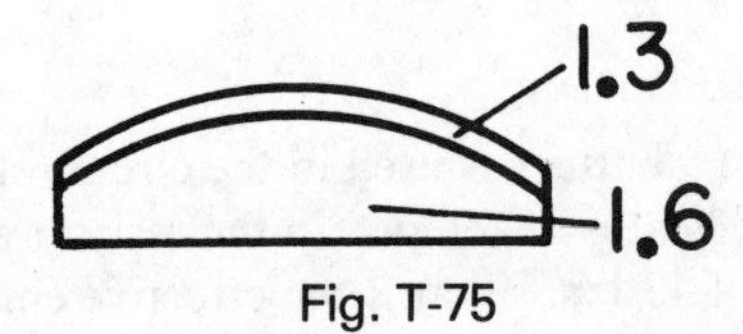

Fig. T-75

23. The graph shown in Fig. T-76 represents the intensity distribution in the light that passes through a narrow slit. The diffraction pattern is observed on a screen 0.80 m from a slit whose width is 2×10^{-5} m. The wavelength of the light is 500 nm. What is the distance from the center of the pattern to the first minimum?

   a. 0.02 m
   b. 32 m
   c. 0.16 m
   d. 0.025 m
   e. 5×10^{-7} m

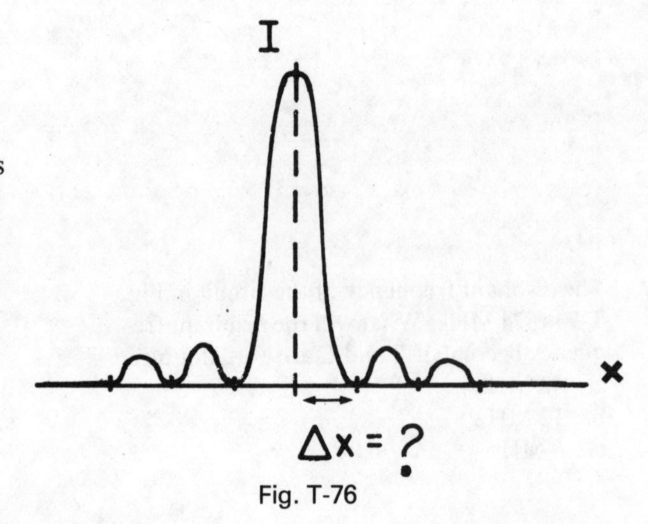

Fig. T-76

24. A simple lens, whose surfaces are spherical, does not form a perfect image. Such a lens is said to have *aberrations*. When a simple double convex lens is used to form an image using the highly monochromatic light of a helium-neon laser (λ = 633 nm), which aberration(s) will be absent?
a. astigmatism
b. chromatic aberration
c. spherical aberration
d. astigmatism and spherical aberration
e. astigmatism, chromatic aberration, and spherical aberration.

25. A simple magnifier consists of a plano-convex lens whose convex surface has a radius of curvature of 5 cm. What is the maximum angular magnification that can be obtained with this lens? (Use n = 1.50.)
a. 6
b. 5
c. 4
d. 3.5
e. 2.5

26. The objective lens of the telescope in Fig. T-77 has a diameter of 10 cm and a focal length of 80 cm. The focal length of the eyepiece is chosen so that the overall angular magnification will be 25. The required focal length is
a. 2.5 cm
b. 3.20 cm
c. 3.33 cm
d. 3.125 cm
e. 2.0 cm

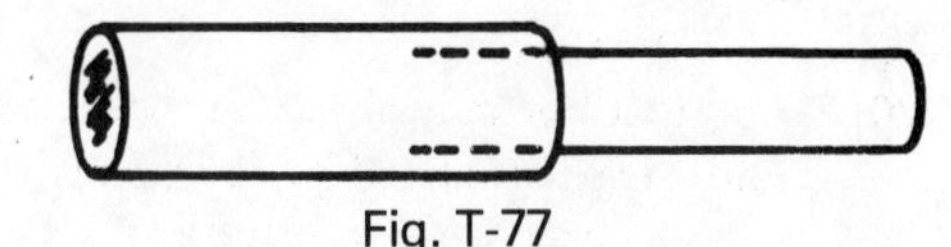
Fig. T-77

27. Which of the following quantities associated with a moving particle has a definite upper limit to the value that it may attain?
a. energy
b. linear momentum
c. velocity
d. angular momentum
e. mass

28. A high-energy cosmic ray particle is observed to have a mass equal to three times its rest mass. What is the speed of this particle in terms of c, the speed of light?
a. 0.33 c
b. 0.889 c
c. 3.0 c
d. 0.943 c
e. 0.90 c

29. The radius of the circular path for a charge q in a magnetic field is R. What is the radius of the circular path for a charge $5q$ of the same mass and same speed in that same magnetic field?
a. R
b. $5R$
c. $R/5$
d. $\sqrt{5}\,R$
e. $R/\sqrt{5}$

30. A certain metal has a work function w of 1.38 eV. If light of wavelength λ = 413 nm is incident upon this metal, what is the KE of the ejected electrons?
a. 3.00 eV
b. 4.38 eV
c. 1.63 eV
d. 2.21×10^{-19} J
e. no photoelectric effect takes place

31. In the Compton effect a photon is incident upon an electron. After the collision has taken place a scattered photon is observed as well as the recoiling electron. In this process, which of the following quantities are conserved (or remain constant): momentum, frequency of photon, or energy?
a. only energy
b. momentum and frequency
c. energy and frequency
d. momentum and energy
e. only momentum

32. The third Bohr postulate concerns the angular momentum of a hydrogen atom. This postulate states that
a. all orbits have the same angular momentum.
b. the angular momentum is conserved.
c. the angular momentum is a continuous function of E, the energy.
d. the angular momentum has discrete values given by integral multiples of $h/2\pi$.
e. angular momentum is a vector quantity.

33. A simplified energy level diagram for the hydrogen atom is shown in Fig. T-78. The energy level E_6 is not shown in Fig. T-78. It will have an energy of
a. -0.38 eV
b. -2.27 eV
c. -0.23 eV
d. -0.27 eV
e. 0.00 eV

Total E		Excitation Energy (eV)
0.00	E_∞	13.60
-0.54	E_5	13.06
-0.85	E_4	12.75
-1.51	E_3	12.09
-3.40	E_2	10.20
-13.60	E_1	0.00

Fig. T-78

34. The M-shell corresponds to a principal quantum number $n = 3$. The possible values of l are 0, 1, 2. What is the total number of electrons permitted in the M-shell?
a. 2
b. 6
c. 8
d. 12
e. 18

35. An x-ray tube operated at an accelerating voltage of 18,600 V will produce a range of wavelengths with a
a. minimum value of 1.50×10^{-10} m.
b. minimum value of 6.68×10^{-11} m.
c. maximum value of 1.06×10^{-29} m.
d. maximum value of 6.68×10^{-11} m.
e. minimum value of 2.82×10^{-11} m.

36. A radioactive sample contains 12×10^{18} atoms, 4×10^{8} of which are decaying each second. What is the disintegration constant of this sample?
a. $3.0 \times 10^{10}\ s^{-1}$
b. 4.8×10^{27}
c. $0.115\ s^{-1}$
d. $3.33 \times 10^{-11}\ s^{-1}$

37. Supposing that a radioactive sample containing 12×10^{18} atoms has a disintegration constant of $2 \times 10^{-8}\ s^{-1}$, how many atoms will remain after 3.47×10^{7} s?
a. 6.0×10^{9}
b. 6.0×10^{18}
c. 3.46×10^{11}
d. 2.4×10^{11}
e. 8.5×10^{11}

38. The particle that triggers the fission reaction in uranium-235 is
a. a beta particle
b. a very high-energy gamma ray (photon)
c. a neutron
d. a proton
e. an alpha particle

39. Pair production is the process in which an energetic photon (γ-ray) produces an electron and a positron. The γ-ray disappears. What is the minimum energy that the photon must have?
a. 0
b. 1.022 MeV
c. 0.511 MeV
d. 931 MeV
e. 1862 MeV

40. The nuclide $^{214}_{84}Po$ decays very rapidly to $^{210}_{82}Pb$, the half-life being 1.6×10^{-4} s. At a certain instant, 2 g of ^{214}Po is present. How much will remain 9.6×10^{-4} s later?
a. 0.17 g
b. 0.031 g
c. 0.063 g
d. 0.0078 g
e. none

Solutions to Review Tests

CHAPTERS 1 AND 2

1. (e)
2. 21.6 kg
3. 6 protons
4. 3×10^{-22} g/atom
5. 1.74×10^{19} molecules
6. 40 m
7. -10 m/s^2
8. 0.83 s
9. (d)
10. (d)
11. (d)
12. 500 mi/h
13. 16 m/s
14. 4 s
15. 1.5 s

CHAPTERS 3 AND 4

1. (b)
2. (d)
3. 8 N
4. 2.69 m/s^2
5. 0.98 m/s^2
6. (d)
7. (d)
8. (d)
9. 6 N
10. 2 N
11. 8.43×10^3 N
12. 10°
13. 51 N
14. 4 m
15. 17.3N

CHAPTERS 5 AND 6

1. 1.5×10^3 N
2. 5 kg·m/s
3. 5 J
4. 24 ft·lb
5. (b)
6. 60°
7. zero
8. 173 J
9. 98 J
10. 7.5 N
11. 51.3°
12. 50 kW

CHAPTERS 7–9

1. 30 cm
2. (a)
3. (a)
4. (a)
5. 2.72×10^{-3} m/s^2
6. 3.33×10^2 rad/s
7. $I_1/I_2 = 3$
8. 8.0×10^4 N

9a. 0.2 N/kg
b. toward the center of the planet

10. 8.49 h
11. 1.91×10^{11} N/m^2
12. (c)
13. 0.0 cm/s
14. 8.66 cm/s
15. 25.1 cm/s
16. 4.90 s

CHAPTERS 10–12

1. 6.4×10^3 m/s
2. (c)
3. 0.68 m
4. 2.3 Hz
5. 6.6×10^5 m/s; receding
6. (c)
7. 850 Hz
8. (c)
9. (b)
10. 50 watts
11. (a)
12. 5.44 g/cm^3
13. 4.4
14. 4.3×10^2 N/m^2 The pressure is higher at A.
15. (b)
16. $R = 24$ cm

CHAPTERS 13–15

a. (c)
2. 42.8°C
3. 25°C
4. 2.7×10^{-2} cm^3
5. 605.5 cm^3
6. 80 cal
7. 30°C
8. 30 g
9. 0.8 cal/g·°C
10. radiation
11. (c)
12. 14×10^{21}
13. 59%
14. (a)

CHAPTERS 1–17

1. (e)
2. (e)
3. (b)
4. (b)
5. (a)
6. (d)
7. (c)
8. (c)
9. (c)
10. (b)
11. (e)
12. (e)
13. (b)
14. (c)
15. (c)
16. (c)
17. (b)
18. (c)
19. (b)
20. (d)
21. (a)
22. (c)
23. (b)
24. (a)
25. (b)
26. (b)

CHAPTERS 18–20

1. (e)
2. (c)
3. (c)
4. (d)
5. 346 V/m
6. 1.6×10^{-4} C
7. 81 J/C
8. 1.67×10^{-3} J
9. 2.4×10^{-2} V/m
10. (a)
11. (b)
12. (a)
13. (b)

CHAPTERS 21–23

1. 0.04 Ω
2. 0.25 A
3. 4.0 V
4. 0.10 A
5. (e)
6. (e)
7. (d)
8. (d)
9. (b)
10. (d)
11. 2.5 V
12. 4.0×10^{-3} N • m
13. 0.5 A
14. (c)
15. 558 Ω

CHAPTERS 24–26

1. 1.57×10^{-4} s
2. 2.0 mH
3. 8.0 mH
4. 679 kHz
5. (e)
6. 6.0×10^{15} Hz
7. (d)
8. (d)
9. 5×10^{14} Hz
10. (a)
11. (d)
12. (b)
13. (e)
14. 0.5 m
15. (d)
16. (b)
17. (b)
18. (b)
19. 5.0×10^{-7} m
20. (d)

CHAPTERS 27–29

1. chromatic aberration
2. $I_1/I_2 = 4$
3. 6
4. 113
5. 20
6. 141.4 km/h; 1.41 h
7. (d)
8. (c)
9. 1.8×10^9 J
10. (b)
11. (d)
12. (e)
13. (d)
14. (d)

CHAPTERS 30–32

1. Path b implies a negatively charged nucleus.
2. (d)
3. 656 nm
4. excitation energy = 13.22 eV
5. 3, 2, 1, 0, −1, −2, −3
6. No two electrons in a given atom can have all their quantum numbers the same.
7. 11 electrons
8. 8 electrons
9. 0.50 Å (the lower limit of the bremsstrahlung)
10. α—helium nuclei, β—negative electrons, γ—short wavelength electromagnetic radiation (photons)
11. 3.8×10^{-15} m
12. $^{25}_{11}\text{Na} \rightarrow ^{25}_{12}\text{Mg} + ^{0}_{-1}\text{e} + \upsilon$
13. the neutrino. Its existence is required by energy conservation and by conservation of angular momentum.
14. 6.25×10^{10}
15. All of the listed properties are the same for all four isotopes.

CHAPTERS 18–34

1. (e)
2. (d)
3. (e)
4. (a)
5. (c)
6. (a)
7. (b)
8. (c)
9. (d)
10. (a)
11. (c)
12. (b)
13. (b)
14. (c)
15. (a)
16. (e)
17. (b)
18. (a)
19. (b)
20. (d)
21. (a)
22. (b)
23. (a)
24. (b)
25. (d)
26. (b)
27. (c)
28. (d)
29. (c)
30. (c)
31. (d)
32. (d)
33. (a)
34. (e)
35. (b)
36. (d)
37. (b)
38. (c)
39. (b)
40. (b)

A 6 B 7 C 8 D 9 E 0 F 1 G 2 H 3 I 4 J 5